PHARMACOGENOMICS

Pharmacogenomics: Challenges and Opportunities in Therapeutic Implementation is accompanied by a website featuring an elaboration on discussion points, a discussion of review questions, and patient case scenarios illustrating how pharmacogenetics may be clinically applied. To access these companion resources, please visit Booksite.elsevier.com/9780123919182

PHARMACOGENOMICS

Challenges and Opportunities in Therapeutic Implementation

EDITORS

YUI-WING FRANCIS LAM

Department of Pharmacology, University of Texas Health Science Center at San Antonio, San Antonio, Texas, USA

LARISA H. CAVALLARI

Department of Pharmacy Practice, University of Illinois at Chicago, Chicago, Illinois, USA

AMSTERDAM • BOSTON • HEIDELBERG • LONDON
NEW YORK • OXFORD • PARIS • SAN DIEGO
SAN FRANCISCO • SINGAPORE • SYDNEY • TOKYO

Academic Press is an imprint of Elsevier

Academic Press is an imprint of Elsevier
The Boulevard, Langford Lane, Kidlington, Oxford, OX5 1GB, UK
225 Wyman Street, Waltham, MA 02451, USA

First published 2013

British Library Cataloguing in Publication Data
A catalogue record for this book is available from the British Library

Library of Congress Cataloguing in Publication Data
A catalog record for this book is available from the Library of Congress

ISBN: 978-0-12-391918-2

For information on all Academic Press publications
visit our website at store.elsevier.com

Printed and bound by CPI Group (UK) Ltd, Croydon, CR0 4YY
Transferred to digital print 2012

Working together to grow
libraries in developing countries

www.elsevier.com | www.bookaid.org | www.sabre.org

ELSEVIER   BOOK AID
             International   Sabre Foundation

Dedication

Dr. Lam dedicates this book to Jennifer, Jessica, and Derek; Aunt Chee-Ming and Uncle Po-Hon; Mom and Dad.

Dr. Cavallari dedicates this book to Andy, Clare, and Meg.

Contents

List of Contributors

L. Annemans   University of Ghent & University of Brussels, Belgium

Christina L. Aquilante   Department of Pharmaceutical Sciences, University of Colorado Skaggs School of Pharmacy and Pharmaceutical Sciences, Aurora, Colorado, USA

Denise Avard   Centre of Genomics and Policy, Department of Human Genetics, Faculty of Medicine, McGill University, Montreal, QC, Canada

Cornelis Boersma   Department of Pharmacy, University of Groningen, Netherlands

John A. Bostrom   Department of Pharmacy Practice and Center for Pharmaceutical Biotechnology, College of Pharmacy, University of Illinois at Chicago, Chicago, Illinois, USA

Daniel A. Brazeau   Genomics, Analytics and Proteomics Core, University of New England, Biddeford, Maine, USA

Gayle A. Brazeau   College of Pharmacy, University of New England, Biddeford, Maine, USA

Jürgen Brockmöller   Department of Clinical Pharmacology, University Göttingen, Göttingen, Germany

Lela Buckingham   Department of Medical Laboratory Sciences, College of Health Sciences, Rush University Medical Center, Chicago, Illinois, USA

Larisa H. Cavallari   Department of Pharmacy Practice, University of Illinois at Chicago, Chicago, Illinois, USA

Julio D. Duarte   Department of Pharmacy Practice, University of Illinois at Chicago, Chicago, Illinois, USA

Sisi Feng   Center for Computational Biology

Naoki Fukui   Department of Psychiatry, Niigata University Graduate School of Medical and Dental Sciences, Chuo-ku, Niigata, Japan

Shiew-Mei Huang   Office of Clinical Pharmacology, Food and Drug Administration, Silver Spring, Maryland, USA

Lou Huei-xin   Premarket Division, Health Products Regulation Group, Health Sciences Authority, Singapore

Yann Joly   Centre of Genomics and Policy, Department of Human Genetics, Faculty of Medicine, McGill University, Montreal, QC, Canada

Edmund Jon Deoon Lee   Pharmacogenetics Laboratory, Department of Pharmacology, Yong Loo Lin School of Medicine, National University of Singapore, Singapore

Emily Kirby   Centre of Genomics and Policy, Department of Human Genetics, Faculty of Medicine, McGill University, Montreal, QC, Canada

Teri E. Klein   Department of Genetics, Stanford University, Stanford, California, USA

Bartha M. Knoppers   Centre of Genomics and Policy, Department of Human Genetics, Faculty of Medicine, McGill University, Montreal, QC, Canada

Y.W. Francis Lam   Department of Pharmacology, University of Texas Health Science Center at San Antonio, San Antonio, Texas, USA

Y.W. Francis Lam   Department of Pharmaceutical Sciences, University of Colorado Skaggs School of Pharmacy and Pharmaceutical Sciences, Aurora, Colorado, USA

Xin Li   Center for Computational Biology, Beijing Forestry University, Beijing, China

Yongci Li   College of Science, Beijing Forestry University, Beijing, China

Mengtao Li Department of Rheumatology, Peking Union Medical College Hospital, Chinese Academy of Medical Science & Peking Union Medical College, Beijing, China

Xinjuan Liu Department of Rheumatology, Peking Union Medical College Hospital, Chinese Academy of Medical Science & Peking Union Medical College, Beijing, 100032, China

Yafei Lu Center for Computational Biology, Beijing Forestry University, Beijing, China

Elsa Haniffah Mejia Mohamed Pharmacogenomics Laboratory, Department of Pharmacology, University Malaya, Malaysia; Pharmacogenetics Laboratory, Department of Pharmacology, Yong Loo Lin School of Medicine, National University of Singapore, Singapore

Kathryn Momary Department of Pharmacy Practice, Mercer University, College of Pharmacy and Health Sciences Center, Atlanta, Georgia, USA

Wenbo Mu Department of Bioengineering, University of Illinois at Chicago, Chicago, Illinois, USA

Vural Özdemir Centre of Genomics and Policy, Department of Human Genetics, Faculty of Medicine, McGill University, Montreal, Quebec, Canada; Group on Complex Collaboration, Faculty of Management, McGill University, Montreal, Quebec, Canada; Data-Enabled Life Sciences Alliance International (DELSA Global), Seattle, Washington, USA

J. Suso Platero Janssen Research & Development, Raritan, New Jersey, USA

Jalene Poh Pharmaceuticals and Biologics Branch, Health Products Regulation Group, Health Sciences Authority, Singapore

Maarten J. Postma Department of Pharmacy, University of Groningen, Netherlands

Michael E. Schaffer Janssen Research & Development, Raritan, New Jersey, USA

Monsheel Sodhi Department of Pharmacy Practice and Center for Pharmaceutical Biotechnology, College of Pharmacy, University of Illinois at Chicago, Chicago, Illinois, USA

Toshiyuki Someya Department of Psychiatry, Niigata University Graduate School of Medical and Dental Sciences, Chuo-ku, Niigata, Japan

Julia Stingl (formerly Kirchheiner) Institute of Pharmacology of Natural Products and Clinical Pharmacology, University Ulm, Ulm, Germany

Takuro Sugai Department of Psychiatry, Niigata University Graduate School of Medical and Dental Sciences, Chuo-ku, Niigata, Japan

Yutato Suzuki Department of Psychiatry, Niigata University Graduate School of Medical and Dental Sciences, Chuo-ku, Niigata, Japan

Dorothy Toh Vigilance Branch, Health Products Regulation Group, Health Sciences Authority, Singapore

Dominique Vandijck University of Ghent & University of Hasselt, Belgium

S. Vegter Department of Pharmacy, University of Groningen, Netherlands

Junzo Watanabe Department of Psychiatry, Niigata University Graduate School of Medical and Dental Sciences, Chuo-ku, Niigata, Japan

Yuichiro Watanabe Department of Psychiatry, Niigata University Graduate School of Medical and Dental Sciences, Chuo-ku, Niigata, Japan

Rongling Wu Center for Statistical Genetics, Pennsylvania State University, Hershey, Pennsylvania, USA

Ophelia Yin Oncology Clinical Pharmacology, Novartis Pharmaceutical Corporation, East Hanover, New Jersey, USA

Xiaofeng Zeng Department of Rheumatology, Peking Union Medical College Hospital, Chinese Academy of Medical Science & Peking Union Medical College, Beijing, 100032, China

Wei Zhang Department of Pediatrics, Institute of Human Genetics; Cancer Center, University of Illinois at Chicago, Chicago, Illinois, USA

Preface

Pharmacogenomics and pharmacogenetics are overlapping sciences, and although the two terminologies have been used interchangeably in the literature, pharmacogenomics reflects the progressive transition that has taken place over the years within the broad scope of personalized medicine. As a discipline, pharmacogenomics is envisioned as a major societal benefit from all the scientific and technical advances related to the Human Genome Project. To date, much work remains to address the challenges of translating pharmacogenomics into clinical practice and drug development in order to achieve the ultimate goal envisioned many years ago. Nevertheless, examples of clinical applications of pharmacogenomic knowledge are beginning to emerge at several major academic medical centers.

This book differs from available pharmacogenomics books in several aspects. It neither contains significant materials on molecular genetics nor lists all the theoretical pharmacogenomic applications organized by therapeutic specialties. Rather, the focus of the book is to provide a timely discussion and viewpoints on a broad range of topics, from the academic, regulatory, pharmaceutical, clinical, socioethical, and economic perspectives that are relevant to the complex processes in translating pharmacogenomic findings into therapeutic applications.

From the beginning, our goal has been to provide information that is not readily available in other books covering the same topic. Although the processes and implementation barriers are presented in depth in one chapter, perspectives on such challenges and limitations as well as examples of successful direct therapeutic applications are presented throughout the book. In addition, we have included a chapter on ethnobridging and pharmacogenomics that discusses the complexity of ethnicity in pharmacogenomics studies and global drug development and several chapters that discuss epigenetics as the next important focus for the pharmacogenomics discipline. To facilitate learning, each chapter also contains objectives and discussion points. Additional material, including elaboration on discussion points, and patient case scenarios illustrating how pharmacogenetics may be clinically applied, are available on the website (Booksite.elsevier.com/9780123919182)

The book chapters are organized into several major sections. Section 1 (chapters 1 to 3) provides an introductory chapter on pharmacogenomics, an overview chapter on the challenges of moving the discipline into real-world settings over the last decade, and a chapter on global academic and governmental efforts to advance and apply the relevant genomic knowledge. Section 2 (chapters 4 to 7) primarily focuses on clinical areas where the evidence strongly supports direct pharmacogenetic applications to patient care. Where appropriate, unsuccessful applications are used to illustrate the challenges for the discipline. In addition, this section also contains three "Looking to the Future" subchapters that describe the concepts and challenges of integrating epigenetics into pharmacogenomics studies for the cardiology, oncology, and psychiatry disciplines. Section 3 (chapters 8 and 9) provides perspective and insights on study design, statistical models, as well as molecular approaches and techniques in pharmacogenomics research and drug development. Section 4 (chapters 10 to 12) is a unique section that covers the diverse topics of ethnobridging, health technology assessment, and governance of personalized medicine. Last but not least, the final section (chapter 13) discusses an example of how genomic education has been integrated into a pharmacy curriculum and provides a framework of how this textbook can be useful for teaching pharmacogenomics to students in various health care disciplines and graduate-level students in health and pharmaceutical sciences.

Because the book details viewpoints on the challenges of translating pharmacogenomics, we intentionally did not limit our contributors with organized content for each chapter. In essence, each chapter follows a general approach of including an overview of the potentials or opportunities within the context of the respective chapter, but the emphasis is on discussion of barriers with perspectives on how to move pharmacogenomics forward. Realizing that overlap is inevitable in a book with multiple authors, we took measures to minimize unnecessary duplicated materials and cross-reference chapters whenever appropriate.

This book is intended not only as a reference book for scientists in academia and the pharmaceutical industry involved in pharmacogenomics research but also for health care clinicians working or interested in the field. In addition, this text is useful as a textbook for teaching clinicians and students in different health care disciplines, and specific materials covered in the book would be useful resources for teaching graduate students in academic disciplines such as pharmacology, neuroscience, structural and cellular biology, and molecular medicine. It is our sincere hope that after completing the textbook, the readers will have not only a critical awareness of the value of pharmacogenomic implementation with actual versus potential applications but also a broad knowledge of the pertinent issues and challenges for pharmacogenomics before advances in scientific findings can be broadly and practically applied to patient care.

Y. W. Francis Lam
Larisa H. Cavallari

Principles of Pharmacogenomics: Pharmacokinetic, Pharmacodynamic, and Clinical Implications

Y.W. Francis Lam*, Larisa H. Cavallari†

*Department of Pharmacology, University of Texas Health Science Center at San Antonio, San Antonio, Texas, USA

†Department of Pharmacy Practice, University of Illinois at Chicago, Chicago, Illinois, USA

OBJECTIVES

1. Describe how a single-nucleotide polymorphism (SNP) can affect protein function or expression and consequently influence drug response.

2. Explain how genetic polymorphisms for drug metabolism or drug transporter enzymes may influence drug pharmacokinetics.

3. Contrast phenotypic responses to genetic variation for drug metabolism versus drug target proteins.

4. Describe a novel drug developed based on an understanding of genes involved in disease pathophysiology.

5. Explain how genetic polymorphisms at the drug target site may influence drug pharmacodynamics.

INTRODUCTION

There is significant interpatient variability in drug response, which is largely attributed to innate differences among individuals in their capacity to process and react to medications. Pharmacogenomics involves incorporating information about a person's genotype into drug therapy decisions, with the goal of providing the most effective and safest therapy for that patient. Over the last decade, there have been significant advances in our understanding of the contribution of genetic differences in pharmacokinetics and pharmacodynamics toward interindividual variability in drug response. Pharmacogenomics may lead not only to improved use of existing therapies but also to novel drugs developed based on an improved understanding of genetic control of cellular functions.

The human genome is composed of approximately 20,000 protein-coding genes. By far, the most common type of variation is the single-nucleotide polymorphism (SNP), which is defined as a single-base difference that exists between individuals. More than 22 million SNPs have been reported in the human genome.[1] SNPs that result in amino acid substitution are termed nonsynonymous. Nonsynonymous SNPs occurring in coding regions of the gene (e.g., exons) can affect protein activity and have significant consequences for responses to medications that depend on the protein for metabolism, transport, or eliciting cellular effects. Synonymous polymorphisms do not result in amino acid substitution; however, those occurring in a gene regulatory region (e.g., promoter region, intron) may alter gene expression and the amount of protein that is produced. Two or more SNPs are often inherited together more frequently than would be expected based on chance alone. This effect is referred to as linkage disequilibrium (LD). A haplotype refers to a set of SNPs that are in LD. Other types of variations include insertion-deletion polymorphisms (indels), short tandem repeats, and multiple-base nucleotide variations.

Polymorphisms commonly occur for genes encoding drug metabolism, drug transporters, and drug target proteins (Figure 1.1). Drug metabolism and transporter genotypes can affect drug availability at the target site; drug target genotypes can affect a patient's sensitivity to a drug. In many instances, genes for proteins involved in drug disposition together with genes for proteins at the drug target site jointly influence drug response.

The terms "pharmacogenetics" and "pharmacogenomics" are often used interchangeably. Because drug responses are mostly determined by multiple proteins, rather than single proteins, recent trends of investigations on determinants of drug response have shifted

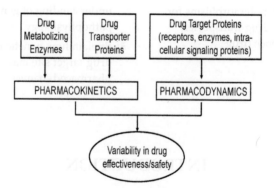

FIGURE 1.1   Location of genetic variations affecting drug response. Those occurring in genes for drug metabolism or transport can affect drug pharmacokinetics, whereas SNPs in genes encoding for drug target proteins can affect drug pharmacodynamics.

from pharmacogenetics to pharmacogenomics. However, for simplicity, this chapter treats pharmacogenetics and pharmacogenomics as synonymous.

Despite the scientific advances made, personalized medicine envisioned many years ago has yet to become a reality in many ways. Exceptions to this delay largely exist in oncology and more recently in cardiology, where genotyping to determine clopidogrel effectiveness is starting to become routine at some large academic medical centers [2,3]. Examples of genotype-guided therapies are beginning to emerge in other therapeutic areas, which are discussed in detail throughout this book. However, significant challenges still exist in ethical, socioeconomic, regulatory, legislative, drug development, and educational issues that need to be addressed and resolved before personalized medicine can be practically and satisfactorily implemented in clinical practice on a broader scale. The goal of this chapter is to review the pharmacokinetic and pharmacodynamics basis of individualized therapy and to briefly discuss the challenges of implementing pharmacogenomics in clinical practice. Further in-depth discussion of specific therapeutic areas and/or disease states—as well as ethical, socioeconomic, regulatory, legislative, drug development, technological, and educational issues—will be the focus of subsequent chapters.

CYTOCHROME P-450 ENZYMES

The cytochrome P-450 (CYP) superfamily of isoenzymes are the most important and studied metabolic enzymes that exhibit clinically relevant genetic polymorphisms. Within this superfamily of isoenzymes, 57 different CYP genes and 58 pseudogenes have been identified and, based on the similarity in their amino acid sequences, grouped into 18 families and 44 subfamilies with increasing extent of sequence similarity. Of these genes and pseudogenes, 42 are involved in the metabolism of exogenous xenobiotics and endogenous substances, such as steroids and prostaglandins, and 15 are known to be involved in the metabolism of drugs in humans.[4]

The genes encoding CYPs are highly polymorphic, resulting in functional genetic polymorphism for several isoenzymes, including CYP2A6, CYP2B6, CYP2C9, CYP2C19, CYP2D6, and CYP3A4/5. These *CYP* polymorphisms cause gene deletions, deleterious mutations resulting in premature stop codon or splicing defects, amino acid changes, and gene duplications. The ensuing differences in the number of functional variants or alleles of the gene encoding a specific CYP and expression of enzyme activity classify patients into four general metabolic phenotypes: the poor metabolizers (PMs), who have defective or deleted alleles and abolished enzyme activity; the intermediate metabolizers (IMs), who carry either one functional and one defective allele, or two partially defective alleles, and in both cases have reduced activity of the enzyme; the extensive metabolizers (EMs) with two functional alleles and normal activity; and the ultra-rapid metabolizers (UMs), who carry a duplicated or amplified gene variant, resulting in two or multiple copies of the functional allele and very high enzyme activity.

The clinical consequences of genetically altered enzyme activity depend on whether the pharmacological activity resides with the parent compound or the metabolite and the relative contribution of the polymorphic isoenzyme to the overall metabolism of the drug. For the majority of the drugs, PMs exhibit a higher risk of adverse drug reactions (ADRs), whereas

UMs experience lower efficacy when administered the standard dosage regimen of a drug that is mostly dependent on the polymorphic enzyme for elimination. In the case of a pro-drug, it would be the UMs who exhibit higher incidence of ADRs and the PMs who experience a lower efficacy, reflecting a difference in the extent of therapeutically active metabolite formed between the two metabolic genotypes.

Among the different *CYP* genetic polymorphisms, those affecting CYP2D6, CYP2C19, and CYP2C9 are currently the most relevant and the ones with the most abundant data, as well as representing most of the revised labeling information. Their potential role in translating the expanding pharmacogenomic knowledge into dose requirements and therapeutic decisions will be discussed first. An overview of the other major CYP isoenzymes will also be presented.

CYP2D6

CYP2D6 is the only drug metabolizing CYP enzyme that is not inducible, and the significant interindividual differences in enzyme activity are largely attributed to genetic variations. In addition, the *CYP2D6* gene polymorphisms are also the best characterized among all of the *CYP* variants, with at least 100 gene variants and 120 alleles identified (www.cypalleles.ki.se/cyp2d6.htm). Nevertheless, Sistonen et al.[5] showed that even with the extensive number of alleles, determining 20 different haplotypes by genotyping 12 SNPs could predict the real phenotype with 90 to 95% accuracy.

Among the multiple alleles listed in Table 1.1, both *CYP2D6*1* and *CYP2D6*2* exhibit normal enzyme activity, whereas the two most important null variants are *CYP2D6*4* (c.1846G>A, rs3892097) and *CYP2D6*5* (gene deletion), resulting in an inactive enzyme and absence of enzyme, respectively. Significant reduction in enzyme activity is commonly associated with *CYP2D6*10* (c.100C>T, rs1065852), *CYP2D6*17* (c.1023C>T, rs28371706, c.2850C>T, rs16947), and *CYP2D6*41* (c.2988G>A, rs28371725), and phenotypically expressed as IM. The IM phenotype has also been associated with the *CYP2D6*9, *29,* and *36* variants.[4] CYP2D6 is also the first CYP isoenzyme for which copy number variants (CNVs) were reported. A CNV represents a DNA segment ($\geq$1 kb) with a variable number of copies of that segment as a result of duplications, deletions, or rearrangement and constitutes a major source of interindividual variation in the human genome. Individuals carrying up to 13 functional gene copies of the *CYP2D6*2* allele[6] have been reported to exhibit variation in response to different drugs [7,8]. After these initial reports, gene duplication has also been documented for the *CYP2D6*1, *4, *6, *10, *17, *29, *35, *41, *43,* and *45* variants.[9] Therefore, although UM can result from duplication or multiduplication of the active *CYP2D6* gene, duplication of partially functional and nonfunctional genes can also occur, resulting in different levels of gene expression and phenotypes of metabolic importance (Table 1.1).

Significant interethnic variations in *CYP2D6* allele and phenotype distributions have also been well documented. *CYP2D6*4* and *CYP2D6*5* (allelic frequency of about 20 to 25% and 4 to 6%, respectively) are predominantly found in Caucasian PMs, whereas the predominant variants in people of Asian and African heritage are *CYP2D6*10* (allelic frequency of about 50%) and *CYP2D6*17* (allelic frequency of about 20 to 34%), respectively, both resulting in the IM phenotype. Therefore, even though the classic PM frequencies determined in Asians

TABLE 1.1 Functional *CYP2D6* polymorphisms, expected enzyme activity, and corresponding metabolic phenotypes for selected common variants

Allelic Variants	Polymorphism/ Substitution	Functional Effect on Enzyme Activity (Metabolic Phenotypes)
CYP2D6*1		Normal (EM)
CYP2D6*1 x N ≥ 2	Copy number variation	Increased (UM)
CYP2D6*2		Normal (EM)
CYP2D6*2 x N ≥ 2	Copy number variation	Increased (UM)
CYP2D6*3	Frameshift	Abolished (PM)
CYP2D6*4	Splicing defect	Abolished (PM)
CYP2D6*4 x2	Copy number variation	Abolished (PM)
CYP2D6*5	Gene deletion	Abolished (PM)
CYP2D6*6, CYP2D6*7, CYP2D6*8, CYP2D6*14	Frameshift (CYP2D6*6)	Abolished (PM)
CYP2D6*9, CYP2D6*29, CYP2D6*36		Decreased (IM)
CYP2D6*10	P34S	Decreased (IM)
CYP2D6*10 x N	Copy number variation	Decreased (IM)
CYP2D6*17	T107I, R296C	Decreased (IM)
CYP2D6*17 x N	Copy number variation	Normal (EM) for N = 2
CYP2D6*41	Splicing defect	Decreased (IM)
CYP2D6*41 x2	Copy number variation	Normal (EM)

(about 0 to 1% of the population) and Africans (0 to 5% of the population) are reported to be lower than that of Caucasians (5 to 14% of the population), the high prevalence of *CYP2D6*10* and *CYP2D6*17* in these two IM populations provides a biologic and molecular explanation for reported higher drug concentrations and/or the practice of prescribing lower dosage requirements in people of Asian and African heritage [10,11,12,13]. On the other hand, the UM frequency is much higher in Northeast Africa and Oceania, including the Saudi Arabian (20%) and black Ethiopian (29%) populations, when compared to Caucasians (1 to 10%).

Even though accounting for only 2% of total CYP content in the liver, CYP2D6 mediates the metabolism of approximately 20 to 30% of currently marketed drugs, and *CYP2D6* polymorphism significantly affects the elimination of 50% of these drugs,[14] which include antidepressants, antipsychotics, analgesics, antiarrhythmics, antiemetics, and anticancer drugs. Although differences in pharmacokinetic parameters (elimination half-lives, clearances, and area under the plasma concentration time curves) for CYP2D6 substrates could be demonstrated among the different metabolic phenotypes, the significant overlap in CYP2D6 activities in EMs and IMs result in therapeutic implication mostly for the PM and UM phenotypes. In the past, the clinical relevance of *CYP2D6* polymorphism primarily concerned the increased prevalence of ADRs in PMs administered standard doses of drugs that rely significantly on CYP2D6 for elimination. These drugs include the antianginal agent perhexiline (neuropathy),[15] the antiarrhythmic agent propafenone (proarrhythmic events),[16] and neuroleptic agents such as perphenazine (sedation and parkinsonism) [17,18]. More recently, occurrences of ADRs have also been highlighted in UMs, primarily a result of a 10- to 30-fold increase in metabolite concentrations. Codeine is converted by CYP2D6 to morphine, which is pharmacologically more active. UMs administered the usual therapeutic dose of codeine have been reported to exhibit symptoms of narcotic overdose associated with significantly elevated morphine concentration. This toxicity potential had been highlighted in several case reports, including a fatal case of a breast-fed infant attributed to extensive formation of morphine from codeine taken by the mother who was a UM [19,20,21]. Similarly, tramadol cardiotoxicity was reported recently in a UM with a high level of the active O-Desmethyltramadol,[22] which has been reported to exhibit a high correlation with increased plasma epinephrine level.[23]

Efficacy of prodrugs (such as codeine and hydrocodone) would be reduced in PMs because less parent drug is converted by CYP2D6 to its respective active metabolite: morphine or hydromorphone, resulting in little analgesic relief.[24] There are similar reports of lower efficacy in PMs with venlafaxine.[25] Another example is tamoxifen, in which CYP2D6 plays a major role in the formation of the abundant and pharmacologically more active metabolite, endoxifen.[26] Because endoxifen possesses greater affinity for the estrogen receptor than tamoxifen, PMs with the *CYP2D6*4/*4* genotype have been shown to have an increased risk of breast cancer recurrence and worse relapse-free survival, as well as a much lower incidence of moderate or severe hot flashes.[26] Even though conflicting data currently complicate the adoption of *CYP2D6* genotyping in patients treated with tamoxifen, available evidence strongly supports a role for CYP2D6 in pharmacological activation of tamoxifen and possibly a likelihood of lesser therapeutic benefit in PMs, with the ultimate impact on patient outcome to be tested in prospective clinical studies.

Inadequate therapeutic response with implications for dosage adjustment has also been demonstrated for UMs administered CYP2D6 substrates. The best evidence describes two patients with multiple copies of *CYP2D6*2* requiring 500 mg daily tricyclic antidepressant nortriptyline (versus the usual recommended daily dose of 100 to 150 mg) in one patient [27] and 300 mg per day clomipramine (versus 25−150 mg) in another patient[6] to achieve adequate therapeutic response. Similarly, lower efficacy in UMs has been reported with other antidepressants [28,29] and antiemetics such as ondansetron.[30]

Furthermore, drug interactions involving competitive inhibition of CYP2D6 also have clinical implications in patients with different metabolic phenotypes. It is not uncommon that tamoxifen-treated patients are also taking antidepressants such as selective serotonin reuptake inhibitors (SSRIs), both for their antidepressant effect as well as their off-label use to manage hot flashes. In view of the abundance and greater antiestrogenic activity of endoxifen, concurrent administration of SSRIs that are potent inhibitors of CYP2D6 (such as fluoxetine and paroxetine) should best be avoided, and SSRIs with a lesser extent of CYP2D6 inhibition (such as citalopram and venlafaxine) would be better alternative antidepressants if there is a need for concurrent administration with tamoxifen. As shown by Hamelin and colleagues,[31] the pharmacological consequences of drug–drug interaction via CYP2D6 inhibition are of greater magnitude in EMs, with pronounced and prolonged hemodynamic responses to metoprolol, than in PMs. Potent CYP2D6 inhibitors have been shown to reduce the metabolic capacity of EMs significantly so that an individual EM could appear metabolically to be a PM during concurrent administration of the potent SSRI.[32] Nevertheless, it is important to realize that the potential for drug interaction via CYP2D6 inhibition could also be affected by the basal metabolic activity of the individual patient. We have shown that the UM phenotype could affect the potential for drug interaction with paroxetine, a CYP2D6 substrate as well as a potent CYP2D6 inhibitor, whence an UM with three functional *CYP2D6* gene copies had undetectable paroxetine concentration with standard dosing and showed no inhibitory effect at CYP2D6.[33]

CYP2C19

Compared to the *CYP2D6* polymorphism, polymorphisms in the *CYP2C19* gene do not affect as many drugs, and their clinical implication has not been extensively evaluated. However, studies involving the proton pump inhibitors (PPIs) provide extensive pharmacokinetic and clinical evidence, as well as economic impact of the importance of taking into consideration *CYP2C19* polymorphism in the management of gastroesophageal diseases.

Over the years, more than 15 *CYP2C19* alleles, including *2, *3, *4, *5, *6, *7, and *8, have been identified (www.cypalleles.ki.se/cyp2c19.htm). The principal null alleles are *2 (c.19154G>A, rs4244285) and *3 (c.17948G>A, rs4986893), resulting in inactive CYP2C19 enzyme, and accounting for the vast majority of the PM phenotype in Caucasians (1 to 6%), black Africans (1 to 7.5%), and Asians (10 to 25%). Genotyping these two defective alleles has been shown to detect about 84%, greater than 90%, and about 100% of PMs in Caucasians, Africans, and Asians, respectively. The detection rate for Caucasians' PMs could be increased to about 92% by including the less common *CYP2C19*4* and *CYP2C19*6* in the genotyping assay. Similar to *CYP2D6* polymorphism, a recently identified *CYP2C19*17* allele (c.-806C>T rs12248560 SNP) in the 5'-flanking region of *CYP2C19* with increased transcription is associated with high enzyme activity and a more rapid metabolism phenotype.[34]

*CYP2C19*2* and *3 are commonly found in Asians, with allele frequencies of about 30% and approximately 10%, respectively. In contrast, *2 occurs at a frequency of about 13% in Caucasians and approximately 18% in African Americans, whereas the allele frequency of *3 is less than 1% in these ethnic groups. About 50% of the Chinese population possess either the *1/*2 or *1/*3 genotype, and 24% have the *2/*2, *2/*3, or *3/*3 genotype.[35] In contrast, only about 2 to 5% and 30 to 40% of the Caucasian population, respectively, have the *2/*2 and *1/*2 genotypes. Similar frequencies of the heterozygous and homozygous variant genotypes are reported in persons of African descent. The higher prevalence of both PMs and heterozygotes for the defective *CYP2C19* alleles in Asians likely accounts for reports of slower rates of metabolism of CYP2C19 substrates and the practice of prescribing lower diazepam dosages for patients of Chinese heritage [36,37]. An opposite direction in ethnic variation was observed in the prevalence of *CYP2C19*17* (18% in Swedes and Ethiopians versus 4% in Chinese), with the *1/*17 and *17/*17 genotypes occurring in more Caucasians and Ethiopians (up to 36%) than Asians (8% of Chinese and 1% of Japanese).[34]

CYP2C19 accounts for about 3% of total hepatic CYP content, and *CYP2C19* polymorphism affects the metabolism of PPIs (omeprazole, lansoprazole, pantoprazole, rabeprazole), antidepressants (citalopram, sertraline, moclobemide, amitriptyline, clomipramine), the antiplatelet agent clopidogrel, the antifungal drug voriconazole, the benzodiazepine diazepam, and the anticancer drug cyclophosphamide. Currently, the PPIs and clopidogrel provide the best examples of clinical relevance of *CYP2C19* polymorphism. When compared to EMs, PMs showed 5- to 12-fold increases in the area under the curve (AUC) of omeprazole, lanzoprazole, and pantoprazole,[38,39] whereas homozygous carriers of the *CYP2C19*17* were shown to have a modest 2.1-fold lower AUC than EMs.[40] In addition, the *CYP2C19* genotype significantly affects the achievable intragastric pH with PPI therapy. In subjects who took a single 20 mg dose of omeprazole, Furuta et al. showed a good relationship not only between *CYP2C19* genotype and AUC, but also between the genotype and achievable intragastric pH: 4.5 in PMs, 3.3 in heterozygous EMs, and 2.1 in homozygous EMs.[41] Given the smaller dependency of esomeprazole and rabeprazole on CYP2C19 for metabolism, the pharmacological action of these two PPIs is less affected by the *CYP2C19* polymorphism [42,43].

An important treatment strategy in the management of patients with peptic ulcer disease is eradication of *Helicobacter pylori* with a regimen of PPI and antibiotics. *CYP2C19* genotype-related pharmacological effects have also been associated with improved eradication rate of *Helicobacter pylori* after dual[44] or triple therapy including omeprazole,[45] lansoprazole,[46] or pantoprazole.[47] The cure rate achieved with dual- and triple-therapy regimens was 100% in PMs compared with 29 to 84% in EMs [44,45,46,47]. Furuta et al. also reported a much higher eradication rate of 97% in EMs who failed initial triple therapy (lansoprazole, clarithromycin, and amoxicillin) and subsequently were retreated with high-dose lansoprazole (30 mg four times daily) and amoxicillin.[48] In addition to showing a gene—dose effect in achieving desirable ranges of intragastric pH and *Helicobacter pylori* cure rates for lansoprazole, Furuta et al. also demonstrated the cost effectiveness of pharmacogenomic-guided dosing when compared to conventional dosing.[49] On the other hand, despite increased metabolism of PPI in carriers of *CYP2C19*17* and the potential of therapeutic failure,[34,50] eradication rates of *Helicobacter pylori* have so far not been shown to be associated with the *CYP2C19*17* allele, at least for patients with peptic ulcer disease who are receiving the triple regimen of pantoprazole, amoxicillin, and metronidazole [47,51].

Similarly, despite the presence of pharmacokinetic differences, the impact of *CYP2C19*17* on therapeutic outcomes with other CYP2C19 substrates is not known. In healthy volunteers given a single 200 mg dose of voriconazole, Wang et al. demonstrated a 48% lower AUC in heterozygous carriers of the *CYP2C19*17* allele as compared to homozygous carriers of *CYP2C19*1*.[52] Investigators have also shown 42% lower escitalopram concentrations and 21% lower AUC in patients who are homozygous carriers of *CYP2C19*17* when compared to *CYP2C19*1* homozygotes.[53] Clearly, more studies are needed to determine whether *CYP2C19*17* homozygotes would require higher doses of most CYP2C19 substrates, including PPIs,[40,50] antidepressants, and voriconazole.[52]

Clopidogrel is an antiplatelet prodrug that requires CYP2C19-mediated conversion to its active metabolite for therapeutic effect, with most pharmacokinetic and pharmacodynamic evidence related to the *CYP2C19*2* allele [54,55,56,57,58]. Shuldiner et al.[59] conducted a genome-wide association study (GWAS) in which *ex vivo* adenosine diphosphate (ADP)—induced platelet aggregation at baseline and after seven days of clopidogrel were measured in a genetically homogeneous cohort of 429 healthy Amish subjects. In addition, 400,230 SNPs were evaluated in each subject for association with platelet activity. They reported that the SNP on chromosome 10q24 with the greatest association signal is in strong LD with CYP2C19*2, accounting for 12% of the interindividual variation in platelet aggregation during clopidogrel treatment. Also important is that there was no association between the CYP2C19 polymorphism and baseline platelet aggregation.[59] The results from this GWAS confirmed results from previous candidate gene studies regarding the role of CYP2C19 as a major genetic determinant of clopidogrel response [54–58]. In a follow-up study of 227 patients undergoing percutaneous coronary intervention (PCI), the investigators also reported a higher incidence of cardiovascular death in carriers of the *2 allele (20.9% versus 10%) at 1-year follow-up. No association with response was found for other CYP2C19 alleles, including *3, *5, and *17, that were also genotyped in the study.[59] Although the increased production of the active clopidogrel metabolite in carriers of the *17 allele has been associated with greater inhibition of platelet aggregation[60,61] and better clinical outcomes,[62] there is also the potential of increased bleeding risk.[63] In addition, the increased response of the *17 allele has been suggested not as a direct effect but rather attributed to that of the *1 allele.[64]

The increased risks of major adverse cardiovascular events and stent thrombosis in carriers of at least one CYP2C19*2 allele were also confirmed in two meta-analyses that included almost 22,000 patients [63,65]. Differences in patient selection for analysis likely account for the lack of association reported in two other recent meta-analyses, which included a significant number of low-risk patients, such as those with acute coronary syndrome managed medically or patients with atrial fibrillation [66,67]. The meta-analysis of Hulot et al.[65] also evaluated the drug interaction potential of PPIs as a result of their inhibitory effect towards CYP2C19, resulting in a metabolic phenotype of CYP2C19 PM similar to that of carriers of the *2 allele. Both Hulot et al. and another study[65,68] suggest that the detrimental effects of PPIs on cardiovascular outcomes with clopidogrel likely occur at a higher frequency in high-risk patients receiving both drugs. Current data do not provide sufficient information to determine whether the observed adverse effects of PPI usage in high-risk patients (e.g., patients undergoing PCI) are related to CYP2C19 inhibition or yet-to-be-discovered mechanisms.

Based on the increasing amount of literature data supporting an association between CYP2C19*2 and poor clopidogrel response, the FDA approved revision of the product label of clopidogrel in March 2010. Although the revised label specifically addresses the implication for homozygotes, there is no guidance on the implication for heterozygotes. In addition, as with other revised labels with addition of genetic information, there is little guidance on clinical management of carriers of CYP2C19*2. In light of the scientific and clinical evidences as well as the regulatory decision, several recent clinical studies addressing alternative antiplatelet agents have been initiated and are discussed in chapter 5 of this book.

CYP2C9

To date, more than 35 CYP2C9 alleles (www.cypalleles.ki.se/cyp2c9.htm) have been identified in the regulatory and coding regions of CYP2C9, with CYP2C9*2 (c.3608C>T, rs1799853) and CYP2C9*3 (c.42614A>C, rs1057910) being the most common in persons of

European descent and also the most extensively studied. Both reduced-function alleles exhibit single-amino-acid substitutions (p.R144C and p.I359L, respectively) in the coding region, accounting for lower enzyme activity by approximately 30% for *2 and 80% for *3.[69] Other reduced-function alleles of potential importance included *5, *6, *8, and *11 [70,71,72]. In addition, a "gain-of-function" CYP2C9 c.18786A>T variant in intron 3 has been identified.[70]

Significant variations in CYP2C9 allele and genotype frequencies exist among different ancestry groups. Both CYP2C9*2 and CYP2C9*3 are more common in Caucasians (11% and 7%, respectively) than in Asians and Africans. In fact, CYP2C9*2 has not been detected in Asians, in whom CYP2C9*3 is the most common allele. On the other hand, CYP2C9*8, as well as *5, *6, and *11 (albeit all at a lower frequency than *8), are present almost exclusively in African Americans. The novel CYP2C9 c.18786A>T variant was reported to occur in about 40% of the African American population.[70] Approximately 1% and 0.4% of Caucasians have the *2/*2 and *3/*3 genotypes. The *1/*3 genotype occurs at a frequency of 4% in the Chinese and Japanese populations, with almost complete absence of the other genotypes (*2/*2, *2/*3, *1/*2, and *3/*3).

CYP2C9 accounts for about 20% of total hepatic CYP content and is involved in the metabolism of about 10% of currently marketed drugs. These CYP2C9 substrates include the nonsteroidal anti-inflammatory drugs such as celecoxib, ibuprofen, and flurbiprofen; oral anticoagulants such as warfarin, acenocoumarol, and phenprocoumon; oral antidiabetic agents such as glyburide and tolbutamide; antiepileptic agents such as phenytoin; and antihypertensive agents such as candesartan and losartan. The enzyme reduction associated with the *3 allele is greater than that with the *2 allele, with a 5- to 10-fold reduction in homozygous *3 carriers and 2-fold reduction in heterozygous *3 carriers when compared to homozygous *1 carriers. For example, clearance of warfarin is reduced by 90%, 75%, and 40% in subjects with the corresponding CYP2C9 genotypes of *3/*3, *1/*3, and *1/*2.[73] Interestingly, the effects of several reduced-function alleles appear to be substrate-dependent. For the *2 allele, a significant effect was shown for clearances of acenocoumarol, celecoxib, tolbutamide, and warfarin but not for other substrates. Similarly, although the *8 allele has no effect on clearance of losartan, it decreases enzyme activity of warfarin and phenytoin and exhibits an increased activity towards tolbutamide.[74]

Of all of the CYP2C9 substrates, warfarin is the most extensively studied, with dosing implications for different metabolic phenotypes. CYP2C9 polymorphism, together with the literature information regarding the gene that encodes the warfarin target, vitamin K epoxide reductase complex (VKORC1),[75] provide promising translational use of the pharmacogenomic data [76,77]. CYP2C9 mediates the conversion of the active S-enantiomer of warfarin to an inactive metabolite. Most of the data document that the *2 and *3 alleles are associated with greater difficulty with warfarin induction therapy, increased time to achieve stable dosing, lower mean dose requirement (e.g., as low as ≤1.5 mg/day with *3/*3), as well as increased risks of elevated international normalized ratios (INRs) and bleeding [76,78,79]. Giving the 30% and 80% difference in enzyme activity reduction between the *2 and 3 alleles, the warfarin dose requirements differ between carriers of these two alleles. Compared to homozygous carriers of the *1 allele, data suggest a dose reduction of 30% and 47% for patients with the heterozygous genotypes of CYP2C9*1/*2 and CYP2C9*1/*3, respectively, and up to 80% for patients with the homozygous CYP2C9*3/*3 genotype [77,78,80,81]. In addition, with the difference in allele prevalence among different ancestral groups, the strength of association between the *2 and *3 alleles and genotypes is stronger in Caucasians [82,83]. Other recently identified alleles (*5, *6, *8, and *11) have been reported to better predict dose requirement (20% lower for *8 carrier) and adverse outcomes in African Americans [70,71,72,74,82,84]. On the other hand, the gain-of-function CYP2C9 c.18786A>T allele was reported to contribute a higher dose requirement (3.7 mg/week/allele).[70] The effect of CYP4F2 and VKORC1 genotypes on warfarin pharmacokinetics and pharmacodynamics will be discussed in later sections of this chapter.

CYP2C8

In addition to *CYP2C9* and *CYP2C19*, the other clinically relevant member of the highly homologous genes (*CYP2C18-CYP2C19-CYP2C9-CYP2C8*) that cluster on chromosome 10q24^{59} is *CYP2C8*. To date, several SNPs within the coding region of the *CYP2C8* gene have been identified (www.cypalleles.ki.se/cyp2c8.htm). The more common variants are *2 (c.11054A>T, rs11572103, resulting in p.I269F), *3 with two amino acid substitutions (c.2130G>A, rs11572080 with p.R139K, and c.30411A>G, rs10509681 with p.K399R) reported to be in total LD, and *4 (p.I264M). Both *3 and *4 alleles are more common in Caucasians, whereas *2 is present primarily in Africans [85,86]. To date, no null allele has been identified for *CYP2C8*.

Accounting for about 7% of total hepatic content, the hepatic expression level of CYP2C8 lies between that of CYP2C19 and CYP2C9,[87] and it plays an important role in the metabolism of different drugs, primarily the antidiabetic agents (pioglitazone, repaglinide, rosiglitazone, and troglitazone), the anticancer agents (paclitaxel), the antiarrhythmic drug amiodarone, and the antimalarial agents amodiaquine and chloroquine. The smaller number of substrates as compared to CYP2C9 and CYP2C19 presumably leads to the lesser interest in studying *CYP2C8* polymorphism. As a result, the molecular mechanisms underlying interindividual variations in CYP2C8 activity remain unclear. Decreased elimination of R-ibuprofen has been reported in carriers of *CYP2C8*3 [88,89]. However, with the presence of a strong LD between *CYP2C8*3 and *CYP2C9*2,[88,90] the individual contribution of *CYP2C8*3 remains to be elucidated. In contrast, increased metabolism of repaglinide was reported in heterozygous carriers of *CYP2C8*3 when compared to carriers of either *1 or *4.[91] Though this finding is interesting, other reports show that genetic polymorphism of the hepatic uptake transporter plays a more important role in determining repaglinide pharmacokinetics.[92] The identification of two *CYP2C8* haplotypes—a high-activity allele associated with *CYP2C8*1B* and a low-activity allele associated with *CYP2C8*4* [93]—highlights the need to further characterize the different *CYP2C8* variants, including any functional relevance.

CYP3A4/5/7

A total of four *CYP3A* genes have been described in humans: *CYP3A4*, *CYP3A5*, *CYP3A7*, and *CYP3A43*, with *CYP3A7* primarily important in fetal CYP3A metabolism and *CYP3A43* exhibiting little functional or clinical relevance. More than 20 variants in the coding region of *CYP3A4*, most of them associated with reduced activity, have been identified to date.[94] However, the current consensus is that *CYP3A4* polymorphism is mostly of minor clinical relevance, and by itself unlikely to account for the 10- to 40-fold interindividual variations in CYP3A4 activities. This is likely a result of low variant allele frequencies and only small changes in enzyme activity in the presence of a variant allele, as well as the overlapping substrate specificity between CYP3A4 and CYP3A5. The most common variant for CYP3A5 polymorphism is the *3 allele (c.6986A>G, rs776746) in intron 3 that results in a splicing defect and absence of enzyme activity. Other CYP3A5 variants with associated reduced enzyme activity include the *6 and *7 alleles.[95]

In general, CYP3A4 polymorphism is more common in Caucasians, with *2 and *7 being the more prevalent alleles, whereas Asians have higher frequencies of *16 and *18 variants. The allele frequency of CYP3A5*3 is much higher in Caucasians and Asians, occurring in 90% and 75% of the populations, respectively, versus a relatively low frequency of 20% in Africans. On the other hand, both CYP3A5*6 and *7 are absent in Caucasians and Asians but present in Africans with frequencies up to 17% [95,96,97].

Although CYP3A4 accounts for about 40% of the total hepatic CYP content and mediates the metabolism of more than 50% of currently used drugs, the clinical relevance of genetic polymorphism is primarily associated with CYP3A5. The pharmacokinetics of the immunosuppressive agent tacrolimus are dependent on the CYP3A5 genotype, with a higher dosage requirement in homozygous or heterozygous carriers of CYP3A5*1 (also known as CYP3A5 expressor in the literature) [98,99]. In addition, results from a randomized controlled trial showed that pharmacogenetic-guided dosing based on CYP3A5 genotype was associated with greater achievement of target tacrolimus concentrations when compared to standard dosing based on body weight.[100] Nevertheless, the overall clinical relevance of CYP3A5 polymorphism is limited by its small contribution (2 to 3%) to the total CYP3A metabolism [101,102].

CYP4F2

The importance of the newly discovered CYP4F2 is related to the recent report of its role in mediating the conversion of vitamin K_1 to hydroxyvitamin K_1. Increased CYP4F2 activity results in decreased activation of vitamin K-dependent clotting factors, reflecting the consequence of reduced availability and reduction of vitamin K_1 to vitamin KH_2 necessary for carboxylation and activation of the clotting factors. On the other hand, a p.V433M SNP in exon 2 of the CYP4F2 gene results in lower protein expression and enzyme activity and consequently greater vitamin K_1 availability [103,104]. Some ethnic differences in the V433M SNP have been reported, with the M433 allele occurring at a much lower frequency in African Americans,[84] which contrasts with its high occurrence in Indonesians and Egyptians [105,106].

In contrast to the CYP2C9 genotype, which accounts for approximately 10 to 12% of the variability in warfarin dose requirement, the CYP4F2 genotype accounts for only 1 to 3% of the overall variability [104,107]. Homozygous carriers of the M allele of the V433M SNP have been shown to require an approximate 1 mg per day higher dose of warfarin than homozygous carriers of the V allele.[108] However, additional studies have demonstrated the association between the CYP4F2 genotypes and dose requirements in Caucasians and Asians [104,107,109,110], but not in Africans Americans, Egyptians, or Indonesians [84,105, 106]. This finding could reflect ethnic differences in CYP4F2 allele and genotype distribution and the minor contribution of CYP4F2, as well as the modulating effects of other more important dose requirement variables such as CYP2C9 and VKORC1.

CYP2B6

Although several variant alleles with low enzyme expression—including CYP2B6*6, *16, and *18—have been identified, to date there have not been any reports of the presence of an important loss-of-function allele.[111] Of these three variant alleles, CYP2B6*6 carrying two SNPs (c.516G>T, rs3745274 and c.785A>G, rs2279343, causing two amino acid changes, p.Q172H and p.K262R, respectively) is the most common and occurs commonly in Caucasians and Asians (16 to 26% allele frequency) whereas *16 and *18 are more common in African subjects with allele frequencies of 7 to 9%.[112] Interestingly, the 785A>G SNP resulting in the K262R amino acid change also occurs as a separate allele, CYP2B6*4, and results in increased expression and enzyme activity [113,114]. Whether the 516G>T and 785A>G mutations are linked to additional mutations creating specific haplotypes causing either high or low CYP2B6 activities is not known. Gatanaga et al. also report a new *26 allele containing 499G for the c.499C>G SNP, and 499G always coexists with 516G>T and 785A>G, thus representing a novel haplotype containing the 499C>G, 516G>T, and 785A>G SNPs.[115]

CYP2B6 accounts for up to 6 to 10% of total CYP content in the liver,[116,117] and known substrates include anticancer drugs such as cyclophosphamide and ifosfamide, the smoking cessation agent bupropion, the antiretroviral agents efavirenz and nevirapine, as well as methadone. The potential clinical relevance of CYP2B6 has been evaluated mostly with the non-nucleoside reverse transcriptase inhibitor efavirenz. Increased central nervous system (CNS) side effects associated with variable systemic exposure of efavirenz could be the result of patients being carriers of the *6, *16, or *18 alleles [111,115]. Incorporating determination of additional less frequent alleles such as *26 and *29 could further improve the prediction of elevated plasma efavirenz concentrations [111,115,118].

In a prospective study of the effect of CYP2B6 polymorphism on efavirenz concentrations and exposure, 456 patients infected with the human immunodeficiency virus type 1 (HIV-1) were genotyped for different SNPs, including the 499C>G, 15631G>T, and 18053A>G polymorphisms.[115] All patients received the standard dosage regimen of 600 mg per day, and extremely high concentrations (9,500 ± 2,580 ng/mL) were obtained in all 14 patients with the CYP2B6*6/*6 genotype and in both patients with the CYP2B6*6/*26 genotype. In contrast, only two patients with other CYP2B6 genotypes had similarly high efavirenz concentrations, and both are heterozygous carrier of either the *6 allele (7,140 ng/mL) or the *26 allele (9,710 ng/mL). Therefore, the *6 and *26 alleles were both associated with high efavirenz concentrations, and patients with the CYP2B6*6/*6 or the CYP2B6*6/*26 genotype had the highest concentrations with standard dosage regimen of 600 mg per day.

To investigate the feasibility of dose reduction in patients with high efavirenz concentrations secondary to CYP2B6 polymorphism, the investigators then reduced the efavirenz dosage regimen to 400 mg per day in 5 patients and to 200 mg per day in another 7 patients. The genotypes in these 12 patients included nine *6/*6 homozygotes, two *6/*26 heterozygotes, and one *1/*26 heterozygote. The plasma concentrations decreased proportionally with the dose reductions. Despite receiving the lower dosage regimens for more than 6 months, the 12 patients were able to maintain therapeutically effective anti-HIV-1 activity with HIV-1 load continuously less than 50 copies/ml. CNS side effects were reported to be much less frequent at the lower dosage regimens. Similar therapeutic success with persistent suppressed HIV-1 load was also demonstrated in efavirenz-naïve patients (*6/*6 and *6/*26) who were administered the lower dosage regimen of 400 mg per day. The overall study results demonstrated the feasibility of genotype-based efavirenz dose reduction in patients with CYP2B6 *6/*6 and *6/*26 genotypes, with additional advantages of less CNS side effects and lower treatment cost.

CYP2A6

Significant variations in CYP2A6 activity are primarily a result of genetic influence. The primary variants for CYP2A6 polymorphism (www.cypalleles.ki.se) include CYP2A6*2 (single-amino-acid substitution), CYP2A6*4 (gene deletion), CYP2A6*5 (gene conversion), and CYP2A6*20 (frameshift), all of which are associated with abolished enzyme activity. Additional alleles associated with reduced enzyme activity include *6, *7, *10, *11, *12, *17, *18, and *19. As with other CYP polymorphisms, there are substantial interethnic differences in allele frequency. Deletion of the CYP2A6 gene is very common in Asian patients,[119] which likely accounts for the dramatic difference in the high occurrence of PMs in Asian (20%) versus Caucasian populations (≤1%).

Nicotine is metabolized by CYP2A6 to cotinine, and the clinical relevance of the CYP2A6 polymorphism has been primarily investigated in managing tobacco abuse. Nonsmokers were found to be more likely to carry the defective CYP2A6 allele than were smokers. In addition, smokers with the defective CYP2A6 allele smoked fewer cigarettes and were more likely to quit. These results likely reflect higher nicotine concentrations, enhanced nicotine tolerance, and increased adverse effects from nicotine in CYP2A6 PMs. Based on these observations, CYP2A6 inhibition may have a role in the management of tobacco dependency.[119]

CYP1A2

Although polymorphisms of the CYP1 family of genes have been studied for association with cancer susceptibility, to date there has been no report of functional CYP1A2 alleles that result in important changes in gene expression and enzyme activity. The CYP1A2*1F allele (rs762551) containing a c.-163C>A mutation in intron 1 has been shown to affect CYP1A2 inducibility and the magnitude of increased caffeine metabolism in smokers [120,121]. However, conflicting findings have been reported for other CYP1A2 substrates [122,123,124]. In general, promoter variation is less likely to result in substrate-dependent effects, and the functional importance of increased CYP1A2 inducibility is currently unknown.

CYP1A2 contributes up to 10% of the total hepatic P-450 content. However, unlike other CYP isoenzymes, it only mediates the metabolism of several commonly used drugs such as olanzapine, clozapine, duloxetine, and theophylline. Although pharmacokinetic studies evaluating CYP1A2 inducibility by smoking or omeprazole have been performed, none of the studies have produced consensus information.

NON-CYP-450 DRUG METABOLIZING ENZYMES

Genetic polymorphisms in many non-P-450 enzymes also play a role in influencing metabolism and elimination of many drugs. Among these enzymes, UDP-glucuronosyl transferase (UGT), thiopurine-S-methyltransferase (TPMT), dihydropyrimidine dehydrogenase (DPD), N-acetyltransferase (NAT), and glutathione-S-transferase (GST) have been characterized and their clinical relevance studied.

UDP-Glucuronosyl Transferase (UGT)

Among the nine UGT1A enzymes, UGT1A1 has been the most extensively investigated. A polymorphism in the promoter region of the *UGT1A1* gene results in the *(TA)[7] TAA* allele or *UGT1A1*28*, with a 35% decrease in transcriptional activity of *UGT1A1* and lower enzyme activity than the wild-type *(TA)[6] TAA* allele.[125] Another *UGT1A1* polymorphism, *UGT1A1*6* carrying the c.211G>A SNP and p.G71R substitution in exon 1, has been associated with lower enzymatic activity.[126] Although the *28* variant is more common in Caucasians (29 to 40%) and Africans (36 to 43%) than in Asians (13 to 16%), the *6* variant is found only in Asians with a frequency of 16 to 23%.

UGT contributes about 35% of phase II drug metabolism and is involved in glucuronidation of endogenous compounds and xenobiotics. For UGT1A1, the substrates include bilirubin, SN-38 (active metabolite of the anticancer drug irinotecan), raltegravir (inhibitor of the HIV integrase enzyme), bazedoxifene (an investigational selective estrogen receptor modulator for prevention and treatment of postmenopausal osteoporosis), and eltrombopag (a thrombopoietin receptor agonist for the management of thrombocytopenia). Irinotecan, also known as CPT-11, is a prodrug that requires metabolic activation via carboxylesterase to SN-38, a potent inhibitor of topoisomerase I. SN-38 is inactivated via glucuronidation by the polymorphic UGT1A1 enzyme. Both *UGT1A1*28* and *6* have been associated with impaired SN-38 glucuronidation, especially in patients who are homozygous carriers [126,127]. The ensuing high SN-38 concentrations lead to increased SN-38 excretion into the gut lumen, predisposing patients to severe diarrhea even with standard irinotecan dosage regimens. Abnormally high SN-38 concentrations have also been reported in patients with severe neutropenia.[128] These pharmacogenetic-related adverse reactions have also been demonstrated in a prospective clinical trial[129] that led to FDA approval of the Invader® UGT1A1 Molecular Assay (Third Wave Technologies) for genotyping *UGT1A1* alleles and revision of the irinotecan product label to include recommendation of adjustment for individuals who are homozygous for *UGT1A1*28*. With the involvement of UGT1A1 in bilirubin glucuronidation and the prevalence of *UGT1A1*6* among Asian populations, UGT1A1 may play a role in the high incidence of neonatal hyperbilirubinemia in those populations.[130]

Thiopurine-S-Methyltransferase (TPMT)

Although more than 20 variants of the *TPMT* gene have been identified, the three most studied are *TPMT*3A* (abolished activity), *TPMT*2* (reduced activity), and *TPMT*3C* (reduced activity), with *3A* being the most common. Approximately 10% and 0.3% of the

Caucasian population is heterozygous and homozygous, respectively, of these mutant alleles.

TPMT mediates the inactivation of thiopurine drugs, including thioguanine, 6-mercapto-purine, and the precursor of 6-mercaptopurine, azathioprine. Compared to patients who possess the wild-type alleles, homozygotes or heterozygotes for the *TPMT* mutant alleles have much higher levels of the cytotoxic thiopurine nucleotides and are at higher risk for developing serious hematological toxicities during treatment with standard dosage regimens of the thiopurine drugs.[131] As an example, those patients with absent and low TPMT activity can tolerate only 5% and 50% of the standard 6-mercaptopurine regimen.

Dihydropyrimidine Dehydrogenase (DPD)

The phase I metabolizing enzyme DPD mediates the metabolism of 5-fluorouracil (5-FU), and genetic polymorphisms in the *DPD* gene result in DPD-deficiency phenotypes with an overall frequency of about 3 to 5%.[132] Among the known SNPs associated with grade-3 and grade-4 toxicities in 5-FU treated patients (c.2846A>T, c.1679 T>G, c.85T>C, and c.IVS14+1G>A), the G>A mutation within intron 14 results in a protein with no catalytic activity [133,134]. Homozygous and heterozygous carriers of the variant IVS14+1G>A allele have complete absence and 50%, respectively, of normal DPD activity, and significant, sometimes life-threatening 5-FU-related toxicities.[135]

N-Acetyltransferase (NAT)

Genetic polymorphism in acetylation capacity was reported more than fifty years ago, when two distinct phenotypes of rapid acetylator (RA) and slow acetylator (SA) were noted in patients enrolled in a clinical trial of the antituberculosis drug isoniazid.[136] Subsequently, the phenotype differences were associated with NAT1 and NAT2 enzyme activities, which are encoded by the *NAT1* and *NAT2* genes, respectively. Nevertheless, the NAT2 enzyme is primarily responsible for acetylation of aromatic amines and hydrazines. Polymorphism in *NAT2* results in more than ten *NAT2* alleles, with *NAT2*4* reported as the wild-type allele, and *NAT2*5* carrying the c.341T>C SNP that results in the p.I114T amino acid change, *NAT2*6* with c.590G>A SNP and p.R197Q substitution, as well as *NAT2*7* with c.857G>A SNP and corresponding p.G286E substitution as the primary variant alleles.[137] These three variant alleles account for the majority of the SA phenotype. The prevalence of SAs varies significantly in different ethnic groups: 90% of Arab populations, 40 to 60% of Caucasians, and 5 to 25% of East Asians.

Substrates for NAT include numerous arylamine- and hydrazine-containing drugs such as sulfamethoxazole, hydralazine, isoniazid, and procainamide. High blood levels of these and similarly acetylated drugs in SAs have been associated with lupus-like syndrome (hydralazine and procainamide), peripheral neuropathy (isoniazid), and liver damage (sulfapyridine). In addition to drug therapy, *NAT2* polymorphism has also been implicated in susceptibility to developing different types of cancer, with SA having an increased risk after prolonged exposure to carcinogenic arylamines and other industrial chemicals.[138]

Glutathione-S-Transferase (GST)

The human GST family of cytosolic enzymes contains at least 17 genes divided into sevenclasses: α, μ, π, σ, θ, ζ, and ω. Of these, the most important genes are *GSTM1* of the μ class, *GSTT1* of the θ class, *GSTP1* of the π class, and *GSTA1* of the α class. Both homozygous deletion polymorphisms and SNPs exist for *GST* genes. Gene deletion results in *GSTM1*0* and *GSTT1*0* and loss of GSTM1 and GSTT1 enzyme function, respectively. The frequency of occurrence is reported to be 42 to 58% of Caucasians and 27 to 41% of Africans lacking the *GSTM1* gene and 2 to 42% of Caucasians lacking the *GSTT1* gene.[139] Two polymorphisms of the *GSTP1* gene have been described: rs947894 carrying the exon 5 c.A1404G SNP and p.I105V substitution at codon 105, and rs1799811 carrying the exon 6 c.C2294T SNP and p.A114V substitution at codon 114. Four different haplotypes have been described for the population: *GSTP1*A* ([105]Ile-[114]Ala), *GSTP1*B* ([105]Val-[114]Ala), *GSTP1*C* ([105]Val-[114]Val), and *GSTP1*D* ([105]ILe-[114]Val).[140] A point mutation in the promoter of the *GSTA1* gene results in lower promotor activity associated with the *GSTA1*B* allele. In contrast to deleted *GST* genotypes, the *GSTP1* and *GSTA1* polymorphisms result in low-activity genotypes. The *GSTP1* and *GSTA1* polymorphisms have been reported to occur in up to 40% of Caucasians and 54% of Africans and 40% of Caucasians and 41% of Africans, respectively.

GSTs are detoxification enzymes that mediate the conjugation of reduced glutathione with different substrates that include carcinogens and chemotherapeutic agents such as oxaliplatin-based chemotherapy and chlorambucil,[139,141,142,143] with poorer response and reduced overall survival in patients with *GSTM1*0* or *GSTT1*0* genotypes and treated with oxaliplatin-based chemotherapy or anthracycline-based induction therapy [143,144,145]. Because GSTs are detoxification enzymes, the shortened survival in patients with reduced GST activity might be related to severe drug-related toxicity, as evidenced by a higher frequency of grade 4 neutropenia in homozygous carriers of *GSTM1*0* treated with oxaliplatin-based chemotherapy for their metastatic colorectal cancer.[146]

POLYMORPHISMS IN DRUG TRANSPORTER GENES

Membrane transporters are present at many endothelial and epithelial barriers, including the blood—brain barrier (BBB), the intestinal epithelial cells, the hepatocytes, and the renal tubular cells. By facilitating drug excretion into the gastrointestinal tract, bile, liver, and kidney, as well as limiting the amount of drug crossing the BBB, they provide an important physiological role of protecting humans against toxic xenobiotics and have recently been recognized as important determinants of drug disposition and response.[147] These drug transporters can be broadly classified into two groups: the efflux adenosine triphosphate-binding cassette (ABC, formerly known as multidrug resistance [MDR]) family of transporters, and the uptake solute carrier (SLC) family of transporters. A total of 49 members are present within the human ABC transporter family. Based on the homology of their amino acid sequences, they are further classified into seven subfamilies. Of all the ABC transporters, the better known examples are ABCB1 (P-glycoprotein [Pgp] or MDR1), ABCC1 (multidrug resistance 1 [MRP1]), ABBC2 (multidrug resistance [MRP2]), and ABCG2 (breast cancer resistance protein [BCRP]). In the SLC family are 360 members that are subdivided into 46 subfamilies. The better known SLC transporters are organic anion transporting polypeptide (OATP), organic cation transporter (OCT), and organic anion transporter (OAT). Genetic variants of the genes encoding these drug transport proteins (www.pharmGKB.org) have been discovered that affect their expression, substrate specificity, and/or intrinsic transport activity, and ultimately the disposition, efficacy, and safety of many drug substrates.

The ABC Transporters

ABCB1

ABCB1 was the first recognized and the most studied ABC transporter. It is encoded by the highly polymorphic *ABCB1*, with more than 50 SNPs and three insertion/deletion polymorphisms reported. The most common SNPs are the c.C1236T (rs1128503) silent polymorphism in exon 12, the rs2032583 polymorphism in exon 21, and the c.C3435T (rs1045642) silent polymorphism in exon 26. The c.G2677A/T (rs2032583) polymorphism results in a change in amino acid sequence p.A893S (G2677T) SNP or p.A893T (G2677A) SNP. Ethnic variations in allelic variant distribution are well known [148,149]. In addition, strong LD among these SNPs had been reported to create haplotypes consisting of 1236C>T, 2677G>A/T, and 3435C>T. The three *ABCB1* SNPs and their haplotypes (Table 1.2) are important in the expression and function of ABCB1.

The functional and clinical implication of the *ABCB1* polymorphism was first evaluated for the C3435T SNP with digoxin as the substrate, demonstrating a relationship between lower expression of ABCB1 and increased digoxin concentration after oral administration in TT homozygotes.[150] The polymorphism also affects plasma concentrations and clinical effects of protease inhibitors. After 6-month therapy with efavirenz or nelfinavir, patients with the TT genotype had a greater rise in CD4 cell counts than patients with the CC genotype.[151] Therefore, *ABCB1* genotyping may have a role in predicting hematologic responses to protease inhibitors. Nevertheless, conflicting results have been reported regarding the functional and clinical significance of the polymorphism for different substrates including psychotropics (see chapter 6), antiretroviral protease inhibitors, immunosuppressants, and anticancer drugs. This conflict may be due to the use of different assays and study designs to identify ABCB1 substrates; the overlapping substrate specificity between ABCB1 and other enzymes and transporters—for example, CYP3A4 for cyclosporine and OATP transporters for fexofenadine;[152] and/or the existence of strong LD necessitating a haplotype approach rather than individual SNPs in association studies or clinical evaluations. In addition, the 1236C>T/2677G>T/3435C>T haplotype was shown to affect the inhibition of substrate transport, not the transport process per se.[153] Thus the functional effect of *ABCB1* polymorphism may be more modest than previously thought. Whether additional mutations resulting in loss of function or significant change in substrate specificity or functionality would have a bigger impact is not known and awaits further studies for clarification.

ABCC1 and ABCC2

Both ABCC1 and ABCC2 are involved in the biliary excretion of conjugated drugs such as glucuronides or sulfates of tamoxifen and SN-38 glucuronide,[154,155] organic anions, and

TABLE 1.2 Selected ABC transporter polymorphisms indicating allele variants and frequency and drug substrates

Genes	Allele Variants, Amino Acid Change	Frequency (%)	Drug Substrate Examples
ABCB1	3435C>T	48–59% in Caucasians 37–66% in Asians 10–27% in Africans	Protease inhibitors (ritonavir, saquinavir, nelfinavir, efavirenz)
	1236C>T	34–42% in Caucasians 60–72% in Asians 15–21% in Africans	Anticancer drugs (anthracyclines, taxanes, vinca alkaloids, imatinib) Immunosuppressants (cyclosporine, tacrolimus)
	2677G>T, A893S	38–47% in Caucasians 32–62% in Asians ≤15% in Africans	Antibiotics (erythromycin, levofloxacin) Calcium channel blockers (diltiazem, verapamil)
	2677G>A, A893T	1–10% in Caucasians 3–22% in Asians	Digoxin, simvastatin
	1236C>T/2677G >A/T/3435C>T haplotype	23–42% in Caucasians 28–56% in Asians 4.5–8.7% in Africans	
ABCC2	1249G>A, V417I	22–26% in Caucasians 13–19% in Asians 14% in Africans	Reverse transcriptase inhibitors (tenofovir), Anticancer drugs (anthracyclines, vinca alkaloids, methotrexate, SN-38 glucuronide), pravastatin, rifampin
ABCG2	421C>A, Q141K	6–14% in Caucasians 15–36% in Asians 0–5% in Africans	Anticancer drugs (methotrexate, imatinib, gefitinib, SN-38, SN-38 glucuronide, topotecan), rosuvastatin, glyburide

some nonconjugated drugs such as methotrexate and pravastatin (Table 1.2), and exhibit overlapping substrate specificities for a variety of drugs. Genetic variation in the *ABCC1* gene is rare, whereas polymorphism of the *ABCC2* gene is more common, including the c.1249G>A SNP (rs2273697) in exon 10 resulting in a p.V417I substitution and lower protein expression. Another identified polymorphism is the c.3972C>T SNP (rs3470066) in exon 28 with an p.1324I amino acid substitution.[156]

Patients with the 1249G>A variant and receiving tenofovir were reported to have a higher risk of drug-induced renal proximal tubulopathy, possibly a result of reduced renal drug excretion.[157] In an exploratory study of an association between *ABCC2* polymorphisms and haplotypes with irinotecan disposition in a cohort of 167 Caucasian patients with solid tumors, a total of 15 *ABCC2* haplotypes were constructed from 6 variants of the *ABCC2* gene. The *ABCC2*2* haplotype (low activity) was found to be associated with lower irinotecan clearance of 28.3 L/hr in 48 patients compared with 31.6 L/hr in 75 patients not carrying the haplotype ($P = 0.02$). Interestingly, patients carrying the *ABCC2*2* haplotype but not

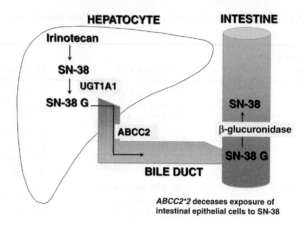

FIGURE 1.2   Schematic representation of potential protective effect of ABCC2 polymorphism against irinotecan-induced diarrhea.

the *UGT1A1*28* allele experienced lower incidence of severe grade 3—4 diarrhea (odds ratio of 0.15) compared to patients carrying at least one *UGT1A1*28* allele (odds ratio of 1.87), suggesting a protective effect of *ABCC2*2* haplotype against diarrhea occurrence.[158] Because ABCC2 mediates the secretion of SN-38 glucuronide into the bile, the protective effect might reflect a lower exposure of intestinal epithelial cells to SN-38 that is formed after cleavage of SN-38 glucuronide by β-glucuronidase within the intestine (Figure 1.2).

ABCG2

The *ABCG2* gene encodes the BCRP, which is also known as mitoxantrone-resistant protein (MXR) or placenta-specific ATP binding cassette transporter (ABCP). More than 80 polymorphisms in *ABCG2* have been reported, with the most studied being the c.421C>A SNP in exon 5, which results in a p.Q141K substitution and lower protein expression (Table 1.2).[159] The 421C>A variant with K141 is commonly present in different ethnic groups [148,149,160]. Patients carrying this SNP were reported to have increased concentrations of gefitinib and topotecan,[161,162] resulting in higher incidence of gefitinib-induced diarrhea.[163] Increased risk of diarrhea was also associated with the *ABCG2* polymorphism in patients with cancer and receiving rituximab plus cyclophosphamide/doxorubicin/vincristine/prednisone (R-CHOP) therapy.[164] ABCG2 also plays a role in disposition of other drugs, with the 421C>A variant reducing biliary excretion of rosuvastatin.[165] In 305 Chinese patients with hypercholesterolemia treated with 10 mg of rosuvastatin per day, a gene—dose dependent reduction in low-density lipoprotein cholesterol levels was observed in carriers of the 421C>A variant.[166]

The SLC Transporters

Organic Anion Transporting Polypeptides (OATPs)

In contrast to ABCB1, OATPs are influx transporters. In addition to facilitating hepatic uptake of drugs from the blood into hepatocytes for further metabolism or biliary secretion,

OATPs also mediate the transport of several endogenous compounds, including bile salts, across the cell membrane. A total of 11 OATP transporters have been identified and classified into 6 families.[167] Of the human OATPs, OATP1A2, OATP1B1, OATP1B3, and OATP2B1 are the best characterized.

OATP1B1

The human *SLCO1B1* gene encodes OATP1B1, which is also known as OATP-C. Since the discovery of the first c.521T>C SNP,[168] multiple SNPs have been reported for *SLCO1B1*, with 17 different *SLCO1B1* alleles identified.[169] The 521T>C SNP (rs4149056) with

TABLE 1.3 Selected SLC transporter polymorphisms indicating allele variants and frequency and drug substrates

Genes	Allele Variants, Amino Acid Change	Frequency (%)	Drug Substrate Examples
SLCO1B1	521T>C, V174A	8–22% in Caucasians 1–19% in Asians 1–5% in Africans	HMG-CoA reductase inhibitors (atorvastatin, simvastatin acid, pravastatin, rosuvasatatin), anticancer drugs (SN-38, methotrexate), antibacterials (rifampicin, cefazolin), repaglinide, valsartan
	388A>G, N130D	30–46% in Caucasians 54–84% in Asians 72–81% in Africans	
SLCO2B1	1457C>T, S486F	1–6% in Caucasians 25–36% in Asians 10–41% in Africans	HMG-CoA reductase inhibitors (atorvastatin, fluvastatin, pravastatin, rosuvasatatin), glibenclamide, fexofenadine
	935G>A, R312Q	2–14% in Caucasians 21–40% in Asians 7–15% in Africans	
SLCO1B3	334T>G, S112A	74–89% in Caucasians 64–83% in Asians 35–41% in Africans	Anticancer drugs (docetaxel, paclitaxel), digoxin
	699G>A, M233I	71–90% in Caucasians 64–84% in Asians 34–48% in Africans	
SLC22A1	1201G>A, G401S	1% in Caucasians 0% in Asians 1% in Africans	Metformin
	1393G>A, G465R	4% in Caucasians 0% in Asians, Africans	
	1256delATG, M420del	60% in Caucasians 74–81% in Asians 74% in Africans	
SLC22A2	808G>T, A270S	16% in Caucasians 14–17% in Asians 11% in Africans	Metformin

p.V174A substitution results in lower expression of the OATP1B1 protein and transport activity. The 521T>C SNP is more common in Caucasians and Asians than in Africans (Table 1.3). Another very common mutation in all ethnic groups studied is the c.388A>G SNP (rs2306283) resulting in p.N130D substitution, although conflicting results exist regarding associated changes in transport activity. More important, though, the 521T>C SNP and 388A>G SNP are in LD, resulting in several known haplotypes, such as OATP1B1*5 carrying the 388A/521C, OATP1B1*15 carrying the 388G/521C, and OATP1B1*17 carrying the 388G/ 521C with -11187A/-10499A of two additional SNPs in the promoter region of SLCO1B1 [169,170].

OATP1B1 plays an important role in hepatic uptake of the 3-hydroxy-3-methylglutaryl-CoA (HMG-CoA) reductase inhibitors such as pravastatin and rosuvastatin, as well as simvastatin acid, the active metabolite of simvastatin. The 521T>C variant has been associated with altered pharmacokinetics of simvastatin acid, with the CC homozygotes having greater than 2- to 3-fold increased systemic exposure compared to the other two genotypes,[171] potentially resulting in increased toxicity,[172] and with decreased intracellular concentration of simvastatin acid for inhibiting HMG-CoA reductase in hepatocytes, a lower efficacy for cholesterol reduction. In a GWAS, 316,184 SNPs were compared between 96 patients treated with 80 mg/day of simvastatin and suffering from myopathy and 96 control subjects without the adverse drug effect. A noncoding rs4363657 SNP within intron 11 of SLCO1B1, found to be in nearly complete LD with the rs4149056 polymorphism (521T>C, V174A) (r[2] > 0.95), was identified as the only strong SNP marker associated with simvastatin-induced myopathy. The odds ratio for myopathy was reported as 4.3 per copy of the C allele, and 17.4 in CC homozygotes compared with TT homozygotes.[173] The magnitude of the clinical significance suggests potential value of genotyping to screen out patients with abnormal OATP1B1 activity to improve the therapeutic index of simvastatin, which may also apply for other HMG-CoA reductase inhibitors such as pravastatin that are also OATP1B1 substrates [174,175,176]. Indeed, both the 521T>C SNP and SLCO1B1*17 haplotype had been shown to be associated with increased pravastatin concentrations and decreased efficacy [177,178,179].

OATP2B1

OATP2B1, also known as OATP-B, possesses substrate selectivity similar to that of OATP1B1.[180] OATP2B1 has also been found to be expressed in the luminal membrane of the small intestinal enterocytes,[181] and hence would have a role in drug absorption. Since the first discovery of genetic polymorphism, several sequence mutations of OATP2B1 have been described, including the c.1457C>T SNP (rs2306168), c.601G>A SNP (rs35199625), c.935G>A SNP (rs12422149), c.43C>T SNP (rs56837383), and a nine-nucleotide deletion of three amino acids 26—28 (26—28, p.QNT) of OATP2B1.[182] Although decreased transport activity had been shown mostly in vitro for most of these SNPs, the results are not consistent among all studies. In addition, significant ethnic variabilities exist in allele frequency of these SNPs.

A recent study evaluated the impact of the 1457C>T SNP on fexofenadine pharmacokinetics in Japanese subjects and found similar pharmacokinetic parameters among the three genotype groups.[183] Although the same SNP did not affect the absorption of the leukotriene receptor antagonist motelukast, patients who carry the 935A variant allele of the

935G>A SNP were reported to show lower plasma concentration and lesser pharmacological response.[184] This finding might suggest that the effect of *OATP2B1* polymorphism on drug absorption could be substrate dependent, but additional studies with other substrates would need to be performed for clarification.

OATP1B3

In humans, the *SLCO1B3* gene encodes OATP1B3, which was previously also known as OATP8 and LST-2. Several sequence variations exist for the *SLCO1B3* gene. The c.334T>G SNP (rs4149117) and the c.699G>A SNP (rs7311358) occur at a high frequency in Caucasian populations. Although OATP1B3 mediates the hepatic uptake of several drugs, including taxanes,[185] a study in 90 patients with cancer from 6 different ethnic groups reported that there were no associations between paclitaxel clearance and the two *OATP1B3* SNPs.[186] Similarly, no associations were found between docetaxel pharmacokinetics and *OATP1B3* SNPs [187,188]. The role of *OATP1B3* polymorphisms in drug disposition and response awaits further clarification from future studies.

In summary, *OATP* polymorphisms can affect disposition and possibly response for a large number of drugs. Current evidence strongly suggests a vital role of specific SNPs of *SLCO1B1* gene (e.g., 521T>C) for statin efficacy and adverse effects. Similar data for other OATP1B1 substrates from future clinical studies would provide further evidence of value of prospective genotyping for *SLCO1B1* variants in individualizing drug therapy. Much more work is needed to clarify the role of genetic polymorphisms for the other uptake transporters, including OATP2B1 and OATP1B3.

Organic Cation Transporter (OCT)

Three OCTs have been identified in humans: OCT1, OCT2, and OCT3, all of which are members of the SLC22A family and are encoded by the corresponding *SLC22A1*, *SLC22A2*, and *SLC22A3* genes, respectively (Table 1.3). OCT1 and OCT2 are primarily expressed in the hepatocyte and the kidney, respectively. Of the different SNPs that have been identified for the *SLC22A2* gene, the most relevant one is the c.808G>T SNP that results in the p.A270S substitution. The antidiabetic drug metformin is primarily renally eliminated by active tubular secretion via OCT2. Homozygotes of the low-activity 270S variant had been shown to have lower renal clearance and higher plasma concentrations of metformin when compared to homozygous carriers of the wild-type 270A [189,190]. Interestingly, Tzvetkov et al. demonstrated that OCT1 is also expressed in the distal tubule and may play a role in tubular reabsorption of metformin. They reported that homozygous and heterozygous carriers of various haplotypes of low-activity alleles of several *SLC22A1* polymorphisms (c.1201G>A SNP with p.G401S substitution, c.1393G>A SNP with p.G465R substitution, and a deletion resulting in M420del) were associated with increases in metformin renal clearance by about 20 to 30%.[191] Nevertheless, OCT1 is primarily expressed in the hepatocyte, the major site of action of metformin. The same low-activity *OCT1* variant alleles of these polymorphisms have also been reported to decrease hepatic uptake of metformin and result in lower blood glucose response.[192] The effects of genetic polymorphisms in other transporters such as the multidrug and toxin extrusion transporters as well as pharmacological targets for metformin are further discussed in chapter 7 of this book.

Organic Anion Transporter (OAT)

In contrast to the OCT belonging to the same SLC22 family, the OATs primarily mediate the transport of organic anions. Four OATs have been studied regarding their tissue location: OAT1, OAT2, and OAT3 are primarily expressed in the basolateral membrane of the renal proximal tubule, whereas OAT4 is located at the apical side. Therefore, OAT1, OAT2, and OAT3 are responsible for uptake of drug substrates into the tubular cells and OAT4 mediates their secretion into the renal tubule. Although several polymorphisms have been reported for *SLC22A6* encoding OAT1, *SLC22A7* encoding OAT2, *SLC22A8* encoding OAT3, and *SLC22A11* encoding OAT4, the allele frequency of these SNPs are all $\leq 1\%$ and their functional significance have not been clarified [193,194].

DRUG TARGET GENES

The study of pharmacodynamics encompasses the biochemical and physiological effects of drugs on the body and the relationship between drug concentration and drug effect. Drugs exert their effects through interaction with numerous protein types, including cell surface receptors (e.g., β-adrenergic and 5-hydroxytryptamine receptors), enzymes (e.g., vitamin K epoxide reductase complex 1, adenosine monophosphate-activated protein kinase), and ion channel proteins (e.g., sodium and potassium channels, epithelial sodium channel). Additionally, numerous intracellular signaling proteins downstream from the target protein are involved in eliciting drug response. Genetic variation affecting either the activity or expression of a drug target or intracellular signaling protein can have significant consequences for pharmacodynamic drug response.

Phenotypic response to genetic variation for drug target proteins generally differs from that of drug metabolizing enzymes (Table 1.4). As illustrated in Figure 1.3, variation in drug metabolizing enzymes results in distinct phenotypes (e.g., PMs, EMs, or UMs), as described in the previous section. With the exception of pharmacogenetics in oncology, where the expression of drug target receptor gene for tumor cells predicts drug efficacy, there are a limited number of examples of genetic variants in drug target proteins in germline cells that result in distinct pharmacodynamic effects. However, one example that exists involves mutations in the vitamin K epoxide complex subunit 1 (*VKORC1*) gene, where rare nonsynonymous mutations result in warfarin resistance in which exceptionally high doses (30 mg/day or higher) are required to achieve therapeutic anticoagulation. Most polymorphisms that affect drug pharmacodynamics tend to be more subtle and help explain response variability across a single distribution curve. For example, commonly occurring variations in the *VKORC1* regulatory regions help explain the significant interpatient variability in the warfarin dose required to produce optimal anticoagulation, as described in detail in chapter 5. The remainder of this section discusses examples of genes for various types of drug target proteins that contribute to the interpatient variability in pharmacodynamic drug responses.

Drug Target Receptor Genes in Oncology

Several cancer chemotherapy agents have been developed based on findings that overexpression of certain tumor cell surface receptors drives tumor cell growth and proliferation.

TABLE 1.4 Consequences of genetic variation in drug disposition and drug target proteins

Gene	Drug	Clinical Consequence
CYP2D6	Atomoxetine	PMs may have 10-fold greater atomoxetine exposure
	Codeine	UMs are at increased risk for morphine toxicity
CYP2C9	Warfarin	CYP2C9 deficiency increases bleeding risk
CYP2C19	Clopidogrel	CYP2C19 deficiency reduces drug effectiveness
G6PD	Rasburicase	G6PD deficiency increases risk for hemolytic anemia
TPMT	Azathioprine, 6-mercaptopurine, Thioguanine	Nonfunctional genotype increases the risk of serious, life-threatening myelosuppression with conventional drug doses
UGT1A1	Irinotecan	Reduced function genotype increases risk for drug-induced neutropenia
DPD	Capecitabine, 5-fluorouracil	DPD deficiency may lead to severe diarrhea, neutropenia, neurotoxicity
SLCO1B1	Simvastatin	Increased risk for myopathy
Drug target genes		
EGFR	Cetuximab, panitumumab	Determines drug effectiveness
HER2	Trastuzumab	Determines drug effectiveness
ADRB1	β-blockers	Influences variability in blood pressure response and possible mortality reduction
VKORC1	Warfarin	Determines dose needed for optimal anticoagulation
KCNJ11 and ABCC8	Sulfonylureas	Drug effectiveness
KCNMB1	Verapamil	Possibly determines reduction in blood pressure
DRD3	Antipsychotics	Risk for tardive dyskinesia
GRK5	β-blockers	Drug effect on clinical outcomes in heart failure
ATM	Metformin	Antidiabetic response
SLC6A4	SSRIs	Drug effectiveness
HTR2A	Clozapine	Drug effectiveness

These are described in detail in chapter 4. Briefly, one example of a receptor whose expression influences disease prognosis and predicts drug response is the epidermal growth factor receptor type 2 (HER2), also known as Her2/neu and ErbB2. Overexpression of HER2 occurs in approximately 20% of metastatic breast cancers and is associated with more aggressive

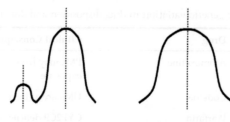

Bimodal phenotype distribution typically resulting for drug metabolizing enzyme polymorphisms

Unimodal phenotype distribution typical of drug target gene polymorphism

FIGURE 1.3 Many drug metabolizing enzyme polymorphisms are inactivating, resulting in distinct phenotypes, such as the poor metabolizer and extensive metabolizer phenotypes. In contrast, drug receptor polymorphisms tend to be more subtle and help explain variability across a single distribution curve.

cancer and poor prognosis.[195] Trastuzumab is a recombinant monoclonal antibody that was developed to target HER2 and block growth and survival of HER2-dependent tumors. The addition of trastuzumab to breast cancer chemotherapy significantly slows the progression of breast cancer in women with HER2-positive tumors, with treatment effects positively correlated with the degree of HER2 overexpression.[196] Thus, testing to confirm HER2 overexpression is necessary before trastuzumab use.

The epidermal growth factor receptor (EGFR), also known as HER1 or ErbB1, is overexpressed in head and neck, colon, and rectal cancer. EGFR overexpression is associated with cancer growth and invasion and portends a poor clinical prognosis. The discovery of the *EGFR* gene and its role in cancer prognosis led to the development of EGFR antagonists, including cetuximab, panitumumab, erlotinib, and gefitinib. Cetuximab is a recombinant monoclonal antibody that binds to the extracellular domain of the EGFR, thus preventing epidermal growth factor and other ligands from activating the receptor. Cetuximab is indicated in the treatment of metastatic colorectal cancers that overexpress EGFR, where it has been shown to improve survival [197,198]. Similar to cetuximab, panitumumab is a monoclonal antibody that blocks activation of the EGFR and is indicated in metastatic colorectal cancer that progresses despite chemotherapy with fluoropyrimidine-, oxaliplatin-, and irinotecan-containing regimens. Erlotinib and gefitinib also target the EGFR and are indicated in non–small cell lung cancer.

Other examples of targeted chemotherapy developed based on genetic abnormalities include:

- rituximab, a monoclonal antibody used to treat CD20-positive, B-cell non-Hodgkin's lymphoma and chronic lymphocytic leukemia;
- imatinib mesylate, a kinase inhibitor developed to block the product of a reciprocal translocation between chromosomes 9 and 22, occurring in 95% of patients with chronic myeloid leukemia; and
- crizotinib, an anaplastic lymphoma kinase (ALK) and c-ros oncogene1, receptor tyrosine kinase (ROS-1) inhibitor that targets the *EML4-ALK* gene fusion product in non–small cell lung cancer.

Drug Target Receptor Genes in Cardiology

β_1-receptors are located in the heart and kidney, where they are involved in the regulation of heart rate, cardiac contractility, and plasma renin release. B_1-receptor mediated effects contribute importantly to the pathophysiology of numerous cardiovascular diseases, including hypertension, coronary artery disease, and heart failure. In particular, plasma renin release and activation of the renin-angiotensin-aldosterone system lead to increased blood volume and vasoconstriction in hypertension. Increases in heart rate and cardiac contractility increase myocardial oxygen demand, thus contributing to myocardial ischemia in patients with coronary heart disease. Furthermore, increased sympathetic nervous system activity is one of the primary mechanisms contributing to cardiac remodeling and heart failure progression. Consequently, β-blockers exert beneficial effects across cardiovascular diseases, resulting in blood pressure reduction in hypertension, lowering of myocardial oxygen demand in ischemic heart disease, and attenuation of cardiac remodeling in heart failure. There is evidence that genetic variation for the β-1 adrenergic receptor (ADRB1) may influence the effectiveness of β-blocker therapy.

The ADRB1 is encoded by an intronless gene located on chromosome 10q24-26. There are two common nonsynonymous SNPs in the *ADRB1*, p.S49G and p.R389G. The S49G SNP is located in the extracellular region of the receptor near the amino terminus, and the R389G variant is located in the cytoplasmic tail in the G-protein coupling domain of the ADRB1. *In vitro* studies show lesser receptor downregulation with the S49 form of the receptor and both greater receptor coupling to the G-protein and greater adenylyl cyclase activity with the R389 form [199,200]. There are ethnic differences in the S49G and R389G allele frequencies, with a G49 frequency of 12 to 16% in Caucasians and 23 to 28% in African Americans and a G389 frequency of 24 to 34% in Caucasians and 39 to 46% in African Americans.[201] The S49G and R389G SNPs are in strong LD such that the G49 allele is rarely inherited with G389.

The *ADRB1* gene has been the primary focus of research into genetic determinants of responses to β-blockers in hypertension, coronary heart disease, and heart failure. In each case, the R389 allele or S49-R389 haplotype has been associated with greater response to β-blockade, presumably because of greater adrenergic activity with this allele and haplotype. For example, treatment of hypertension with metoprolol produced greater blood pressure reduction in patients who were homozygous for the S49-R389 haplotype than in carriers of the G49 or G389 allele.[202] Among patients with coronary heart disease, the S49-R389 haplotype was associated with an increased risk for death compared to other haplotypes, an effect negated by treatment with atenolol.[203] In patients with heart failure, the homozygous R389 genotype was associated with greater improvements in left ventricular ejection fraction with carvedilol or metoprolol and greater survival benefits with bucindolol [204,205,206]. These clinical data are consistent with the *in vitro* data implying greater agonist-mediated effects (e.g., greater sympathetic nervous system-driven hemodynamic effects) with the S49 and R389 alleles and suggest that *ADRB1* genotype is an important determinant of blood pressure and cardiac responses to β-blockers.

The *ADRB1* genotype is also associated with β-blocker tolerability in heart failure. β-blockers are indicated for patients with heart failure because they attenuate the detrimental effects of the sympathetic nervous system on heart failure progression. However, because β-blockers have negative inotropic effects (i.e., reduce cardiac contractility), they can worsen heart failure when first started. For this reason, they must be started in very low doses with careful up-titration. Although most heart failure patients tolerate β-blocker initiation at low doses and slow up-titration, some experience significant heart failure exacerbation. The influence of *ADRB1* genotype on tolerability to β-blocker initiation and up-titration has been examined, and it was found that carriers of the G389 allele or the S49 homozygotes more frequently require increases in concomitant heart failure therapy (predominately diuretics) for symptoms of worsening heart failure during β-blocker titration than patients with other genotypes.[207]

The gene for alpha 2C-adrenergic receptor (ADRA2C), which helps regulate adrenergic activity, has also been correlated with β-blocker response. Stimulation of the ADRA2C regulates sympathetic response by inhibiting norepinephrine release. The *ADRA2C* Del322-325 polymorphism causes an in-frame deletion of 12 nucleic acids, resulting in the loss of 4 amino acids in the ADRA2C protein and loss of protein function. Loss of ADRA2C function would be expected to result in less inhibition of norepinephrine release, and consequently increased norepinephrine levels and sympathetic tone. The frequency of the Del322-325 variant exhibits marked variability by ancestry, with a frequency of approximately 40% in African Americans and <5% in those of European descent.[208] In a large, multicenter, randomized, placebo-controlled heart failure trial, investigators found that individuals with the Del322-325 allele had greater reductions in sympathetic activity with bucindolol, a nonselective β-blocker with α1-receptor blocker properties. However, individuals with the wild-type (Ins322-325) *ADRA2C* genotype derived significant survival benefits from bucindolol, whereas Del322-325 allele carriers did not.[209] The mechanism underlying this association was not determined. However, it was hypothesized that the significant sympatholytic activity with bucindolol in Del322-325 allele carriers caused detrimental clinical effects. These findings might explain the negative association between bucindolol use and heart failure survival in the study population overall. Specifically, whereas carvedilol, metoprolol, and bisoprolol were all shown to improve survival in heart failure, bucindolol was not [210,211]. However, compared to other β-blocker trials, the trial with bucindolol enrolled a large number of African Americans, in whom the Del322-325 allele, associated with lack of benefit with bucindolol, is 10 times more common.

Drug Target Genes in Psychiatry

Antidepressants target 5-hydroxytryptamine (5-HT) receptors, and a number of studies have examined the association between antidepressant treatment response and 5-HT genotype, as described in chapter 6. However, results of these studies are largely inconsistent and even conflicting. For example, in a large-scale association study of 68 candidate genes, only the synonymous IVS2 A/G (rs7997012) SNP within intron 2 of the *HTR2A* gene, which codes for the postsynaptic 5-HT_{2A} receptor, was associated with response to citalopram.[212] Although a large study in European Caucasians confirmed the association between the rs7997012 SNP and antidepressant response, the findings were opposite of those in the initial study.[213]

The majority of drug target genetic associations discussed so far relate to drug effectiveness. Variation in the *DRD3* gene, encoding for the dopamine D3 receptor, is an example of a drug target genotype linked to adverse drug effects. Specifically, the *DRD3* p.S9G variant has been implicated in risk for developing tardive dyskinesia, an irreversible movement disorder that develops after long-term antipsychotic treatment, particularly with typical antipsychotics. In a meta-analysis, the G9 allele was significantly overrepresented among 317 patients with tardive dyskinesia compared to 463 patients without this adverse drug effect. [214] Further, G9 allele homozygotes had higher abnormal involuntary movement scores compared to both heterozygotes and S9 allele homozygotes. This association was confirmed in another meta-analysis.[215]

Signal Transduction Proteins

Signal transduction encompasses the cascade of events following drug binding to a receptor that ultimately lead to a change in cellular response. G-protein receptor kinase 5 (GRK5) is an example of a signal transduction protein linked to drug response. The ADRB1 and other adrenergic receptors are coupled to Guanosine triphosphate (GTP)-binding proteins also called G-proteins. Upon ligand binding, the receptor couples to the intracellular G-protein to elicit a cellular response. GRKs phosphorylate cardiac receptors, essentially inhibiting receptor-mediated signaling, thus serving in a manner analogous to natural β-blockade. The *GRK5* p.Q41L polymorphism occurs commonly in African Americans, with an allele frequency >30%. However, it rarely occurs in Caucasians. The L41 allele has been found to more effectively uncouple agonist-mediated receptor signaling and has been associated with increased transplant-free survival in African Americans with heart failure.[216] Patients with the L41 allele derived no benefit from β-blocker therapy, presumably because they already have inherent downregulation of ADRB1 receptor signaling.[217] However, in patients with the *GRK5* 41QQ genotype, which is associated with a poor prognosis, treatment with β-blocker therapy significantly improved transplant-free survival.[216]

The dopamine and serotonin receptors targeted by antipsychotics are also G-protein-coupled receptors. The regulator of G-protein signaling 4 (RGS4) shortens the duration of neurotransmitter-mediated receptor signaling, and variants of *RSG4* have been studied as predictors for antipsychotic treatment response. The rs951439 SNP has been associated with response to perphenazine among patients of African descent and with response to risperidone among patients of European descent.[218] Further associations between *RGS4* variants and antipsychotic response are described in chapter 6.

The *ADD1* gene encodes for α-adducin, a cytoskeletal protein involved in signal transduction and renal sodium transport. The *ADD1* p.G460W variant is associated with greater renal sodium—potassium pump activity, renal sodium retention, and salt-sensitive hypertension [219,220]. Given its role in regulating sodium reabsorption and potentially mediating increased hypertension risk, the *ADD1* gene has been studied for its contribution to diuretic response. Though the W460 allele has been linked to greater blood pressure reduction with thiazide diuretics, the data are inconsistent [219,221]. The *ADD1* gene appears to interact with other genes involved in renal sodium reabsorption, including the neural precursor cell expressed, developmentally downregulated 4-like (*NEDD4L*), and lysine-deficient protein kinase 1 (*WNK*) genes.[219] This finding may explain the inconsistencies in the data when *ADD1* is analyzed alone rather than in the context of other genes involved in renal sodium handling and illustrates the likely contribution of multiple genes to the efficacy of many drugs.

Enzyme Genes

VKORC1 is the target site for warfarin. Specifically, warfarin inhibits VKORC1 to prevent regeneration of a reduced form of vitamin K necessary for clotting factor activation. A common variant, c.−1639G>A, occurs in the *VKORC1* gene promoter region, with reduced gene expression with the −1639A allele.[222] The frequency of the −1639A allele is highest in Asians (~90%) and lowest in persons of African descent (10%), with an intermittent frequency in populations of European descent (~40%).[223]

Numerous studies have documented the association between *VKORC1* genotype and warfarin dose requirements. The −1639AA, AG, and GG genotypes are associated with average warfarin dose requirements of approximately 3 mg/day, 5 mg/day, and 6 mg/day, respectively. Recent data suggest that dosing warfarin based on *VKORC1*, in addition to *CYP2C9*, genotype leads to more accurate dose prediction and may reduce the risk for adverse clinical outcomes early in the course of warfarin therapy [224,225]. The *VKORC1* genotype is described in detail in chapter 5.

The angiotensin-converting enzyme (ACE) gene has been widely studied for its effects on ACE inhibitor response. An insertion/deletion (I/D) polymorphism in intron 16 of the *ACE* gene results in the presence or absence of a 287-base-pair fragment. The *ACE* D allele has been linked consistently to higher plasma concentrations of ACE, the enzyme responsible for the conversion of angiotensin I to the potent vasoconstrictor angiotensin II.[226] Given its association with ACE concentrations, a number of investigators have examined whether the I/D polymorphism contributes to the interpatient variability in ACE inhibitor response. However, much of the data with the I/D polymorphism and blood pressure response to ACE inhibitors are inconsistent and even conflicting, with some studies demonstrating greater response with the D/D genotype, others showing greater response with the I/I genotype, and further studies showing no association. In one of the largest pharmacogenetic studies to date, including nearly 38,000 patients, there was no association between the *ACE* I/D genotype and either blood pressure response or cardiovascular or renal outcomes with antihypertensive therapy.[227]

Numerous polymorphic proteins are involved in the complex signaling pathway of the renin−angiotensin system, including renin, angiotensinogen, the angiotensin II type 1 receptor, bradykinin, and aldosterone synthase. Thus, a likely explanation for the inconsistent data with the *ACE* gene and ACE inhibitor response in hypertension is that a single polymorphism provides minimal contribution to ACE inhibitor response. Rather, ACE inhibitor response may be best determined by a combination of multiple polymorphisms occurring in multiple genes involved in the renin−angiotensin pathway.

The data with the *ACE* I/D genotype and ACE inhibitor response in patients with heart failure are more compelling. In this population, the *ACE* D allele has been associated with an increased risk for cardiac transplant or death [228,229,230,231,232]. As described in detail in chapter 5, the detrimental effect of the *ACE* D allele on transplant-free survival appears greatest among patients who are taking lower than recommended doses of ACE inhibitors. These data suggest that maximizing the ACE inhibitor dose may be necessary in *ACE* D allele carriers to attenuate the harmful effects of this allele [232,233].

Metformin is an anti diabetic drug that works in part by activating adenosine monophosphate-activated protein kinase (AMPK), which is a master regulator of cell and body energy homeostasis and glucose uptake in skeletal muscle.[234] A GWAS identified a significant association between metformin response in patients with type 2 diabetes and a polymorphism in a locus containing the ataxia-telangiectasia mutated (ATM) gene.[235] ATM is a DNA repair gene that acts upstream of AMPK and appears to be necessary for metformin action.[235] This topic is further described in chapter 7.

Ion Channel Genotype

One of the most often cited examples of ion channel genes with consequences for drug response are genes for the pore-forming channel proteins that affect potassium and sodium transport across the cardiac cell membrane. Mutations in cardiac ion channel genes predispose individuals to congenital long-QT syndrome. Moreover, there is evidence that these mutations may increase the risk for drug-induced torsades de pointes [236,237]. This subject is discussed in detail in chapter 5.

The large-conductance calcium and voltage-dependent potassium (BK) channel is another example of an ion channel with genetic contributions to drug response. The BK channel is found in vascular smooth muscle and consists of pore-forming-α and regulatory-β1 subunits. The β1 subunit enhances calcium sensitivity and decreases smooth muscle cell excitability, thus attenuating smooth muscle contraction. The KCNMB1 gene encodes for the BK channel β1 subunit. A common SNP in the KCNMB1 gene, p.E65K, results in a gain of function of the channel and increased calcium sensitivity compared to the wild type.[238] Given its role in mediating calcium sensitivity, the KCNMB1 gene was examined for its effect on response to the calcium channel blocker verapamil. Among patients with hypertension and coronary heart disease who were started on verapamil, K65 allele carriers achieved blood pressure control more rapidly than homozygotes, suggesting that the E65 SNP enhances response to calcium channel blockers and contributes to the interpatient variability in blood pressure reduction during calcium channel blocker therapy.[239]

The epithelial sodium channel (ENaC) is another example of an ion channel with genetic contributions to drug response. The ENaC is located in the distal renal tubule and collecting duct of the nephron and serves as the final site for sodium reabsorption. The channel is composed of α, β, and γ subunits, encoded by the SCNN1A, SCNN1B, and SCNN1G genes, respectively. In a healthy volunteer study, SNPs in the SCNN1B and SCNN1G genes were associated with natriuretic and diuretic responses to single oral doses of loop diuretics. Loop diuretics are commonly prescribed for managing symptoms of fluid overload in heart failure. Whether genes encoding for ENaC subunits influence response to loop diuretics in heart failure remains to be determined. But given the significant consequences of under- or overdosing loop diuretics in this disease, such information could have significant clinical value.

The potassium inwardly rectifying channel, subfamily J, member 11 gene (KCNJ11) and the sulfonylurea receptor gene (ABCC8) encode the Kir6.2 and sulfonylurea receptor-1 (SUR1) subunits of pancreatic ATP-sensitive potassium (K_{ATP}) channels, respectively. Activating mutations in the KCNJ11 and ABCC8 cause K_{ATP} channels to remain open, which promotes hyperpolarization of the pancreatic β cell membrane and impaired insulin release [240,241]. Sulfonylurea drugs promote K_{ATP} channel closure, thereby attenuating the effects of activating mutations in *KCNJ11* and *ABCC8*. As such, sulfonylureas are especially effective in patients with *KCNJ11* or *ABCC8* activating mutations [241,242]. Chapter 7 includes a detailed discussion of these genetic variations and their effects on response to antidiabetic agents.

CONCLUSION

Variations in genes influencing drug pharmacokinetics and pharmacodynamics often jointly influence drug response, as is the case with warfarin, whose dose requirements are influenced by both the *CYP2C9* and *VKORC1* genotypes. Thus, when taking a candidate gene approach to discovery of variants impacting drug response, genes encoding proteins involved in determining drug bioavailability (transporter proteins, drug metabolizing enzymes) and response (receptor, enzyme, ion channel, and/or intracellular signaling proteins) should be considered. Genome-wide approaches to identifying determinants of drug response may reveal previously unknown proteins involved in eliciting drug response that represent potential biomarkers for predicting drug effectiveness or risk for toxicity. In addition, proteins involved in disease pathophysiology may represent attractive targets for drug development, as most often demonstrated in the area of oncology.

QUESTIONS FOR DISCUSSION

1. What are examples of drug metabolism and drug transporter genotypes that affect drug response?
2. What are examples of drugs developed based on an understanding of genes involved in disease pathophysiology?
3. What are examples of drug target genes with implications for drug response?
4. How might genes for drug metabolism, drug transport, and/or drug target sites jointly influence drug response?

References

[1] Database of Single Nucleotide Polymorphisms (dbSNP). Bethesda (MD): National Center for Biotechnology Information, National Library of Medicine. (dbSNP Build ID: 137). Vol. 2012.
[2] Nelson DR, Conlon M, Baralt C, et al. University of Florida Clinical and Translational Science Institute: transformation and translation in personalized medicine. Clin Transl Sci 2011;4(6):400–2.
[3] Pulley JM, Denny JC, Peterson JF, Bernard GR, Vnencak-Jones CL, Ramirez AH, et al. Operational implementation of prospective genotyping for personalized medicine: the design of the Vanderbilt PREDICT Project. Clin Pharmacol Ther 2012;92:87–95.
[4] Zanger UM, Turpeinen M, Klein K, Schwab M. Functional pharmacogenetics/genomics of human cytochromes P450 involved in drug biotransformation. Anal Bioanal Chem 2008;392:1093–108.
[5] Sistonen J, Sajantila A, Lao O, Corander J, Barbujani G, Fuselli S. CYP2D6 worldwide genetic variation shows high frequency of altered activity variants and no continental structure. Pharmacogenet Genomics 2007;17:93–101.
[6] Bertilsson L, Dahl ML, Sjoqvist F, Aberg-Wistedt A, Humble M, Johansson I, et al. Molecular basis for rational megaprescribing in ultrarapid hydroxylators of debrisoquine. Lancet 1993;341:63.

[7] Dalen P, Dahl ML, Bernal Ruiz ML, Nordin J, Bertilsson L. 10-Hydroxylation of nortriptyline in white persons with 0, 1, 2, 3, and 13 functional CYP2D6 genes. Clin Pharmacol Ther 1998;63:444−52.

[8] Johansson I, Lundqvist E, Bertilsson L, Dahl ML, Sjoqvist F, Ingelman-Sundberg M. Inherited amplification of an active gene in the cytochrome P450 CYP2D locus as a cause of ultrarapid metabolism of debrisoquine. Proc Natl Acad Sci U S A 1993;90:11825−9.

[9] Gaedigk A, Ndjountche L, Divakaran K, Dianne Bradford L, Zineh I, Oberlander TF, et al. Cytochrome P4502D6 (CYP2D6) gene locus heterogeneity: characterization of gene duplication events. Clin Pharmacol Ther 2007;81:242−51.

[10] Lin KM, Finder E. Neuroleptic dosage for Asians. Am J Psychiatry 1983;140:490−1.

[11] Mihara K, Otani K, Tybring G, Dahl ML, Bertilsson L, Kaneko S. The CYP2D6 genotype and plasma concentrations of mianserin enantiomers in relation to therapeutic response to mianserin in depressed Japanese patients. J Clin Psychopharmacol 1997;17:467−71.

[12] Droll K, Bruce-Mensah K, Otton SV, Gaedigk A, Sellers EM, Tyndale RF. Comparison of three CYP2D6 probe substrates and genotype in Ghanaians, Chinese and Caucasians. Pharmacogenetics 1998;8:325−33.

[13] Masimirembwa C, Persson I, Bertilsson L, Hasler J, Ingelman-Sundberg M. A novel mutant variant of the CYP2D6 gene (CYP2D6*17) common in a black African population: association with diminished debrisoquine hydroxylase activity. Br J Clin Pharmacol 1996;42:713−9.

[14] Evans WE, Relling MV. Pharmacogenomics: translating functional genomics into rational therapeutics. Science 1999;286:487−91.

[15] Barclay ML, Sawyers SM, Begg EJ, Zhang M, Roberts RL, Kennedy MA, et al. Correlation of CYP2D6 genotype with perhexiline phenotypic metabolizer status. Pharmacogenetics 2003;13:627−32.

[16] Lee JT, Kroemer HK, Silberstein DJ, Funck-Brentano C, Lineberry MD, Wood AJ, et al. The role of genetically determined polymorphic drug metabolism in the beta-blockade produced by propafenone. N Engl J Med 1990;322:1764−8.

[17] Dahl-Puustinen ML, Liden A, Alm C, Nordin C, Bertilsson L. Disposition of perphenazine is related to polymorphic debrisoquin hydroxylation in human beings. Clin Pharmacol Ther 1989;46:78−81.

[18] Spina E, Ancione M, Di Rosa AE, Meduri M, Caputi AP. Polymorphic debrisoquine oxidation and acute neuroleptic-induced adverse effects. Eur J Clin Pharmacol 1992;42:347−8.

[19] Gasche Y, Daali Y, Fathi M, Chiappe A, Cottini S, Dayer P, et al. Codeine intoxication associated with ultrarapid CYP2D6 metabolism. N Engl J Med 2004;351:2827−31.

[20] Koren G, Cairns J, Chitayat D, Gaedigk A, Leeder SJ. Pharmacogenetics of morphine poisoning in a breastfed neonate of a codeine-prescribed mother. Lancet 2006;368:704.

[21] Ciszkowski C, Madadi P, Phillips MS, Lauwers AE, Koren G. Codeine, ultrarapid-metabolism genotype, and postoperative death. N Engl J Med 2009;361:827−8.

[22] Elkalioubie A, Allorge D, Robriquet L, Wiart JF, Garat A, Broly F, et al. Near-fatal tramadol cardiotoxicity in a CYP2D6 ultrarapid metabolizer. Eur J Clin Pharmacol 2011;67:855−8.

[23] Garcia-Quetglas E, Azanza JR, Sadaba B, Munoz MJ, Gil I, Campanero MA. Pharmacokinetics of tramadol enantiomers and their respective phase I metabolites in relation to CYP2D6 phenotype. Pharmacol Res 2007;55:122−30.

[24] Sindrup SH, Brosen K, Bjerring P, Arendt-Nielsen L, Larsen U, Angelo HR, et al. Codeine increases pain thresholds to copper vapor laser stimuli in extensive but not poor metabolizers of sparteine. Clin Pharmacol Ther 1990;48:686−93.

[25] Wijnen PA, Limantoro I, Drent M, Bekers O, Kuijpers PM, Koek GH. Depressive effect of an antidepressant: therapeutic failure of venlafaxine in a case lacking CYP2D6 activity. Ann Clin Biochem 2009;46:527−30.

[26] Goetz MP, Rae JM, Suman VJ, Safgren SL, Ames MM, Visscher DW, et al. Pharmacogenetics of tamoxifen biotransformation is associated with clinical outcomes of efficacy and hot flashes. J Clin Oncol 2005;23:9312−8.

[27] Bertilsson L, Aberg-Wistedt A, Gustafsson LL, Nordin C. Extremely rapid hydroxylation of debrisoquine: a case report with implication for treatment with nortriptyline and other tricyclic antidepressants. Ther Drug Monit 1985;7:478−80.

[28] Kawanishi C, Lundgren S, Agren H, Bertilsson L. Increased incidence of CYP2D6 gene duplication in patients with persistent mood disorders: ultrarapid metabolism of antidepressants as a cause of nonresponse. A pilot study. Eur J Clin Pharmacol 2004;59:803−7.

[29] Rau T, Wohlleben G, Wuttke H, Thuerauf N, Lunkenheimer J, Lanczik M, et al. CYP2D6 genotype: impact on adverse effects and nonresponse during treatment with antidepressants—a pilot study. Clin Pharmacol Ther 2004;75:386—93.

[30] Candiotti KA, Birnbach DJ, Lubarsky DA, Nhuch F, Kamat A, Koch WH, et al. The impact of pharmacogenomics on postoperative nausea and vomiting: do CYP2D6 allele copy number and polymorphisms affect the success or failure of ondansetron prophylaxis? Anesthesiology 2005;102:543—9.

[31] Hamelin BA, Bouayad A, Methot J, Jobin J, Desgagnes P, Poirier P, et al. Significant interaction between the nonprescription antihistamine diphenhydramine and the CYP2D6 substrate metoprolol in healthy men with high or low CYP2D6 activity. Clin Pharmacol Ther 2000;67:466—77.

[32] Alfaro CL, Lam YW, Simpson J, Ereshefsky L. CYP2D6 status of extensive metabolizers after multiple-dose fluoxetine, fluvoxamine, paroxetine, or sertraline. J Clin Psychopharmacol 1999;19:155—63.

[33] Lam YW, Gaedigk A, Ereshefsky L, Alfaro CL, Simpson J. CYP2D6 inhibition by selective serotonin reuptake inhibitors: analysis of achievable steady-state plasma concentrations and the effect of ultrarapid metabolism at CYP2D6. Pharmacotherapy 2002;22:1001—6.

[34] Sim SC, Risinger C, Dahl ML, Aklillu E, Christensen M, Bertilsson L, et al. A common novel CYP2C19 gene variant causes ultrarapid drug metabolism relevant for the drug response to proton pump inhibitors and antidepressants. Clin Pharmacol Ther 2006;79:103—13.

[35] Yamada S, Onda M, Kato S, Matsuda N, Matsuhisa T, Yamada N, et al. Genetic differences in CYP2C19 single nucleotide polymorphisms among four Asian populations. J Gastroenterol 2001;36:669—72.

[36] Ghoneim MM, Korttila K, Chiang CK, Jacobs L, Schoenwald RD, Mewaldt SP, et al. Diazepam effects and kinetics in Caucasians and Orientals. Clin Pharmacol Ther 1981;29:749—56.

[37] Kumana CR, Lauder IJ, Chan M, Ko W, Lin HJ. Differences in diazepam pharmacokinetics in Chinese and white Caucasians—relation to body lipid stores. Eur J Clin Pharmacol 1987;32:211—5.

[38] Furuta T, Sugimoto M, Shirai N, Ishizaki T. CYP2C19 pharmacogenomics associated with therapy of Helicobacter pylori infection and gastro-esophageal reflux diseases with a proton pump inhibitor. Pharmacogenomics 2007;8:1199—210.

[39] Andersson T, Holmberg J, Rohss K, Walan A. Pharmacokinetics and effect on caffeine metabolism of the proton pump inhibitors, omeprazole, lansoprazole, and pantoprazole. Br J Clin Pharmacol 1998;45:369—75.

[40] Baldwin RM, Ohlsson S, Pedersen RS, Mwinyi J, Ingelman-Sundberg M, Eliasson E, et al. Increased omeprazole metabolism in carriers of the CYP2C19*17 allele; a pharmacokinetic study in healthy volunteers. Br J Clin Pharmacol 2008;65:767—74.

[41] Furuta T, Ohashi K, Kosuge K, Zhao XJ, Takashima M, Kimura M, et al. CYP2C19 genotype status and effect of omeprazole on intragastric pH in humans. Clin Pharmacol Ther 1999;65:552—61.

[42] Lou HY, Chang CC, Sheu MT, Chen YC, Ho HO. Optimal dose regimens of esomeprazole for gastric acid suppression with minimal influence of the CYP2C19 polymorphism. Eur J Clin Pharmacol 2009;65:55—64.

[43] Qiao HL, Hu YR, Tian X, Jia LJ, Gao N, Zhang LR, et al. Pharmacokinetics of three proton pump inhibitors in Chinese subjects in relation to the CYP2C19 genotype. Eur J Clin Pharmacol 2006;62:107—12.

[44] Furuta T, Ohashi K, Kamata T, Takashima M, Kosuge K, Kawasaki T, et al. Effect of genetic differences in omeprazole metabolism on cure rates for Helicobacter pylori infection and peptic ulcer. Ann Intern Med 1998;129:1027—30.

[45] Tanigawara Y, Aoyama N, Kita T, Shirakawa K, Komada F, Kasuga M, et al. CYP2C19 genotype-related efficacy of omeprazole for the treatment of infection caused by Helicobacter pylori. Clin Pharmacol Ther 1999;66:528—34.

[46] Kawabata H, Habu Y, Tomioka H, Kutsumi H, Kobayashi M, Oyasu K, et al. Effect of different proton pump inhibitors, differences in CYP2C19 genotype and antibiotic resistance on the eradication rate of Helicobacter pylori infection by a 1-week regimen of proton pump inhibitor, amoxicillin and clarithromycin. Aliment Pharmacol Ther 2003;17:259—64.

[47] Gawronska-Szklarz B, Siuda A, Kurzawski M, Bielicki D, Marlicz W, Drozdzik M. Effects of CYP2C19, MDR1, and interleukin 1-B gene variants on the eradication rate of Helicobacter pylori infection by triple therapy with pantoprazole, amoxicillin, and metronidazole. Eur J Clin Pharmacol 2010;66:681—7.

[48] Furuta T, Shirai N, Takashima M, Xiao F, Hanai H, Sugimura H, et al. Effect of genotypic differences in CYP2C19 on cure rates for Helicobacter pylori infection by triple therapy with a proton pump inhibitor, amoxicillin, and clarithromycin. Clin Pharmacol Ther 2001;69:158—68.

[49] Furuta T, Shirai N, Kodaira M, Sugimoto M, Nogaki A, Kuriyama S, et al. Pharmacogenomics-based tailored versus standard therapeutic regimen for eradication of H. pylori. Clin Pharmacol Ther 2007;81:521–8.

[50] Hunfeld NG, Mathot RA, Touw DJ, van Schaik RH, Mulder PG, et al. Effect of CYP2C19*2 and *17 mutations on pharmacodynamics and kinetics of proton pump inhibitors in Caucasians. Br J Clin Pharmacol 2008;65:752–60.

[51] Kurzawski M, Gawronska-Szklarz B, Wrzesniewska J, Siuda A, Starzynska T, Drozdzik M. Effect of CYP2C19*17 gene variant on Helicobacter pylori eradication in peptic ulcer patients. Eur J Clin Pharmacol 2006;62:877–80.

[52] Wang G, Lei HP, Li Z, Tan ZR, Guo D, Fan L, et al. The CYP2C19 ultra-rapid metabolizer genotype influences the pharmacokinetics of voriconazole in healthy male volunteers. Eur J Clin Pharmacol 2009;65:281–5.

[53] Rudberg I, Mohebi B, Hermann M, Refsum H, Molden E. Impact of the ultrarapid CYP2C19*17 allele on serum concentration of escitalopram in psychiatric patients. Clin Pharmacol Ther 2008;83:322–7.

[54] Collet JP, Hulot JS, Pena A, Villard E, Esteve JB, Silvain J, et al. Cytochrome P450 2C19 polymorphism in young patients treated with clopidogrel after myocardial infarction: a cohort study. Lancet 2009;373:309–17.

[55] Hulot JS, Bura A, Villard E, Azizi M, Remones V, Goyenvalle C, et al. Cytochrome P450 2C19 loss-of-function polymorphism is a major determinant of clopidogrel responsiveness in healthy subjects. Blood 2006;108:2244–7.

[56] Kim KA, Park PW, Hong SJ, Park JY. The effect of CYP2C19 polymorphism on the pharmacokinetics and pharmacodynamics of clopidogrel: a possible mechanism for clopidogrel resistance. Clin Pharmacol Ther 2008;84:236–42.

[57] Mega JL, Close SL, Wiviott SD, Shen L, Hockett RD, Brandt JT, et al. Cytochrome p-450 polymorphisms and response to clopidogrel. N Engl J Med 2009;360:354–62.

[58] Umemura K, Furuta T, Kondo K. The common gene variants of CYP2C19 affect pharmacokinetics and pharmacodynamics in an active metabolite of clopidogrel in healthy subjects. J Thromb Haemost 2008;6: 1439–41.

[59] Shuldiner AR, O'Connell JR, Bliden KP, Gandhi A, Ryan K, Horenstein RB, et al. Association of cytochrome P450 2C19 genotype with the antiplatelet effect and clinical efficacy of clopidogrel therapy. JAMA 2009;302:849–57.

[60] Frere C, Cuisset T, Gaborit B, Alessi MC, Hulot JS. The CYP2C19*17 allele is associated with better platelet response to clopidogrel in patients admitted for non-ST acute coronary syndrome. J Thromb Haemost 2009;7:1409–11.

[61] Sibbing D, Koch W, Gebhard D, Schuster T, Braun S, Stegherr J, et al. Cytochrome 2C19*17 allelic variant, platelet aggregation, bleeding events, and stent thrombosis in clopidogrel-treated patients with coronary stent placement. Circulation 2010;121:512–8.

[62] Tiroch KA, Sibbing D, Koch W, Roosen-Runge T, Mehilli J, Schomig A, et al. Protective effect of the CYP2C19 *17 polymorphism with increased activation of clopidogrel on cardiovascular events. Am Heart J 2010; 160:506–12.

[63] Mega JL, Simon T, Collet JP, Anderson JL, Antman EM, Bliden K, et al. Reduced-function CYP2C19 genotype and risk of adverse clinical outcomes among patients treated with clopidogrel predominantly for PCI: a meta-analysis. JAMA 2010;304:1821–30.

[64] Gurbel PA, Tantry US, Shuldiner AR. Letter by Gurbel et al. regarding article "Cytochrome 2C19*17 allelic variant, platelet aggregation, bleeding events, and stent thrombosis in clopidogrel-treated patients with coronary stent placement". Circulation 2010;122. e478; author reply e479.

[65] Hulot JS, Collet JP, Silvain J, Pena A, Bellemain-Appaix A, Barthelemy O, et al. Cardiovascular risk in clopidogrel-treated patients according to cytochrome P450 2C19*2 loss-of-function allele or proton pump inhibitor coadministration: a systematic meta-analysis. J Am Coll Cardiol 2010;56:134–43.

[66] Bauer T, Bouman HJ, van Werkum JW, Ford NF, ten Berg JM, et al. Impact of CYP2C19 variant genotypes on clinical efficacy of antiplatelet treatment with clopidogrel: systematic review and meta-analysis. BMJ 2011; 343:d4588.

[67] Holmes MV, Perel P, Shah T, Hingorani AD, Casas JP. CYP2C19 genotype, clopidogrel metabolism, platelet function, and cardiovascular events: a systematic review and meta-analysis. J Am Med Assoc 2011;306: 2704–14.

[68] Kwok CS, Loke YK. Meta-analysis: the effects of proton pump inhibitors on cardiovascular events and mortality in patients receiving clopidogrel. Aliment Pharmacol Ther 2010;31:810–23.

[69] Stubbins MJ, Harries LW, Smith G, Tarbit MH, Wolf CR. Genetic analysis of the human cytochrome P450 CYP2C9 locus. Pharmacogenetics 1996;6:429—39.

[70] Perera MA, Gamazon E, Cavallari LH, Patel SR, Poindexter S, Kittles RA, et al. The missing association: sequencing-based discovery of novel SNPs in VKORC1 and CYP2C9 that affect warfarin dose in African Americans. Clin Pharmacol Ther 2011;89:408—15.

[71] Allabi AC, Gala JL, Horsmans Y, Babaoglu MO, Bozkurt A, Heusterspreute M, et al. Functional impact of CYP2C95, CYP2C96, CYP2C98, and CYP2C911 in vivo among black Africans. Clin Pharmacol Ther 2004;76:113—8.

[72] Dickmann LJ, Rettie AE, Kneller MB, Kim RB, Wood AJ, Stein CM, et al. Identification and functional characterization of a new CYP2C9 variant (CYP2C9*5) expressed among African Americans. Mol Pharmacol 2001;60:382—7.

[73] Takahashi H, Kashima T, Nomoto S, Iwade K, Tainaka H, Shimizu T, et al. Comparisons between in-vitro and in-vivo metabolism of (S)-warfarin: catalytic activities of cDNA-expressed CYP2C9, its Leu359 variant and their mixture versus unbound clearance in patients with the corresponding CYP2C9 genotypes. Pharmacogenetics 1998;8:365—73.

[74] Liu Y, Jeong H, Takahashi H, Drozda K, Patel SR, Shapiro NL, et al. Decreased warfarin clearance associated with the CYP2C9 R150H (*8) polymorphism. Clin Pharmacol Ther 2012;91:660—5.

[75] Rieder MJ, Reiner AP, Gage BF, Nickerson DA, Eby CS, McLeod HL, et al. Effect of VKORC1 haplotypes on transcriptional regulation and warfarin dose. N Engl J Med 2005;352:2285—93.

[76] Anderson JL, Horne BD, Stevens SM, Grove AS, Barton S, Nicholas ZP, et al. Randomized trial of genotype-guided versus standard warfarin dosing in patients initiating oral anticoagulation. Circulation 2007; 116:2563—70.

[77] Sconce EA, Khan TI, Wynne HA, Avery P, Monkhouse L, King BP, et al. The impact of CYP2C9 and VKORC1 genetic polymorphism and patient characteristics upon warfarin dose requirements: proposal for a new dosing regimen. Blood 2005;106:2329—33.

[78] Higashi MK, Veenstra DL, Kondo LM, Wittkowsky AK, Srinouanprachanh SL, Farin FM, et al. Association between CYP2C9 genetic variants and anticoagulation-related outcomes during warfarin therapy. JAMA 2002;287:1690—8.

[79] Aithal GP, Day CP, Kesteven PJ, Daly AK. Association of polymorphisms in the cytochrome P450 CYP2C9 with warfarin dose requirement and risk of bleeding complications. Lancet 1999;353:717—9.

[80] Lindh JD, Holm L, Andersson ML, Rane A. Influence of CYP2C9 genotype on warfarin dose requirements—a systematic review and meta-analysis. Eur J Clin Pharmacol 2009;65:365—75.

[81] Steward DJ, Haining RL, Henne KR, Davis G, Rushmore TH, Trager WF, et al. Genetic association between sensitivity to warfarin and expression of CYP2C9*3. Pharmacogenetics 1997;7:361—7.

[82] Limdi NA, Arnett DK, Goldstein JA, Beasley TM, McGwin G, Adler BK, et al. Influence of CYP2C9 and VKORC1 on warfarin dose, anticoagulation attainment and maintenance among European-Americans and African-Americans. Pharmacogenomics 2008;9:511—26.

[83] Limdi NA, McGwin G, Goldstein JA, Beasley TM, Arnett DK, Adler BK, et al. Influence of CYP2C9 and VKORC1 1173C/T genotype on the risk of hemorrhagic complications in African-American and European-American patients on warfarin. Clin Pharmacol Ther 2008;83:312—21.

[84] Cavallari LH, Langaee TY, Momary KM, Shapiro NL, Nutescu EA, Coty WA, et al. Genetic and clinical predictors of warfarin dose requirements in African Americans. Clin Pharmacol Ther 2010;87:459—64.

[85] Cavaco I, Stromberg-Norklit J, Kaneko A, Msellem MI, Dahoma M, Ribeiro VL, et al. CYP2C8 polymorphism frequencies among malaria patients in Zanzibar. Eur J Clin Pharmacol 2005;61:15—8.

[86] Soyama A, Saito Y, Komamura K, Ueno K, Kamakura S, Ozawa S, et al. Five novel single nucleotide polymorphisms in the CYP2C8 gene, one of which induces a frame-shift. Drug Metab Pharmacokinet 2002; 17:374—7.

[87] Edwards RJ, Adams DA, Watts PS, Davies DS, Boobis AR. Development of a comprehensive panel of antibodies against the major xenobiotic metabolising forms of cytochrome P450 in humans. Biochem Pharmacol 1998;56:377—87.

[88] Garcia-Martin E, Martinez C, Tabares B, Frias J, Agundez JA. Interindividual variability in ibuprofen pharmacokinetics is related to interaction of cytochrome P450 2C8 and 2C9 amino acid polymorphisms. Clin Pharmacol Ther 2004;76:119—27.

[89] Martinez C, Garcia-Martin E, Blanco G, Gamito FJ, Ladero JM, Agundez JA. The effect of the cytochrome P450 CYP2C8 polymorphism on the disposition of (R)-ibuprofen enantiomer in healthy subjects. Br J Clin Pharmacol 2005;59:62—9.

[90] Lundblad MS, Stark K, Eliasson E, Oliw E, Rane A. Biosynthesis of epoxyeicosatrienoic acids varies between polymorphic CYP2C enzymes. Biochem Biophys Res Commun 2005;327:1052—7.

[91] Niemi M, Leathart JB, Neuvonen M, Backman JT, Daly AK, Neuvonen PJ. Polymorphism in CYP2C8 is associated with reduced plasma concentrations of repaglinide. Clin Pharmacol Ther 2003;74:380—7.

[92] Niemi M, Backman JT, Kajosaari LI, Leathart JB, Neuvonen M, Daly AK, et al. Polymorphic organic anion transporting polypeptide 1B1 is a major determinant of repaglinide pharmacokinetics. Clin Pharmacol Ther 2005;77:468—78.

[93] Rodriguez-Antona C, Niemi M, Backman JT, Kajosaari LI, Neuvonen PJ, Robledo M, et al. Characterization of novel CYP2C8 haplotypes and their contribution to paclitaxel and repaglinide metabolism. Pharmacogenomics J 2008;8:268—77.

[94] Miyazaki M, Nakamura K, Fujita Y, Guengerich FP, Horiuchi R, Yamamoto K. Defective activity of recombinant cytochromes P450 3A4.2 and 3A4.16 in oxidation of midazolam, nifedipine, and testosterone. Drug Metab Dispos 2008;36:2287—91.

[95] Kuehl P, Zhang J, Lin Y, Lamba J, Assem M, Schuetz J, et al. Sequence diversity in CYP3A promoters and characterization of the genetic basis of polymorphic CYP3A5 expression. Nat Genet 2001;27:383—91.

[96] Lamba JK, Lin YS, Thummel K, Daly A, Watkins PB, Strom S, et al. Common allelic variants of cytochrome P4503A4 and their prevalence in different populations. Pharmacogenetics 2002;12:121—32.

[97] Sata F, Sapone A, Elizondo G, Stocker P, Miller VP, Zheng W, et al. CYP3A4 allelic variants with amino acid substitutions in exons 7 and 12: evidence for an allelic variant with altered catalytic activity. Clin Pharmacol Ther 2000;67:48—56.

[98] Hesselink DA, van Schaik RH, van der Heiden IP, van der Werf M, Gregoor PJ, et al. Genetic polymorphisms of the CYP3A4, CYP3A5, and MDR-1 genes and pharmacokinetics of the calcineurin inhibitors cyclosporine and tacrolimus. Clin Pharmacol Ther 2003;74:245—54.

[99] Zheng H, Zeevi A, Schuetz E, Lamba J, McCurry K, Griffith BP, et al. Tacrolimus dosing in adult lung transplant patients is related to cytochrome P4503A5 gene polymorphism. J Clin Pharmacol 2004;44:135—40.

[100] Thervet E, Loriot MA, Barbier S, Buchler M, Ficheux M, Choukroun G, et al. Optimization of initial tacrolimus dose using pharmacogenetic testing. Clin Pharmacol Ther 2010;87:721—6.

[101] Stevens JC, Hines RN, Gu C, Koukouritaki SB, Manro JR, Tandler PJ, et al. Developmental expression of the major human hepatic CYP3A enzymes. J Pharmacol Exp Ther 2003;307:573—82.

[102] Westlind-Johnsson A, Malmebo S, Johansson A, Otter C, Andersson TB, Johansson I, et al. Comparative analysis of CYP3A expression in human liver suggests only a minor role for CYP3A5 in drug metabolism. Drug Metab Dispos 2003;31:755—61.

[103] McDonald MG, Rieder MJ, Nakano M, Hsia CK, Rettie AE. CYP4F2 is a vitamin K1 oxidase: an explanation for altered warfarin dose in carriers of the V433M variant. Mol Pharmacol 2009;75:1337—46.

[104] Takeuchi F, McGinnis R, Bourgeois S, Barnes C, Eriksson N, Soranzo N, et al. A genome-wide association study confirms VKORC1, CYP2C9, and CYP4F2 as principal genetic determinants of warfarin dose. PLoS Genet 2009;5:e1000433.

[105] Shahin MH, Khalifa SI, Gong Y, Hammad LN, Sallam MT, El Shafey M, et al. Genetic and nongenetic factors associated with warfarin dose requirements in Egyptian patients. Pharmacogenet Genomics 2011;21:130—5.

[106] Suriapranata IM, Tjong WY, Wang T, Utama A, Raharjo SB, Yuniadi Y, et al. Genetic factors associated with patient-specific warfarin dose in ethnic Indonesians. BMC Med Genet 2011;12:80.

[107] Chan SL, Suo C, Lee SC, Goh BC, Chia KS, Teo YY. Translational aspects of genetic factors in the prediction of drug response variability: a case study of warfarin pharmacogenomics in a multi-ethnic cohort from Asia. Pharmacogenomics J 2012;12:312—8.

[108] Caldwell MD, Awad T, Johnson JA, Gage BF, Falkowski M, Gardina P, et al. CYP4F2 genetic variant alters required warfarin dose. Blood 2008;111:4106—12.

[109] Cha PC, Mushiroda T, Takahashi A, Kubo M, Minami S, Kamatani N, et al. Genome-wide association study identifies genetic determinants of warfarin responsiveness for Japanese. Hum Mol Genet 2011;19:4735—44.

[110] Choi JR, Kim JO, Kang DR, Yoon SA, Shin JY, Zhang X, et al. Proposal of pharmacogenetics-based warfarin dosing algorithm in Korean patients. J Hum Genet 2011;56:290—5.

[111] Rotger M, Tegude H, Colombo S, Cavassini M, Furrer H, Decosterd L, et al. Predictive value of known and novel alleles of CYP2B6 for efavirenz plasma concentrations in HIV-infected individuals. Clin Pharmacol Ther 2007;81:557—66.

[112] Mehlotra RK, Bockarie MJ, Zimmerman PA. CYP2B6 983T>C polymorphism is prevalent in West Africa but absent in Papua New Guinea: implications for HIV/AIDS treatment. Br J Clin Pharmacol 2007;64:391—5.

[113] Jinno H, Tanaka-Kagawa T, Ohno A, Makino Y, Matsushima E, Hanioka N, et al. Functional characterization of cytochrome P450 2B6 allelic variants. Drug Metab Dispos 2003;31:398—403.

[114] Kirchheiner J, Klein C, Meineke I, Sasse J, Zanger UM, Murdter TE, et al. Bupropion and 4-OH-bupropion pharmacokinetics in relation to genetic polymorphisms in CYP2B6. Pharmacogenetics 2003;13:619—26.

[115] Gatanaga H, Hayashida T, Tsuchiya K, Yoshino M, Kuwahara T, Tsukada H, et al. Successful efavirenz dose reduction in HIV type 1-infected individuals with cytochrome P450 2B6 *6 and *26. Clin Infect Dis 2007; 45:1230—7.

[116] Hofmann MH, Blievernicht JK, Klein K, Saussele T, Schaeffeler E, Schwab M, et al. Aberrant splicing caused by single nucleotide polymorphism c.516G>T [Q172H], a marker of CYP2B6*6, is responsible for decreased expression and activity of CYP2B6 in liver. J Pharmacol Exp Ther 2008;325:284—92.

[117] Wang H, Tompkins LM. CYP2B6: new insights into a historically overlooked cytochrome P450 isozyme. Curr Drug Metab 2008;9:598—610.

[118] Rotger M, Saumoy M, Zhang K, Flepp M, Sahli R, Decosterd L, et al. Partial deletion of CYP2B6 owing to unequal crossover with CYP2B7. Pharmacogenet Genomics 2007;17:885—90.

[119] Malaiyandi V, Sellers EM, Tyndale RF. Implications of CYP2A6 genetic variation for smoking behaviors and nicotine dependence. Clin Pharmacol Ther 2005;77:145—58.

[120] Ghotbi R, Christensen M, Roh HK, Ingelman-Sundberg M, Aklillu E, Bertilsson L. Comparisons of CYP1A2 genetic polymorphisms, enzyme activity and the genotype-phenotype relationship in Swedes and Koreans. Eur J Clin Pharmacol 2007;63:537—46.

[121] Sachse C, Brockmoller J, Bauer S, Roots I. Functional significance of a C->A polymorphism in intron 1 of the cytochrome P450 CYP1A2 gene tested with caffeine. Br J Clin Pharmacol 1999;47:445—9.

[122] Mihara K, Kondo T, Suzuki A, Yasui-Furukori N, Ono S, Otani K, et al. Effects of genetic polymorphism of CYP1A2 inducibility on the steady-state plasma concentrations of trazodone and its active metabolite m-chlorophenylpiperazine in depressed Japanese patients. Pharmacol Toxicol 2001;88:267—70.

[123] Shimoda K, Someya T, Morita S, Hirokane G, Yokono A, Takahashi S, et al. Lack of impact of CYP1A2 genetic polymorphism (C/A polymorphism at position 734 in intron 1 and G/A polymorphism at position -2964 in the 5'-flanking region of CYP1A2) on the plasma concentration of haloperidol in smoking male Japanese with schizophrenia. Prog Neuropsychopharmacol Biol Psychiatry 2002;26:261—5.

[124] van der Weide J, Steijns LS, van Weelden MJ. The effect of smoking and cytochrome P450 CYP1A2 genetic polymorphism on clozapine clearance and dose requirement. Pharmacogenetics 2003;13:169—72.

[125] Beutler E, Gelbart T, Demina A. Racial variability in the UDP-glucuronosyltransferase 1 (UGT1A1) promoter: a balanced polymorphism for regulation of bilirubin metabolism? Proc Natl Acad Sci U S A 1998;95:8170—4.

[126] Jinno H, Tanaka-Kagawa T, Hanioka N, Saeki M, Ishida S, Nishimura T, et al. Glucuronidation of 7-ethyl-10-hydroxycamptothecin (SN-38), an active metabolite of irinotecan (CPT-11), by human UGT1A1 variants, G71R, P229Q, and Y486D. Drug Metab Dispos 2003;31:108—13.

[127] Ando Y, Saka H, Asai G, Sugiura S, Shimokata K, Kamataki T. UGT1A1 genotypes and glucuronidation of SN-38, the active metabolite of irinotecan. Ann Oncol 1998;9:845—7.

[128] Wasserman E, Myara A, Lokiec F, Goldwasser F, Trivin F, Mahjoubi M, et al. Severe CPT-11 toxicity in patients with Gilbert's syndrome: two case reports. Ann Oncol 1997;8:1049—51.

[129] Iyer L, Das S, Janisch L, Wen M, Ramirez J, Karrison T, et al. UGT1A1*28 polymorphism as a determinant of irinotecan disposition and toxicity. Pharmacogenetics J 2002;2:43—7.

[130] Kadakol A, Ghosh SS, Sappal BS, Sharma G, Chowdhury JR, Chowdhury NR. Genetic lesions of bilirubin uridine-diphosphoglucuronate glucuronosyltransferase (UGT1A1) causing Crigler-Najjar and Gilbert syndromes: correlation of genotype to phenotype. Hum Mutat 2000;16:297—306.

[131] Relling MV, Hancock ML, Rivera GK, Sandlund JT, Ribeiro RC, Krynetski EY, et al. Mercaptopurine therapy intolerance and heterozygosity at the thiopurine S-methyltransferase gene locus. J Natl Cancer Inst 1999;91:2001—8.

[132] Lu Z, Zhang R, Diasio RB. Dihydropyrimidine dehydrogenase activity in human peripheral blood mono-nuclear cells and liver: population characteristics, newly identified deficient patients, and clinical implication in 5-fluorouracil chemotherapy. Cancer Res 1993;53:5433—8.

[133] Morel A, Boisdron-Celle M, Fey L, Soulie P, Craipeau MC, Traore S, et al. Clinical relevance of different dihydropyrimidine dehydrogenase gene single nucleotide polymorphisms on 5-fluorouracil tolerance. Mol Cancer Ther 2006;5:2895—904.

[134] Vreken P, Van Kuilenburg AB, Meinsma R, Smit GP, Bakker HD, et al. A point mutation in an invariant splice donor site leads to exon skipping in two unrelated Dutch patients with dihydropyrimidine dehydrogenase deficiency. J Inherit Metab Dis 1996;19:645—54.

[135] Van Kuilenburg AB, Vreken P, Beex LV, Meinsma R, Van Lenthe H, et al. Heterozygosity for a point mutation in an invariant splice donor site of dihydropyrimidine dehydrogenase and severe 5-fluorouracil related toxicity. Eur J Cancer 1997;33:2258—64.

[136] Evans DA, Manley KA, Mc KV. Genetic control of isoniazid metabolism in man. Br Med J 1960;2:485—91.

[137] Hein DW. Molecular genetics and function of NAT1 and NAT2: role in aromatic amine metabolism and carcinogenesis. Mutat Res 2002;506-507:65—77.

[138] Cartwright RA, Glashan RW, Rogers HJ, Ahmad RA, Barham-Hall D, Higgins E, et al. Role of N-acetyl-transferase phenotypes in bladder carcinogenesis: a pharmacogenetic epidemiological approach to bladder cancer. Lancet 1982;2:842—5.

[139] Davies SM, Robison LL, Buckley JD, Tjoa T, Woods WG, Radloff GA, et al. Glutathione S-transferase poly-morphisms and outcome of chemotherapy in childhood acute myeloid leukemia. J Clin Oncol 2001; 19:1279—87.

[140] Ali-Osman F, Akande O, Antoun G, Mao JX, Buolamwini J. Molecular cloning, characterization, and expression in Escherichia coli of full-length cDNAs of three human glutathione S-transferase Pi gene variants. Evidence for differential catalytic activity of the encoded proteins. J Biol Chem 1997;272:10004—12.

[141] Ciaccio PJ, Tew KD, LaCreta FP. Enzymatic conjugation of chlorambucil with glutathione by human gluta-thione S-transferases and inhibition by ethacrynic acid. Biochem Pharmacol 1991;42:1504—7.

[142] Petros WP, Hopkins PJ, Spruill S, Broadwater G, Vredenburgh JJ, Colvin OM, et al. Associations between drug metabolism genotype, chemotherapy pharmacokinetics, and overall survival in patients with breast cancer. J Clin Oncol 2005;23:6117—25.

[143] Goekkurt E, Al-Batran SE, Hartmann JT, Mogck U, Schuch G, Kramer M, et al. Pharmacogenetic analyses of a phase III trial in metastatic gastroesophageal adenocarcinoma with fluorouracil and leucovorin plus either oxaliplatin or cisplatin: a study of the arbeitsgemeinschaft internistische onkologie. J Clin Oncol 2009; 27:2863—73.

[144] Voso MT, D'Alo F, Putzulu R, Mele L, Scardocci A, Chiusolo P, et al. Negative prognostic value of glutathione S-transferase (GSTM1 and GSTT1) deletions in adult acute myeloid leukemia. Blood 2002;100:2703—7.

[145] Xiao Z, Yang L, Xu Z, Zhang Y, Liu L, Nie L, et al. Glutathione S-transferases (GSTT1 and GSTM1) genes polymorphisms and the treatment response and prognosis in Chinese patients with de novo acute myeloid leukemia. Leuk Res 2008;32:1288—91.

[146] McLeod HL, Sargent DJ, Marsh S, Green EM, King CR, Fuchs CS, et al. Pharmacogenetic predictors of adverse events and response to chemotherapy in metastatic colorectal cancer: results from North American Gastro-intestinal Intergroup Trial N9741. J Clin Oncol 2010;28:3227—33.

[147] Giacomini KM, Huang SM, Tweedie DJ, Benet LZ, Brouwer KL, Chu X, et al. Membrane transporters in drug development. Nat Rev Drug Discov 2010;9:215—36.

[148] Gradhand U, Kim RB. Pharmacogenomics of MRP transporters (ABCC1-5) and BCRP (ABCG2). Drug Metab Rev 2008;40:317—54.

[149] Marzolini C, Paus E, Buclin T, Kim RB. Polymorphisms in human MDR1 (P-glycoprotein): recent advances and clinical relevance. Clin Pharmacol Ther 2004;75:13—33.

[150] Sakaeda T, Nakamura T, Horinouchi M, Kakumoto M, Ohmoto N, Sakai T, et al. MDR1 genotype-related pharmacokinetics of digoxin after single oral administration in healthy Japanese subjects. Pharm Res 2001;18:1400—4.

[151] Fellay J, Marzolini C, Meaden ER, Back DJ, Buclin T, Chave JP, et al. Response to antiretroviral treatment in HIV-1-infected individuals with allelic variants of the multidrug resistance transporter 1: a pharmacogenetics study. Lancet 2002;359:30—6.

[152] Wacher VJ, Wu CY, Benet LZ. Overlapping substrate specificities and tissue distribution of cytochrome P450 3A and P-glycoprotein: implications for drug delivery and activity in cancer chemotherapy. Mol Carcinog 1995;13:129–34.

[153] Kimchi-Sarfaty C, Oh JM, Kim IW, Sauna ZE, Calcagno AM, Ambudkar SV, et al. A "silent" polymorphism in the MDR1 gene changes substrate specificity. Science 2007;315:525–8.

[154] Lien EA, Solheim E, Lea OA, Lundgren S, Kvinnsland S, Ueland PM. Distribution of 4-hydroxy-N-desmethyltamoxifen and other tamoxifen metabolites in human biological fluids during tamoxifen treatment. Cancer Res 1989;49:2175–83.

[155] Sugiyama Y, Kato Y, Chu X. Multiplicity of biliary excretion mechanisms for the camptothecin derivative irinotecan (CPT-11), its metabolite SN-38, and its glucuronide: role of canalicular multispecific organic anion transporter and P-glycoprotein. Cancer Chemother Pharmacol 1998;42 Suppl:S44–9.

[156] Itoda M, Saito Y, Soyama A, Saeki M, Murayama N, Ishida S, et al. Polymorphisms in the ABCC2 (cMOAT/MRP2) gene found in 72 established cell lines derived from Japanese individuals: an association between single nucleotide polymorphisms in the 5′-untranslated region and exon 28. Drug Metab Dispos 2002;30:363–4.

[157] Izzedine H, Hulot JS, Villard E, Goyenvalle C, Dominguez S, Ghosn J, et al. Association between ABCC2 gene haplotypes and tenofovir-induced proximal tubulopathy. J Infect Dis 2006;194:1481–91.

[158] de Jong FA, Scott-Horton TJ, Kroetz DL, McLeod HL, Friberg LE, et al. Irinotecan-induced diarrhea: functional significance of the polymorphic ABCC2 transporter protein. Clin Pharmacol Ther 2007;81:42–9.

[159] Imai Y, Nakane M, Kage K, Tsukahara S, Ishikawa E, Tsuruo T, et al. C421A polymorphism in the human breast cancer resistance protein gene is associated with low expression of Q141K protein and low-level drug resistance. Mol Cancer Ther 2002;1:611–6.

[160] Bosch TM, Kjellberg LM, Bouwers A, Koeleman BP, Schellens JH, Beijnen JH, et al. Detection of single nucleotide polymorphisms in the ABCG2 gene in a Dutch population. Am J Pharmacogenomics 2005; 5:123–31.

[161] Li J, Cusatis G, Brahmer J, Sparreboom A, Robey RW, Bates SE, et al. Association of variant ABCG2 and the pharmacokinetics of epidermal growth factor receptor tyrosine kinase inhibitors in cancer patients. Cancer Biol Ther 2007;6:432–8.

[162] Sparreboom A, Loos WJ, Burger H, Sissung TM, Verweij J, Figg WD, et al. Effect of ABCG2 genotype on the oral bioavailability of topotecan. Cancer Biol Ther 2005;4:650–8.

[163] Cusatis G, Gregorc V, Li J, Spreafico A, Ingersoll RG, Verweij J, et al. Pharmacogenetics of ABCG2 and adverse reactions to gefitinib. J Natl Cancer Inst 2006;98:1739–42.

[164] Kim IS, Kim HG, Kim DC, Eom HS, Kong SY, Shin HJ, et al. ABCG2 Q141K polymorphism is associated with chemotherapy-induced diarrhea in patients with diffuse large B-cell lymphoma who received frontline rituximab plus cyclophosphamide/doxorubicin/vincristine/prednisone chemotherapy. Cancer Sci 2008;99: 2496–501.

[165] Zhang W, Yu BN, He YJ, Fan L, Li Q, Liu ZQ, et al. Role of BCRP 421C>A polymorphism on rosuvastatin pharmacokinetics in healthy Chinese males. Clin Chim Acta 2006;373:99–103.

[166] Tomlinson B, Hu M, Lee VW, Lui SS, Chu TT, Poon EW, et al. ABCG2 polymorphism is associated with the low-density lipoprotein cholesterol response to rosuvastatin. Clin Pharmacol Ther 2010;87:558–62.

[167] Hagenbuch B, Meier PJ. Organic anion transporting polypeptides of the OATP/ SLC21 family: phylogenetic classification as OATP/ SLCO superfamily, new nomenclature and molecular/functional properties. Pflugers Arch 2004;447:653–65.

[168] Tamai I, Nezu J, Uchino H, Sai Y, Oku A, Shimane M, et al. Molecular identification and characterization of novel members of the human organic anion transporter (OATP) family. Biochem Biophys Res Commun 2000;273:251–60.

[169] Tirona RG, Leake BF, Merino G, Kim RB. Polymorphisms in OATP-C: identification of multiple allelic variants associated with altered transport activity among European- and African-Americans. J Biol Chem 2001;276: 35669–75.

[170] Pasanen MK, Backman JT, Neuvonen PJ, Niemi M. Frequencies of single nucleotide polymorphisms and haplotypes of organic anion transporting polypeptide 1B1 SLCO1B1 gene in a Finnish population. Eur J Clin Pharmacol 2006;62:409–15.

[171] Pasanen MK, Neuvonen M, Neuvonen PJ, Niemi M. SLCO1B1 polymorphism markedly affects the pharmacokinetics of simvastatin acid. Pharmacogenet Genomics 2006;16:873–9.

[172] Voora D, Shah SH, Spasojevic I, Ali S, Reed CR, Salisbury BA, et al. The SLCO1B1*5 genetic variant is associated with statin-induced side effects. J Am Coll Cardiol 2009;54:1609—16.

[173] Link E, Parish S, Armitage J, Bowman L, Heath S, Matsuda F, et al. SLCO1B1 variants and statin-induced myopathy—a genomewide study. N Engl J Med 2008;359:789—99.

[174] Niemi M, Pasanen MK, Neuvonen PJ. SLCO1B1 polymorphism and sex affect the pharmacokinetics of pravastatin but not fluvastatin. Clin Pharmacol Ther 2006;80:356—66.

[175] Nishizato Y, Ieiri I, Suzuki H, Kimura M, Kawabata K, Hirota T, et al. Polymorphisms of OATP-C (SLC21A6) and OAT3 (SLC22A8) genes: consequences for pravastatin pharmacokinetics. Clin Pharmacol Ther 2003; 73:554—65.

[176] Pasanen MK, Fredrikson H, Neuvonen PJ, Niemi M. Different effects of SLCO1B1 polymorphism on the pharmacokinetics of atorvastatin and rosuvastatin. Clin Pharmacol Ther 2007;82:726—33.

[177] Niemi M, Neuvonen PJ, Hofmann U, Backman JT, Schwab M, Lutjohann D, et al. Acute effects of pravastatin on cholesterol synthesis are associated with SLCO1B1 (encoding OATP1B1) haplotype *17. Pharmacogenet Genomics 2005;15:303—9.

[178] Niemi M, Schaeffeler E, Lang T, Fromm MF, Neuvonen M, Kyrklund C, et al. High plasma pravastatin concentrations are associated with single nucleotide polymorphisms and haplotypes of organic anion transporting polypeptide-C (OATP-C, SLCO1B1). Pharmacogenetics 2004;14:429—40.

[179] Tachibana-Iimori R, Tabara Y, Kusuhara H, Kohara K, Kawamoto R, Nakura J, et al. Effect of genetic polymorphism of OATP-C (SLCO1B1) on lipid-lowering response to HMG-CoA reductase inhibitors. Drug Metab Pharmacokinet 2004;19:375—80.

[180] Tamai I, Nozawa T, Koshida M, Nezu J, Sai Y, Tsuji A. Functional characterization of human organic anion transporting polypeptide B (OATP-B) in comparison with liver-specific OATP-C. Pharm Res 2001;18:1262—9.

[181] Kobayashi D, Nozawa T, Imai K, Nezu J, Tsuji A, Tamai I. Involvement of human organic anion transporting polypeptide OATP-B (SLC21A9) in pH-dependent transport across intestinal apical membrane. J Pharmacol Exp Ther 2003;306:703—8.

[182] Nozawa T, Nakajima M, Tamai I, Noda K, Nezu J, Sai Y, et al. Genetic polymorphisms of human organic anion transporters OATP-C (SLC21A6) and OATP-B (SLC21A9): allele frequencies in the Japanese population and functional analysis. J Pharmacol Exp Ther 2002;302:804—13.

[183] Imanaga J, Kotegawa T, Imai H, Tsutsumi K, Yoshizato T, Ohyama T, et al. The effects of the SLCO2B1 c.1457C>T polymorphism and apple juice on the pharmacokinetics of fexofenadine and midazolam in humans. Pharmacogenet Genomics 2011;2011:84—93.

[184] Mougey EB, Feng H, Castro M, Irvin CG, Lima JJ. Absorption of montelukast is transporter mediated: a common variant of OATP2B1 is associated with reduced plasma concentrations and poor response. Pharmacogenet Genomics 2009;19:129—38.

[185] Smith NF, Acharya MR, Desai N, Figg WD, Sparreboom A. Identification of OATP1B3 as a high-affinity hepatocellular transporter of paclitaxel. Cancer Biol Ther 2005;4:815—8.

[186] Smith NF, Marsh S, Scott-Horton TJ, Hamada A, Mielke S, Mross K, et al. Variants in the SLCO1B3 gene: interethnic distribution and association with paclitaxel pharmacokinetics. Clin Pharmacol Ther 2007;81: 76—82.

[187] Baker SD, Verweij J, Cusatis GA, van Schaik RH, Marsh S, et al. Pharmacogenetic pathway analysis of docetaxel elimination. Clin Pharmacol Ther 2009;85:155—63.

[188] Chew SC, Singh O, Chen X, Ramasamy RD, Kulkarni T, Lee EJ, et al. The effects of CYP3A4, CYP3A5, ABCB1, ABCC2, ABCG2 and SLCO1B3 single nucleotide polymorphisms on the pharmacokinetics and pharmacodynamics of docetaxel in nasopharyngeal carcinoma patients. Cancer Chemother Pharmacol 2011;67:1471—8.

[189] Song IS, Shin HJ, Shim EJ, Jung IS, Kim WY, Shon JH, et al. Genetic variants of the organic cation transporter 2 influence the disposition of metformin. Clin Pharmacol Ther 2008;84:559—62.

[190] Wang ZJ, Yin OQ, Tomlinson B, Chow MS. OCT2 polymorphisms and in-vivo renal functional consequence: studies with metformin and cimetidine. Pharmacogenet Genomics 2008;18:637—45.

[191] Tzvetkov MV, Vormfelde SV, Balen D, Meineke I, Schmidt T, Sehrt D, et al. The effects of genetic polymorphisms in the organic cation transporters OCT1, OCT2, and OCT3 on the renal clearance of metformin. Clin Pharmacol Ther 2009;86:299—306.

[192] Shu Y, Sheardown SA, Brown C, Owen RP, Zhang S, Castro RA, et al. Effect of genetic variation in the organic cation transporter 1 (OCT1) on metformin action. J Clin Invest 2007;117:1422—31.

[193] Bhatnagar V, Xu G, Hamilton BA, Truong DM, Eraly SA, Wu W, et al. Analyses of 5′ regulatory region polymorphisms in human SLC22A6 (OAT1) and SLC22A8 (OAT3). J Hum Genet 2006;51:575—80.

[194] Xu G, Bhatnagar V, Wen G, Hamilton BA, Eraly SA, Nigam SK. Analyses of coding region polymorphisms in apical and basolateral human organic anion transporter (OAT) genes [OAT1 (NKT), OAT2, OAT3, OAT4, URAT (RST)]. Kidney Int 2005;68:1491—9.

[195] Hudis CA. Trastuzumab—mechanism of action and use in clinical practice. N Engl J Med 2007;357:39—51.

[196] Goldenberg MM. Trastuzumab, a recombinant DNA-derived humanized monoclonal antibody, a novel agent for the treatment of metastatic breast cancer. Clin Ther 1999;21:309—18.

[197] Jonker DJ, O'Callaghan CJ, Karapetis CS, Zalcberg JR, Tu D, Au HJ, et al. Cetuximab for the treatment of colorectal cancer. N Engl J Med 2007;357:2040—8.

[198] Bokemeyer C, Bondarenko I, Makhson A, Hartmann JT, Aparicio J, de Braud F, et al. Fluorouracil, leucovorin, and oxaliplatin with and without cetuximab in the first-line treatment of metastatic colorectal cancer. J Clin Oncol 2009;27:663—71.

[199] Mason DA, Moore JD, Green SA, Liggett SB. A gain-of-function polymorphism in a G-protein coupling domain of the human beta1-adrenergic receptor. J Biol Chem 1999;274:12670—4.

[200] Levin MC, Marullo S, Muntaner O, Andersson B, Magnusson Y. The myocardium-protective Gly-49 variant of the beta 1-adrenergic receptor exhibits constitutive activity and increased desensitization and down-regulation. J Biol Chem 2002;277:30429—35.

[201] Pacanowski MA, Johnson JA. PharmGKB submission update: IX. ADRB1 gene summary. Pharmacol Rev 2007;59:2—4.

[202] Johnson JA, Zineh I, Puckett BJ, McGorray SP, Yarandi HN, Pauly DF. Beta 1-adrenergic receptor polymorphisms and antihypertensive response to metoprolol. Clin Pharmacol Ther 2003;74:44—52.

[203] Pacanowski MA, Gong Y, Cooper-Dehoff RM, Schork NJ, Shriver MD, Langaee TY, et al. Beta-adrenergic receptor gene polymorphisms and beta-blocker treatment outcomes in hypertension. Clin Pharmacol Ther 2008;84:715—21.

[204] Chen L, Meyers D, Javorsky G, Burstow D, Lolekha P, Lucas M, et al. Arg389Gly-beta1-adrenergic receptors determine improvement in left ventricular systolic function in nonischemic cardiomyopathy patients with heart failure after chronic treatment with carvedilol. Pharmacogenet Genomics 2007;17:941—9.

[205] Terra SG, Hamilton KK, Pauly DF, Lee CR, Patterson JH, Adams KF, et al. Beta1-adrenergic receptor polymorphisms and left ventricular remodeling changes in response to beta-blocker therapy. Pharmacogenet Genomics 2005;15:227—34.

[206] Liggett SB, Mialet-Perez J, Thaneemit-Chen S, Weber SA, Greene SM, Hodne D, et al. A polymorphism within a conserved beta(1)-adrenergic receptor motif alters cardiac function and beta-blocker response in human heart failure. Proc Natl Acad Sci U S A 2006;103:11288—93.

[207] Terra SG, Pauly DF, Lee CR, Patterson JH, Adams KF, Schofield RS, et al. Beta-adrenergic receptor polymorphisms and responses during titration of metoprolol controlled release/extended release in heart failure. Clin Pharmacol Ther 2005;77:127—37.

[208] Small KM, Forbes SL, Rahman FF, Bridges KM, Liggett SB. A four amino acid deletion polymorphism in the third intracellular loop of the human alpha 2C-adrenergic receptor confers impaired coupling to multiple effectors. J Biol Chem 2000;275:23059—64.

[209] Bristow MR, Murphy GA, Krause-Steinrauf H, Anderson JL, Carlquist JF, Thaneemit-Chen S, et al. An alpha2C-adrenergic receptor polymorphism alters the norepinephrine-lowering effects and therapeutic response of the beta-blocker bucindolol in chronic heart failure. Circ Heart Fail 2010;3:21—8.

[210] Hunt SA, Abraham WT, Chin MH, Feldman AM, Francis GS, Ganiats TG, et al. Focused update incorporated into the ACC/AHA 2005 Guidelines for the Diagnosis and Management of Heart Failure in Adults. J Am Coll Cardiol 2009;53:e1—e90.

[211] Beta-Blocker Evaluation of Survival Trial Investigators. A trial of the beta-blocker bucindolol in patients with advanced chronic heart failure. N Engl J Med 2001;344:1659—67.

[212] Peters EJ, Slager SL, Jenkins GD, Reinalda MS, Garriock HA, Shyn SI, et al. Resequencing of serotonin-related genes and association of tagging SNPs to citalopram response. Pharmacogenet Genomics 2009;19:1—10.

[213] Horstmann S, Lucae S, Menke A, Hennings JM, Ising M, Roeske D, et al. Polymorphisms in GRIK4, HTR2A, and FKBP5 show interactive effects in predicting remission to antidepressant treatment. Neuropsychopharmacology 2010;35:727—40.

[214] Lerer B, Segman RH, Fangerau H, Daly AK, Basile VS, Cavallaro R, et al. Pharmacogenetics of tardive dyskinesia: combined analysis of 780 patients supports association with dopamine D3 receptor gene Ser9Gly polymorphism. Neuropsychopharmacology 2002;27:105–19.

[215] Bakker PR, van Harten PN, van Os J. Antipsychotic-induced tardive dyskinesia and the Ser9Gly polymorphism in the DRD3 gene: a meta analysis. Schizophr Res 2006;83:185–92.

[216] Liggett SB, Cresci S, Kelly RJ, Syed FM, Matkovich SJ, Hahn HS, et al. A GRK5 polymorphism that inhibits beta-adrenergic receptor signaling is protective in heart failure. Nat Med 2008;14:510–7.

[217] Cresci S, Kelly RJ, Cappola TP, Diwan A, Dries D, Kardia SL, et al. Clinical and genetic modifiers of long-term survival in heart failure. J Am Coll Cardiol 2009;54:432–44.

[218] Campbell DB, Ebert PJ, Skelly T, Stroup TS, Lieberman J, Levitt P, et al. Ethnic stratification of the association of RGS4 variants with antipsychotic treatment response in schizophrenia. Biol Psychiatry 2008;63:32–41.

[219] Manunta P, Lavery G, Lanzani C, Braund PS, Simonini M, Bodycote C, et al. Physiological interaction between alpha-adducin and WNK1-NEDD4L pathways on sodium-related blood pressure regulation. Hypertension 2008;52:366–72.

[220] Cusi D, Barlassina C, Azzani T, Casari G, Citterio L, Devoto M, et al. Polymorphisms of alpha-adducin and salt sensitivity in patients with essential hypertension. Lancet 1997;349:1353–7.

[221] Davis BR, Arnett DK, Boerwinkle E, Ford CE, Leiendecker-Foster C, Miller MB, et al. Antihypertensive therapy, the alpha-adducin polymorphism, and cardiovascular disease in high-risk hypertensive persons: the Genetics of Hypertension-Associated Treatment Study. Pharmacogenomics J 2007;7:112–22.

[222] Wang D, Chen H, Momary KM, Cavallari LH, Johnson JA, Sadee W. Regulatory polymorphism in vitamin K epoxide reductase complex subunit 1 (VKORC1) affects gene expression and warfarin dose requirement. Blood 2008;112:1013–21.

[223] Limdi NA, Wadelius M, Cavallari L, Eriksson N, Crawford DC, Lee MT, et al. Warfarin pharmacogenetics: a single VKORC1 polymorphism is predictive of dose across 3 racial groups. Blood 2010;115:3827–34.

[224] Epstein RS, Moyer TP, Aubert RE, DJ OK, Xia F, et al. Warfarin genotyping reduces hospitalization rates results from the MM-WES (Medco-Mayo Warfarin Effectiveness study). J Am Coll Cardiol 2010;55: 2804–12.

[225] Anderson JL, Horne BD, Stevens SM, Woller SC, Samuelson KM, Mansfield JW, et al. Randomized and clinical effectiveness trial comparing two pharmacogenetic algorithms and standard care for individualizing warfarin dosing: CoumaGen-II. Circulation 2012;125:1997–2005.

[226] Scharplatz M, Puhan MA, Steurer J, Bachmann LM. What is the impact of the ACE gene insertion/deletion (I/D) polymorphism on the clinical effectiveness and adverse events of ACE inhibitors?—Protocol of a systematic review. BMC Med Genet 2004;5:23.

[227] Arnett DK, Davis BR, Ford CE, Boerwinkle E, Leiendecker-Foster C, Miller MB, et al. Pharmacogenetic association of the angiotensin-converting enzyme insertion/deletion polymorphism on blood pressure and cardiovascular risk in relation to antihypertensive treatment: the Genetics of Hypertension-Associated Treatment (GenHAT) study. Circulation 2005;111:3374–83.

[228] Tiret L, Rigat B, Visvikis S, Breda C, Corvol P, Cambien F, et al. Evidence, from combined segregation and linkage analysis, that a variant of the angiotensin I-converting enzyme (ACE) gene controls plasma ACE levels. Am J Hum Genet 1992;51:197–205.

[229] Danser AH, Derkx FH, Hense HW, Jeunemaitre X, Riegger GA, Schunkert H. Angiotensinogen (M235T) and angiotensin-converting enzyme (I/D) polymorphisms in association with plasma renin and prorenin levels. J Hypertens 1998;16:1879–83.

[230] Winkelmann BR, Nauck M, Klein B, Russ AP, Bohm BO, Siekmeier R, et al. Deletion polymorphism of the angiotensin I-converting enzyme gene is associated with increased plasma angiotensin-converting enzyme activity but not with increased risk for myocardial infarction and coronary artery disease. Ann Intern Med 1996;125:19–25.

[231] Andersson B, Sylven C. The DD genotype of the angiotensin-converting enzyme gene is associated with increased mortality in idiopathic heart failure. J Am Coll Cardiol 1996;28:162–7.

[232] McNamara DM, Holubkov R, Postava L, Janosko K, MacGowan GA, Mathier M, et al. Pharmacogenetic interactions between angiotensin-converting enzyme inhibitor therapy and the angiotensin-converting enzyme deletion polymorphism in patients with congestive heart failure. J Am Coll Cardiol 2004;44:2019–26.

[233] Wu CK, Luo JL, Tsai CT, Huang YT, Cheng CL, Lee JK, et al. Demonstrating the pharmacogenetic effects of angiotensin-converting enzyme inhibitors on long-term prognosis of diastolic heart failure. Pharmacogenomics J 2010;10:46−53.

[234] Zolk O. Disposition of metformin: variability due to polymorphisms of organic cation transporters. Ann Med 2011;44:119−29.

[235] Zhou K, Bellenguez C, Spencer CC, Bennett AJ, Coleman RL, Tavendale R, et al. Common variants near ATM are associated with glycemic response to metformin in type 2 diabetes. Nat Genet 2011;43:117−20.

[236] Kannankeril P, Roden DM, Darbar D. Drug-induced long QT syndrome. Pharmacol Rev 2010;62:760−81.

[237] Roden DM. Personalized medicine and the genotype-phenotype dilemma. J Interv Card Electrophysiol 2011;31:17−23.

[238] Fernandez-Fernandez JM, Tomas M, Vazquez E, Orio P, Latorre R, Senti M, et al. Gain-of-function mutation in the KCNMB1 potassium channel subunit is associated with low prevalence of diastolic hypertension. J Clin Invest 2004;113:1032−9.

[239] Beitelshees AL, Gong Y, Wang D, Schork NJ, Cooper-Dehoff RM, Langaee TY, et al. KCNMB1 genotype influences response to verapamil SR and adverse outcomes in the INternational VErapamil SR/Trandolapril STudy (INVEST). Pharmacogenet Genomics 2007;17:719−29.

[240] Gloyn AL, Pearson ER, Antcliff JF, Proks P, Bruining GJ, Slingerland AS, et al. Activating mutations in the gene encoding the ATP-sensitive potassium-channel subunit Kir6.2 and permanent neonatal diabetes. N Engl J Med 2004;350:1838−49.

[241] Babenko AP, Polak M, Cave H, Busiah K, Czernichow P, Scharfmann R, et al. Activating mutations in the ABCC8 gene in neonatal diabetes mellitus. N Engl J Med 2006;355:456−66.

[242] Pearson ER, Flechtner I, Njolstad PR, Malecki MT, Flanagan SE, Larkin B, et al. Switching from insulin to oral sulfonylureas in patients with diabetes due to Kir6.2 mutations. N Engl J Med 2006;355:467−77.

2

Translating Pharmacogenomic Research to Therapeutic Potentials

Y.W. Francis Lam

Department of Pharmacology, University of Texas Health Science Center at San Antonio, San Antonio, Texas, USA

OBJECTIVES

1. List the major steps involved in implementing pharmacogenomic testing in a clinical environment.

2. Discuss the challenges associated with translating pharmacogenomic research findings to clinical practice.

3. Delineate the differences between analytical validity, clinical validity, and clinical utility of a pharmacogenomic diagnostic test.

4. Explain the potential roles of pharmacogenomics in all phases of drug development.

INTRODUCTION

The genetic basis of altered drug pharmacokinetics and pharmacodynamics, as well as how these interindividual variabilities can potentially help optimize drug therapy in different disease states, are highlighted in chapter 1 and chapters 4 to 7 of this book. However, most clinicians and researchers would agree that although pharmacogenomic research findings are now used to different extents in clinical practice at many major academic medical institutions,[1,2] we are still some years away from achieving the goal of broadly offered personalized therapy in health care that was envisioned decades ago.

Application of genomic data in clinical practice requires the use and interpretation of biomarker-based pharmacogenomic diagnostic tests. Although there are established genetic biomarkers that clinicians can use to predict drug efficacy and/or toxicity (e.g., Her2neu testing for trastuzumab, *HLA-B**5701 for abacavir), the challenge is to address several issues, including whether the biomarkers should be used in patient assessment, as well as when and in whom to use the diagnostic tests. Establishing the clinical utility of the biomarker has been advocated to ensure that the use of the biomarker is appropriate in patients and that the testing is cost-effective and ultimately improves clinical outcome.

Translation of scientific knowledge into practice and integration within the health care system has been further hampered by commercial, economical, educational, legal, and societal barriers, each of which is fueled by stakeholders with different interests and goals. In addition, there is constant debate within the scientific community, with little agreement, as to how much data (replication studies, sample size) and what quality of data (retrospective cohort versus randomized controlled clinical trial) are scientifically appropriate but at the same time realistically achievable.

This chapter provides a perspective on the existing steps and challenges (Table 2.1) in the complex process for translating a pharmacogenetic biomarker that has been discovered to its

TABLE 2.1   Major steps and challenges involved in the implementation of pharmacogenomic testing in clinical practice

---

DISCOVERY AND VALIDATION OF PHARMACOGENOMIC BIOMARKERS IN WELL-CONTROLLED STUDIES

Challenge: Clinical validity and clinical utility: how to meaningfully translate a statistical genomic association between single-nucleotide polymorphisms SNPs (or haplotypes) and drug response in a controlled, but usually underpowered, study to the real-world clinical environment.

REPLICATION OF GENE/DRUG ASSOCIATION

Challenge: Multiple nongenetic variables make it difficult to identify the most appropriate response phenotype for a specific biomarker in most replication studies.

DEVELOPMENT AND APPROVAL OF COMPANION DIAGNOSTIC TESTS OR VALIDATION OF GENOTYPE RESULTS IN A CLINICAL LABORATORY PER QUALITY STANDARDS

Challenge: Financial incentive and resources for most small diagnostic companies. Approval process, especially in global markets, not completely delineated.

IDENTIFICATION OF APPROPRIATE PATIENT POPULATIONS FOR CLINICAL IMPLEMENTATION

Challenge: Continued debate over routine genotyping for all patients versus reserving the test for selected patients.

DESIGNING AN EFFICIENT WORKFLOW STRATEGY FOR ORDERING AND RECEIVING GENETIC TEST RESULTS

Challenge: Reasonable turnaround time for point-of-care utility. Electronic medical record not universally adopted in hospitals and clinics.

INTERPRETATION OF RESULTS

Challenge: Not a simple normal versus abnormal interpretation. Proper education crucial to appropriate clinical use.

EDUCATION OF CLINICIANS

Challenge: Educating current and future clinicians at a level such that the genomic information can be efficiently utilized.

ASSESSING CLINICIAN ACCEPTANCE OF GENETIC TESTING AND ADDRESSING BARRIERS TO CLINICAL IMPLEMENTATION

Challenge: Most appropriate ways to address issues of lack of reimbursement from payers for most tests, privacy and discrimination concerns from patients, ownership of genetic information, health disparity, and potential legal liability.

---

clinical implementation as a test, as well as incorporating pharmacogenomics into drug development.

IMPLEMENTATION OF BIOMARKERS IN CLINICAL PRACTICE

Complexity of Genetic Variabilities and Non genetic Influences

Although many biomarkers have been identified over the last decade, most of them have not advanced beyond the discovery phase. Exceptions to this state are primarily in the area of oncology; however, there are also examples of biomarker use in neurology (*HLA-B**1502 testing to predict risk for severe skin reactions to carbamazepine), infectious disease (*HLA-B**5701 testing for hypersensitivity risk with abacavir), and cardiology (*CYP2C19* to predict clopidogrel effectiveness). The major issue has been the inconsistent results for replication of the genetic associations for most biomarkers, whether alone or in combination. The scientific challenge for study replication is magnified by our understanding that drug disposition and response phenotypes are more accurately predicted by multiple gene variations and not single-nucleotide polymorphisms (SNPs), population differences in prevalence of most genetic variants, as well as the accompanying sample size requirement for statistical power in most pharmacogenomic studies. The atypical antipsychotic clozapine, with its complex pharmacological effects via the dopaminergic, serotonergic, adrenergic, and histaminergic receptors within the central nervous system, is a good example to illustrate the challenge of multiple gene variants each accounting for a portion of the response variability. Conflicting study results exist in the literature for association between clozapine response with either SNPs of each receptor subtype[3–6] or combinations of polymorphisms,[7] suggesting that yet-to-be identified genes could account for additional variability in patients' responses to clozapine. The presence of different allele variants of *CYP2D6*, *HLA-B*, *UGT1A1*, and *SLC6A4* among different ethnic groups (discussed in chapters 1 and 6) reminds investigators of the importance of ethnicity in pharmacogenomic studies.

The drug disposition and response phenotypes can further be affected by patient-specific and environmental variables. Concurrent therapy with a potent CYP2D6 inhibitor such as fluoxetine or paroxetine could significantly reduce the CYP2D6 metabolic capacity of a genotypic extensive metabolizer to that of a poor metabolizer,[8] thereby creating a genotype-phenotype discordance and affecting the ability to predict possible drug response based on genotype-guided dosing and achievable drug concentration. Another example is inflammation-mediated down-regulation of drug metabolizing enzymes. Using a transgenic mouse model of human *CYP3A4* regulation, Robertson et al. showed that presence of extrahepatic tumors elicited an inflammatory response, including release of cytokines such as interluekin-6, and resulted in transcriptional down-regulation of the human *CYP3A4* gene [9]. Therefore, literature reporting lower docetaxel clearance in cancer patients could be related to tumor-associated inflammation and subsequent transcriptional repression of *CYP3A4*, potentially leading to unanticipated toxicity despite normal enzymatic activity in the patient. On the pharmacodynamic side, excessive vitamin K intake can override the effects of *VKORC1* genotype on warfarin dose requirements.

Currently, much less is known about the influence of environmental variables and gene-environment interactions on drug disposition and response phenotypes. There is an increasing appreciation that genetic heterogeneity alone cannot explain interindividual variations in drug responses. The three "Looking to the Future" subchapters (chapters 4a, 5a, and 6a) in this book provide an overview and discuss potential implications of epigenetics, which in its simplest form refers to changes in phenotype without DNA sequence alteration. In the not-too-distant future, pharmacoepigenetics could provide the basis of studying the interaction among drugs, environment, and genes and provide additional explanation of drug response variations beyond the level of genetic polymorphisms.

Analytical Validity, Clinical Validity, and Clinical Utility

For evaluation of a pharmacogenomic biomarker test, regardless of whether it is to be developed as a companion diagnostic, both analytical and clinical validity of the test have to be considered. For the purpose of personalized therapy, a companion diagnostic for a drug can be defined as a biomarker that is critical to the safe and effective use of the drug. Analytical validity defines how well a diagnostic test measures what it is intended to measure, regardless of whether it is a mutation, protein, or an expression pattern. Clinical validity measures the ability of the test to differentiate responders from nonresponders or to identify patients who are at risk for adverse drug reactions. For practical implementation of the validated pharmacogenomic biomarker test, the clinical utility of the test also has to be determined. Clinical utility measures the ability of the test result to predict outcome in a clinical environment and what value would be obtained compared to nontesting, that is, standard empirical treatment. The ACCE (analytical validity, clinical validity, clinical utility and associated ethical, legal, and social implications [ELSI]) Model Project [10] sponsored by the Office of Public Health Genomics, Centers for Disease Control and Prevention (CDC), has been recently advocated by some investigators to be the basis for evaluation of pharmacogenomic biomarker tests.

In 2004, the CDC launched the Evaluation of Genomic Applications in Practice and Prevention (EGAPP) initiative, which aims to establish an evidence-based process for evaluating genetic tests and genomic technology that are being translated from research to clinical practice. In 2007, the EGAPP Working Group, incorporating the three levels of evaluations (analytical validity, clinical validity, and clinical utility), published their evidence-based review of the literature on the use of cytochrome P-450 (CYP) genotyping for clinical management of depressed patients with the selective serotonin reuptake inhibitors (SSRIs). Based on strong evidence of analytical validity, possible demonstration of clinical validity, and lack of study data to support evaluation of potential clinical utility, the working group does not recommend the application of the CYP2D6 test for SSRI pharmacotherapy [11].

The analytical validity of most CYP genotyping tests in detecting CYP450 gene variants is strong and to be expected, as their approval by the Food and Drug Administration (FDA) is dependent on technical performance. Nevertheless, there is no universal agreement as to which allele variant should be tested routinely, and ethnic variations in allele importance and distribution further complicate the picture. The weak evidence of association between the gene variants and SSRI metabolism, efficacy, and response are more likely related to most SSRIs relying on multiple enzymes for metabolism, some of which are not polymorphic; a flat dose-response relationship; and wide therapeutic index. The clinical validity of the test to differentiate response phenotypes is further limited by the CYP genotype—metabolic phenotype discordance that can occur as a result of drug—drug interactions or environmental influences. Given these pharmacokinetic and pharmacodynamic limitations as well as the lack of cost-effectiveness data, it is not surprising that the SSRIs are not good candidates for genotype-based pharmacogenomic therapy, and hence the recommendation of the EGAPP Working Group.

Traditional clinical studies aim to gather evidence of drug efficacy and safety in large patient cohorts in an attempt to overcome statistical issues related to disease and population heterogeneities, placebo effects, inadequate understanding of disease etiologies, and finally, drug response variabilities per se. All too often, such studies result in the achievement of a small average benefit in the entire heterogeneous patient cohort. Nevertheless, given the current evidence-based driven clinical environment, it is expected that any clinical trial to validate the clinical utility of pharmacogenomic biomarkers would have to be not only hypothesis driven but also extensive in terms of time and sample size, and therefore costly. Although prospective double-blind randomized clinical trials would provide the ideal evidence-based approach advocated by many investigators, a balance between the scientific demand of randomized clinical trials and the practical value of genotyping for patient care seems appropriate. In contrast to evidence-based practice, the emphasis and value of pharmacogenomics are more geared towards the outliers (the nonresponders, the poor metabolizers, or the ultra-rapid metabolizers). Therefore, to generate more robust evidence of efficacy, prospective enrichment design clinical trials (discussed in chapter 9a) have been advocated by many investigators and sponsors to include patients who are more likely to respond or at least be stratified according to disease subtypes, and/or exclude patients who are highly susceptible to adverse drug reactions. Even with the assumption of (and sometimes proven) association between genetic variabilities and drug response, both advantages and disadvantages exist for this study design (Table 2.2).

TABLE 2.2   Advantages and disadvantages for prospective enrichment design studies

Advantages
- Substantial reduction in response (efficacy, side effects, disposition) variabilities
- Possibly can explore a greater dose range than otherwise achieved with entire population
- Smaller number of patients needed in pivotal Phase III trial
- Possible reduction of patient discontinuance (from less efficacy or increased side effects)
- Greater probability of a successful trial
- Possible shorter duration of a clinical trial
- Possible reduction in safety monitoring, including plasma concentration

Disadvantages
- No opportunity to study excluded subjects in the pivotal trial
- Information available only for a much narrower range of response variabilities
- Potential of overestimating drug efficacy in a highly selective group
- Possibly less inclination to monitor for safety in genotyped patients
- Resultant less information on short-term and long-term safety

Additional Concern
- Any regulatory requirement (and may be ethical reasons) for studying safety in excluded patients?

For patient care, a good example for the need of balance between evidence-based medicine and personalized medicine is clopidogrel. As discussed in chapters 1 and 5, there is extensive evidence of clopidogrel efficacy linked to CYP2C19 genetic polymorphism. However, the continued debate over the routine use of CYP2C19 genotyping to guide clopidogrel therapy prevents more widespread use of the biomarker in individualized therapy, despite the significantly higher rates of stent thrombosis and the associated mortality rates in carriers of the reduced-function CYP2C19*2 allele. Based on lack of outcomes data, the joint clinical alert issued in 2010 by the American College of Cardiology and the American Heart Association did not recommend routine genotyping and suggested the need for large, prospective, controlled trials [12]. In the 2011 Practice Guideline for Percutaneous Coronary Intervention (PCI), routine clinical use of genetic testing to screen patients treated with clopidogrel who are undergoing PCI is not recommended. However, the guideline did suggest that genetic testing might be considered for patients at high risk for poor clinical outcomes [13]. The Pharmacogenomics of Anti platelet Intervention-2 (PAPI-2) trial is an ongoing multisite, randomized, controlled clinical trial designed to evaluate the effect of genotype-guided antiplatelet therapy versus standard care on cardiovascular events among 7,200 patients undergoing PCI (clinicaltrials.gov, NCT01452152). However, the results will likely not be available until 2015. In the meantime, the questions then become: Are we sacrificing patient care on the insistence of waiting for proof of value via the evidence-based approach? If no such proof is available in the near future, should we focus on steps that can facilitate the genotyping implementation in the clinical setting and examine the cost effectiveness of genotypes-guided antiplatelet therapy with a variety of different approaches?

Another example of taking an alternative approach to evidence-based evaluation is tolbutamide. Based on a pharmacokinetic study in subjects genotyped for the CYP2C9 polymorphisms, the elimination of tolbutamide in carriers of the CYP2C9*2 and CYP2C9*3 variants were 50% and 84% lower, respectively, than in subjects with the CYP2C9*1/*1 genotype [14]. However, there has not been any prospective, controlled clinical study to evaluate whether dosage reduction on the order of 50% and 90% in patients with these two genotypes would be appropriate in clinical practice. Because tolbutamide efficacy can be easily monitored in the clinical setting, implementing these dosage reductions in clinics or physician offices in lieu of expensive and time-consuming large-scale clinical trials could constitute the first step to obtain information regarding the clinical utility of CYP2C9 genotype in optimizing tolbutamide therapy.

Evaluation of Cost-Effectiveness

For many health care facilities and systems, demonstration of cost-effectiveness of any test or procedure is critical prior to its implementation. Ideally, the pharmacogenomic biomarker not only will result in cost-effective improved clinical care in patients who will benefit from individualized therapy with the drug, but also will lead to avoidance of cost-ineffective treatment for patients who likely will not benefit from the drug, either as a result of lack of response or increased adverse drug reactions. Given the differences in revenue generation between a pharmacogenomic diagnostic companion test and a drug, there could conceivably be far fewer incentives for pharmaceutical companies to include thorough cost-effectiveness analyses as part of drug development. However, demonstration of cost-effectiveness of pharmacogenomic-based therapeutic approaches can range the spectrum of comparing per-patient cost for specific clinical outcome between genotype-based regimen and standard regimen, as shown by the study of Furuta et al.,[15] to decision model—based study using a simulated patient cohort similar to the recent report of Reese et al. for antiplatelet therapy [16].

The economic impact and cost-effectiveness of screening can be affected by different variables. To study the potential clinical and economic outcomes for pharmacogenomic-guided dosing of warfarin, two studies utilized modeling techniques in separate simulated patient cohorts. Despite the conclusion that the relatively high cost of a *CYP2C9* and *VKORC1* bundled test ($326 to $570) resulted in only modest improvements (quality-adjusted life-years, survival rates, and total adverse rates), the investigators also suggested that the cost-effectiveness can be improved in several ways, including cost reduction of the genotyping test by 50% and applying a genotype-guided warfarin dosing algorithm in outliers (patients with out of range international normalized ratios (INRs) and/or those who are at high risk for hemorrhage) [17,18]. Other variables such as prevalence of a specific variant in a population and cost of alternative treatment approaches should also be taken into consideration.

Regulatory Approval of Pharmacogenomic Diagnostic Tests

Over the last seven years, the FDA has progressively acknowledged the importance of biomarkers and provided new recommendations on pharmacogenomic diagnostic tests and data submission. These efforts included the publication of FDA Guidance for Pharmacogenomic Data Submission in 2005, the introduction of the Voluntary Data Submission Program in 2007, and the issuance of the draft guidance for "In Vitro Diagnostic Multivariate Index Assays" in 2008. These and other FDA initiatives are further described in chapter 3. Pharmacogenomic biomarker information for more than 120 drugs has been previously classified by the FDA into three categories: test required before the drug is prescribed, either for predicting efficacy or toxicity; test recommended; and test available only for information purposes. Table 3.4 lists up-to-date examples of drugs with pharmacogenetic-related information in their FDA-approved labeling.

Within the United States, tests for a pharmacogenomic biomarker are performed either as a test developed by a clinical laboratory or as an *in vitro* diagnostic device, each with its own regulatory oversight. Quality standards for clinical laboratory tests are governed by the Clinical Laboratory Improvement Amendments (CLIA). In addition, the laboratories are accredited by either the College of American Pathologists, the Joint Commission on Accreditation of Healthcare Organizations, or the Health Department of each individual state, which takes into consideration CLIA compliance and laboratory standard practices that are in line with Good Laboratory Practice (GLP) regulations enforced by the FDA.

In contrast to clinical laboratories, the GLP regulations govern the testing of *in vitro* medical diagnostic devices. Although currently there is no formal regulatory process for submission of companion diagnostic tests, the well-established medical test and device regulatory process within the Office of *In Vitro* Diagnostic Devices seems amenable for application to biomarker approval. Despite original attempts to perform and market the

AmpliChip CYP450 Test under CLIA regulations, the FDA decided that evaluation and approval as an *in vitro* diagnostic device was required. It is also of note that additional historical precedence had been set with the FDA fast track approval of trastuzumab and the accompanying Hercep Test for detecting over-expression of HER2 protein in breast cancer tissue in 2001, and more recently for tests that utilize fluorescence in situ hybridization (FISH). Last year, both vemurafenib and crizotinib were approved by the FDA with their respective companion diagnostic tests. Other drugs for which pharmacogenomic markers have been developed and approved by the FDA are listed in Table 3.4. In the European Union and the United Kingdom, there is no similar framework for premarketing regulatory review and approval of pharmacogenomic biomarkers, and most regulations apply in the postmarketing phase. An example is approval of gefitinib by the European Medicines Agency (EMA) in June 2009. Subsequently, the EMA last year approved a companion diagnostic test for *HER1* mutations. These efforts by the FDA and foreign regulatory agencies not only increase the availability of companion diagnostic tests but also provide an impetus for pharmacogenomic data submission for drug approval and additional research to address the debate over the utility of the information incorporated in the revised labels, such as for clopidogrel [19].

Integration of Testing within the Health Care Environment

There are two practical aspects that need to be addressed before pharmacogenomic markers can be successfully utilized in any health care setting. Low test volume for the diagnostic test may not justify in-house testing in institutional clinical laboratories. The ideal point-of-care performance for rapid decision making at the bedside or within the clinic is not available at most hospitals. The inevitable outcome is longer turnaround time for test results coming from external clinical laboratories or research institutions. Although this outcome might be acceptable in relatively less "urgent" settings, such as *HER2* expression or CYP2C19 genotyping prior to scheduled PCI, the contrary would be true for on-the-spot warfarin dosing adjustment or when there is a need for emergency PCI. Nevertheless, progress has been made in this aspect. A point-of-care *CYP2C19* genotyping device capable of detecting the *CYP2C19*2* allele with a buccal swab has been developed and recently used to explore the feasibility of bedside genetic testing [20]. With a turnaround time of about an hour, incorporation of *CYP2C19* testing into the clinical protocol for antiplatelet dosing is realistic and could pave the way for more widespread use of clopidogrel pharmacogenetics in clinical practice. Ideally, all pharmacogenomic information will be available in a robust system of electronic medical records (EMR) for use by all physicians taking care of the patients. However, the adoption of EMR is not universal [21].

With the availability of test results comes the need for interpretation and education of clinicians. Most pharmacogenomic diagnostic tests report the genotype result, the interpretation of which is usually not difficult, especially for deciding the appropriateness of a specific drug for a patient. Examples include the presence of the *HLA-B*5701* variant for exclusion of abacavir therapy in patients with HIV-1 infection and the use of gefitinib in patients with the epidermal growth factor receptor mutation. The interpretation is more complicated and challenging when the test result is used for dosing adjustment (e.g., with warfarin), given the many genetic and nongenetic variables that can affect drug disposition and response. Other examples include drugs that rely on the P-450 enzyme system for elimination, given the significant interindividual variabilities in activities of most of the isoenzymes and the possibility of phenocopying with change in metabolic phenotype in the presence of a drug–drug interaction. This difference in interpretation complexity related to the intended use of the test is likely one of the reasons for the FDA having previously separated pharmacogenomic biomarkers into three categories.

Currently, warfarin represents an example with some of the greatest potential for clinical implementation of pharmacogenomic biomarker. As described in chapters 1 and 5, the availability of pharmacogenetic dosing algorithms helps utilize the patient's *CYP2C9* and *VKORC1* genotypes and other nongenetic factors (e.g., age, body size, concurrent interacting drugs) in determining dosage. The algorithms based on the work of Gage et al [22] and the International Warfarin Pharmacogenetics Consortium (IWPC)[23] are publicly available via the Internet [24,25]. Although these algorithms can be useful, they are not without limitations. As with other CYP genotyping, the clinical utility is mostly demonstrated in the outliers. For example, the IWPC showed that a pharmacogenetic dosing algorithm was most predictive of therapeutic anticoagulation in 46% of the patient cohort who required less than 25 mg/week or greater than 49 mg/week of warfarin. Although the algorithm approach has been successfully used in inpatients receiving multiple drug therapy,[26] most data have been primarily derived from outpatients receiving stable warfarin dosage regimens. Not surprisingly, different—albeit not statistically significant—dosage requirements were obtained with various algorithms,[27] likely reflecting the inconsistency in the choice of specific nongenetic variables among these algorithms. Most algorithms also do not include detection of *CYP2C9*8* or assessment of *CYP4F2* genotype. Although the contribution of *CYP4F2* genetic polymorphism accounts for only 1 to 3% of the variability of warfarin dosing requirement, the exclusion of *CYP2C9*8* commonly found in African Americans likely would account for lower successful dose prediction associated with the use of these algorithms in this ethnic group. This limitation is similar to the challenge discussed earlier of deciding which *CYP* alleles or which *UGT1A1* alleles should be included in the diagnostic tests for these pharmacogenomic biomarkers.

Much like other clinical diagnostic tests, patients expect clinicians to be able to explain the pharmacogenomic diagnostic test results and answer their questions. However, a survey of more than 10,000 physicians conducted in 2008 found that only 10% felt adequately trained to apply genetic information in clinical practice and only 26% had received pharmacogenomic education during their medical school or postgraduate training. This lack of training is unfortunate, given that 98% of those surveyed agreed that patients' genetic profiles could influence drug therapy decisions [28]. In their 2011 update to accreditation guidelines (version 2, January 23, 2011), the Accreditation Council for Pharmaceutical Education listed pharmacogenomics as part of professional curriculum course work [29]. Although most pharmacy schools have a pharmacogenomic courses or materials in place, this is not the case with medical schools. The gap in knowledge can at the moment be addressed through clinical guidelines and algorithms such as the guidelines available through the Clinical Pharmacogenetics Implementation Consortium (CPIC) for abacavir, clopidogrel, codeine, simvastatin, and warfarin [30–34]. Hopefully, additional clinical practice guidelines from diverse groups of organizations and expert panels will pave the way to greater extent of implementation. To that end, it is of note that regulatory guidance [35] has been published to support the recommendation of the clinical practice guidelines.

Reimbursement Issues

The successful implementation of pharmacogenomic biomarkers in clinical practice not only involves multidisciplinary coordination from physicians, pharmacists, and clinical laboratories but also requires efforts from the payer. With the high cost of providing health care, whether any particular test is reimbursable plays a significant role in its implementation status in clinical practice. Reimbursement for diagnostic tests in the United States is primarily linked to the Current Procedural Terminology (CPT) codes. Although the cost of testing for thiopurine S-methyltransferase is reimbursed according to CPT codes in some hospitals, that is not the case for most pharmacogenomic biomarker tests. Even with the revised product labeling information regarding the impact of CYP variants for drugs such as warfarin and clopidogrel, insurers are reluctant to reimburse the cost of the tests on the basis that either (1) such tests are not medically necessary (because they have never been classified by the FDA as a required test), (2) there is no evidence of clinical utility (clinical utility is usually associated with endorsement by professional organizations), or (3) lack of cost-effectiveness analysis and/or comprehensive comparative effectiveness analysis. It should be noted that even for trastuzumab, which is reimbursed by most insurers, there have been few cost-effectiveness analyses of HER2 protein expression and treatment with trastuzumab [36]. For most pharmacogenomic biomarkers, the ideal analysis might not be available until years after the diagnostic test is marketed.

In their survey of 12 payers, Cohen et al [37]. found that 67 to 75% provide reimbursements to cover the costs of the required companion diagnostic tests for trastuzumab and cetuximab, even though a lower percentage (42 to 50%) require documentation of testing prior to reimbursement. Likewise, 33% of payers will reimburse *CYP2C9* and *VKORC1* tests, but none requires test documentation. Six of the eight study drugs included in the survey were antineoplastics. Reimbursements for the corresponding diagnostic tests are provided by most payers for these antineoplastics, with the exception of irinotecan, for which only two payers will provide reimbursement for *UGT1A1* testing. Most payers consider conclusive evidence of a link between the diagnostic test and health outcome to be much more important than evidence of test accuracy in identifying subpopulations of interest. Interestingly, the authors reported that most payers indicate that test cost, medication adherence, and off-label use are not factors in their consideration for reimbursement.

It is clear that currently payers are reluctant to pay for the diagnostic tests (most costing up to $500), even though they will pay for the more expensive drugs. Such a stance would pose much less incentive for diagnostic companies to develop biomarkers, as they usually have less financial resources than pharmaceutical companies. An obvious solution to this would be the codevelopment of a proprietary drug and diagnostic test that would be rewarding to both parties,[38] and the recent FDA approval of crizotinib with a codeveloped diagnostic test might pave the way for further parallel development of drug and companion diagnostics.

Trastuzumab provides a good example of the paradigm shift in thinking about market share: the manufacturer's development of the drug along with the diagnostic device has resulted in the capture of 100% of the market share associated with breast cancer drug treatment in women overexpressing the HER2 protein. Therefore, trastuzumab is used and reimbursed only for patients with HER2 protein overexpression. Likewise, a similar paradigm shift might be applicable for reimbursement of companion diagnostics. Instead of reimbursing the same rate for every patient tested for a pharmacogenomic biomarker, it might be less of a financial burden for the payer to institute a differential reimbursement based on indication (e.g., for *CYP2C19* testing, a higher rate for high-risk PCI, a lower rate for a PCI that is not high risk, and none if no PCI is performed). This approach could provide additional incentive to use pharmacogenomic biomarkers as the equivalent of a differential diagnostic test to identify patients who will benefit the most from genomic-guided personalized drug therapy. The financial cost of the one-time test should be easily covered through cost saving associated with not using the drug when it is ineffective or harmful in specific patient populations. Adopting this approach may provide a workaround to some payers' insisting on conclusive evidence of linking diagnostic tests to health outcomes,[37] which would take many years and cost much to achieve.

Ethical, Legal, and Social Issues

The public is in general receptive to genetic-based prescribing [39]. Although the benefit of pharmacogenomic testing lies in identifying individual patients with unanticipated responses and/or adverse drug reactions, it also provides an opportunity to reveal information about an individual's disease or medical condition to other parties, however unintended. As a result, concerns have been raised regarding the individual's right to privacy, as well as the potential for discrimination and ineligibility for employment and insurance [39]. In response, the U.S. Congress passed the Genetic Information Nondiscrimination Act (GINA) in 2008. The GINA specifically prohibits the misuse of genetic information in determining employment decisions, insurance coverage and premiums, as well as payers requiring individuals to submit genetic results prior to underwriting decisions. Another intention of the GINA is to encourage individuals to participate in genetic research, although the issues of information sharing and confidentiality have not been addressed to the satisfaction of stakeholders—primarily patients. Of paramount importance are concerns regarding ownership of genetic materials, who has the right to access the information (reported as laboratory test values and sometimes stored in the EMR), and patients' awareness of the consequences of storing genetic materials and phenotypic data. These concerns have been amplified by different efforts to facilitate research collaboration among investigators, as outlined in chapter 11.

When discussing the ethical, legal, and social implications of genetic technology, pharmacogenomic biomarkers are usually grouped with tests predicting disease likelihood into the "generic" category of genetic tests. In terms of patient care, does consent need to be obtained from patients for tests designed to individualize their drug therapy (choice and/or dosage regimen), or should consent be reserved for only those tests that disclose disease susceptibility? It would seem appropriate that pharmacogenomic biomarker tests not be treated with the same degree of scrutiny and protection as genetic testing for disease susceptibility, which carries a much greater potential for abuse. A lessening of regulation and consent requirements for pharmacogenomic markers would make it easier for their implementation. However, this issue of the need for consent is very much open for further discussion and debate. The bioethical implications of the ever-expanding genomic data are further addressed in chapter 12 of this book.

Social concerns also arise regarding potential challenges for health care systems. It is not unusual under insurance coverage for patients to be required to pay for some of the cost of the medical service. Therefore, potential beneficial pharmacogenomic test information might be excluded as a result of an individual patient's socioeconomic status, thus exacerbating health care disparities. In addition, for those identified by a pharmacogenomic test as either nonresponders or at high risk of adverse drug reaction to a specific drug, the use of pharmacogenomic test as a "gatekeeper" of accessibility to drug treatment might pose a problem if there is no suitable alternative drug available. How should those patients be advised and treated? Is it ethical or appropriate if the patient and/or the physician opt for off-label use of a drug regardless of the unfavorable response and/or risk associated with a specific genotype? These are relevant questions because the clinical validity and utility of most pharmacogenomic tests have not been universally accepted in clinical practice. Another potential concern is liability on the part of the provider. If a genetic test (e.g., *CYP2C9*) is ordered to guide therapy with one drug (e.g., warfarin) but the patient is later prescribed another drug that is also affected by the gene previously tested (e.g., phenytoin), would the clinician be responsible for acting on the genotype results when dosing the second agent? If so, some point-of-care mechanism must be in place, such as in an EMR containing the pharmacogenomic information, for the clinician to readily determine that genetic test results relevant to the prescribed drug are available.

Pharmacogenomic biomarker tests are a subset of the ever-increasing genetic tests available as a result of advances in genomic technology. Most of these genetic tests are "home-brewed" and advertised over the Internet directly to the consumer, primarily for determining inherited susceptibility of different diseases. The majority of these direct-to-consumer (DTC) genetic tests are not subject to regulatory oversight by the FDA and/or CLIA compliance for test quality standards and proficiency. In addition, companies selling DTC genetic tests can develop and market the tests without establishing clinical utility, which contrasts significantly to what is demanded for pharmacogenomic biomarkers discussed earlier in this chapter as well as in the relevant sections of chapters 3, 5, 6, 9, and 12. These concerns of oversight gap and test validity likely contribute to the conclusion by Janssens et al. that most DTC genetic tests are not useful in predicting disease risk [40]. Current knowledge suggests that genomic profiling based on a single SNP—a feature common to most DTC genetic tests—is not necessarily clinically accurate or useful. In this regard, the recent availability of a DTC genome-wide platform [41] could provide a useful example of how genomic profiling can have potential impact on patient care. Finally, the increased availability of DTC genetic tests and the increased consumer desire for health-related information underscores the importance of educating clinicians and preparing them to provide the appropriate test interpretation for clinical decision making.

INCORPORATING PHARMACOGENOMICS INTO DRUG DEVELOPMENT

The scientific rationale and the applicable technical challenges for incorporating pharmacogenomics into drug development is discussed in chapter 9b and, more specifically for clinical development of anticancer drugs, in chapter 4. This section summarizes the additional obstacles and considerations for the pharmaceutical industry.

The concept of the blockbuster drug and its financial impact on revenue have played a major role in pharmaceutical drug development. As such, the concept of pharmacogenomics and the resultant segmented (and smaller) market tailored to a subpopulation with a specific genotype have been viewed unfavorably because of lower revenue and decreased profit. However, as discussed earlier in this chapter, trastuzumab represents a paradigm shift in such a perspective with regard to revenue. With little or no competition, the perception of a smaller market share in the entire population of patients with breast cancer can be overcome by 100% market share of all—albeit a smaller number—of the patients with HER2 overexpression. There are additional drug development advantages associated with this paradigm shift of product differentiation instead of market segmentation. Identifying patients likely to participate in clinical trials would enable benefits to be shown in smaller number of patients, resulting in shorter Phase II and III studies and reducing the cost of development. It could also screen out patients likely to have unfavorable side effects that come to light only in Phase IV postmarketing surveillance studies, events which sometimes lead to the inevitable and unfavorable outcomes of postmarketing product recall and litigation. The litigation and financial burden could be even smaller if the pharmaceutical company could work with regulatory agencies to incorporate the pharmacogenomic information into a drug label that more accurately describes contra-indications, precautions, and warnings. Finally, as indicated earlier in this chapter, proper partnerships to develop and market a companion diagnostic test can also lead to additional revenue streams.

Nevertheless, relevant drug efficacy and safety data and issues that are important for regulatory decision making were developed long before the era of pharmacogenomics, and it is unclear how traditional regulatory review would approach the inclusion of any pharmacogenomic data in a new drug application (NDA) package. As described earlier in this chapter, the FDA has developed multiple initiatives to encourage the use and submission of pharmacogenomic data by the pharmaceutical industry. However, concerns and questions remain regarding what type of pharmacogenomic data are necessary and when it should be incorporated in the NDA process. Such issues are summarized in Table 2.3.

The high attrition rate in drug development is a well-known fact for the pharmaceutical industry, and a much less discussed and explored role of pharmacogenomics is the potential of "rescuing" drugs that fail clinical trials during drug development. The prime example for this benefit is gefitinib, which originally was destined to failure because only a small number of patients with small cell lung cancer responded to the drug. However, in 2004, published results showed that tumor response to the drug was linked to mutations in the epidermal growth factor receptor (EGFR). Subsequently, development of pharmacogenomic biomarker tests for EGFR mutations enabled identification of responders for gefitinib. This example showed that investigational drugs found to be ineffective or unsafe during Phase II or III clinical trials might deserve a second look from the perspective of pharmacogenomics. Another example is lumiracoxib, a selective cyclooxygenase-2 inhibitor that was withdrawn from most global pharmaceutical markets because of hepatotoxicity. Recently, a strong association between patients with an HLA haplotype and lumiracoxib-related liver injury has been

TABLE 2.3   Issues for pharmacogenomic data submission

Voluntary genomic data submission and voluntary exploratory data submission:
- How much data (all biomarkers and all genes or only those of interest) should be included?
- What is the requirement (type and extent) for biomarker validation?
- Are genome-wide analyses required?
- Would approval of biomarkers be required before they can be used in studies?
- Are data from retrospective subgroup analyses appropriate and sufficient?
- Are prospective case/control designs required?
- How should the issue of ethnicity be addressed?

Regulatory review:
- How would exploratory data be evaluated?
- What are the criteria that could change a voluntary submission to a required one?
- How would the extra time for additional studies affect the duration of exclusivity?

Specific concerns:
- If required (changed from voluntary) data are not sufficient, then what will happen?
- Are there unifying approaches to reviewing the data among regulatory agencies?
- If not, how would global drug development be affected?
- How would a company's intellectual property be affected by *voluntary* submission?

identified [42]. Therefore, "failing" drugs can be further developed with a smaller target population with the genetic profile predictive of improved efficacy and/or reduced toxicity. This result can then be used for approval with appropriate product labeling containing the pharmacogenomic information. In reality, the possibility of such "drug rescue" with potential drug approval might not have sufficient incentive for the pharmaceutical company to spend additional cost to conduct another clinical trial, albeit in a smaller number of patients. Such incentive likely has to come from regulatory changes in the form of conditional drug approval with subsequent requirement of a Phase IV trial. Without such changes, another incentive could take an approach to drug development and approval similar to that described under the Orphan Drug Act, intended primarily for therapeutic agents in treatment of rare diseases.

CONCLUSION

Over the last few decades, significant achievements have been made in identifying variants in (or haplotypes linked to) genes that regulate the disposition and target pathways of drugs. Despite these advances, translating the pharmacogenomic results into clinical practice has been met with continued scientific debates, as well as commercial, economical, educational, legal, and societal barriers. Much work remains to address the logistics and challenges for fully incorporating pharmacogenomics into clinical practice and drug development. There are urgent needs to improve drug efficacy and safety for patient care, as well as the efficiency of the drug development process, and that can only be achieved with all stakeholders in the field working together and occasionally accepting a paradigm change in their current approach.

QUESTIONS FOR DISCUSSION

1. Why is the evidence-based approach not necessarily the most appropriate means to evaluate the clinical utility of pharmacogenomic data?
2. What specific aspects of drug development can benefit from the incorporation of pharmacogenomic evaluation?
3. How much pharmacogenomic information should be included in electronic medical record systems?
4. How important is the cost-effectiveness of a pharmacogenomic biomarker in the decision regarding its implementation in clinical practice?

References

[1] Nelson DR, Conlon M, Baralt C, et al. University of Florida Clinical and Translational Science Institute: transformation and translation in personalized medicine. Clin Transl Sci

[2] Pulley JM, Denny JC, Peterson JF, Bernard GR, Vnencak-Jones CL, Ramirez AH, et al. Operational implementation of prospective genotyping for personalized medicine: the design of the Vanderbilt PREDICT Project. Clin Pharmacol Ther 2012;92:87—95.

[3] Malhotra AK, Goldman D, Buchanan RW, Rooney W, Clifton A, Kosmidis MH, et al. The dopamine D3 receptor (DRD3) Ser9Gly polymorphism and schizophrenia: a haplotype relative risk study and association with clozapine response. Mol Psychiatry 1998;3:72—5.

[4] Malhotra AK, Goldman D, Mazzanti C, Clifton A, Breier A, Pickar D. A functional serotonin transporter (5-HTT) polymorphism is associated with psychosis in neuroleptic-free schizophrenics. Mol Psychiatry 1998;3:328—32.

[5] Malhotra AK, Goldman D, Ozaki N, Breier A, Buchanan R, Pickar D. Lack of association between polymorphisms in the 5-HT2A receptor gene and the antipsychotic response to clozapine. Am J Psychiatry 1996;153:1092—4.

[6] Masellis M, Basile V, Meltzer HY, Lieberman JA, Sevy S, Macciardi FM, et al. Serotonin subtype 2 receptor genes and clinical response to clozapine in schizophrenia patients. Neuropsychopharmacology 1998;19:123—32.

[7] Arranz MJ, Munro J, Birkett J, Bolonna A, Mancama D, Sodhi M, et al. Pharmacogenetic prediction of clozapine response. Lancet 2000;355:1615—6.

[8] Alfaro CL, Lam YW, Simpson J, Ereshefsky L. CYP2D6 status of extensive metabolizers after multiple-dose fluoxetine, fluvoxamine, paroxetine, or sertraline. J Clin Psychopharmacol 1999;19:155—63.

[9] Robertson GR, Liddle C, Clarke SJ. Inflammation and altered drug clearance in cancer: transcriptional repression of a human CYP3A4 transgene in tumor-bearing mice. Clin Pharmacol Ther 2008;83:894—7.

[10] Centers for Disease Control and Prevention. Genetic testing: ACCE model system for collecting, analyzing and disseminating information on genetic tests. http://www.cdc.gov/genomics/gtesting/ACCE/FBR/index.htm. Accessed March 26, 2012.

[11] EGAPP Working Group. Recommendations from the EGAPP Working Group: testing for cytochrome P450 polymorphisms in adults with nonpsychotic depression treated with selective serotonin reuptake inhibitors. Genet Med 2007;9:819—25.

[12] Holmes Jr DR, Dehmer GJ, Kaul S, Leifer D, O'Gara PT, Stein CM. ACCF/AHA Clopidogrel clinical alert: approaches to the FDA "boxed warning": a report of the American College of Cardiology Foundation Task Force on Clinical Expert Consensus Documents and the American Heart Association. Circulation 2010;122:537—57.

[13] Levine GN, Bates ER, Blankenship JC, Bailey SR, Bittl JA, Cercek B, et al. ACCF/AHA/SCAI Guideline for Percutaneous Coronary Intervention: executive summary: a report of the American College of Cardiology Foundation/American Heart Association Task Force on Practice Guidelines and the Society for Cardiovascular Angiography and Interventions. Circulation 2011;124:2574—609.

[14] Kirchheiner J, Bauer S, Meineke I, Rohde W, Prang V, Meisel C, et al. Impact of CYP2C9 and CYP2C19 polymorphisms on tolbutamide kinetics and the insulin and glucose response in healthy volunteers. Pharmacogenetics 2002;12:101—9.

[15] Furuta T, Shirai N, Kodaira M, Sugimoto M, Nogaki A, Kuriyama S, et al. Pharmacogenomics-based tailored versus standard therapeutic regimen for eradication of H. pylori. Clin Pharmacol Ther 2007;81:521–8.

[16] Reese ES, Daniel Mullins C, Beitelshees AL, Onukwugha E. Cost-effectiveness of cytochrome P450 2C19 genotype screening for selection of antiplatelet therapy with clopidogrel or prasugrel. Pharmacotherapy 2012;32:323–32.

[17] Eckman MH, Rosand J, Greenberg SM, Gage BF. Cost-effectiveness of using pharmacogenetic information in warfarin dosing for patients with nonvalvular atrial fibrillation. Ann Intern Med 2009;150:73–83.

[18] You JH, Tsui KK, Wong RS, Cheng G. Potential clinical and economic outcomes of CYP2C9 and VKORC1 genotype-guided dosing in patients starting warfarin therapy. Clin Pharmacol Ther 2009;86:540–7.

[19] Johnson JA, Roden DM, Lesko LJ, Ashley E, Klein TE, Shuldiner AR. Clopidogrel: a case for indication-specific pharmacogenetics. Clin Pharmacol Ther 2012;91:774–6.

[20] Roberts JD, Wells GA, Le May MR, Labinaz M, Glover C, Froeschl M, et al. Point-of-care genetic testing for personalisation of antiplatelet treatment (RAPID GENE): a prospective, randomised, proof-of-concept trial. Lancet 2012;379:1705–11.

[21] Jha AK, DesRoches CM, Campbell EG, Donelan K, Rao SR, Ferris TG, et al. Use of electronic health records in U.S. hospitals. N Engl J Med 2009;360:1628–38.

[22] Gage BF, Eby C, Johnson JA, Deych E, Rieder MJ, Ridker PM, et al. Use of pharmacogenetic and clinical factors to predict the therapeutic dose of warfarin. Clin Pharmacol Ther 2008;84:326–31.

[23] Owen RP, Altman RB, Klein TE. PharmGKB and the International Warfarin Pharmacogenetics Consortium: the changing role for pharmacogenomic databases and single-drug pharmacogenetics. Hum Mutat 2008;29: 456–60.

[24] IWPC Pharmacogenetic Dosing Algorithm. http://pharmgkb.org/drug/PA451906. Accessed April 17, 2012.

[25] WarfarinDosing. www.warfarindosing.org. Accessed April 17, 2012.

[26] Michaud V, Vanier MC, Brouillette D, Roy D, Verret L, Noel N, et al. Combination of phenotype assessments and CYP2C9-VKORC1 polymorphisms in the determination of warfarin dose requirements in heavily medicated patients. Clin Pharmacol Ther 2008;83:740–8.

[27] Wu AH, Wang P, Smith A, Haller C, Drake K, Linder M, et al. Dosing algorithm for warfarin using CYP2C9 and VKORC1 genotyping from a multi-ethnic population: comparison with other equations. Pharmacogenomics 2008;9:169–78.

[28] Stanek EJ, Sanders CL, Taber KA, Khalid M, Patel A, Verbrugge RR, et al. Adoption of pharmacogenomic testing by US physicians: results of a nationwide survey. Clin Pharmacol Ther 2012;91(3):450–8.

[29] Accreditation Council for Pharmaceutical Education Accreditation standards and guidelines for the professional program in pharmacy leading to the doctor of pharmacy degree. www.acpe-accredit.org/deans/standards.asp. Accessed April 17, 2012.

[30] Crews KR, Gaedigk A, Dunnenberger HM, Klein TE, Shen DD, Callaghan JT, et al. Clinical Pharmacogenetics Implementation Consortium (CPIC) guidelines for codeine therapy in the context of cytochrome P450 2D6 (CYP2D6) genotype. Clin Pharmacol Ther 2012;91:321–6.

[31] Johnson JA, Gong L, Whirl-Carrillo M, Gage BF, Scott SA, Stein CM, et al. Clinical Pharmacogenetics Implementation Consortium Guidelines for CYP2C9 and VKORC1 genotypes and warfarin dosing. Clin Pharmacol Ther 2011;90:625–9.

[32] Martin MA, Klein TE, Dong BJ, Pirmohamed M, Haas DW, Kroetz DL. Clinical pharmacogenetics implementation consortium guidelines for HLA-B genotype and abacavir dosing. Clin Pharmacol Ther 2012;91:734–8.

[33] Scott SA, Sangkuhl K, Gardner EE, Stein CM, Hulot JS, Johnson JA, et al. Clinical Pharmacogenetics Implementation Consortium guidelines for cytochrome P450-2C19 (CYP2C19) genotype and clopidogrel therapy. Clin Pharmacol Ther 2011;90:328–32.

[34] Wilke RA, Ramsey LB, Johnson SG, Maxwell WD, McLeod HL, Voora D, et al. The Clinical Pharmacogenomics Implementation Consortium: CPIC Guideline for SLCO1B1 and Simvastatin-Induced Myopathy. Clin Pharmacol Ther 2012;92:112–7.

[35] Food and Drug Administration. Guidance on pharmacogenetic tests and genetic tests for heritable markers. www.fda.gov/MedicalDevices/DeviceRegulationsandGuidance/GuidanceDocuments/ucm077862.htm. Accessed April 16, 2012.

[36] Elkin EB, Weinstein MC, Winer EP, Kuntz KM, Schnitt SJ, Weeks JC. HER-2 testing and trastuzumab therapy for metastatic breast cancer: a cost-effectiveness analysis. J Clin Oncol 2004;22:854–63.

[37] Cohen J, Wilson A, Manzolillo K. Clinical and economic challenges facing pharmacogenomics. Pharmacogenomics J 2012. doi:10.1038/tpj.2011.63 [Epub date is Jan 11, 2012].

[38] Blair ED. Assessing the value-adding impact of diagnostic-type tests on drug development and marketing. Mol Diagn Ther 2008;12:331–7.

[39] Bevan JL, Lynch JA, Dubriwny TN, Harris TM, Achter PJ, Reeder AL, et al. Informed lay preferences for delivery of racially varied pharmacogenomics. Genet Med 2003;5:393–9.

[40] Janssens AC, Gwinn M, Bradley LA, Oostra BA, van Duijn CM, et al. A critical appraisal of the scientific basis of commercial genomic profiles used to assess health risks and personalize health interventions. Am J Hum Genet 2008;82:593–9.

[41] Ashley EA, Butte AJ, Wheeler MT, Chen R, Klein TE, Dewey FE, et al. Clinical assessment incorporating a personal genome. Lancet 2010;375:1525–35.

[42] Singer JB, Lewitzky S, Leroy E, Yang F, Zhao X, Klickstein L, et al. A genome-wide study identifies HLA alleles associated with lumiracoxib-related liver injury. Nat Genet 2010;42:711–4.

[37] Cohen J, Wilson A, Manzolillo K. Clinical and economic challenges facing pharmacogenomics. Pharmacogenomics J 2012; doi:10.1038/tpj.2011.63 [Epub date is Jan 31, 2012].

[38] Blair ED. Assessing the value-adding impact of diagnostic-type tests on drug development and marketing. Mol Diagn Ther 2008;12:331-7.

[39] Bevan JL, Lynch JA, Dubriwny TN, Harris TM, Achter PJ, Reeder AL, et al. Informed lay preferences for delivery of reality varied pharmacogenomics. Genet Med 2003;5:393-8.

[40] Janssens AC, Gwinn M, Bradley LA, Oostra BA, van Duijn CM et al. A critical appraisal of the scientific basis of commercial genomic profiles used to assess health risks and personalize health interventions. Am J Hum Genet 2008;82:593-9.

[41] Ashley EA, Butte AJ, Wheeler MT, Chen R, Klein TE, Dewey FE, et al. Clinical assessment incorporating a personal genome. Lancet 2010;375:1525-35.

[42] Singer JB, Lewitzky S, Leroy E, Yang F, Zhao X, Klickstein L, et al. A genome-wide study identifies HLA alleles associated with lumiracoxib-related liver injury. Nat Genet 2010;42:711-4.

Governmental and Academic Efforts to Advance the Field of Pharmacogenomics

Larisa H. Cavallari[*], *Teri E. Klein*[†], *Shiew-Mei Huang*[‡]*

[*]Department of Pharmacy Practice, University of Illinois at Chicago, Chicago, Illinois, USA
[†]Department of Genetics, Stanford University, Stanford, California, USA
[‡]Office of Clinical Pharmacology, Food and Drug Administration, Silver Spring, Maryland, USA

OBJECTIVES

1. Describe scientific and regulatory challenges associated with the clinical implementation of pharmacogenomic data.

2. Describe efforts by the National Institutes of Health to promote pharmacogenetic research discoveries.

3. Discuss pharmacogenetic initiatives by the Food and Drug Administration in regard to drug development and drug use.

4. Provide examples of regulatory activities by non-U.S. agencies in the area of personalized medicine.

INTRODUCTION

In 2010, leaders of the National Institutes of Health (NIH) and U.S. Food and Drug Administration (FDA) announced their shared vision of personalized medicine [1]. Together, they outlined the scientific and regulatory structure necessary to address the challenges in advancing personalized medicine. Examples of such challenges are listed in Table 3.1 and include identifying optimal genetic markers for drug response and encouraging the discovery of novel genetic targets for therapeutic intervention. In addition, personalized medicine is contingent on the accurate identification of patients who are likely to respond favorably to a particular drug. This identification may require the development of a diagnostic product that is approved for use with a drug, similar to that available for predicting response to the anticancer drug trastuzumab. Thus, another challenge is to identify the optimal means for coordinated approval of drug therapy and companion diagnostics.

The NIH and FDA are investing significant resources to address the challenges with advancing the field of pharmacogenomics from scientific discovery to clinical implementation. For example, the NIH is supporting pharmacogenomic discoveries through the Pharmacogenomics Research Network and promoting development of genotype-based therapies for rare inherited diseases through the Therapeutics for Rare and Neglected Diseases (TRND) Program [1]. The NIH also offers several funding opportunities for pharmacogenomic analysis of biologic specimens from large NIH-sponsored epidemiologic studies and clinical trials. Pharmacogenomic-related activities of the FDA include the Voluntary Exploratory Data Submission program, which is designed to encourage pharmaceutical companies to integrate pharmacogenomic data into the drug development process and the possible inclusion of pharmacogenetic information in drug labeling. Through these and other activities, the NIH and the FDA hope to ease the transition from the identification of a genetic marker for drug response or a potential target for therapeutic intervention to the clinical implementation of novel therapies and strategies for improved disease management. This chapter will discuss specific initiatives by the NIH and FDA in coordination with external researchers to advance the field of pharmacogenomics. The chapter concludes with a brief discussion of pharmacogenomics-related activities of the European Medicines Agency (EMA) and the Pharmaceutical and Medical Devices Agency (PMDA) in Japan.

TABLE 3.1   Challenges in advancing personalized medicine and efforts to address these challenges

Challenge	Effort to Address Challenge
Identifying genetic markers correlated with drug response	NIH-funded Pharmacogenomics Research Network (PGRN) NIH-supported analysis of tissue and sample banks from large epidemiologic studies
Identifying novel molecular targets for therapeutic intervention	NIH-funded Pharmacogenomics Research Network (PGRN)
Encouraging the development of novel therapies targeting gene-based disease pathways	NIH-funded Therapeutics for Rare and Neglected Diseases (TRND)
Delivering personalized medicine to patients	Clinical Pharmacogenetics Implementation Consortium (CPIC)
Defining the process for coordinated approval of drug therapy and companion diagnostics	FDA Voluntary Exploratory Data Submission (VXDS) Program
Ensuring high-quality diagnostic tests to predict drug response	FDA standards for the efficient review and oversight of genetic diagnostic tests and manufacturer claims are being established; NIH-funded voluntary genetic testing registry

THE ROLE OF THE NATIONAL INSTITUTES OF HEALTH

Historical Efforts

Pharmacogenomic discoveries date back at least to the 1950s, when an inherited deficiency in the glucose-6 phosphate dehydrogenase enzyme was identified as the cause for primaquine-induced anemia [2]. There has been a resurgence of interest in pharmacogenomics since the completion of the Human Genome Project, which began in 1990 as a collaborative effort between the NIH National Human Genome Research Institute and the U.S. Department of Energy Human Genome Project. The goal of the Human Genome Project was to sequence the entire human genome by 2005. The project was completed in 13 years, 2 years ahead of schedule. The working draft of the human genome was published in companion papers in 2001 [3,4]. The final sequence includes 3 billion DNA base pairs and contains 99% of the gene-containing sequence, with 99.9% accuracy. Sequence data from the project were deposited into a freely accessible database run by the National Center for Biotechnology Information (NCBI; www.ncbi.nlm.nih.gov) in order to encourage genetic research and ultimately improve human health and well-being [5].

Medical advances from the Human Genome Project are many. Early pharmacogenetic discoveries were largely in the area of oncology, for which drugs have been developed to target specific cancer mutations. For example, trastuzumab inhibits proliferation of cancer cells that overexpress the human epidermal receptor type 2 (*HER* or *ERBB2*) gene. Later discoveries were in the areas of cardiology, neurology, and infectious disease. In cardiology, genetic testing may influence the choice of antiplatelet therapy or dosing of the anticoagulant warfarin. In neurology, genetic testing for the major histocompatibility complex, class I, B (*HLA-B*)*1502 allele, is indicated in persons of southern Asian ancestry to determine risk for life-threatening skin reactions to the antiepileptic drug carbamazepine. Similarly, screening for the *HLA-B*57:01 allele is recommended prior to initiation of the antiretroviral agent abacavir to predict risk for serious hypersensitivity reactions.

According to data from a recent independent analysis, the Human Genome Project has also had a considerable economic impact [6]. Specifically, the $3.8 billion ($5.6 billion in 2010 dollars) invested in the project over 13 years generated an economic output of $796 billion. In other words, every $1 invested in the Human Genome Project by the U.S. government generated a return of $141 to the U.S. economy. The cumulative economic impact of human genome sequencing includes 3.8 million job-years of employment (i.e., one person employed full time for one year) and $244 billion in personal income.

International HapMap Project

The International HapMap Project followed the Human Genome Project. The purpose of the HapMap project was to create a publicly accessible database of common patterns of heritability in the human genome in order to facilitate genetic studies of common human diseases, including genome-wide association studies (GWAS). The HapMap Project is based on the occurrence of linkage disequilibrium (LD) among single-nucleotide polymorphisms (SNPs) in the genome, whereby SNPs are inherited together in sets or blocks more often than would be expected based on chance alone. A single SNP within the group is representative of SNPs within the haplotype block and thus may serve as a "tag SNP" for the haplotype. Based on patterns of LD in a given chromosomal region, only a few carefully chosen SNPs within the region need to be included to identify association at that locus with disease or drug response.

Patterns of LD may vary across ancestral groups, particularly for populations of recent African ancestry [7]. Thus, the HapMap Project was a collaborative effort among researchers from different countries (the United States, the United Kingdom, Canada, China, Japan, and Nigeria) to capture the LD patterns in various populations. For Phase I of the project, a total of 270 samples from populations with African, European, and Asian ancestry were genotyped. These included 30 mother–father–adult child trios from the Yoruba in Nigeria (YRI), 30 trios from the Centre d'Etude du Polymorphisme Humain (CEPH) collection of Utah residents of Northern and Western European ancestry, 45 unrelated Han Chinese individuals in Beijing (CHB), and 45 unrelated Japanese individuals in Tokyo (JPT). These four populations, with different ancestral geographies, were chosen in order to identify the major common haplotypes that occur worldwide. Specifically, the YRI, CEPH, CHB, and JPT populations include samples from at least three Old World continents exhibiting a range of haplotype frequencies [8]. Inclusion criteria for individuals providing YRI, CHB, and JPT HapMap samples were generally limited to adult age and the ability to provide informed consent. In addition, the YRI and Asian populations had to have evidence of ancestry from their respective race group (e.g., grandparents of the same race). The samples for the CEPH population were collected in 1980; therefore, inclusion criteria for this population were not specified. All participants provided informed consent for genetic sampling, and no medical or identifying information was obtained for privacy reasons. The sample size allowed for the detection of most haplotypes with a frequency of 5% or higher. Centers in each country involved in the HapMap Project were assigned different chromosomal regions to interrogate in HapMap samples using various high-throughput genotyping technologies.

The results of Phase I describing the LD patterns of approximately 1.1 million SNPs were published in 2005 [9]. The results of the Phase II second-generation human haplotype map, involving over 3.1 million SNPs, were published two years later [10]. In order to better define tag SNPs across geographic regions, samples genotyped for Phase II included samples from the original HapMap populations plus samples from seven additional populations:

- Luhya in Webuye, Kenya (LWK)
- Maasai in Kinyawa, Kenya (MKK)
- Tuscans in Italy (TSI)
- Gujarati Indians in Houston, Texas (GIH)
- Chinese in metropolitan Denver, Colorado (CHD)
- Persons of Mexican ancestry in Los Angeles, California (MXL)
- Persons of African ancestry in the Southwestern United States (ASW)

The populations were added to determine whether haplotype patterns differ substantially by specific ancestry or geographic region, with haplotypes thus requiring different tag SNPs for disease or phenotype association studies. Investigators have genotyped 1.6 million SNPs in an expanded set of HapMap I and II samples and sequenced ten 100-kilobase regions in 692 of these samples to create an integrated data set of both common and rare alleles [11].

Data from the HapMap Project are freely available through the International HapMap Project website (www.hapmap.org). These data have enabled numerous GWASes, in which tag SNPs for haplotypes across the genome are examined for their association with disease risk. According to the catalog of published GWASes maintained by the National Human Genome Research Institute, more than 170 studies of genetic associations with complex diseases and traits have been published based on HapMap data [12]. Genome-wide association studies of drug response have also emerged in recent years and include studies of genetic associations with warfarin dose requirements [13], gemcitabine response in pancreatic cancer [14], antibiotic-induced liver injury [15], statin-induced myopathy [16], and metformin response in diabetes [17]. In many cases, GWASes have led to the discovery of variants in genes not previously suspected to have a role in disease pathology or drug response. For example, a GWAS identified the *SLCO1B1* gene that encodes the organic anion–transporting polypeptide OATP1B1 as associated with risk of simvastatin-induced myopathy [16]. This association has since been confirmed by other investigators [18]. Per recommendations by the NIH, GWAS data from NIH-sponsored or conducted studies is made available to the scientific community through the NCBI Database of Genotype and Phenotype (DBGaP) [19].

1000 Genomes Project

The goal of a GWAS is to discover regions of the genome that are associated with an outcome of interest (e.g., drug response). The associated variant often serves as a marker (tag SNP) for the actual causal variant (the variant that underlies the observed association). Further studies of variants within the candidate region are required to identify the causal variant(s) and elucidate the mechanism underlying its effects. In many cases, rare variants within the candidate locus contribute to disease risk or phenotype. However, these rare variants are not included in the HapMap database or captured with available genotyping platforms and are thus missed in GWASes [12]. Resequencing of candidate gene regions is historically necessary after GWASes to identify potentially causal, yet rare, variants. This is a timely and costly process. Data provided by the 1000 Genomes Project are expected to limit the sequencing efforts necessary to identify rare variants underlying genetic associations from GWASes.

The 1000 Genomes Project began in 2008 with the goal of developing a comprehensive catalog of less common genetic variation through a DNA sequencing approach. The project involves sequencing short pieces of DNA randomly distributed across the genome for 2,500 individuals representing European, East Asian, South Asian, West African, and American populations. In addition to the HapMap samples described previously, the 1000 Genomes Project includes samples from the following groups, among others, in order to capture most of the genetic variation occurring in populations worldwide:

- British from England and Scotland (GBR)
- Finnish from Finland (FIN)
- Iberian populations in Spain (IBS)
- Han Chinese South (CHS)

- Chinese Dai in Xishuangbanna (CDX)
- Kinh in Ho Chi Minh City, Vietnam (KHV)
- Gambain in Western Division, The Gambia (GWD)
- African American in Southwest US (ASW)
- Puerto Rican (PUR)
- Columbian (CLM)
- Peruvian (PEL)
- Punjabi in Lahore, Pakistan (PJL)

Each sample in 1000 Genomes will be sequenced an average of four times, thus providing 4X coverage, which is insufficient to provide a statistical assurance of whole genome coverage but is expected to allow detection of variants with frequencies of 1% or greater. Deeper sequencing would be considerably more costly, and the use of a high number of samples (2,500) at lesser depth should provide accurate data on genotypes and variants through computational processes [20].

Sequencing approximately 28 times would be necessary to identify the complete genomic sequence. The high cost associated with providing such deep coverage precludes deep sequencing with a large number of samples. However, combining data from 4X sequencing of a large number of samples allows for detection of most variants in a given region with a frequency of at least 1% in the represented populations.

Results from the initial pilot phase of the 1000 Genome Project were reported in 2010 [21]. The authors carried out three projects with the goal of developing and assessing various strategies for genome-wide sequencing with high-throughput technology. These projects included low-coverage (2X−6X) sequencing of approximately 180 individuals; deep sequencing (42X) of two parent−child trios; and exome sequencing of nearly 700 individuals. The data set resulting from these efforts contained over 95% of accessible variants in any individual and included more than 15 million SNPs, 1 million short insertions and deletions, and 20,000 structural variants.

Similar to the HapMap Project, data from 1000 Genomes are made available through a publicly accessible database (www.1000genomes.org). Investigators conducting GWASes may use computational approaches to impute variants from 1000 Genomes into their data set of genotyped SNPs, thus significantly expanding the number of variants interrogated for association with the phenotype of interest. This expansion increases the likelihood of capturing the causal variant underlying the observed association. The samples from 1000 Genomes are available to researchers from the nonprofit Coriell Institute for Medical Research.

The ENCyclopedia of DNA Elements (ENCODE) Project

The ENCyclopedia of DNA Elements (ENCODE) Project began in 2003 as an initiative of the National Human Genome Research Institute to develop a catalogue of functional elements in the human genome. Functional elements refer to discrete genomic regions that encode for a particular product, such as a protein, or for a biochemical signature, such as transcription or a chromatin structure [22]. The initial pilot phase of the project was a global, multidisciplinary effort of scientists from academia, industry, and governmental institutions to identify and analyze functional elements contained within a targeted 1% (about 30,000 kilobases, or kb) of the genome. Scientists involved included those from several American universities; the Municipal Institute of Medical Research, Barcelona, Spain; the Wellcome Trust Sanger Institute, Hinxton, U.K; and Affymetrix, Inc. Additional investigators were funded to develop new technologies or to improve existing technologies to allow for efficient and cost-effective discovery and analysis of functional elements.

Results from the pilot project were published in 2007 [23]. Overall, the project showed that the organization and function of the human genome is much more complex than many had expected. Specific observations of the pilot project include the following [23,24]:

- The majority of the human genome is found in primary transcripts, which overlap and include non-protein-coding regions.
- There are numerous unannotated transcription start sites.
- Many protein-coding genes have alternative transcription start sites that may be located > 100 kb upstream of the annotated start site.
- There is more alternative splicing than originally thought.
- Regulatory sites for a given gene may be located at a chromosomal position quite distal to the gene.
- Noncoding RNA genes are involved in gene regulation (e.g., microRNAs (miRNAs)) and processing (e.g., small nucleolar RNAs (sno-RNAs)), in addition to protein synthesis.
- Noncoding "pseudogenes" can influence the structure and function of the human genome.
- The majority (approximately 60%) of bases under evolutionary constraint are related to functional sites rather than protein-coding exons or their associated untranslated regions.

In 2007, the ENCODE project was expanded to cover the entire genome, with a focus on completing annotations for protein-coding genes, noncoding pseudogenes, noncoding transcripts, and their RNA transcripts and transcriptional regulatory regions [22]. Other areas of focus include DNA binding proteins that interact with *cis*-regulatory regions, such as transcription factors and histones; DNA methylation patterns; DNaseI footprints; long-range chromatin interactions; protein:RNA interactions; transcriptional silencer elements; and promoter sequence architecture. Similar to other NIH initiatives, all data from the ENCODE project are available through a freely accessible database (genome.ucsc .edu/ENCODE/).

Genotype-Tissue Expression (GTEx) Project

The Genotype-Tissue Expression (GTEx) Project was launched in 2010 by the NIH Common Fund with the goal of mapping genetic variation that affects gene expression. The project involves correlating genetic variation with tissue-specific gene expression levels through expression quantitative trait loci (eQTL) analysis. This information will be useful to researchers in selecting variants to interrogate for the effects on drug response. For example, variants found to affect expression of the ATP-binding cassette, subfamily B (*ABCB1*) gene, which codes for P-glycoprotein, are candidates for affecting disposition of drugs that are P-glycoprotein substrates. A database, developed by NCBI, will allow researchers to view and download GTEx data. Tissue samples will also be made available for research use.

Therapeutics for Rare and Neglected Diseases (TRND) Program

Rare diseases are generally defined as diseases affecting fewer than 200,000 people in the U.S. Examples of rare diseases are listed on the NIH Office of Rare Diseases Research website (rarediseases.info.nih.gov/default.aspx) and include Marfan syndrome, Gaucher disease, and severe combined immunodeficiency, all of which have a genetic cause. Because a rare disease affects a relatively small population, drug development is challenging and lacks the financial incentive inherent in drug development for common diseases.

The Therapeutics for Rare and Neglected Diseases (TRND) Program is a congressionally mandated program for preclinical and early clinical development of new drug entities for rare diseases. TRND is overseen by the NIH Office of Rare Diseases Research with laboratory operations administered by the National Human Genome Research Institute. Through TRND, the NIH supports development of gene-based therapies targeting rare, inherited diseases. Examples of diseases targeted by the program include Duchenne muscular dystrophy, Niemann–Pick type C, and fibrodysplasia ossificans progressive. TRND may also enable development of compounds targeting molecularly distinct subtypes of some common diseases [1].

Pharmacogenomics Research Network (PGRN)

TRND targets rare diseases; work by investigators within the NIH-sponsored Pharmacogenomics Research Network (PGRN) primarily targets common diseases, such as hypertension, asthma, cancers, and nicotine addiction. The PGRN consists of multiple research groups across the nation with varied yet complementary expertise and the common broad objective of elucidating genetic contribution to drug response. The vision of the PGRN is "to lead discovery and advance translation in genomics in order to enable safer and more effective drug therapies."[25] The ultimate goal of the PGRN is to allow personalized medicine, whereby therapy is tailored based on patient-specific information, including genotype.

The PGRN was initially funded in 2000, with funding renewed in 2005 and 2010. Groups within the network have an interest in pharmacogenomics related to various disease states (e.g., cardiovascular disease, pulmonary disease, cancer), proteins (e.g., drug transporters), and/or specific ancestral groups (e.g., Native Americans). Specific projects funded in the most recent renewal period are listed in Table 3.2. Funding is also provided to support resources that enhance the endeavors of PGRN investigators. These network resources include the Pharmacogenomic Ontology Network Resource, which aims to develop standardized phenotype definitions and research terminology; coordination sites for high-throughput genotyping, exome resequencing, and deep DNA sequencing; statistical workshops and expert statistical analysis; resources for pharmacogenomic discovery in large populations identified through use of the de-identified electronic medical record; and a global alliance for pharmacogenomics with the RIKEN Center for Genomic Medicine in Japan.

Researchers within the network use both phenotype-to-genotype and genotype-to-phenotype strategies to identify and characterize genetic influences of drug response [26]. In the former, investigators search for variants predicting a phenotypic response in a well-characterized population (e.g., blood pressure response in a hypertensive patient cohort). In the latter, individuals with known genotypes may be exposed to a drug to

TABLE 3.2   Pharmacogenetics Research Network (PGRN) projects funded in 2010

Project Name	Goal
Pharmacogenomics of Membrane Transporters (PMT)	To develop a functional map of genetic variants in the ATP binding cassette and solute carrier membrane transporters; find membrane transporter variants that contribute to variable drug response
Pharmacogenomic Evaluation of Antihypertensive Response (PEAR)	Identify and define the genetic basis of antihypertensive effects, metabolic responses, and clinical outcomes with thiazide diuretics and β-blockers
Pharmacogenomics of Mood Stabilizer Response to Bipolar Disorders (PGBD)	Identify genes involved in lithium response
Pharmacogenomics and Risk of Cardiovascular Disease (PARC)	Identify genetic contributors to inter individual variability in phenotypic and clinical response to statins
Pharmacogenetics of Nicotine Addiction Treatment (PNAT)	Generate the evidence base to optimize smoking cessation therapy
Pharmacogenomics of Anticancer Agents (PAAR)	Discover and validate functional polymorphisms contributing to the pharmacokinetics and/or pharmacodynamics of anticancer agents, especially for solid tumors
Pharmacogenomics of Anticancer Agents Research in Children (PAAR4Kids)	Define the pharmacogenomics of childhood acute lymphoblastic leukemia
Pharmacogenomics of Arrhythmia Therapy (PAT)	Define the genomic basis for susceptibility to drug-induced arrhythmias and variability in drug response in arrhythmias; identify variants contributing to warfarin-related bleeding using a DNA repository linked to electronic health records
Expression Genetics in Drug Therapy (XGEN)	Use functional genomics to discover genetic and epigenetic factors regulating gene expression
The Amish Pharmacogenomics of Anti-Platelet Intervention Study (PAP)	Compare CYP2C19 genotype-directed versus standard of care anti-platelet therapy in coronary heart disease; identify novel common and rare variants for clopidogrel response
Pharmacogenetics in Rural and Underserved Populations (NWAP)	Pharmacogenetics in Northwest American Indian/Alaska Native people
Pharmacogenomics of Phase II Drug Metabolizing Enzymes (PPII)	Discover and characterize variants influencing response to therapy of cancer and depression
Pharmacogenomics of Asthma Treatment (PHAT)	Identify novel genetic determinants of response to anti-asthmatic agents
Pharmacogenomics of Rheumatoid Arthritis Therapy (PhRAT)	Search for variants contributing to response to anti-TNF agents

determine response. Investigational approaches include *in vitro* mechanistic studies, sometimes using cell lines from the International HapMap Project; GWASes; large population studies; and clinical trials. In addition to individual efforts, PGRN investigators work cooperatively through data and resource sharing and formation of cross-disciplinary teams to discover and disseminate new findings. Furthermore, individual groups collaborate with each other on network-wide projects, such as the Translational Pharmacogenomics Project (TPP), which aims to translate actionable pharmacogenetic discoveries to clinical practice.

Numerous NIH components provide support for the PGRN, with efforts led by the National Institute of General Medical Sciences. The governing structure of the PGRN consists of a rotating leader who works with a steering committee composed of all principal investigators and NIH staff. The network also utilizes several standing committees and interest-based working groups, such as the coordinating committee and systems biology working group. An external panel, consisting of individuals from the scientific community selected by PGRN members, serves an advisory role, especially for network-wide activities.

The PGRN is expected to be a champion of pharmacogenomics and leader in translating pharmacogenomic research to the clinical setting. More than 1,000 publications have resulted so far from PGRN-funded activities [25]. These describe important genetic contributions to drug response, and in some cases have led to the inclusion of pharmacogenomic information in FDA-approved drug labeling. Individuals and groups outside the PGRN who conduct research in pharmacogenomics or related fields are welcome to apply for PGRN affiliate membership and participate in PGRN webinars and scientific discussions [25].

Pharmacogenomics Knowledge Base (PharmGKB)

The Pharmacogenomics Knowledge Base (PharmGKB) is a centralized resource serving the entire research community to provide the most comprehensive and up-to-date knowledge and tools in pharmacogenomics [27]. It is a comprehensive resource that curates and disseminates knowledge about the impact of genetic variation on drug response for researchers and clinicians. The mission of PharmGKB is to advance research and facilitate clinical implementation of pharmacogenomics. Originated in 2000 as an NIH-sponsored pharmacogenomics knowledge base for the scientific community [28], the PharmGKB now encompasses clinical information including genotype-guided dosing guidelines and drug labels, potentially clinically actionable gene-drug associations, and genotype–phenotype relationships [27].

The PharmGKB is publicly available through the www.pharmgkb.org website (Figure 3.1). The PharmGKB website serves as a premier source for pharmacogenomic information. The website includes annotations of genetic variations and their relationship to disease and drug response, drug-centered pathways, and a clinical interpretation of the pharmacogenomic data. Very important pharmacogene (VIP) summaries are available for key genes that are of significant pharmacogenomic importance and include detailed description for individual variants and haplotypes that have been associated with drug response.

As outlined in Figure 3.2, viewers may search the website by drug/small molecule, gene, gene variant (using the reference SNP ID number (rsID)) drug-centric pathways, or phenotype. The data are organized such that the same information can be retrieved from various starting points (e.g., searches by gene or by drug) [29]. Examples of information available when searching by drug include variants related to drug response, FDA drug labeling information, any pharmacogenomic dosing guidelines, and any genetic tests available to predict drug response. A search by gene leads to information on gene location, alternative gene names, variants within the gene, related genes, and any PharmGKB-curated pathways for the gene. The viewer may also access available

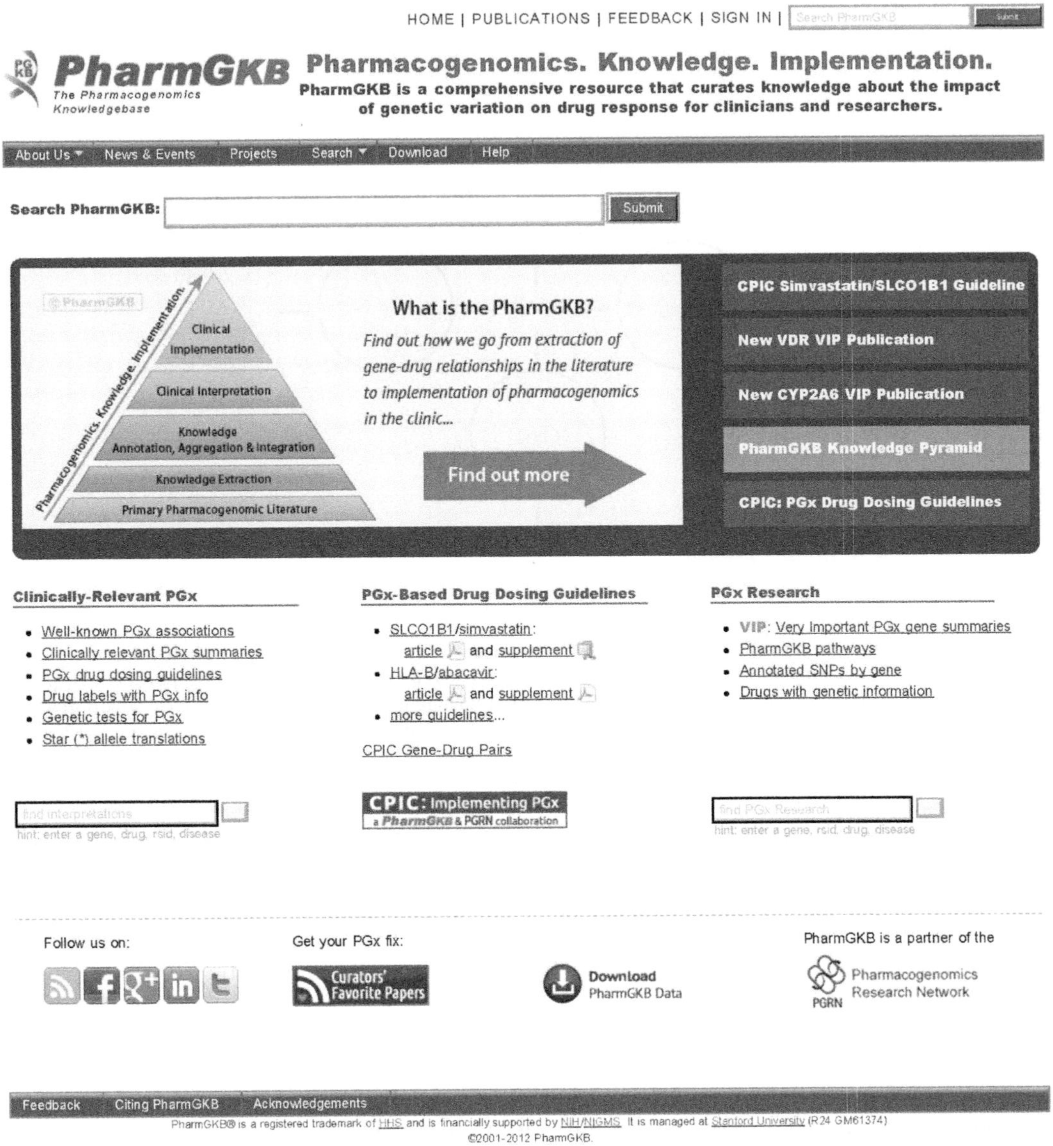

FIGURE 3.1 **The Pharmacogenomics Knowledge Base home page [27].** (See color plate 1.) *Copyright to PharmGKB with permission given by PharmGKB and Stanford University for reproduction.*

pharmacogenomic dosing guidelines, labeling information, and genetic tests through the search by gene. The dosing guidelines currently available through PharmGKB are those published by the CPIC (described in detail in an upcoming section) or the Royal Dutch Association for the Advancement of Pharmacy Pharmacogenetics Working Group (DPWG)

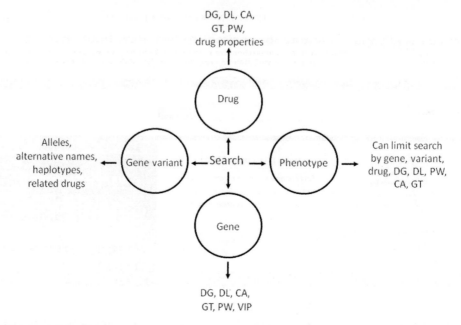

FIGURE 3.2   Information available through the Pharmacogenomics Knowledge Base website. As shown, the data are organized such that the same information can be retrieved from different starting points. DG, pharmacogenetic drug dosing guideline; DL, drug label with pharmacogenomic information; CA, clinical annotation; GT, genetic test for pharmacogenomics; PW, pharmacokinetic and pharmacodynamic pathways; VIP, very important pharmacogenes.

and manually curated by the PharmGKB staff. Examples of these guidelines are shown in Table 3.3. Like the CPIC, the DPWG is composed of a small group of clinical pharmacology experts in the Netherlands who provide consensus guidelines for integrating genetic information into therapeutic decisions [30]. Although none of these guidelines are specifically FDA-approved, the drugs included in the guidelines have genetic information in their FDA-approved labeling.

Clinical annotations related to specific gene variants are accessible to registered PharmGKB users. These annotations provide information about variants linked to drug response based on data from one or more research manuscripts, with links to access the original article and ratings for the strength of evidence. The drug pathway view shows diagrams of proteins related to the pharmacokinetics or pharmacodynamics of a drug. The pathway for clopidogrel is shown in Figure 3.3 [31]. Readers may click on individual proteins to view information about related genes and drugs. The VIPs and pathways on PharmGKB are peer-reviewed and published monthly in the *Pharmacogenetics and Genomics Journal* [32].

Another goal of the PharmGKB is to enable consortia examining important pharmacogenomic questions that are beyond the scope of individual research groups [33]. In this regard, the PharmGKB serves to curate (collect, format, and subject to quality control) data from disparate groups, facilitate communication among groups, actively participate in data

TABLE 3.3 Examples of pharmacogenomics dosing guidelines curated by the Pharmacogenomics Knowledge Base [30,38,40,41]

Gene(s)	Drug(s)	Publishing body
HLA-B	Abacavir	CPIC DPWG
CYP2D6	Amitriptyline Atomoxetine Nortriptyline Flecainide Haloperidol Tamoxifen Venlafaxine	DPWG
CYP2D6	Codeine	CPIC, DPWG
CYP2C9, VKORC1	Warfarin	CPIC, DPWG
CYP2C9	Phenytoin	DPWG
CYP2C19	Citalopram Lansoprazole Imipramine Omeprazole	DPWG
CYP2C19	Clopidogrel	CPIC, DPWG
FVL	Estrogen-containing oral contraceptives	DPWG
UGT1A1	Irinotecan	DPWG
TPMT	Azathioprine Mercaptopurine Thioguanine	CPIC, DPWG
DPYD	Fluorouracil/capecitabine Tegafur/uracil	DPWG

HLA-B, major histocompatibility complex, class I, B; CYP, cytochrome P450; VKORC1, vitamin K epoxide reductase complex 1; FVL, factor V Leiden; UGT, UDP glucuronosyltransferase 1 family polypeptide A1; TPMP, thiopurine-S-methyltransferase; DPYD, dihydropyrimidine dehydrogenase; CPIC, Clinical Pharmacogenetics Implementation Consortium; DPWG, Dutch Pharmacogenetics Working Group.

analyses, and publish and disseminate the final data and research results to the community at large. There are both data-centric and knowledge-centric consortia ongoing, and these are described in the following subsections.

International Warfarin Pharmacogenetics Consortium (IWPC)

The International Warfarin Pharmacogenetics Consortium (IWPC) was the initial consortium convened by the PharmGKB. The IWPC is an international collaboration of at least 21 research groups representing at least 11 countries and 4 continents who have combined genotype and phenotype data from over 5,500 warfarin-treated patients. The initial aim of the IWPC was to enable research groups to share warfarin pharmacogenetic data and create

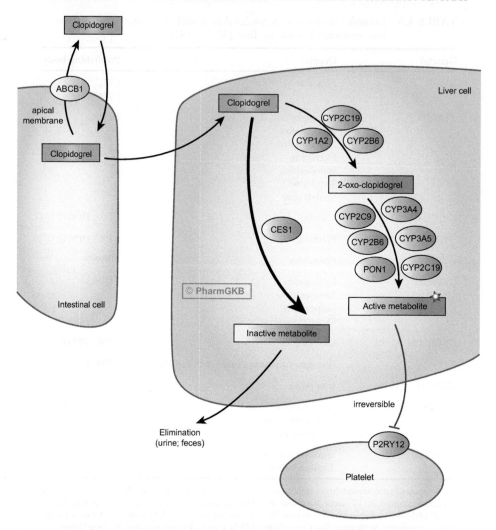

FIGURE 3.3 Clopidogrel pathway on the Pharmacogenomics Knowledge Base [31]. (See color plate 2.) *Copyright to PharmGKB with permission given by PharmGKB and Stanford University for reproduction.*

a dosing equation with global clinical utility [33,34]. Results of the second effort of the IWPC, which was to examine genetic influences of warfarin dose requirements across racial groups, were published in 2010 [35].

IWPC—Genome Wide Association Studies (IWPC-GWAS)

The IWPC-GWAS consortium consists of research groups with the shared interest of discovering novel genetic determinants of warfarin dose response. One aim of the consortium is to conduct a GWAS of warfarin dose requirements in African Americans. A second aim is to perform a combined analysis of GWAS data from multiple sites and populations of warfarin-treated patients. Preliminary results of the African American GWAS were presented at the 2011 American Heart Association Scientific Sessions [36].

International Tamoxifen Pharmacogenomic Consortium (ITPC)

The International Tamoxifen Pharmacogenomics Consortium (ITPC) was established to amass worldwide data on genetic determinants of tamoxifen response. The ITPC seeks to identify genetic determinants of clinical outcomes and adverse effects with tamoxifen in women receiving the drug as adjuvant therapy for breast cancer. One specific objective of the ITPC is to resolve discordant data with the *CYP2D6* genotype and clinical outcomes with tamoxifen [37].

As of 2009, data from more than 3,400 patients with early-stage estrogen-receptor-positive invasive breast cancer treated with tamoxifen had been deposited at PharmGKB. These data were contributed by investigators from the United States, Sweden, the United Kingdom, Germany, Belgium, and Japan. Results of the primary analysis of these data were presented at the 2009 San Antonio Breast Cancer Symposium [37].

International SSRI Pharmacogenomics Consortium (ISPC)

Depression is a major psychiatric disease worldwide, and the International SSRI Pharmacogenomics Consortium (ISPC) seeks to aggregate samples from the majority of currently available selective serotonin reuptake inhibitor (SSRI) pharmacogenetic studies worldwide to perform GWASes, as well as large replication studies. Their goal is to elucidate the genetic determinants of variable response to SSRIs. This effort focuses on finding new genetic variants associated with SSRI pharmacodynamics (PD) by adjusting for genotypes associated with SSRI metabolizing enzymes via covariates in the analysis to make it possible to aggregate data for multiple SSRI agents.

International Clopidogrel Pharmacogenomics Consortium (ICPC)

The International Clopidogrel Pharmacogenomics Consortium (ICPC) came about as a result of the excitement and controversy regarding the role of *CYP2C19* loss-of-function variants on clinical outcomes in clopidogrel-treated patients. Here, the ICPC's goal is to amass very large sample sizes and expertise from across the globe to dissect the genetic underpinnings of variable response to clopidogrel. The ICPC seeks to further increase sample size through *de novo* genotyping as well as by performing additional analyses of platelet function and clinical outcomes to better understand the role of *CYP2C19* polymorphism in clopidogrel response. They also hope to identify novel variants for clopidogrel response through candidate gene and genome-wide approaches.

Clinical Pharmacogenetics Implementation Consortium (CPIC)

The Clinical Pharmacogenetics Implementation Consortium (CPIC) is a joint project of the PharmGKB and PGRN to address barriers associated with transitioning data from the laboratory into clinical applications [38]. The CPIC is an international collaboration of PGRN members, PharmGKB staff, and other individuals from academic centers, clinical institutions, and pharmacy benefits management with expertise in pharmacogenomics or laboratory medicine. The consortium scores evidence linking drug dosing decisions to genetic tests and provides consensus-based guidelines on how to use genetic test results to optimize pharmacotherapy. Their goal is not to recommend whether or not genetic testing should be done, but rather, provide guidelines on how to use existing genetic information. Guidelines have been published related to the pharmacogenomics of thiopurine dosing [39], clopidogrel [40], warfarin [41], codeine-containing analgesics [42], abacavair [43], and simvastatin [44]. All guidelines are posted, integrated, and updated on the PharmGKB. Guidelines related to the pharmacogenomics of several anticancer agents (e.g., capecitabine), antihyperuricemia (e.g., allopurinol), and antiseizure drugs (e.g., carbamazepine) are forthcoming.

Electronic Medical Records and Genomics (eMERGE) Network

The Electronic Medical Records and Genomics (eMERGE) network was formed in 2007 with the goal of exploring the use of electronic medical records (EMRs) coupled to DNA repositories for large-scale genomic research [45]. An additional focus is on the social and ethical aspects (e.g., privacy and confidentiality) related to merging genomic information with the medical record. The eMERGE network is administered by the National Human Genome Research Institute's Office of Population Genomics, with additional funding from the National Institute of General Medical Sciences. The network is composed of experts in clinical medicine, genomics, health information technology, statistics, and ethics. Vanderbilt University Medical Center serves as the coordinating center and supports and facilitates the work by network investigators.

The first phase of eMERGE included five institutions, each with a unique DNA biobank linked to the EMRs. Each site examined genome-wide associations in specific diseases, such as cataracts, type 2 diabetes, peripheral artery disease, Alzheimer's disease, and cardiac conduction defects [45]. A sixth GWAS was conducted using samples accrued across sites in the network. Results from eMERGE Phase I show that linking data from the EMRs with patient genotypes is feasible for identifying genetic contributors to disease phenotype and for large-scale GWASes across multiple clinical sites [46–49]. Phase II of eMERGE began in 2011 with the goal of exploring the incorporation of genetic information into the EMR for use in patient care. Additional sites, with focuses including abdominal aortic aneurysm, obesity, antipsychotic-induced weight gain, dementia, and infectious disease susceptibility, among others, have been added to Phase II. Future funding will be geared toward pediatric eMERGE studies.

THE ROLE OF THE FOOD AND DRUG ADMINISTRATION AND OTHER INTERNATIONAL GOVERNMENT AGENCIES

Recognizing limitations of the trial-and-error approach to drug prescribing and the high attrition rates in drug development, the FDA has engaged in a number of activities to optimize drug use and enhance drug development through a better understanding of genetic determinants of drug response. Examples of FDA's initiatives towards personalized medicine include:

- publication of the critical path initiative [50,51];
- establishment of an Interdisciplinary Pharmacogenomics Review Group to evaluate voluntary submissions [52–55];
- formation of a pharmacogenomics group in the Office of Clinical Pharmacology, Center for Drug Evaluation and Research;[56] and
- formation of a personalized medicine group in the Office of In Vitro Diagnostic Devices, Center for Device Evaluation and Radiological Health[57] within the FDA.

Voluntary Exploratory Data Submission (VXDS) Program

The FDA has introduced the "Safe Harbor" concept [52,58] and has encouraged submission of exploratory genomic data from drug and device sponsors and research investigators for the purpose of scientific discussion without regulatory implications. A guide was published to facilitate the process and to encourage pharmaceutical companies to integrate genetic information into the development of new drugs [59]. The initial Voluntary Genomic Data Submission process met with significant success [60,61]. The concept was therefore extended to a Voluntary Exploratory Data Submission (VXDS) program to reflect the expanded scope to both nongenomic and genomic biomarkers [62]. Although pharmaceutical companies were accruing genetic data prior to the start of the voluntary submission program, they were reluctant to share these data with the FDA for fear that the FDA may make premature regulatory decisions based on the data [58]. Thus, the program was designed to allay these fears and provide an opportunity for companies to discuss exploratory genetic information pertaining to their product with FDA staff and obtain valuable informal feedback, in a process that is independent of the drug review. By the end of 2009, the FDA had received more than 40 voluntary data submissions.

The Interdisciplinary Pharmacogenomics Review Group (IPRG) was created to ensure high-quality review of the voluntary data submissions [61]. The IPRG engages in face-to-face meetings with industry sponsors, during which the IPRG provides their scientific perspective on the use of the genetic biomarker for a particular drug. In turn, the IPRG gains perspective on how the sponsor will generate and analyze genetic data. In addition, the IPRG conducts joint meetings with the European Medicines Agency (EMA). Such meetings have resulted in an international consensus on both the opportunities and limitations of incorporating genomic data into drug development. An example of this collaboration has been the review of biomarkers for acute kidney toxicity [61]. Many of the initial voluntary submissions have resulted in regulatory submissions.

Guidance Development

The FDA provides guidance to drug developers on incorporating pharmacogenomic principles in the drug development process. The reviewers at the FDA ensure that the appropriate biomarkers, including genetic markers, are in place at the Investigational New Drug (IND) stage and assesses the benefit risk profile of the drug according to genetic and other biomarkers in the

New Drug Application (NDA) and Biologic License Application (BLA) review stages [63,64]. In addition, the FDA also recommends pharmacogenomic-based postapproval studies, as appropriate, to better understand the drug-related benefits and risks in certain patient subsets.

In the draft guide, "Clinical Pharmacogenomics: Premarketing Evaluation in Early Phase Clinical Studies," the FDA provides drug developers with recommendations on when to consider genetic information during the drug development process [65]. Specifically, the document outlines the application of genetic data to individualize drug dosing, prioritize drug–drug interaction studies, assess the molecular basis for adverse drug effects or inadequate drug response, and identify subgroups to examine in clinical trials. With regard to the latter, genetics can help to identify individuals most likely to respond to a particular therapy, thus enabling enrichment strategies in Phase III studies. An enrichment design strategy has been discussed [66]. In addition, if a genetic predisposition to a serious adverse drug effect is identified, the development of the drug may continue with the exclusion of high-risk subjects from clinical trials. The guide also provides recommendations regarding appropriate collection and storage of genetic samples in clinical trial populations.

Clinical Utility

Clinical utility is often test- (genetic or other biomarker), drug-, and context-dependent, and it is not easily quantified. In the context of a drug–test pair, clinical utility has historically been a composite measure of benefit, risk, and perceived value in medical decisions [67]. Although the most robust and efficient method for translating biomarker test information may be a randomized, controlled clinical trial, there are many complementary sources of evidence (mechanistic, pharmacological, and observational studies) that can all contribute to establishing clinical utility [68]. Earlier publications have addressed the clinical utility of measuring metabolizing enzymes[69] or the integration of biomarker evaluation in drug development, regulatory review, and clinical practice [70,71]. Improved understanding of the mechanistic basis of adverse drug events and the availability of biomarker tests, including genetic tests [72], have enabled development of tools to select patients or dosing regimens for individual patients. Recent studies, including pharmacogenomic studies, have suggested that transporters play an important part *in vivo* in drug disposition, therapeutic efficacy, and adverse drug reactions [73,74]. To determine the clinical utility of pharmacogenomic tests, considerations to improve the design of association studies have been outlined [75].

Pharmacogenomic Biomarkers in Drug Labels

The FDA continues to update labeling to reflect current knowledge on the safe and effective use of drug products. A request to update the labeling of an approved drug may be made by regulatory scientists at the FDA, the manufacturer of the drug in question, or an external researcher [63]. Based on the strength of evidence and potential impact of a gene–drug response association, FDA scientists from various offices, together with individuals from the drug manufacturing company, determine when a pharmacogenomic update to drug labeling is required.

Pharmacogenomic information may relate to alterations in gene structure (e.g., genetic polymorphism) or expression or to chromosomal abnormalities. Pharmacogenomic information may be included in various sections of the drug label (e.g., clinical pharmacology, dosage and administration, warnings and precautions). Pharmacogenomic associations with significant implications for drug safety may be included in a boxed warning. Currently, the labeling for more than 80 FDA-approved drugs contains content related to genetic biomarkers with implications for drug exposure, clinical response, risk for adverse effects, and or dose optimization. Examples of drugs with pharmacogenomic labeling are shown in Table 3.4. The FDA maintains a Table of Pharmacogenomic Biomarkers in Drug Labels on its website [76].

Pharmacogenomic updates to the labeling of existing drugs are usually based on data from the literature. In many cases, the data are based on retrospective assessment of genetic associations with drug response. For example, the data informing the decision to update the warfarin labeling were largely from pharmacogenomic studies using existing data 50 years after the drug was first approved for marketing [77]. Further, scientists at the FDA do not typically have access to the raw data from pharmacogenomic studies carried out on drugs that have already been approved, because these data are not usually included in publications reporting study results. In contrast, prospective data generated in properly designed drug clinical trials are available to FDA to guide pharmacogenomic labeling decisions for newly approved drugs. Pharmacogenomics is only one of many factors to consider in individualization of drug therapy [78]. Therefore, to assess a drug response (beneficial or adverse) in the context of multiple patient factors (e.g., age, gender, genetics, organ impairment, concomitant medications), a systems-based approach [79], including physiologically based pharmacokinetic modeling, is needed [80].

Diagnostic Approval/Clearance

An additional role of the FDA in advancing personalized medicine involves the approval of companion diagnostics to predict drug response, or to select a population for whom the drug safety and effectiveness are known. The availability of accurate diagnostic testing is key for successful personalized medicine. To date, the FDA has approved companion diagnostics for a few oncology medications including cetuximab, crizotinib, imatinib, panitumumab, trastuzumab, and vemurafenib (Table 3.4). However, there are also laboratories performing and marketing non-FDA-approved tests, commonly referred to as "laboratory-developed" tests, with claims that such tests predict drug response. Although clinicians may assume that such tests provide accurate information on which to base treatment decisions, this might not always be the case. In order to ensure the quality of diagnostic tests and protect those ultimately affected by test results (i.e., the patients), the FDA is establishing standards for the efficient review and oversight of companion diagnostic tests and manufacturer claims.

ACTIVITIES OF NON-U.S. AGENCIES

European Medicines Agency (EMA)

The EMA is a decentralized body of the European Union responsible for evaluating drugs for human and veterinary use. Similar to the FDA, the EMA has engaged in several activities

TABLE 3.4   Examples of drugs with pharmacogenomic-related information in their FDA-approved labeling

Drug	Gene	Context
Abacavir	HLA-B	Boxed warning that patients with the HLA-B*57:01 allele are at increased risk for hypersensitivity to abacavir. Genetic screening is recommended before starting abacavir.
Azathioprine and 6-mercaptopurine	TPMT	Description of increased risk for myelotoxicity with conventional azathioprine or 6-mercaptopurine doses in patients with a nonfunctional TPMT allele in the clinical pharmacology section. Consideration of TPMT genetic testing is recommended.
Atomoxetine	CYP2D6	Warning that dose adjustment may be necessary in CYP2D6 poor metabolizers to avoid adverse drug effects.
Capecitabine	DPD	Warning about an increased risk for severe toxicity (e.g., diarrhea, stomatitis, neutropenia, and neurotoxicity) in patients with dihydropyrimidine dehydrogenase deficiency.
Carbamazepine	HLA-B	Boxed warning of increased risk for serious dermatologic reactions (e.g., toxic epidermal necrolysis, Stevens—Johnson syndrome) in patients with the HLA-B*1502 variant. Patients from genetically at-risk regions (e.g., Southeast Asia) should be screened for the HLA-B*1502 allele prior to starting carbamazepine.
Cetuximab and Panitumumab	EGRF, KRAS	These drugs are indicated for EGRF-expressing colorectal cancer and may be ineffective in patients whose tumors have a KRAS mutation in codon 12 or 13.
Codeine	CYP2D6	Warning about greater conversion to morphine and higher than expected morphine concentrations in patients who are ultra-rapid metabolizers secondary to the CYP2D6*2x*2 genotype. These individuals are at increased risk for symptoms of overdose (e.g., extreme sleepiness, confusion, respiratory depression) with conventional doses of codeine.
Clopidogrel	CYP2C19	Boxed warning of possible reduced drug effectiveness in CYP2C19 poor metabolizers with 2 loss-of-function alleles.
Crizotinib	ALK	Confirmation of the lymphoma kinase (ALK)-positive mutation is required using an FDA-approved test prior to drug use. A clinical trial to explore responses in ALK-negative patients is being conducted postmarketing.
Irinotecan	UGT1A1	Recommendation to reduce irinotecan dosage by one level in homozygotes for the UGT1A1*28 allele because of an increased risk for neutropenia.
Lenalidomide	Chromosome 5q	Boxed warning to monitor complete blood counts weekly for the first 8 weeks of therapy for patients with del 5q myelodysplastic syndrome.
Tetrabenazine	CYP2D6	Recommendation to test CYP2D6 genotype if the patient needs a higher than 50 mg dose.
Trastuzumab	Her2/neu	The drug is indicated for HER2-overexpressing cancers. Detection of HER2 protein overexpression or gene amplification is necessary prior to treatment.

TABLE 3.4 Examples of drugs with pharmacogenomic-related information in their FDA-approved labeling—cont'd

Drug	Gene	Context
Vemurafenib	*BRAF*	Confirmation of the *BRAF* V600E mutation is required using an FDA-approved test prior to drug use. Drug efficacy and safety have not been studied in patients without the mutation.
Warfarin	*CYP2C9* *VKORC1*	Dosing recommendations are provided according to *CYP2C9* and *VKORC1* genotypes.

Note: For FDA-cleared or FDA-approved tests, see the following website: www.accessdata.fda.gov/scripts/cdrh/cfdocs/cfPMN/pmn.cfm; for information related to *in vitro* companion diagnostic devices, see the following website: www.fda.gov/MedicalDevices/ProductsandMedicalProcedures/InVitroDiagnostics/ucm301431.htm.

related to pharmacogenomics. These include formation of a Pharmacogenomics Working Party (PgWP) and publication of scientific guidelines on pharmacogenomics.

The PgWP is composed of experts in the field of genetics or pharmacogenomics who are nominated by the Committee for Medicinal Products for Human Use (CHMP) and charged with providing recommendations to the CHMP on issues related to pharmacogenomics. Drug developers may seek informal advice on pharmacogenomic-related matters from the PgWP. Other activities of the PgWP include preparing guidelines for the evaluation of pharmacogenomic information in regulatory submissions and organizing pharmacogenomics workshops.

In early 2012, the CHMP published a guideline for pharmaceutical companies on the use of pharmacogenomic methodologies in the evaluation of medicinal products [81]. The guideline focuses on genetic variation affecting drug pharmacokinetics and provides guidance on both required and recommended studies at different stages of drug development. The guidelines specifically address when pharmacogenomic studies are appropriate, how to design and conduct pharmacogenomic studies, how to evaluate the clinical impact of genetic variability, and how to study drug dosing and both drug and disease interactions in genetic subpopulations.

Pharmaceutical and Medical Devices Agency (PMDA), Japan

The Pharmaceutical and Medical Devices Agency (PMDA) is responsible for regulatory drug review and approval in Japan. In 2009, the PMDA introduced a pilot scientific consultation program that focuses on pharmacogenomics and biomarker qualification [82]. This program involves consultation with drug sponsors to identify strategies for utilizing pharmacogenomics and biomarkers in drug development. The purpose of this program is to improve efficiency in drug development as well as enable development of personalized medicines.

Similar to FDA-approved drug labeling, pharmacogenomic information has been added to the package inserts for many drugs marketed in Japan. In a recent review, the authors found that the majority of drugs with FDA-approved pharmacogenomic labeling also contained pharmacogenomic information in their PMDA-approved package insert [83]. However, there are fewer instances of pharmacogenomic information included in the warning or contraindication sections of labeling for drugs marketed in Japan versus the United States. In the case of carbamazepine, the pharmacogenomic information is specific for the Japanese population. Specifically, the *HLA-B*1502* allele, associated with an increased of carbamazepine-induced Stevens—Johnson syndrome (SJS) and toxic epidermal necrolysis (TEN) in Han Chinese, is rare in persons from Japan (frequency <0.1%). In a recent GWAS conducted in Japanese patients, the *HLA-A*3101*, but not the *HLA-B*1502* allele, was associated with serious carbamazepine-induced cutaneous adverse reactions, including SJS and TEN [84]. The PMDA-approved package insert for carbamazepine has since been amended to include information about the *HLA-A*3101* allele [83].

CONCLUSION

The Human Genome Project and subsequent efforts in the field have led to an improved understanding of the structure and function of the human genome. This understanding has enabled numerous discoveries of genetic contributions to drug response in addition to discovery of novel therapeutic targets. Although there are examples of pharmacogenomic applications to patient care, broad application of personalized medicine has yet to be realized.

In 2011, the National Human Genome Research Institute published its updated vision for the future of genomic medicine. This vision focuses on prevention and treatment of disease based on an understanding of human biology and diagnosis [85]. Opportunities for genomic medicine are many. As outlined by the National Human Genome Research Institute [85], these include enabling routine use of genomic-based diagnostic panels; better characterizing genetic contributors to disease phenotype and drug response, thus allowing for improved therapeutic strategies and revealing sites for novel drug development; and developing practical systems for applying genomic information to patient care. Ongoing efforts by the NIH and FDA are addressing each of these areas and their related challenges in order to further advance personalized medicine. Non-U.S. agencies, including the EMA and PMDA, are also engaged in regulatory activities with the goal of safer medication use through individualized therapy.

QUESTIONS FOR DISCUSSION

1. What are the scientific and regulatory challenges of advancing personalized medicine?
2. What are some examples of efforts by the National Institutes of Health and the FDA to address the challenges associated with advancing personalized medicine?
3. What resources are available to assist clinicians with implementing pharmacogenomics into patient care? How do you gain access to these resources?
4. How is the FDA encouraging the incorporation of genetic information in drug development?
5. What efforts are non-U.S. agencies taking to advance pharmacogenomics?

References

[1] Hamburg MA, Collins FS. The path to personalized medicine. N Engl J Med 2010;363:301–4.

[2] Alving AS, Carson PE, Flanagan CL, Ickes CE. Enzymatic deficiency in primaquine-sensitive erythrocytes. Science 1956;124:484–5.

[3] Venter JC, Adams MD, Myers EW, et al. The sequence of the human genome. Science 2001;291:1304–51.

[4] Lander ES, Linton LM, Birren B, et al. Initial sequencing and analysis of the human genome. Nature 2001; 409:860–921.

[5] Collins FS, Green ED, Guttmacher AE, Guyer MS. A vision for the future of genomics research. Nature 2003; 422:835–47.

[6] Battelle Technology Partnership Practice. Economic Impact of the Human Genome Project. 2011 (Accessed February 14, 2012, at, http://www.battelle.org/docs/default-document-library/economic_impact_of_the_human_genome_project.pdf.).

[7] de Bakker PI, Burtt NP, Graham RR, et al. Transferability of tag SNPs in genetic association studies in multiple populations. Nat Genet 2006;38:1298–303.

[8] The International HapMap Project: Ethical, Social, and Cultural Issues. (Accessed April 30, 2012, at hapmap.ncbi.nlm.nih.gov/downloads/presentations/HapMapELSI.en.ppt)

[9] The International HapMap Consortium. A haplotype map of the human genome. Nature 2005; 437:1299–320.

[10] Frazer KA, Ballinger DG, Cox DR, et al. A second generation human haplotype map of over 3.1 million SNPs. Nature 2007;449:851–61.

[11] Altshuler DM, Gibbs RA, Peltonen L, et al. Integrating common and rare genetic variation in diverse human populations. Nature 2010;467:52–8.

[12] Manolio TA, Collins FS. The HapMap and genome-wide association studies in diagnosis and therapy. Ann Rev Med 2009;60:443–56.

[13] Cooper GM, Johnson JA, Langaee TY, et al. A genome-wide scan for common genetic variants with a large influence on warfarin maintenance dose. Blood 2008;112:1022–7.

[14] Innocenti F, Owzar K, Cox NL, et al. A genome-wide association study of overall survival in pancreatic cancer patients treated with gemcitabine in CALGB 80303. Clin Cancer Res 2012;18:577–84.

[15] Lucena MI, Molokhia M, Shen Y, et al. Susceptibility to amoxicillin-clavulanate-induced liver injury is influenced by multiple HLA class I and II alleles. Gastroenterology 2011;141:338–47.

[16] Link E, Parish S, Armitage J, et al. SLCO1B1 variants and statin-induced myopathy—a genomewide study. N Engl J Med 2008;359:789–99.

[17] Jablonski KA, McAteer JB, de Bakker PI, et al. Common variants in 40 genes assessed for diabetes incidence and response to metformin and lifestyle intervention in the diabetes prevention program. Diabetes 2010; 59:2672–81.

[18] Voora D, Shah SH, Spasojevic I, et al. The SLCO1B1*5 genetic variant is associated with statin-induced side effects. J Am Coll Cardiol 2009;54:1609–16.

[19] Mailman MD, Feolo M, Jin Y, et al. The NCBI dbGaP database of genotypes and phenotypes. Nat Genet 2007; 39:1181–6.

[20] 1000 Genomes—A deep catalog of human genetic variation (Accessed February 15, 2012, at, http://www.1000genomes.org/about.).

[21] A map of human genome variation from population-scale sequencing. Nature 2010;467:1061–73.

[22] Myers RM, Stamatoyannopoulos J, Snyder M, et al. A user's guide to the encyclopedia of DNA elements (ENCODE). PLoS Biol 2011;9:e1001046.

[23] Birney E, Stamatoyannopoulos JA, Dutta A, et al. Identification and analysis of functional elements in 1% of the human genome by the ENCODE pilot project. Nature 2007;447:799–816.

[24] Gerstein MB, Bruce C, Rozowsky JS, et al. What is a gene, post-ENCODE? History and updated definition. Genome Res 2007;17:669–81.

[25] Long RM, Berg JM. What to expect from the Pharmacogenomics Research Network. Clin Pharmacol Ther 2011; 89:339–41.

[26] Giacomini KM, Brett CM, Altman RB, et al. The pharmacogenetics research network: from SNP discovery to clinical drug response. Clin Pharmacol Ther 2007;81:328–45.

[27] McDonagh EM, Whirl-Carrillo M, Garten Y, Altman RB, Klein TE. From pharmacogenomic knowledge acquisition to clinical applications: the PharmGKB as a clinical pharmacogenomic biomarker resource. Biomark Med 2011;5:795—806.

[28] Klein TE, Chang JT, Cho MK, et al. Integrating genotype and phenotype information: an overview of the PharmGKB project. Pharmacogenetics Research Network and Knowledge Base. Pharmacogenomics J 2001; 1:167—70.

[29] Hewett M, Oliver DE, Rubin DL, et al. PharmGKB: the Pharmacogenetics Knowledge Base. Nucleic Acids Res 2002;30:163—5.

[30] Swen JJ, Nijenhuis M, de Boer A, et al. Pharmacogenetics: from bench to byte—an update of guidelines. Clin Pharmacol Ther 2011;89:662—73.

[31] Sangkuhl K, Klein TE, Altman RB. Clopidogrel pathway. Pharmacogenet Genomics 2010;20:463—5.

[32] Eichelbaum M, Altman RB, Ratain M, Klein TE. New feature: pathways and important genes from PharmGKB. Pharmacogenet Genomics 2009;19:403.

[33] Owen RP, Altman RB, Klein TE. PharmGKB and the International Warfarin Pharmacogenetics Consortium: the changing role for pharmacogenomic databases and single-drug pharmacogenetics. Hum Mutat 2008; 29:456—60.

[34] Klein TE, Altman RB, Eriksson N, et al. Estimation of the warfarin dose with clinical and pharmacogenetic data. N Engl J Med 2009;360:753—64.

[35] Limdi NA, Wadelius M, Cavallari L, et al. Warfarin pharmacogenetics: a single VKORC1 polymorphism is predictive of dose across 3 racial groups. Blood 2010;115:3827—34.

[36] Perera MA, Limdi NA, Cavallari L, et al. Novel SNPs associated with warfarin dose in a large multicenter cohort of African Americans: genome wide association study and replication results. Circulation 2011;124: #15518.

[37] Goetz M, Berry D, Klein T, and the International Tamoxifen Pharmacogenomics Consortium. Adjuvant tamoxifen treatment outcome according to cytochrome P450 2D6 (CYP2D6) phenotype in early stage breast cancer: findings from the International Tamoxifen Pharmacogenomics Consortium Cancer Res 2009;69:33.

[38] Relling MV, Klein TE. CPIC: Clinical Pharmacogenetics Implementation Consortium of the Pharmacogenomics Research Network. Clin Pharmacol Ther 2011;89:464—7.

[39] Relling MV, Gardner EE, Sandborn WJ, et al. Clinical Pharmacogenetics Implementation Consortium guidelines for thiopurine methyltransferase genotype and thiopurine dosing. Clin Pharmacol Ther 2011;89:387—91.

[40] Scott SA, Sangkuhl K, Gardner EE, et al. Clinical Pharmacogenetics Implementation Consortium guidelines for cytochrome P450-2C19 (CYP2C19) genotype and clopidogrel therapy. Clin Pharmacol Ther 2011;90:328—32.

[41] Johnson JA, Gong L, Whirl-Carrillo M, et al. Clinical Pharmacogenetics Implementation Consortium Guidelines for CYP2C9 and VKORC1 genotypes and warfarin dosing. Clin Pharmacol Ther 2011;90:625—9.

[42] Crews KR, Gaedigk A, Dunnenberger HM, et al. Clinical Pharmacogenetics Implementation Consortium (CPIC) guidelines for codeine therapy in the context of cytochrome P450 2D6 (CYP2D6) genotype. Clin Pharmacol Ther 2011;91:321—6.

[43] Martin MA, Klein TE, Dong BJ, Pirmohamed M, Haas DW, Kroetz DL. Clinical Pharmacogenetics Implementation Consortium guidelines for hla-B genotype and abacavir dosing. Clin Pharmacol Ther 2012; 91:734—8.

[44] Wilke RA, Ramsey LB, Johnson SG, et al. The Clinical Pharmacogenomics Implementation Consortium CPIC guideline for SLCO1B1 and simvastatin-induced myopathy. Clin Pharmacol Ther 2012 [In press].

[45] McCarty CA, Chisholm RL, Chute CG, et al. The eMERGE Network: a consortium of biorepositories linked to electronic medical records data for conducting genomic studies. BMC Med Genomics 2011;4:13.

[46] Peissig PL, Rasmussen LV, Berg RL, et al. Importance of multi-modal approaches to effectively identify cataract cases from electronic health records. J Am Med Inform Assoc 2012;19:225—34.

[47] Denny JC, Ritchie MD, Crawford DC, et al. Identification of genomic predictors of atrioventricular conduction: using electronic medical records as a tool for genome science. Circulation 2010;122:2016—21.

[48] Bielinski SJ, Chai HS, Pathak J, et al. Mayo Genome Consortia: a genotype-phenotype resource for genome-wide association studies with an application to the analysis of circulating bilirubin levels. Mayo Clin Proc 2011; 86:606—14.

[49] Crosslin DR, McDavid A, Weston N, et al. Genetic variants associated with the white blood cell count in 13,923 subjects in the eMERGE Network. Hum Genet 2012;131:639—52.

[50] Buckman S, Huang SM, Murphy S. Medical product development and regulatory science for the 21st century: the critical path vision and its impact on health care. Clin Pharmacol Ther 2007;81:141—4.

[51] Barratt RA, Bowens SL, McCune SK, Johannessen JN, Buckman SY. The critical path initiative: leveraging collaborations to enhance regulatory science. Clin Pharmacol Ther 2012;91:380—3.

[52] Lesko LJ, Woodcock J. Translation of pharmacogenomics and pharmacogenetics: a regulatory perspective. Nat Rev 2004;3:763—9.

[53] Ruano G, Collins JM, Dorner AJ, Wang SJ, Guerciolini R, Huang SM. Pharmacogenomic data submissions to the FDA: clinical pharmacology case studies. Pharmacogenomics 2004;5:513—7.

[54] Huang SM, Goodsaid F, Rahman A, Frueh F, Lesko LJ. Application of pharmacogenomics in clinical pharmacology. Toxicol Mech Methods 2006;16:89—99.

[55] Hinman LM, Huang SM, Hackett J, et al. The drug diagnostic co-development concept paper: commentary from the 3rd FDA-DIA-PWG-PhRMA-BIO Pharmacogenomics Workshop. Pharmacogenomics J 2006;6:375—80.

[56] Zineh I, Woodcock J. The clinical pharmacogeneticist: an emerging regulatory scientist at the US Food and Drug Administration. Hum Genomics 2010;4:221—5.

[57] Sapsford KE, Tezak Z, Kondratovich M, Pacanowski MA, Zineh I, Mansfield E. Biomarkers to improve the benefit/risk balance for approved therapeutics: a US FDA perspective on personalized medicine. Therapeutic Delivery 2010;1:631—41.

[58] Lesko LJ, Zineh I. DNA, drugs and chariots: on a decade of pharmacogenomics at the US FDA. Pharmacogenomics 2010;11:507—12.

[59] Food and Drug Administration Draft Guidance for Industry and Food and Drug Administration Staff - In Vitro Companion Diagnostic Devices; published in July 2011 (Accessed May 18, 2012, at http://www.fda.gov/downloads/MedicalDevices/DeviceRegulationandGuidance/GuidanceDocuments/UCM262327.pdf.)

[60] Food and Drug Administration, guidance for industry. Pharmacogenomic data submission. (Accessed April 15, 2012, at http://www.fda.gov/downloads/Drugs/GuidanceComplianceRegulatoryInformation/Guidances/UCM079849.pdf.)

[61] Orr MS, Goodsaid F, Amur S, Rudman A, Frueh FW. The experience with voluntary genomic data submissions at the FDA and a vision for the future of the voluntary data submission program. Clin Pharmacol Ther 2007;81:294—7.

[62] Goodsaid FM, Amur S, Aubrecht J, et al. Voluntary exploratory data submissions to the US FDA and the EMA: experience and impact. Nat Rev 2010;9:435—45.

[63] Zineh I, Pacanowski MA. Pharmacogenomics in the assessment of therapeutic risks versus benefits: inside the United States Food and Drug Administration. Pharmacotherapy 2011;31:729—35.

[64] Zineh I, Huang SM. Biomarkers in drug development and regulation: a paradigm for clinical implementation of personalized medicine. Biomark Med 2011;5:705—13.

[65] Food and Drug Administration. Guidance for Industry. Clinical Pharmacogenomics: Premarketing Evaluation in Early Phase Clinical Studies. 2011 (Accessed February 27, 2012, at, http://www.fda.gov/downloads/Drugs/GuidanceComplianceRegulatoryInformation/Guidances/UCM243702.pdf.).

[66] Temple R. Enrichment of clinical study populations. Clin Pharmacol Ther 2010;88:774—8.

[67] Lesko LJ, Zineh I, Huang SM. What is clinical utility and why should we care? Clin Pharmacol Ther 2010;88:729—33.

[68] Woodcock J. Assessing the clinical utility of diagnostics used in drug therapy. Clin Pharmacol Ther 2010;88:765—73.

[69] Andersson T, Flockhart DA, Goldstein DB, et al. Drug-metabolizing enzymes: evidence for clinical utility of pharmacogenomic tests. Clin Pharmacol Ther 2005;78:559—81.

[70] Amur S, Frueh FW, Lesko LJ, Huang SM. Integration and use of biomarkers in drug development, regulation and clinical practice: a US regulatory perspective. Biomark Med 2008;2:305—11.

[71] Amur S, Zineh I, Abernethy DR, Huang SM, Lesko L. Pharmacogenomics and adverse drug reactions (ADRs). Per Med 2010;7:633—42.

[72] Giacomini KM, Huang S-M, Tweedie DJ, et al, for the International Transporter Consortium. Membrane Transporters in Drug Development: Report form the FDA Critical Path Initiative Sponsored Workshop. Nat Rev 2010;9:215—36.

[73] Huang SM, Woodcock J. Transporters in drug development: advancing on the Critical Path. Nat Rev 2010;9:175—6.

[74] Huang SM, Chen LG, Giacomini KM. Pharmacogenomic mechanisms of drug toxicity. In: Atkinson AJ, Huang SM, Lertora JL, Markey SP, editors. Principles of Clinical Pharmacology. 3rd ed. Burlington: Elsevier; 2012. pp. 285–308.

[75] Ware J, Zhang L, Huang SM. Mechanisms and genetics of drug transport. In: Atkinson AJ, Huang SM, Lertora JL, Markey SP, editors. Principles of Clinical Pharmacology. 3rd ed. Burlington: Elsevier; 2012. pp. 217–238.

[76] Center for Drug Evaluation and Research, Food and Drug Administration. Genomics at the FDA website (Accessed April 15, 2012, at, http://www.fda.gov/drugs/scienceresearch/researchareas/pharmacogenetics/default.htm.).

[77] Kim MJ, Huang SM, Meyer UA, Rahman A, Lesko LJ. A regulatory science perspective on warfarin therapy: a pharmacogenetic opportunity. J Clin Pharmacol 2009;49:138–46.

[78] Huang SM, Temple R. Is this the drug or dose for you? Impact and consideration of ethnic factors in global drug development, regulatory review, and clinical practice. Clin Pharmacol Ther 2008;84:287–94.

[79] Abernethy DR, Woodcock J, Lesko LJ. Pharmacological mechanism-based drug safety assessment and prediction. Clin Pharmacol Ther 2011;89:793–7.

[80] Huang SM, Rowland M. The role of physiologically based pharmacokinetic modeling in regulatory review. Clin Pharmacol Ther 2012;91:542–9.

[81] European Medicines Agency. Guideline on the use of pharmacogenetic methodologies in the pharmacokinetic evaluation of medicinal products. Available through 2012 (Accessed April 30, 2012, at, http://www.ema.europa.eu/docs/en_GB/document_library/Scientific_guideline/2012/02/WC500121954.pdf.).

[82] Ichimaru K, Toyoshima S, Uyama Y. PMDA's challenge to accelerate clinical development and review of new drugs in Japan. Clin Pharmacol Ther 2010;88:454–7.

[83] Otsubo Y, Asahina Y, Noguchi A, Sato Y, Ando Y, Uyama Y. Similarities and differences between US and Japan as to pharmacogenomic biomarker information in drug labels. Drug Metab Pharmacokinet 2012;27:142–9.

[84] Ozeki T, Mushiroda T, Yowang A, et al. Genome-wide association study identifies HLA-A*3101 allele as a genetic risk factor for carbamazepine-induced cutaneous adverse drug reactions in Japanese population. Hum Mol Genet 2011;20:1034–41.

[85] Green ED, Guyer MS. Charting a course for genomic medicine from base pairs to bedside. Nature 2011; 470:204–13.

Pharmacogenomics in Cancer Therapeutics

Michael E. Schaffer, J. Suso Platero

Janssen Research & Development, Raritan, New Jersey, USA

OBJECTIVES

1. Provide a historical perspective of targeted oncology therapies.

2. Highlight advantages of biomarker-driven drug development.

3. Introduce challenges faced by codevelopment of a companion diagnostic.

INTRODUCTION

An estimated 12.7 million newly diagnosed cancer cases and 7.6 million cancer-related deaths were reported worldwide in 2008 [1]. In the United States alone, nearly 1.6 million new cases and more than half a million related deaths were expected for 2011 [2]. Despite these discouraging figures, striking decreases in U.S. death rates between 1990 and 2007 for breast, prostate, and colorectal cancers highlight promising recent advances in early detection and treatment options. Cancer is a multifactorial disease driven by genetic events that disrupt cellular machinery and signaling, ultimately leading to unregulated proliferation. Because this dysregulation is heterogeneous across malignancies, personalized treatment approaches tailored to a patient's disease are quite attractive. Pharmacogenomics in oncology builds on this notion and aims to increase therapeutic success rates by identifying patients most likely to receive treatment benefit based on their unique genetic makeup and that of their cancer.

Over the past decade, a fundamental paradigm shift transitioned oncology drug development from traditional nonselective cytotoxic agents to targeted therapies designed to

89

ameliorate specific oncogenic transformations [3]. This approach, comprising small molecule and monoclonal antibody therapies, is important in that it is amenable to predictive companion diagnostic tests that empower clinicians to identify patients suitable for a given treatment. This change in development focus, however, requires deeper knowledge and consideration of the molecular pathogenesis of cancer subtypes. Consequently, many advances in our collective understanding are dividends of the Human Genome Project and significant innovation in analytical tool development. One notable example is in the treatment of breast cancer, where hormone receptor status permits disease subtyping and provides clinicians with increasingly important treatment guidance. Moreover, the U.S. Food and Drug Administration (FDA) has approved gene expression—based prognostic assays that predict 10-year risk of recurrence in specific patient populations. This evolution of treatments and related companion diagnostic tools has propelled oncology into the era of personalized medicine.

Past successes notwithstanding, scientific, development, and regulatory hurdles remain in translating basic research findings to concrete clinical value. In this chapter, the role of pharmacogenomics in these processes will be presented, focusing on applications in oncology and enumerating the challenges that lie ahead. Advantages and limitations of such approaches and key examples are highlighted throughout.

ROLE OF ONCOLOGY BIOMARKERS

For almost a century, physicians have relied on noninvasive analyte measurements to assess patient health and organ function. Deviations from normal levels of biological markers frequently indicate underlying pathology or treatment failure. Over the past 20 years, advances in biomarker identification have touched many areas of medicine, including oncology, where detectable marker changes may be systemic or tumor localized. Today, biomarkers provide disease detection with diagnostic markers, prognostication with prognostic markers, patient selection with predictive markers, and treatment response monitoring with pharmacodynamic markers. In addition to protein and metabolic markers, molecular methods now permit investigation of transcriptional alterations, genetic variants, and genomic structural anomalies. Moreover, recent advances have made inexpensive whole-genome and transcriptome sequencing a reality. Together, these technologies form the foundation for emerging pharmacogenomics research and companion diagnostic assay development.

Diagnostic Biomarkers

One of the first diagnostic biomarker assays in oncology measured serum prostate-specific antigen (PSA). Normally produced in small amounts by prostate cells, PSA is a serum marker that may suggest prostate tumor activity when detected at high levels. In conjunction with a digital rectal exam (DRE), the PSA test is FDA approved for prostate cancer detection and monitoring recurrence in patients with a history of the disease. Since 1987, this test has been routinely used and recommended for screening patients over 50 years of age or those with other risk factors, as positive test results often prompt additional diagnostic tests

[4]. However, with a positive predictive value of less than 35% [5], PSA testing has been criticized for producing misleading results in patients with other nonmalignant conditions such as benign prostatic hyperplasia or prostatitis. More important, some prostate tumors do not elicit these characteristic changes in serum PSA levels, and in the Prostate Cancer Prevention Trial, around 15% of patients with consistently normal PSA levels (≤ 4 ng/mL) went on to develop prostate cancer [6].

In the seven-year Prostate, Lung, Colorectal, and Ovarian (PLCO) Cancer Screening Trial in the United States, prostate cancer screening with PSA and DRE demonstrated little effect on death rates, although a 22% increase in diagnosis relative to the control population was observed [7]. Conversely, another study of 182,000 European men identified a benefit of PSA screening with an estimated 20% decrease in death rate, but with collateral overdiagnosis and overtreatment in many subjects [8]. Despite the challenges in interpreting PSA results and the ongoing debate of testing utility, screening more generally demonstrates that patients and physicians alike see value in diagnostic biomarker assays. In an effort to limit overtreatment of the disease, changes have been made to the best practice guidelines for PSA screening [9]. No longer is a single PSA threshold recommended to determine biopsy candidates; rather, a composite of factors including age, PSA change, and family history of the disease is suggested.

Building on evidence that tumors may drive measurable serum analyte changes, the search for other diagnostic biomarkers continued for other cancers. Cancer antigen 125 (CA-125) was identified as a serum marker of cancer risk that is often elevated in patients with ovarian cancer. Unfortunately, this marker suffers similar limitations as PSA with low sensitivity and specificity. Its utility as a diagnostic marker is compromised, as high levels of the protein also occur in kidney, liver, and pancreatic diseases. However, similar to PSA, this marker can provide an early indication of relapse in patients at high risk for recurrence or with familial ovarian cancer. Likewise, another marker, carcinoembryonic antigen (CEA), has poor specificity, but with frequently elevated levels in colorectal, breast, lung, and pancreatic cancers it may still hold value in detecting treatment failure or postsurgical recurrence.

Prognostic and Predictive Biomarkers

Although diagnostic biomarkers are useful for noninvasive assessment of patient health, more consistent with pharmacogenomics is identification of markers for disease prognosis or prediction of treatment response. Accordingly, cytotoxic chemotherapy or targeted treatment regimens can be most effective when markers predictive of sensitivity or resistance are known. Without such indicators, ineffective treatments waste precious time and may unnecessarily expose patients to toxic compounds.

As a tumor progresses, spontaneous somatic mutations and genomic instability distort the genetic content of malignant cells. With this divergence from the original germline sequence, the cancer genome encodes unique changes enabling escape from normal cellular regulation. Moreover, some genetic alterations provide greater fitness to cells within a growing tumor and, through Darwinian selection, become fixed within the tumor cell population. Thus, both germline and somatic genetic variants should be considered for prognostic or predictive markers. In some cases, a patient's genome may contain inherited variation that confers elevated cancer risk or can serve as a predictor of drug sensitivity. Although familial cancers make up less than 15% of all cancers, germline mutations can inform decisions about preventative therapy or surgical intervention before any signs of malignancy.

Examples of such markers are polymorphisms identified in the *BRCA1* and *BRCA2* tumor suppressor genes. In the early 1990s, these mutations were associated with elevated lifetime risk of breast and ovarian cancers [10-12]. Because these gene products are involved in DNA repair, inheriting one defective parental copy predisposes individuals to develop malignancies following spontaneous mutations in the functional wild-type copy. Patients with risk factors, including *BRCA1* and *BRCA2* mutations, may proactively opt for tamoxifen chemoprevention or prophylactic mastectomies to lessen the chance of developing breast cancer. Such risk mutations have also been suggested for colorectal, kidney, and pancreatic cancers.

More often, however, the cancer genome itself provides essential genetic cues to determine disease aggressiveness or to predict clinical response to a drug. When considering targeted therapies, prognostic and predictive biomarkers are frequently interrelated, with the drug target occasionally assuming the role of either or both. Thus, these markers serve important functions in the development and clinical application of targeted therapies. The following section explores how current cancer therapies relate to prognostic and predictive biomarkers.

CONCEPTS IN TARGETED CANCER THERAPY

For more than 60 years, primary approaches to cancer treatment included nonspecific cytotoxic agents that disrupt cellular proliferation. Pioneering work in the late 1940s introduced the folate analogue aminopterin, which possessed antineoplastic properties and induced remission of pediatric acute lymphoblastic leukemia. Since then, the related antimetabolite methotrexate has supplanted aminopterin use and was joined by other classes of chemotherapeutic drugs including alkylating agents (e.g., nitrogen mustards, nitrosoureas, platinum drugs), mitotic inhibitors (e.g., taxanes, vinca alkaloids), topoisomerase inhibitors, corticosteroids, and antitumor antibiotics (e.g., anthracyclines). Nonspecific cytotoxic agents interfere with mechanisms involved in cell division either by direct inhibition or through depletion of analytes necessary for cellular replication. Often used in combination, these therapies have proven to be useful tools in fighting cancer. Nevertheless, because these agents lack specificity, rapidly dividing nonmalignant cells such as those of the gastrointestinal tract and bone marrow are similarly affected, leading to a range of side effects from undesirable to severe. Consequently, tolerable dose levels are limited by the inherent toxicity of such compounds. Despite typical initial responses, disease progression often occurs as tumors acquire resistance mutations to the nonspecific therapies [13].

The discovery of tumor suppressors and oncogenes in the 1970s fundamentally reshaped cancer research by demonstrating the oncogenic potential in key genes. Association of Rb mutations to retinoblastoma and characterization of src, the cellular counterpart to v-src in the Rous sarcoma virus, were pivotal discoveries that underscored the cellular origin of cancer. Dysregulation of receptor and cytoplasmic tyrosine kinases, regulatory guanosine 5'-triphosphatese (GTPases), and tumor suppressors were shown to drive uncontrolled proliferation (Figure 4.1). Furthermore, an explanation for poorly understood familial cancers emerged with the identification of heritable germline genetic polymorphisms that

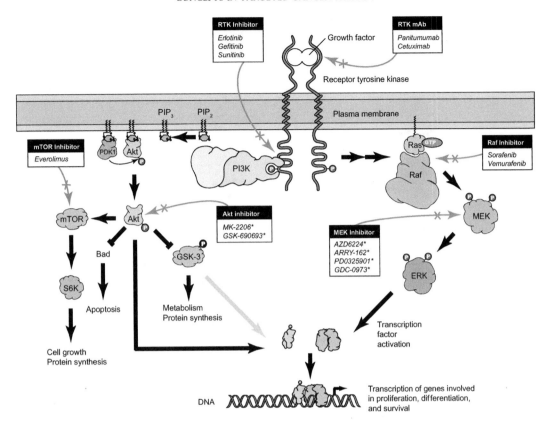

FIGURE 4.1   Canonical receptor tyrosine kinase signaling pathway. Simplified diagram of a canonical biological pathway representing key intracellular signaling cascades following binding of a receptor tyrosine kinase (RTK) by its growth factor ligand. Noted in black rectangles are small molecule or monoclonal antibody (mAb) interventions that may block aberrant signaling that drives cancer cell survival and proliferation. Asterisks indicate compounds in development.

increase cancer risk in carriers, such as mutations in *BRCA1* and *BRCA2*. Although these advances foreshadowed the inevitable era of personalized medicine, decades would separate these seminal discoveries from proven targeted therapies.

Breast Cancer: Hormonal Therapy

The role of pharmacogenomics in oncology has in many ways been shaped by early discoveries and investments in breast cancer research. From efforts in 1932 that first identified hormonally induced breast cancer and subsequent studies in the four decades that followed, estrogen and its cognate binding partner, the estrogen receptor (ER), emerged as key drivers of breast tumor growth [14, 15]. The ER is a nuclear receptor responsible for transducing estrogenic signals and modulating transcription of targets, including genes involved in proliferation and cell growth. In a recent study of almost 6,000 breast cancer cases, 75% of

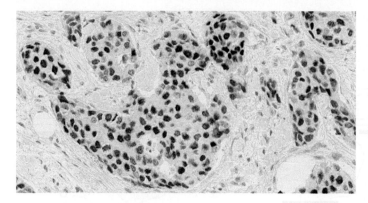

FIGURE 4.2 Estrogen receptor staining of breast tumor tissue section. Image of ER-positive breast cancer tissue section with immuno-histochemical staining for ER (brown) and hematoxylin counter-staining (blue). (See color plate 3.)

all invasive breast cancers and all grade 1 ductal carcinomas expressed the ER (ER-positive) [16]. The active metabolite of the drug tamoxifen antagonizes the ER in breast tissue. Since the FDA approved tamoxifen in 1977 for advanced breast cancer, this prodrug has been used to treat and prevent recurrence in other breast cancer populations by potently blocking ER ligand binding.

A meta-analysis of 55 clinical studies demonstrated striking benefit of adjuvant tamoxifen treatment in 10-year survival rates of ER-positive patients [17]. Conversely, patient tumors expressing low or no ER protein (ER-poor or ER-negative) showed little tamoxifen treatment effects. The progesterone receptor (PR) is a related hormone receptor with expression regulated by estrogen. Binding of PR by its ligand, progesterone, induces similar protumorigenic transcriptional responses that may be blocked by restricting estrogen signaling. Accordingly, the presence of ER and PR define breast tumor subtypes with distinct targeted therapy options including selective ER modulators (SERMs), such as tamoxifen, or aromatase inhibitors (AIs) that disrupt estrogen biosynthesis in postmenopausal patients. To infer a patient's likely response to these therapies, immunohistochemical staining is routinely used to detect presence of ER (Figure 4.2) and PR (Figure 4.3).

Because tamoxifen is a prodrug, a liver enzyme, cytochrome P450 (CYP) 2D6, is required for conversion to its most potent metabolite, endoxifen. Patients treated with tamoxifen and

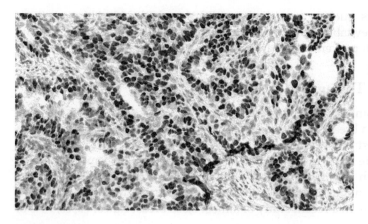

FIGURE 4.3 Progesterone receptor staining of breast tumor tissue section. Image of PR-positive breast cancer tissue section with immuno-histochemical staining for PR (brown) and hematoxylin counterstaining (blue). (See color plate 4.)

concomitant CYP2D6 inhibitors, such as selective serotonin reuptake inhibitor antidepressants, have significantly lower endoxifen levels [18]. Moreover, about 7% of Caucasians have polymorphisms in hepatic CYP2D6, which reportedly reduce the efficiency of tamoxifen metabolism [19]. In this poor metabolizer population, AIs are a viable alternative to tamoxifen for postmenopausal women; however, premenopausal patients lack this option, as AIs do not effectively block ovarian estrogen production [20].

Several retrospective studies investigating the clinical significance of CYP2D6 polymorphisms have suggested a significantly greater risk of poor outcome in poor metabolizers [21, 22]. Although recommendations for routine CYP2D6 screening for breast cancer patients have not been established, in 2006 an FDA advisory subcommittee suggested amending the tamoxifen label to reflect increased risk of recurrence in CYP2D6 poor metabolizers with ER-positive breast cancer [20, 23]. Despite previous reports, however, two large prospective clinical trials, Arimidex, Tamoxifen, Alone or in Combination (ATAC) and Breast International Group (BIG) 1-98, found no significant association between outcome of tamoxifen-treated patients and CYP2D6 polymorphisms [24-26]. The BIG 1-98 results have ignited a controversy, with critics of the study citing genotyping analysis deficiencies that invalidate its conclusions, though the authors maintain the approach was appropriate [27]. Despite the contradictory evidence and lack of specific treatment guidelines, commercial genotyping tests are available to identify CYP2D6 polymorphisms prior to tamoxifen administration [28].

Breast Cancer: Trastuzumab

Along with the ER and PR, the human epidermal growth factor receptor 2 (HER2/neu or ErbB-2) is an additional marker used to stratify breast cancer patients. As a member of the epidermal growth factor receptor family, the orphan receptor is often co-opted in breast cancer and other tumor types to drive cell growth and differentiation. Amplified in about 18 to 22% of breast cancers [29, 30], HER2 overexpression (HER2-positive) (Figure 4.4) is associated with a more aggressive disease and poor clinical outcome [31]. The 1998 FDA approval of the HER2 antagonist, trastuzumab, marked an oncology milestone, as the drug is the first

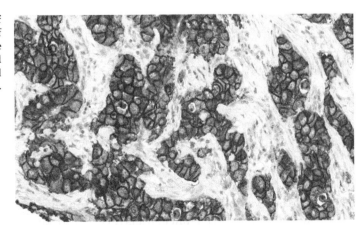

FIGURE 4.4 **HER2 staining of breast tumor tissue section.** Image of HER2-positive breast cancer tissue section with immunohistochemical staining for HER2 (brown) and hematoxylin counterstaining (blue). (See color plate 5.)

humanized monoclonal antibody targeting a specific cancer marker. The Phase III study of trastuzumab plus paclitaxel in HER2-positive metastatic breast cancer patients demonstrated a significant improvement in time to disease progression, overall response rate, and one-year survival compared to paclitaxel alone [32]. Moreover, the treatment effects were positively associated with the degree of HER2 overexpression.

To limit the risk of adverse events in patients not likely to benefit from the drug, trastuzumab treatment is restricted to patients overexpressing HER2. Thus, HER2 serves as both a prognostic and predictive biomarker, with elevated levels suggesting a more aggressive tumor and expression that correlates with trastuzumab treatment outcome. The growing list of FDA-approved tests for HER2 overexpression includes the original immunohistochemistry-based HercepTest™ as well as the fluorescence *in situ* hybridization-based HER2 FISH pharmDx™, HER2/neu ELISA, and the INFORM HER2 dual ISH assay [33].

The American Society of Clinical Oncology (ASCO) and College of American Pathologists (CAP) recommend checking ER, PR, and HER2 status in all invasive breast cancers and provide guidelines for testing [34, 35]. Together, these three markers are used routinely to stratify breast cancers and assist physicians in selecting treatment options with the highest likelihood of success.

B-Cell Malignancies: Rituximab

Among the earliest clinical applications of a targeted therapeutic in oncology was a chimeric monoclonal antibody designed against CD20 cell surface markers present on B-cells and cells of related malignancies, such as lymphoma and leukemia. Although CD20 is not the driver of these diseases, cancer cells can be effectively targeted via this marker with limited side effects occurring as a result of destruction of normal (noncancerous) B-cells. In 1997, rituximab became the first FDA-approved monoclonal antibody cancer therapy, and it is currently used to treat CD20-positive, B-cell non-Hodgkin's lymphoma and chronic lymphocytic leukemia. Rituximab's mechanism of B-cell depletion is primarily from antibody-directed cellular cytotoxicity (ADCC), complement-dependent cytotoxicity (CDC), and apoptosis when bound to CD20 [36]. Because rituximab marshals an immunological response to attack cancer cells, it may be better characterized as a targeted immunotherapy. Interestingly, this antibody has also been used to treat B-cell overproduction in other diseases, such as rheumatoid arthritis and multiple sclerosis.

Chronic Myeloid Leukemia: Imatinib

A key example of early rational drug discovery in oncology was the development of imatinib mesylate. This small molecule inhibitor primarily blocks the product of a reciprocal translocation between chromosomes 9 and 22, which occurs in 95% of patients with the myeloproliferative disorder, chronic myeloid leukemia (CML) [37]. Initially described in 1960, the so-called Philadelphia chromosome is a CML hallmark that was subsequently discovered to produce *BCR-ABL*, a gene fusion encoding a constitutively active oncogenic tyrosine kinase [38-40]. The imatinib precursor, STI571, was identified in 1996 as a multikinase inhibitor of the bcr-abl fusion product, constitutively activated c-Kit, and platelet-derived growth factor receptor [41]. Human trials revealed striking clinical benefit in imatinib-treated

CML patients [42], leading to the 2002 FDA approval of the drug for treating CML, and later for Kit-positive gastrointestinal stromal tumors (GIST). As the first FDA-approved therapy to directly block oncogenic protein signaling, imatinib embodied the targeted drug development paradigm, and its clinical and economic success inspired similarly fruitful research across the pharmaceutical industry. To this end, a number of other tyrosine kinase inhibitors (TKIs) have received approval for additional oncology indications.

Importantly, targeting specific kinases may be complicated by primary or acquired resistance mutations that prevent potent inhibition by TKIs. For example, imatinib typically demonstrates activity in GIST by targeting constitutively active c-Kit; however, specific mutations, such as D816V in the kinase domain, confer drug resistance. Moreover, common genotypes, such as a two codon duplication in exon 9, demonstrate lower drug sensitivity over other mutations and deletions, such as those often found in exon 11. Fortunately, this effect can be modulated by higher doses of imatinib treatment [43]. Thus, mutational testing can provide a basis for selecting drug doses or indicate that imatinib is unsuitable in specific GIST patients and other indications driven by c-Kit.

Metastatic Colorectal Cancer: Cetuximab and Panitumumab

Epidermal growth factor receptor (EGFR) is another receptor tyrosine kinase that signals to downstream cell survival, proliferation, and differentiation pathways and has been associated with cancer growth and invasion. Natural receptor ligands, including epidermal growth factor, normally induce EGFR dimerization and subsequent activation of its intracellular tyrosine kinase domain. Two primary effector pathways are downstream targets of EGFR activation, the mitogen-activated protein kinase (MAPK) pathway and the phosphatidylinositol 3-kinase- (PI3K-) protein kinase B (AKT) pathway. For many years, elevated expression of EGFR was suggested to be a prognostic indicator across a range of malignancies including head and neck, ovarian, lung, pancreatic, and colorectal cancers [44]. Specifically in colorectal cancer (CRC), EGFR is expressed or elevated in 60 to 80% of cases [45] and is associated with poor clinical outcome [46, 47]. With this understanding, antibodies and small molecule inhibitors targeting EGFR were extensively investigated for treating CRC. Prior to targeted therapies, interventions for advanced CRC included fluorouracil, the topoisomerase inhibitor, irinotecan, and platinum-based chemotherapies.

In 2004, the FDA approved the human-mouse chimeric IgG_1 EGFR antibody, cetuximab, for treating advanced metastatic colorectal cancer (mCRC). Cetuximab blocks natural ligands from activating EGFR by competitively binding its external receptor domain. In the pivotal Phase II trial, single-agent cetuximab was tested against combination cetuximab with irinotecan in 329 EGFR-expressing mCRC patients failing first-line irinotecan chemotherapy [45]. The results demonstrated significant improvements in response rate and median time to progression, with a nonsignificant increase in median survival for the combination arm over cetuximab monotherapy. A subsequent study in a similar patient population reported a significant overall survival and progression-free survival benefit in patients treated with cetuximab alone compared with the best supportive care [48]. Additionally, two larger randomized trials, the Phase II Oxaliplatin and Cetuximab in First-Line Treatment of mCRC (OPUS) trial [49] and Phase III Cetuximab Combined with Irinotecan in First-Line Therapy for Metastatic Colorectal Cancer (CRYSTAL) trial [50], reproduced the synergistic effects of cetuximab and chemotherapy, showing reduced risk of mCRC progression compared to chemotherapy alone.

Concurrent investigation of several other EGFR-targeting therapies was underway during cetuximab development. These included panitumumab, a high-affinity, fully human monoclonal IgG$_2$ antibody that, like cetuximab, competes with natural ligands to block EGFR activation. In 2006, this treatment was approved by the FDA for use in EGFR-expressing chemorefractory mCRC.

Importantly, mutations in *KRAS*, which occur in about 40% of CRC patients, were recognized as key prognostic markers in CRC [51, 52]. In addition, *KRAS* mutations were found to be predictive of cetuximab resistance, consistent with tumor escape from EGFR regulation [51, 53]. This key role of *KRAS* was further validated in the CRYSTAL and OPUS trials, where anti-EGFR therapeutic benefit was restricted to patients with wild-type *KRAS* tumors [54, 55]. As these mutations are contraindicated for anti-EGFR antibody therapies, testing for *KRAS* mutations prior to initiating cetuximab or panitumumab therapy is now part of the mCRC treatment guidelines set forth by ASCO and the National Comprehensive Cancer Network (NCCN) [54]. These *KRAS* findings prompted the FDA to require updated labels for these treatments that specifically exclude patients not likely to benefit.

Non—Small Cell Lung Cancer: Gefitinib, Erlotinib, and Crizotinib

An estimated 221,130 new cases of lung cancer and 156,940 deaths attributable to the disease were expected for 2011 [2]. Accounting for about 80% of cases, non—small cell lung cancer (NSCLC) is the most common type of lung cancer and includes adenocarcinoma, squamous cell carcinoma, and large cell carcinoma. Development of several targeted therapies to treat NSCLC has included inhibitors of the EGFR and fusion product of two genes, echinoderm microtubule-associated protein-like 4 (*EML4*) and anaplastic lymphoma kinase (*ALK*). Just as in CRC, aberrant EGFR activation can drive NSCLC tumor growth and has been associated with more aggressive tumors and chemotherapy resistance. Two tyrosine kinase inhibitors targeting EGFR include erlotinib and gefitinib.

In a Phase III study, gefitinib improved disease symptoms and induced radiographic tumor regressions but was not significantly associated with overall survival benefit [56]. In 2005, notwithstanding a small cohort of responders, the overall limited efficacy led the FDA to restrict approval for gefitinib to only patients who previously benefited from the drug. On the other hand, erlotinib demonstrated response in about 9% of patients failing first- or second-line chemotherapy, with small but significant improvements in both progression-free survival and overall survival compared to placebo [57]. Thus, erlotinib first received approval as a single agent for treatment of locally advanced and metastatic NSCLC in 2004 [58].

A landmark follow-up to these early studies was an evaluation of tumors and matched normal samples from patients who did respond to gefitinib. Sequencing the EGFR gene from tumors of study subjects revealed somatic mutations in eight of nine gefitinib responders, none of which were present in the nonresponding patients [59]. Subsequent *in vitro* characterization demonstrated that these EGFR mutations conferred increased tyrosine kinase activity but additionally enhanced sensitivity to gefitinib. Importantly, later studies showed the incidence of these NSCLC EGFR mutations depends on ethnicity and gender, with higher rates in Asian populations (32%) over non-Asians (7%) and in women (38%) versus men (10%) [60]. When selecting patients with these mutations across six independent clinical studies, gefitinib and erlotinib monotherapies demonstrated striking tumor response rates from 55 to 84% compared to 11 to 47% with standard cytotoxic chemotherapy, along with significantly longer progression-free survival [61]. Thus, the earlier retrospective analyses had indeed identified critical predictive markers for gefitinib and erlotinib treatments.

Another recent success story of pharmacogenomics is crizotinib, which targets the *EML4-ALK* gene fusion product for the treatment of NSCLC. Resulting from a small inversion on chromosome 2p, this gene fusion encodes a constitutively active tyrosine kinase and is found in about 4 to 5% of NSCLC tumors [62, 63]. In a Phase I trial of 82 patients with NSCLC tumors harboring ALK-rearrangements, 57% of crizotinib-treated subjects responded with partial or complete responses compared to only 10% with chemotherapy alone [64]. A critical part of this study was the availability of FISH and RT-PCR biomarker assays needed to screen NSCLC patients for *EML4-ALK* expression. Remarkably, only three years elapsed from the initial discovery of the *EML4-ALK* fusion in 2007 to Phase III trial recruitment. In August 2011, crizotinib won FDA approval for the treatment of locally advanced or metastatic late-stage NSCLC. As additional crizotinib clinical trials proceed, other investigations of ALK inhibitor resistance have identified a number of specific mutations that may contribute to treatment failure [65, 66]. The rapid development of crizotinib was a remarkable achievement and represents the streamlined translation of early pharmacogenomics insights into a tangible therapeutic agent for patients.

Gaps in our knowledge of cancer drivers and acquired resistance mechanisms have historically hampered drug target identification and widespread biomarker adoption in oncology. In contrast to CML, which is primarily driven by consistent chromosomal damage, most cancers originate and progress via a unique collection of genetic triggers despite superficial similarities across patients. Even within the same patient, the molecular features of cancer often change during progression, and between primary and secondary tumor sites. If overlooked, this heterogeneity presents a substantial challenge to developing universally effective oncology therapies. However, disease subtypes sharing common drivers have been revealed at the molecular level and can often be represented by distinct protein, transcriptional, or genetic markers. Consequently, the identification of such subtypes can drive development of more effective focused therapies and enable patient stratification with biomarker assays. With the groundwork of basic cancer research laid over the past century, the last decade has produced an explosion of oncology investigations ignited by a deeper understanding of cancer pathways and access to high-throughput molecular assays. Together these advances have produced important insights about key mechanisms of cancer development and molecular subclasses of cancer.

ONCOLOGY IN THE POSTGENOMIC ERA

In 2001, the Human Genome Sequencing Consortium marked a major biological milestone with the release of the draft sequence of the human genome. Over a decade later, this work continues to shape our biological understanding of normal and disease processes, with long-term implications for public health that cannot be overstated. Of course the raw genome sequence itself was not useful without precise accompanying annotations. Thus, work over the last decade focused on filling these gaps, consequently producing a complete list of genes, genomic structure, and initial descriptions of functional roles, regulatory interactions, and disease-causing variants. Although much of the aforementioned therapeutic research predated the draft human genome release, genomic information is now inextricably embedded within contemporary drug development projects and is a central component of pharmacogenomics.

Prior to 2001, initial data from the Human Genome Project motivated the concomitant development of DNA microarray technology. These are grids of cDNA or oligonucleotide probes immobilized on a glass slide or silicon substrate. Washing labeled samples over the array allows hybridization to complementary probes, which provides a readout of gene expression based on the emitted fluorescence intensity and grid position. Leveraging the nascent catalogue of human genes and sequences, microarrays provided snapshots of genome-wide transcriptional activity. For over a decade, microarray assays have been widely used and have produced fundamental insights for many areas of biological research. For oncology, in particular, these tools allowed deeper tumor characterization and identification of distinct molecular subtypes based on gene expression patterns. In addition, this technology inspired the design of other high-throughput genomic assays including single-nucleotide polymorphism (SNP) microarrays for surveying genome-wide genetic variation.

Molecular Characterization of Leukemia and Non-Hodgkin's Lymphoma

One early example of microarray utility in oncology was for molecular classification of acute myeloid leukemia (AML) and acute lymphoblastic leukemia (ALL) samples [67]. In this study, a 50-gene classifier was identified from expression data and applied to classify 24 independent AML and ALL samples with 100% accuracy. Hematopathological means for leukemia classification were already well established, but the authors noted errors that could lead to incorrect treatment decisions with serious consequences for the patient. Thus, these results suggested a potential role for an unbiased expression-based differential diagnostic test in the clinical setting.

A year later, another study demonstrated the gene expression-based subclassification of diffuse large B-cell lymphoma (DLBCL) samples [68]. This aggressive malignancy makes up about 40% of non-Hodgkin's lymphoma cases and is characterized by variable clinical responses to chemotherapy. The study aimed to identify distinct molecular classes of the disease using gene expression data, in part because morphological DLBCL classification was not considered reproducible. Using custom microarrays with nearly 18,000 cDNA clones, 96 normal and cancer lymphocytes from B-cell malignancies were profiled. Through expression analysis, two divergent classes of DLBCL were identified: one with expression indicative of germinal center B-cells (GC B-like) and the other more similar to peripheral blood or activated B-cell expression (ABC). Further analysis revealed that following chemotherapy treatment, DLBCL patients with cancers expressing GC B-like patterns had a significant overall survival benefit over the ABC population. Moreover, low-risk patients, as defined by the International Prognostic Indicator [69], revealed similar stratification by gene expression, with significant overall survival benefit in GC B-like patients. Since this study was published, a third DLBCL molecular subtype has been identified that is primary mediastinal B-cell lymphoma, with strong JAK2 up-regulation and frequent 9p24 amplifications [70].

Together, these early applications of gene expression analysis highlight the potential of microarrays to help define diagnostic and prognostic expression signatures. Since the first microarray studies, oncology has celebrated similar success in describing molecular subtypes of other cancers. However, the challenge remains to translate these signatures into reproducible and validated prognostic or predictive tests.

Breast Cancer Prognostic Signatures

Microarray characterization of breast cancer subtypes was first reported over a decade ago. The earliest studies defined a set of "intrinsic" gene expression signatures that segregated tumors into one of four classes, with the luminal type primarily describing ER-positive tumors and the three remaining classes dominated by ER-negative tumors including basal-like, Her2+, and normal-like subtypes [71, 72]. The luminal and basal classes were aptly named to reflect similarity to normal luminal epithelial and basal breast cells. Subsequent studies revealed that the luminal type could further be subdivided into the luminal A and luminal B subtypes, with luminal A associated with a more favorable prognosis over the other classes, including luminal B [73, 74] (Figure 4.5).

Following these initial studies, other reports explored the association of breast cancer expression patterns to clinical outcomes. One study developed a 70-gene prognostic signature to identify lymph node—negative breast cancer patients with a higher five-year risk of distant metastases [75]. Moreover, the approach identified patients most likely to benefit from adjuvant therapy and those with little to gain from such treatment. Interestingly, some of the genes defining this signature represent broad classes, including ER-response, HER2-response, and proliferation. In an independent validation study of 295 breast cancer patients, the classifier predicted subpopulations with good or poor prognosis and established a significant difference in 10-year survival rates between the groups, even after accounting for lymph node status, a traditional prognostic factor [76].

This work was the basis for the first FDA-cleared *in vitro* diagnostic multivariate index assay (IVDMIA), MammaPrint®. This microarray-based test determines the activity of the 70-gene signature in fresh frozen tumor tissue sections from lymph node—negative breast cancers. The output is a dichotomous result indicating either low or high risk of 5- or 10-year recurrence without adjuvant hormonal therapy or chemotherapy [77]. Further validation of the test is underway in the prospective Phase III Microarray In Node Negative and 1-3 Positive Lymph Node Disease May Avoid Chemotherapy (MINDACT) trial [78]. One primary objective of this trial is to determine whether low-risk patients as identified by MammaPrint® can avoid unnecessary chemotherapy treatment without increased risk of distant metastases.

Another commercially available prognostic test for early-stage breast cancer is the Oncotype DX® quantitative RT-PCR assay. In addition to measuring ER-associated and proliferation-related genes, as in the MammaPrint® assay, this test surveys other factors associated with disease aggressiveness, including invasion components of breast cancer. Based on a 21-gene expression profile, the Oncotype DX® algorithm outputs a weighted score that reportedly predicts 10-year disease recurrence and chemotherapy benefit in ER-positive,

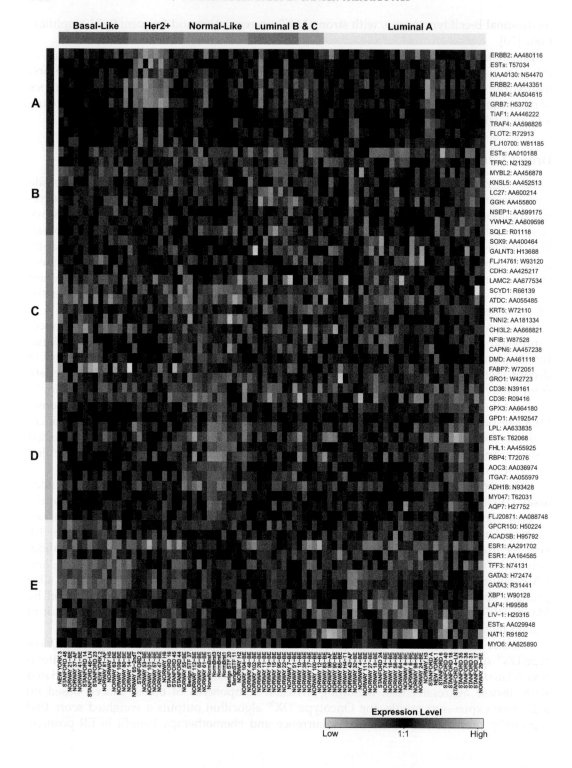

node-negative breast cancer patients [79, 80]. As described shortly, the test produces a recurrence score (RS) by combining expression levels of 16 genes across distinct functional groups normalized against a reference gene panel (*ACTB, GAPDH, RPLPO, GUS,* and *TFRC*) [79].

RS = + 0.47 × HER-2 group (*GRB7* and *HER2*) score
−0.34 × ER group (*ER, PR, Bcl2,* and *SCUBE2*) score
+ 1.04 × proliferation group (*KI67, STK15,* Survivin, *CCNB1,* and *MYBL2*) score
+ 0.10 × invasion group (*CTSL2 and MMP11*) score
+ 0.05 × *CD68*
−0.08 × *GSTM1*
−0.07 × *BAG1*

Within each gene group, individual gene expression levels are adjusted and the weighted composite group scores are combined. The sign of each weight indicates a group's contribution to the overall risk of recurrence, with negative coefficients producing lower recurrence scores and positive coefficients conferring greater risk of recurrence. Consequently, high expression of hormonal response genes in the ER group translates to a lower recurrence risk, and conversely, expression of groups representing proliferation and invasion induce a higher recurrence risk. Once scaled between 0 and 100, the recurrence scores are interpreted as low (<18), intermediate (≥ 18 and <31), and high (≥ 31) risks of recurrence [79].

To assess the prognostic value of the RS, a prospective validation study tested 668 archived paraffin-embedded tissue blocks from tamoxifen-treated ER-positive, lymph node−negative invasive breast cancer patients. Comparing patient tumors generating low and high recurrence scores, the study revealed a significant difference in 10-year distant recurrence rates of 6.8% and 30.5%, respectively ($P < 0.001$) [79]. A follow-up investigation demonstrated that the RS could predict adjuvant chemotherapy benefit in tamoxifen-treated ER-positive, lymph node−negative breast cancer. Patients with low-RS tumors received little benefit, in striking contrast to patients with high-RS tumors [80]. Further validation of Oncotype DX® test is ongoing in the 10-year prospective Phase III Trial Assigning Individualized Options for Treatment (TAILORx) [81]. This study, which has enrolled more than 10,000 women and expects results in 2015, aims to identify chemotherapy benefit in lymph node−negative breast cancer patients that have midrange (11−25) recurrence scores.

◄───────────────────────────────────

FIGURE 4.5   Heat map of breast cancer gene expression data. Heat map illustrating intrinsic breast cancer gene signatures identified by Sørlie et al. by profiling 78 breast tumor samples, 3 benign tumor samples, and 4 normal breast tissue samples [74]. Each row in the heat map represents a gene and is annotated with the Entrez Gene gene symbol, when available, and Genbank accession number. Each sample is depicted as a column and annotated with a unique sample identifier. Gene expression levels are depicted as a color gradient relative to the median expression ratio per row. Based on distinct expression signatures, breast tumors can be grouped into five broad classes: Basal-like, Her2+, Normal-like, or one of the Luminal subtypes (typically now reported as either "Luminal A" or "Luminal B"). These expression patterns have been reproduced across several independent studies and reflect unique drivers of tumor growth [71, 74, 111, 112]. The genes shown are representative of a larger set of 456 intrinsic genes previously reported and are arranged by distinct functional groups noted by labels A-E. Note two considerable divisions in the data derive from expression of ER-related genes (gene group E) and Her2-related genes (gene group A). This figure was created in the R software package from original data deposited in the Stanford Microarray Database [113] and gene lists adapted from Sørlie et al. [74]. (See color plate 6.)

Both the MammaPrint® and Oncotype DX® assays demonstrate the prognostic and predictive utility of multigene assays and reinforce the association of molecular features of tumors and patient outcome. Other assays that leverage unique breast cancer expression patterns have been described (Table 4.1). These include the Breast Cancer Index^{SM} (bio-Theranostics), which combines a five-gene molecular grade index with a ratio of the *HOXB13* and *IL17BR* genes [82-84]. This prognostic test determines 10-year risk of distant recurrence in lymph node—negative, ER-positive breast cancer patients receiving adjuvant tamoxifen. In addition, the Insight™ Dx Breast Cancer Profile (Clarient) assay, though not a gene expression—based prognostic test, combines staging factors such as tumor size, grade, and lymph node status with seven molecular markers including ER, PR, HER2, and EGFR [85]. A risk score is generated using a nonlinear algorithm that reflects high and low risk of 10-year distant metastasis in early-stage hormone-receptor-positive breast cancer patients receiving hormone therapy. Finally, the microarray-based 76-gene "Rotterdam signature" (Veridex, Raritan, NJ) reportedly predicts risk of distant metastases at 5 and 10 years following diagnosis, but the assay is not commercially available [86-88].

TABLE 4.1 Multimarker prognostic molecular assays for breast cancer

Assay Name	Description	Clinical Use
MammaPrint® (also referred to as the "Amsterdam signature")	70-gene microarray assay for fresh or frozen tissue from node negative, ER-negative, or positive breast cancer patients, with tumors of less than 5 cm.	Prognostic tool to determine 5- or 10-year recurrence risk depending on the patient's age.
Oncotype™ DX	21-gene RT-PCR assay for Formalin-fixed, Paraffin-embedded (FFPE) tissue from node-negative, ER-positive breast cancer patients. Weighted expression-based recurrence score (RS) correlates with the recurrence of breast cancer.	Decision-making tool to select nonmetastatic breast cancer patients for adjuvant chemotherapy. Prognostic tool for 10-year recurrence risk.
Breast Cancer Index^{SM}	Simultaneous assessment 5-gene bioTheranostics Molecular Grade Index (MGI^{SM}) and 2-gene (HOXB13:IL17BR) ratio bioTheranostics H/I^{SM} in node-negative, ER-positive breast cancer patients receiving adjuvant tamoxifen therapy.	Prognostic tool to determine 10-year risk of distant recurrence.
Insight™ Dx Breast Cancer Profile	Combines genomic and proteomic markers with clinicopathologic features in early-stage, hormone receptor-positive breast cancer patients receiving hormone therapy.	Prognostic tool with nonlinear algorithm that computes a risk score for high or low risk of 10-year distant metastasis.
76-gene "Rotterdam signature" assay	76-gene microarray assay (using Affymetrix U133a GeneChip) for fresh or frozen tissue from node-negative patients independent of hormone receptor status.	Not commercially available.

Microarray- or IHC-based prognostic assays for breast cancer reported in the literature.

Although such prognostic tests can identify patients at high risk of recurrence, these are not regarded as true predictive assays. Thus, modifying treatment decisions based on the test outcome is quite limited. In contrast to other protein- or genotype-based predictive tests that indicate sensitivity or resistance to a drug, no IVDMIA predictive test that uses pharmacogenomics has yet been approved by the FDA in conjunction with an oncology therapeutic.

Characterization of Somatic Mutations in Cancer

Conventional cytogenetics has contributed greatly to our understanding of cancer-related chromosomal aberrations. For several decades, standard molecular techniques provided detection of chromosomal translocations, copy number variation, and SNPs. However, the availability of the human genome sequence and expression arrays transformed these early assays into efficient high-throughput techniques with unprecedented resolution for studying genome-wide variation. The introduction of tiling arrays with millions of probes, each representing a short overlapping sequence along the genome, suddenly allowed genome-wide detection of copy number changes and chromosomal breakpoints. Similarly, the availability of SNP chips permitted simultaneous survey of 500,000 to 1 million SNPs on a single microarray. Just as gene expression microarrays revolutionized studies of transcriptional alterations, these technologies provided snapshots of the genetic variation within samples.

A significant advance in genomics has been the recent rebirth of DNA sequencing technology. Standard Sanger sequencing has long been a staple of molecular biology laboratories and, through incremental refinements, has consistently produced robust sequencing results for decades. Nevertheless, the cost and speed of this method is prohibitive for large sequencing projects. Thus, the demand for inexpensive alternatives ignited a race among competing technologies that could accommodate genome-scale projects. Today, nearly a dozen commercially available instruments or services are collectively referred to as next-generation sequencing (NGS). Although the underlying methodologies differ considerably among the technologies, a common theme distinct from Sanger sequencing is their massively parallel nature. Each experiment can produce millions of short sequence reads, which are effectively binary readouts of small genomic regions from each DNA sample. Computational tools are required for assembly of the reads into a contiguous consensus sequence, and algorithms are used to detect variants, rearrangements, and copy number variation. These technologies have matured over the past few years and dramatically reduce the time and expense to sequence large genomic intervals, including protein-coding exomes and entire genomes.

Increased access to NGS has produced noteworthy findings outside of oncology. A number of studies identified causal mutations of some rare Mendelian disorders following genomic sequencing of afflicted individuals [89-92]. Cancer, on the other hand, presents substantially greater challenges for interpreting NGS results. Because of our limited understanding of individual germline variation, interpreting a cancer genome requires the corresponding noncancer reference genome from the same individual to differentiate relevant somatic variation. Moreover, tumor samples are often a heterogeneous mixture of different cell types with variable fractions of malignant cells and distinct clones. Such variability across samples can confound identification of driver mutations or structural variants common to a sample set.

Nevertheless, comparative analysis of cancer genomes provides a unique perspective of cancer's genetic underpinnings that has the potential to profoundly affect patient care in years to come. Until recently, cytogenetic aberrations were primarily described in terms of known or suspected mutations and structural variations. With NGS, the entire genome is now completely accessible. The first full cancer genome was reported in a comparison of primary AML cancer cells to normal skin cells [93]. This study identified over 63,000 tumor-specific single-nucleotide variants (SNVs) and 726 small indels. Among these, eight genes were identified with novel nonsynonymous somatic mutations along with two genes containing previously described somatic insertions in AML. Since these initial studies, other cancer genomes and exomes have been reported, including 24 breast cancer genomes [94], seven prostate cancer genomes [95], and two lung cancer genomes [96]. In a recent study of intratumor heterogeneity, renal carcinoma samples from the same tumor were sequenced, and phylogenetic reconstruction allowed inference of an acquired mutation timeline across the tumor [97]. Interestingly, mutations identified from samples of the same tumor varied considerably, with up to 69% of somatic mutations undetectable across all sample sites.

Although initial NGS studies have identified somatic variants that frequently occur within specific tumor types, realizing the full potential of whole-genome NGS will require thousands of cancer genome sequences and integration with other data types. Identification of every major cancer-causing variant is the aim of the Cancer Genome Atlas project (cancergenome.nih.gov) [98]. By combining exome sequencing, gene expression profiling, SNP genotyping, copy number analysis, and methylation analysis, this multidisciplinary effort has proposed to characterize hundreds of diverse tumors samples. The first report from this project analyzed 206 glioblastoma multiforme samples and included an integrated analysis of gene expression, DNA copy number, methylation, and exome sequencing for a sample subset [98]. Most recently, a similar detailed characterization of ovarian adenocarcinomas that included exome sequencing for 316 samples was released [99].

Teasing common threads from fully characterized cancer genomes promises to advance oncology research and translational medicine. As cancer is fundamentally a genetic disease, these technologies are poised to highlight molecular similarities between malignancies that may otherwise go unnoticed. Mutations in *BRAF* commonly found in melanoma or translocations such as the Philadelphia chromosome in CML are excellent examples of cancers with distinct genetic etiologies and equally unique treatment strategies. Identification of such defining characteristics in other cancers assures continued pharmacogenomics investments and motivates development of new targeted interventions.

CHALLENGES IN DRUG DEVELOPMENT AND CANCER TRIALS

Universally applicable "blockbuster" drug development is increasingly being supplanted by more modest and focused strategies. In a relatively short period, pharmacogenomics has transformed the drug development process by both defining patient subpopulations and identifying key predictive markers of successful treatment. Application of these principles has been fruitful for a number of therapies; however, challenges remain in incorporating pharmacogenomics into clinical development. These include technical hurdles regarding diagnostic test development along with logistics and costs associated with integrating exploratory biomarker endpoints into clinical trial plans [100]. Moreover, regulatory guidance is limited for companion diagnostic assays that employ new genomics technologies. Without compelling evidence of robust patient benefit, the additional burden of developing personalized medicines and companion diagnostics may be prohibitive.

Technical Challenges

There are a number of technical challenges with the development of personalized medicines and requisite companion diagnostics. Some are inherent in the technologies used for patient stratification; others reflect the overall difficulty in targeting such a complex disease. To translate research findings into clinically testable hypotheses, robust biomarkers must be identified, along with subsequent development of reproducible, rapid, and convenient biomarker assays. These three assay features are critical for companion diagnostic adoption, as they ensure for caregivers a reliable method with timely impact on treatment decisions.

One problem with reproducibility stems from tumor sampling. Tumors are generally heterogeneous, and biopsies from different regions may contain varying proportions of component cells. Thus, the sample content may differ considerably within and between patients and prevent identification of a meaningful biomarker, which can be especially challenging when genomics technologies, such as gene expression or genotyping analysis, are in use where measurable levels are directly impacted by the tumor content. Moreover, results from assays requiring a pathologist's assessment can vary between experts and complicate the definition of a positive or negative test result.

Another source of variability in biomarker identification is the assay itself, as is particularly evident in whole-genome gene expression studies, where artifacts from sample handling are common and sample sizes are generally very small relative to the number of array features tested. Moreover, in some published studies, models derived from expression analyses may overfit the data, causing exaggeration of the predictive value and clinical relevance of the resulting gene signature. One must be particularly cautious when interpreting reports in which discovery and validation data sets are not independently evaluated [101]. Although these methods have revolutionized pharmacogenomics research, independent prospective validation of retrospectively identified biomarkers is imperative.

In addition, such high-dimensional data requires correction for multiple testing when isolating differentially expressed genes or associating SNP variants with a particular phenotype. When utilizing genomic technologies in a trial, achieving sufficient statistical power for class discrimination can be difficult without very large studies [102, 103]. Thus, these assays tend to be exploratory to help generate hypotheses that can inform the design of later studies. Once a well-defined predictive biomarker is characterized, smaller and faster clinical validation studies may result.

A final technical challenge that affects pharmacogenomics research is the volume of data generated by current high-throughput assays. In recent years, NGS and other assays have outpaced conventional methods for data storage and analysis. As an example, a single human genome sequence, with mapped reads, quality scores, and associated genomic assembly, can easily consume up to a terabyte of disk space. Analysis of datasets on this scale requires investments in computational infrastructure and informatics expertise to accommodate these advances. Just a few years ago, parallel processing with multicore CPUs, graphical processing units, and computational clusters were the principal solutions to handling large data sets. Today, these tools are supplemented by on-demand cloud computing and storage services that rent nearly boundless processing time and data storage over the Internet. Integration of information technology and informatics is essential for extracting the greatest value from these experiments.

Logistical and Regulatory Challenges

One of the primary logistical challenges for therapeutic biomarker development is timing. Unless pharmacogenomics approaches are incorporated early in a drug's development, the effort may not realize returns. Balancing retrospective identification of markers with prospective study timelines can be challenging. Moreover, incorporating exploratory biomarker assays may be perceived to some as a trial hindrance, especially when the primary endpoints are of paramount concern to investigators. One approach to identify predictive biomarkers and prospectively test the markers within the same clinical study is the use of an adaptive clinical trial design [104]. In this design, a classifier is developed to identify subjects most likely to respond in an initial stage of the study. At the conclusion of the trial, the drug efficacy is tested in the entire randomized study population as well as in the subset of randomized patients predicted to respond but not included in the first stage. This adaptive approach can identify significant treatment effects in cases where response is primarily limited to the biomarker-selected population.

Importantly, regulatory requirements for companion diagnostic assays are rigorous, and failure to sufficiently demonstrate the utility of such assays in a prospective study can jeopardize the regulatory filing for a targeted drug. Such was the case in a recent FDA review of omacetaxine mepesuccinate, a drug for imatinib-resistant chronic myeloid leukemia. During this review, the FDA withheld approval for the drug, citing an inadequate assay to identify the target patient population. Thus, approval of a drug targeting a specific patient subpopulation requires the concomitant development and approval of the cognate biomarker assay. Development of these assays as an afterthought or delay in this process can prevent an otherwise efficacious drug from reaching the market in a timely fashion.

Another challenge in incorporating multimarker assays into clinical trials is the limited regulatory guidance for such approaches. In 2007, the FDA released a draft guidance document outlining provisions for IVDMIA [105]. However, no additional provisions have been released, and the guidance is still considered to be in a draft state. Thus, pharmaceutical companies acting as early adopters of these technologies could be unexpectedly penalized during a drug filing in the absence of sufficiently clear expectations.

A final issue pertains to payer coverage and reimbursement for companion diagnostics and prognostic tests. In the United States, every procedure performed by a medical professional requires a Current Procedural Terminology (CPT) code that allows uniform payment from insurers for providing these services. One difficulty is that payers may consider some biomarker assays investigational and not medically necessary. Thus, it is incumbent upon the practicing medical community to evaluate the claims of these assays and incorporate the best approaches into treatment guidelines. Without adoption of biomarker tests into guidelines or sufficient mechanisms for reimbursement, targeted therapeutics with corresponding biomarker assays may be prohibitively expensive for patients to bear out of pocket.

SEEKING PHARMACOGENOMIC VALUE

Patients

All patients are the decisive beneficiaries of targeted therapies and pharmacogenomics research. In years past, particular cancer diagnoses were considered terminal, but because of improvements over the past 20 years in early detection and personalized treatments, some of these same malignancies are now treatable. In addition, the arsenal to fight cancer has grown beyond cytotoxics to include better-tolerated treatments with fewer severe side effects. Pharmacogenomics has helped define patient populations who are better suited to respond to a given therapy. Thus, treating physicians can bypass treatment options that are likely to show little benefit or actually worsen a patient's condition.

One common challenge in cancer management is the propensity of tumors to acquire resistance to a therapy. Initially recognized with chemotherapy treatments, this problem similarly extends to targeted therapeutics. Moreover, relapse of a dormant cancer often results in more aggressive tumors with fewer treatment options. A standard strategy for patient care in these scenarios is to vary or combine treatments as a tumor becomes unresponsive. With pharmacogenomics, a more appropriate therapeutic may be selected to combat the new resistance mutations. In many cases, however, an effective follow-up therapy is not available, and the refractory disease represents an unmet medical need. Understanding the escape mechanisms of a cancer presents an opportunity to fill this niche with a novel targeted treatment. Thus, a successive cycle of pharmacogenomics can further dissect the disease to identify biomarkers of resistance and suggest additional targets that motivate development of complementary therapies.

Healthcare Providers and Payers

Biomarkers allow unbiased assessment of a patient's disease and can assist caregivers by facilitating diagnosis and guiding treatment. Additionally, pharmacogenomics research can improve treatment guidelines and lead to a more consistent adoption of proven therapeutics. As a consequence, unmet medical needs are more clearly defined as diseases are stratified into distinct subtypes. For example, so-called triple-negative (ER^-/PR^-/$HER2^-$) breast cancer expresses none of the histopathological markers predictive of effective targeted treatment and represents a distinct breast cancer subtype with few treatment options beyond chemotherapy. This trend will continue as new technologies, such as NGS, uncover the full spectrum of cancer mutations.

As of 2008, nearly 25% of all prescriptions were for drugs with pharmacogenomics labeling [106]. Currently, there are 13 FDA-approved oncology treatments with biomarkers specified in the Indications and Usage section of the drug label (Table 4.2). From the payers' perspective, incorporating biomarkers presents an opportunity to maximize treatment efficiency and reduce waste [107].

TABLE 4.2 Pharmacogenomic biomarkers in oncology drug labels

Drug Name	Approved Oncology Indications	First FDA Approval	Biomarker
Brentuximab vedotin	Hodgkin's lymphoma; anaplastic large cell lymphoma	August 2011	CD30
Cetuximab	metastatic colorectal cancer; squamous cell carcinoma of the head and neck cancer	February 2004	EGFR; KRAS
Crizotinib	non–small cell lung carcinoma	August 2011	ALK
Dasatinib	chronic myelogenous leukemia; acute lymphoblastic leukemia	June 2006	Philadelphia chromosome
Fulvestrant	metastatic breast cancer	April 2002	Estrogen Receptor
Imatinib	chronic myelogenous leukemia; gastrointestinal stromal tumors; acute lymphoblastic leukemia; chronic eosinophilic leukemia; aggressive systemic mastocytosis; myelodysplastic/ myeloproliferative diseases	May 2001	Philadelphia chromosome; c-Kit; PDGFR; FIP1L1-PDGFRα
Lapatinib	breast cancer	March 2007	HER2/neu
Nilotinib	drug-resistant chronic myelogenous leukemia	October 2007	Philadelphia chromosome
Panitumumab	metastatic colorectal cancer	September 2006	EGFR; KRAS
Tamoxifen	breast cancer	December 1977	Estrogen Receptor
Tositumomab	non-Hodgkin's lymphoma	June 2003	CD20 antigen
Trastuzumab	breast cancer	September 1998	HER2/neu
Vemurafenib	melanoma	June 2011	BRAF (V600E mutation)

FDA-approved oncology drugs and related biomarkers that are included in the drug labels.

Pharmaceutical Industry

With notoriously low success rates in bringing new drugs to market, the pharmaceutical industry has embraced strategies that increase productivity and decrease pipeline attrition. Although development costs are not typically published by the industry, estimates top $883 million for a single drug [108]. Reducing time to market with more focused drug applications has the potential to cut overall development costs and effectively extend intellectual property protection for a therapeutic [109]. Thus, targeted therapies provide a clear economic incentive by enriching clinical populations with expected responders.

The Phase III trastuzumab trial demonstrated the value of limiting the treated population to a subgroup overexpressing HER2. With the biomarker, only 470 patients were required for enrollment in the pivotal trial as opposed to an estimated 2,200 that would have otherwise been needed [100]. Moreover, the trial could have required up to 10 years to complete, as opposed to the actual 1.6 years needed to show benefit [110]. Although the target market for such drugs may be smaller by comparison, product differentiation and the greater likelihood of treatment success can contribute to a net gain for investment in such approaches. Likewise, companion diagnostics can provide additional revenue associated with a product and potentially offset the added expense of bringing the test to market.

Taken together, pharmacogenomics presents a number of key challenges but offers substantially compelling opportunities. Costs associated with additional pharmacogenomics development can be reclaimed by faster targeted drug development, fewer patients required for clinical trials, and marketing of companion diagnostics. The technologies that enable greater understanding of the complexities of cancer are just being harnessed. In decades to come, one can anticipate a myriad of personalized cancer treatments based on pharmacogenomics principles.

QUESTIONS FOR DISCUSSION

1. Differentiate between prognostic versus predictive biomarkers.
2. Discuss, with specific examples, the principles of targeted cancer therapy.
3. Explain the challenges associated with targeted anticancer drug development.

References

[1] Jemal A, Bray F, Center MM, Ferlay J, Ward E, Forman D. Global cancer statistics. CA Cancer J Clin 2011;61(2):69—90.
[2] Siegel R, Ward E, Brawley O, Jemal A. Cancer statistics, 2011: The impact of eliminating socioeconomic and racial disparities on premature cancer deaths. CA Cancer J Clin 2011;61(4):212—36.
[3] Gerber DE. Targeted therapies: a new generation of cancer treatments. Am Fam Physician 2008;77(3):311—9.
[4] American Urological Association. Prostate-specific antigen (PSA) best practice policy. American Urological Association (AUA). Oncology 2000;14(2):267—72, 277—8, 280 passim.
[5] Smith DS, Humphrey PA, Catalona WJ. The early detection of prostate carcinoma with prostate specific antigen: the Washington University experience. Cancer 1997;80(9):1852—6.
[6] Thompson IM, Pauler DK, Goodman PJ, Tangen CM, Lucia MS, Parnes HL, et al. Prevalence of prostate cancer among men with a prostate-specific antigen level < or =4.0 ng per milliliter. N Engl J Med 2004;350(22):2239—46.
[7] Andriole GL, Crawford ED, Grubb 3rd RL, Buys SS, Chia D, Church TR, et al. Mortality results from a randomized prostate-cancer screening trial. N Engl J Med 2009;360(13):1310—9.
[8] Schroder FH, Hugosson J, Roobol MJ, Tammela TL, Ciatto S, Nelen V, et al. Screening and prostate-cancer mortality in a randomized European study. N Engl J Med 2009;360(13):1320—8.
[9] Greene KL, Albertsen PC, Babaian RJ, Carter HB, Gann PH, Han M, et al. Prostate specific antigen best practice statement: 2009 update. J Urol 2009;182(5):2232—41.
[10] Miki Y, Swensen J, Shattuck-Eidens D, Futreal PA, Harshman K, Tavtigian S, et al. A strong candidate for the breast and ovarian cancer susceptibility gene BRCA1. Science 1994;266(5182):66—71.
[11] Wooster R, Bignell G, Lancaster J, Swift S, Seal S, Mangion J, et al. Identification of the breast cancer susceptibility gene BRCA2. Nature 1995;378(6559):789—92.
[12] Antoniou A, Pharoah PD, Narod S, Risch HA, Eyfjord JE, Hopper JL, et al. Average risks of breast and ovarian cancer associated with BRCA1 or BRCA2 mutations detected in case series unselected for family history: a combined analysis of 22 studies. Am J Hum Genet 2003;72(5):1117—30.
[13] Raguz S, Yague E. Resistance to chemotherapy: new treatments and novel insights into an old problem. Br J Cancer 2008;99(3):387—91.

[14] MacGregor JI, Jordan VC. Basic guide to the mechanisms of antiestrogen action. Pharmacol Rev 1998;50(2): 151–96.

[15] Lacassagne A. Apparition de cancers de la mamelle chez la souris mâle soumise á des injections de folliculine. Comptes rendus de l'Académie des Sciences 1932;195:630–2.

[16] Nadji M, Gomez-Fernandez C, Ganjei-Azar P, Morales AR. Immunohistochemistry of estrogen and progesterone receptors reconsidered: experience with 5,993 breast cancers. Am J Clin Pathol 2005;123(1): 21–7.

[17] Early Breast Cancer Trialists' Collaborative Group. Tamoxifen for early breast cancer: an overview of the randomised trials. Lancet 1998;351(9114):1451–67.

[18] Stearns V. Active tamoxifen metabolite plasma concentrations after coadministration of tamoxifen and the selective serotonin reuptake inhibitor paroxetine. J Natl Cancer Inst 2003;95(23):1758–64.

[19] Gaston C, Kolesar J. Clinical significance of CYP2D6 polymorphisms and tamoxifen in women with breast cancer. Clin Adv Hematol Oncol 2008;6(11):825–33.

[20] Hartman AR, Helft P. The ethics of CYP2D6 testing for patients considering tamoxifen. Breast Cancer Res 2007;9(2):103.

[21] Schroth W, Antoniadou L, Fritz P, Schwab M, Muerdter T, Zanger UM, et al. Breast cancer treatment outcome with adjuvant tamoxifen relative to patient CYP2D6 and CYP2C19 genotypes. J Clin Oncol 2007;25(33): 5187–93.

[22] Goetz MP, Knox SK, Suman VJ, Rae JM, Safgren SL, Ames MM, et al. The impact of cytochrome P450 2D6 metabolism in women receiving adjuvant tamoxifen. Breast Cancer Res Treat 2007;101(1):113–21.

[23] Goetz MP, Kamal A, Ames MM. Tamoxifen pharmacogenomics: the role of CYP2D6 as a predictor of drug response. Clin Pharmacol Ther 2008;83(1):160–6.

[24] Rae JM. Personalized tamoxifen: what is the best way forward? J Clin Oncol 2011;29(24):3206–8.

[25] Regan MM, Leyland-Jones B, Bouzyk M, Pagani O, Tang W, Kammler R, et al. CYP2D6 genotype and tamoxifen response in postmenopausal women with endocrine-responsive breast cancer: the breast international group 1-98 trial. J Natl Cancer Inst 2012;104(6):441–51.

[26] Rae JM, Drury S, Hayes DF, Stearns V, Thibert JN, B.HaynesSalter J, et al. CYP2D6 and UGT2B7 genotype and risk of recurrence in tamoxifen-treated breast cancer patients. J Natl Cancer Inst 2012;104(6):452–60.

[27] Golberg P. Experts claim errors in breast cancer study, demand retraction of practice-changing paper. Cancer Lett 2012;38(20):1–6.

[28] Heller T, Kirchheiner J, Armstrong VW, Luthe H, Tzvetkov M, Brockmoller J, et al. AmpliChip CYP450 GeneChip: a new gene chip that allows rapid and accurate CYP2D6 genotyping. Ther Drug Monit 2006;28(5):673–7.

[29] Yaziji H, Goldstein LC, Barry TS, Werling R, Hwang H, Ellis GK, et al. HER-2 testing in breast cancer using parallel tissue-based methods. J Am Med Assoc 2004;291(16):1972–7.

[30] Owens MA, Horten BC, Da Silva MM. HER2 amplification ratios by fluorescence in situ hybridization and correlation with immunohistochemistry in a cohort of 6556 breast cancer tissues. Clin Breast Cancer 2004;5(1):63–9.

[31] Slamon DJ, Clark GM, Wong SG, Levin WJ, Ullrich A, McGuire WL. Human breast cancer: correlation of relapse and survival with amplification of the HER-2/neu oncogene. Science 1987;235(4785):177–82.

[32] Goldenberg MM. Trastuzumab, a recombinant DNA-derived humanized monoclonal antibody, a novel agent for the treatment of metastatic breast cancer. Clin Ther 1999;21(2):309–18.

[33] Allison M. The HER2 testing conundrum. Nat Biotechnol 2010;28(2):117–9.

[34] Wolff AC, Hammond ME, Schwartz JN, Hagerty KL, Allred DC, Cote RJ, et al. American Society of Clinical Oncology/College of American Pathologists guideline recommendations for human epidermal growth factor receptor 2 testing in breast cancer. Arch Pathol Lab Med 2007;131(1):18–43.

[35] Hammond ME, Hayes DF, Wolff AC, Mangu PB, Temin S. American society of clinical oncology/college of american pathologists guideline recommendations for immunohistochemical testing of estrogen and progesterone receptors in breast cancer. J Oncol Pract 2010;6(4):195–7.

[36] van Meerten T, Hagenbeek A. CD20-targeted therapy: a breakthrough in the treatment of non-Hodgkin's lymphoma. Neth J Med 2009;67(7):251–9.

[37] Morel F, Ka C, Le Bris MJ, Herry A, Morice P, Bourquard P, et al. Deletion of the 5′ abl region in Philadelphia chromosome-positive chronic myeloid leukemia. Leukemia 2003;17(2):473–4.

[38] Witte ON, Goff S, Rosenberg N, Baltimore D. A transformation-defective mutant of Abelson murine leukemia virus lacks protein kinase activity. Proc Natl Acad Sci U S A 1980;77(8):4993—7.

[39] Witte ON, Dasgupta A, Baltimore D. Abelson murine leukaemia virus protein is phosphorylated in vitro to form phosphotyrosine. Nature 1980;283(5750):826—31.

[40] Nowell PC, Hungerford DA. A minute chromosome in human chronic granulocytic leukemia. Science 1960;132(3438):1497—501.

[41] Druker BJ, Tamura S, Buchdunger E, Ohno S, Segal GM, Fanning S, et al. Effects of a selective inhibitor of the Abl tyrosine kinase on the growth of Bcr-Abl positive cells. Nat Med 1996;2(5):561—6.

[42] Druker BJ, Talpaz M, Resta DJ, Peng B, Buchdunger E, Ford JM, et al. Efficacy and safety of a specific inhibitor of the BCR-ABL tyrosine kinase in chronic myeloid leukemia. N Engl J Med 2001;344(14):1031—7.

[43] Debiec-Rychter M, Sciot R, Le Cesne A, Schlemmer M, Hohenberger P, van Oosterom AT, et al. KIT mutations and dose selection for imatinib in patients with advanced gastrointestinal stromal tumours. Eur J Cancer 2006;42(8):1093—103.

[44] Nicholson RI, Gee JM, Harper ME. EGFR and cancer prognosis. Eur J Cancer 2001;37(Suppl 4):S9—15.

[45] Cunningham D, Humblet Y, Siena S, Khayat D, Bleiberg H, Santoro A, et al. Cetuximab monotherapy and cetuximab plus irinotecan in irinotecan-refractory metastatic colorectal cancer. N Engl J Med 2004;351(4): 337—45.

[46] Tabernero J, Salazar R, Casado E, Martinelli E, Gomez P, Baselga J. Targeted therapy in advanced colon cancer: the role of new therapies. Ann Oncol 2004;15(Suppl 4):55—62.

[47] Spano JP, Lagorce C, Atlan D, Milano G, Domont J, Benamouzig R, et al. Impact of EGFR expression on colorectal cancer patient prognosis and survival. Ann Oncol 2005;16(1):102—8.

[48] Jonker DJ, O'Callaghan CJ, Karapetis CS, Zalcberg JR, Tu D, Au HJ, et al. Cetuximab for the treatment of colorectal cancer. N Engl J Med 2007;357(20):2040—8.

[49] Bokemeyer C, Bondarenko I, Makhson A, Hartmann JT, Aparicio J, de Braud F, et al. Koralewski, Fluorouracil, leucovorin, and oxaliplatin with and without cetuximab in the first-line treatment of metastatic colorectal cancer. J Clin Oncol 2009;27(5):663—71.

[50] Van Cutsem E, Kohne CH, Hitre E, Zaluski J, Chang Chien CR, Makhson A, et al. Cetuximab and chemotherapy as initial treatment for metastatic colorectal cancer. N Engl J Med 2009;360(14):1408—17.

[51] Lievre A, Bachet JB, Le Corre D, Boige V, Landi B, Emile JF, et al. KRAS mutation status is predictive of response to cetuximab therapy in colorectal cancer. Cancer Res 2006;66(8):3992—5.

[52] Lievre A, Bachet JB, Boige V, Cayre A, Le Corre D, Buc E, et al. KRAS mutations as an independent prognostic factor in patients with advanced colorectal cancer treated with cetuximab. J Clin Oncol 2008;26(3):374—9.

[53] Karapetis CS, Khambata-Ford S, Jonker DJ, O'Callaghan CJ, Tu D, Tebbutt NC, et al. K-ras mutations and benefit from cetuximab in advanced colorectal cancer. N Engl J Med 2008;359(17):1757—65.

[54] Allegra CJ, Jessup JM, Somerfield MR, Hamilton SR, Hammond EH, Hayes DF, et al. American Society of Clinical Oncology provisional clinical opinion: testing for KRAS gene mutations in patients with metastatic colorectal carcinoma to predict response to anti-epidermal growth factor receptor monoclonal antibody therapy. J Clin Oncol 2009;27(12):2091—6.

[55] Amado RG, Wolf M, Peeters M, Van Cutsem E, Siena S, Freeman DJ, et al. Wild-type KRAS is required for panitumumab efficacy in patients with metastatic colorectal cancer. J Clin Oncol 2008;26(10):1626—34.

[56] Kris MG, Natale RB, Herbst RS, Lynch Jr TJ, Prager D, Belani C, Schiller JH, et al. Efficacy of gefitinib, an inhibitor of the epidermal growth factor receptor tyrosine kinase, in symptomatic patients with non-small cell lung cancer: a randomized trial. J Am Med Assoc 2003;290(16):2149—58.

[57] Shepherd FA, Rodrigues Pereira J, Ciuleanu T, Tan EH, Hirsh V, Thongprasert S, et al. Erlotinib in previously treated non-small-cell lung cancer. N Engl J Med 2005;353(2):123—32.

[58] Cohen MH, Johnson JR, Chen YF, Sridhara R, Pazdur R. FDA drug approval summary: erlotinib (Tarceva) tablets. Oncologist 2005;10(7):461—6.

[59] Lynch TJ, Bell DW, Sordella R, Gurubhagavatula S, Okimoto RA, Brannigan BW, et al. Activating mutations in the epidermal growth factor receptor underlying responsiveness of nonsmall-cell lung cancer to gefitinib. N Engl J Med 2004;350(21):2129—39.

[60] Mitsudomi T, Yatabe Y. Mutations of the epidermal growth factor receptor gene and related genes as determinants of epidermal growth factor receptor tyrosine kinase inhibitors sensitivity in lung cancer. Cancer Sci 2007;98(12):1817—24.

[61] Mok TS. Personalized medicine in lung cancer: what we need to know. Nat Rev Clin Oncol 2011.

[62] Soda M, Choi YL, Enomoto M, Takada S, Yamashita Y, Ishikawa S, et al. Identification of the transforming EML4-ALK fusion gene in non-small-cell lung cancer. Nature 2007;448(7153):561–6.

[63] Mano H. Non-solid oncogenes in solid tumors: EML4-ALK fusion genes in lung cancer. Cancer Sci 2008;99(12):2349–55.

[64] Kwak EL, Bang YJ, Camidge DR, Shaw AT, Solomon B, Maki RG, et al. Anaplastic lymphoma kinase inhibition in non-small-cell lung cancer. N Engl J Med 2010;363(18):1693–703.

[65] Heuckmann JM, Holzel M, Sos ML, Heynck S, Balke-Want H, Koker M, et al. ALK mutations conferring differential resistance to structurally diverse ALK inhibitors. Clin Cancer Res 2011;17(23):7394–401.

[66] Katayama R, Khan TM, Benes C, Lifshits E, Ebi H, Rivera VM, et al. Therapeutic strategies to overcome crizotinib resistance in non-small cell lung cancers harboring the fusion oncogene EML4-ALK. Proc Natl Acad Sci U S A 2011;108(18):7535–40.

[67] Golub TR, Slonim DK, Tamayo P, Huard C, Gaasenbeek M, Mesirov JP, et al. Molecular classification of cancer: class discovery and class prediction by gene expression monitoring. Science 1999;286(5439):531–7.

[68] Alizadeh AA, Eisen MB, Davis RE, Ma C, Lossos IS, Rosenwald A, et al. Distinct types of diffuse large B-cell lymphoma identified by gene expression profiling. Nature 2000;403(6769):503–11.

[69] International Non-Hodgkin's Lymphoma Prognostic Factors Project. A predictive model for aggressive non-Hodgkin's lymphoma. The International Non-Hodgkin's Lymphoma Prognostic Factors Project. N Engl J Med 1993;329(14):987–94.

[70] Lenz G, Wright GW, Emre NC, Kohlhammer H, Dave SS, Davis RE, et al. Molecular subtypes of diffuse large B-cell lymphoma arise by distinct genetic pathways. Proc Natl Acad Sci U S A 2008;105(36):13520–5.

[71] Perou CM, Sorlie T, Eisen MB, van de Rijn M, Jeffrey SS, Rees CA, et al. Molecular portraits of human breast tumours. Nature 2000;406(6797):747–52.

[72] Perou CM, Jeffrey SS, van de Rijn M, Rees CA, Eisen MB, Ross DT, et al. Distinctive gene expression patterns in human mammary epithelial cells and breast cancers. Proc Natl Acad Sci U S A 1999;96(16):9212–7.

[73] Sørlie T, Tibshirani R, Parker J, Hastie T, Marron JS, Nobel A, et al. Repeated observation of breast tumor subtypes in independent gene expression data sets. Proc Natl Acad Sci U S A 2003;100(14):8418–23.

[74] Sørlie T, Perou CM, Tibshirani R, Aas T, Geisler S, Johnsen H, et al. Gene expression patterns of breast carcinomas distinguish tumor subclasses with clinical implications. Proc Natl Acad Sci U S A 2001;98(19): 10869–74.

[75] L.J., Dai H, van de Vijver MJ, He YD, Hart AA, Mao M, et al. Gene expression profiling predicts clinical outcome of breast cancer. Nature 2002;415(6871):530–6.

[76] van de Vijver MJ, He YD, van 't Veer LJ, Dai H, Hart AA, Voskuil DW, et al. A gene-expression signature as a predictor of survival in breast cancer. N Engl J Med 2002;347(25):1999–2009.

[77] Buyse M, Loi S, van 't Veer L, Viale G, Delorenzi M, Glas AM, et al. Validation and clinical utility of a 70-gene prognostic signature for women with node-negative breast cancer. J Natl Cancer Inst 2006;98(17): 1183–92.

[78] Cardoso F, van 't Veer L, Rutgers E, Loi S, Mook S, Piccart-Gebhart MJ. Clinical application of the 70-gene profile: the MINDACT trial. J Clin Oncol 2008;26(5):729–35.

[79] Paik S, Shak S, Tang G, Kim C, Baker J, Cronin M, et al. A multigene assay to predict recurrence of tamoxifen-treated, node-negative breast cancer. N Engl J Med 2004;351(27):2817–26.

[80] Paik S, Tang G, Shak S, Kim C, Baker J, Kim W, et al. Gene expression and benefit of chemotherapy in women with node-negative, estrogen receptor-positive breast cancer. J Clin Oncol 2006;24(23):3726–34.

[81] Sparano JA, Paik S. Development of the 21-gene assay and its application in clinical practice and clinical trials. J Clin Oncol 2008;26(5):721–8.

[82] Jerevall PL, Ma XJ, Li H, Salunga R, Kesty NC, Erlander MG, et al. Prognostic utility of HOXB13:IL17BR and molecular grade index in early-stage breast cancer patients from the Stockholm trial. Br J Cancer 2011;104(11): 1762–9.

[83] Ma XJ, Salunga R, Dahiya S, Wang W, Carney E, Durbecq V, et al. A five-gene molecular grade index and HOXB13:IL17BR are complementary prognostic factors in early stage breast cancer. Clin Cancer Res 2008; 14(9):2601–8.

[84] Ma XJ, Wang Z, Ryan PD, Isakoff SJ, Barmettler A, Fuller A, et al. A two-gene expression ratio predicts clinical outcome in breast cancer patients treated with tamoxifen. Cancer Cell 2004;5(6):607–16.

[85] Bremer TM, Jacquemier J, Charafe-Jauffret E, Viens P, Birnbaum D, Linke SP. Prognostic marker profile to assess risk in stage I-III hormone receptor-positive breast cancer patients. International journal of cancer. Int J Cancer 2009;124(4):896—904.

[86] Foekens JA, Atkins D, Zhang Y, Sweep FC, Harbeck N, Paradiso A, et al. Multicenter validation of a gene expression-based prognostic signature in lymph node-negative primary breast cancer. J Clin Oncol 2006; 24(11):1665—71.

[87] Wang Y, Klijn JG, Zhang Y, Sieuwerts AM, Look MP, Yang F, et al. Gene-expression profiles to predict distant metastasis of lymph-node-negative primary breast cancer. Lancet 2005;365(9460):671—9.

[88] Desmedt C, Piette F, Loi S, Wang Y, Lallemand F, Haibe-Kains B, et al. Strong time dependence of the 76-gene prognostic signature for node-negative breast cancer patients in the TRANSBIG multicenter independent validation series. Clin Cancer Res 2007;13(11):3207—14.

[89] Ng SB, Turner EH, Robertson PD, Flygare SD, Bigham AW, Lee C, et al. Targeted capture and massively parallel sequencing of 12 human exomes. Nature 2009;461(7261):272—6.

[90] Rios J, Stein E, Shendure J, Hobbs HH, Cohen JC. Identification by whole-genome resequencing of gene defect responsible for severe hypercholesterolemia. Hum Mol Genet 2010;19(22):4313—8.

[91] Ng SB, Buckingham KJ, Lee C, Bigham AW, Tabor HK, Dent KM, et al. Exome sequencing identifies the cause of a mendelian disorder. Nat Genet 2010;42(1):30—5.

[92] Li M, Wang IX, Li Y, Bruzel A, Richards AL, Toung JM, et al. Widespread RNA and DNA sequence differences in the human transcriptome. Science 2011;333(6038):53—8.

[93] Ley TJ, Mardis ER, Ding L, Fulton B, McLellan MD, Chen K, et al. DNA sequencing of a cytogenetically normal acute myeloid leukaemia genome. Nature 2008;456(7218):66—72.

[94] Stephens PJ, McBride DJ, Lin ML, Varela I, Pleasance ED, Simpson JT, et al. Complex landscapes of somatic rearrangement in human breast cancer genomes. Nature 2009;462(7276):1005—10.

[95] Berger MF, Lawrence MS, Demichelis F, Drier Y, Cibulskis K, Sivachenko AY, et al. The genomic complexity of primary human prostate cancer. Nature 2011;470(7333):214—20.

[96] Campbell PJ, Stephens PJ, Pleasance ED, O'Meara S, Li H, Santarius T, et al. Identification of somatically acquired rearrangements in cancer using genome-wide massively parallel paired-end sequencing. Nat Genet 2008;40(6):722—9.

[97] Gerlinger M, Rowan AJ, Horswell S, Larkin J, Endesfelder D, Gronroos E, et al. Intratumor heterogeneity and branched evolution revealed by multiregion sequencing. N Engl J Med 2012;366(10):883—92.

[98] The Cancer Genome Atlas Research Network. Comprehensive genomic characterization defines human glioblastoma genes and core pathways. Nature 2008;455(7216):1061—8.

[99] The Cancer Genome Atlas Research Network. Integrated genomic analyses of ovarian carcinoma. Nature 2011;474(7353):609—15.

[100] Davis JC, Furstenthal L, Desai AA, Norris T, Sutaria S, Fleming E, et al. The microeconomics of personalized medicine: today's challenge and tomorrow's promise. Nat Rev Drug Discov 2009;8(4):279—86.

[101] Hartmann LC, Lu KH, Linette GP, Cliby WA, Kalli KR, Gershenson D, et al. Gene expression profiles predict early relapse in ovarian cancer after platinum-paclitaxel chemotherapy. Clin Cancer Res 2005;11(6):2149—55.

[102] Simon R. The use of genomics in clinical trial design. Clin Cancer Res 2008;14(19):5984—93.

[103] Simon R. Clinical trial designs for evaluating the medical utility of prognostic and predictive biomarkers in oncology. Per Med 2010;7(1):33—47.

[104] Freidlin B, Simon R. Adaptive signature design: an adaptive clinical trial design for generating and prospectively testing a gene expression signature for sensitive patients. Clin Cancer Res 2005;11(21):7872—8.

[105] Center for Biologics Evaluation and Research, F.a.D.A. Draft guidance for industry, clinical laboratories, and FDA staff: in vitro diagnostic multivariate index assays; 2007. http://www.fda.gov/cdrh/oivd/guidance/1610.pdf.

[106] Frueh FW, Amur S, Mummaneni P, Epstein RS, Aubert RE, DeLuca TM, et al. Pharmacogenomic biomarker information in drug labels approved by the United States food and drug administration: prevalence of related drug use. Pharmacotherapy 2008;28(8):992—8.

[107] Epstein RS, Frueh FW, Geren D, Hummer D, McKibbin S, O'Connor S, et al. Payer perspectives on pharmacogenomics testing and drug development. Pharmacogenomics 2009;10(1):149—51.

[108] Morgan S, Grootendorst P, Lexchin J, Cunningham C, Greyson D. The cost of drug development: a systematic review. Health Policy 2011;100(1):4—17.

[109] Deverka PA, Vernon J, McLeod HL. Economic opportunities and challenges for pharmacogenomics. Ann Rev Pharmacol Toxicol 2010;50:423–37.

[110] Zwierzina H. Biomarkers in drug development. Ann Oncol 2008;19(Suppl 5):v33–7.

[111] Fan C, Oh DS, Wessels L, Weigelt B, Nuyten DS, Nobel AB, et al. Concordance among gene-expression-based predictors for breast cancer. N Engl J Med 2006;355(6):560–9.

[112] Hu Z, Fan C, Oh DS, Marron JS, He X, Qaqish BF, et al. The molecular portraits of breast tumors are conserved across microarray platforms. BMC Genomics 2006;7:96.

[113] Sherlock G, Hernandez-Boussard T, Kasarskis A, Binkley G, Matese JC, Dwight SS, et al. The Stanford Microarray Database. Nucleic Acids Res 2001;29(1):152–5.

A Look to the Future: Cancer Epigenetics

Lela Buckingham

Department of Medical Laboratory Sciences, College of Health Sciences, Rush University Medical Center, Chicago, Illinois, USA

LEARNING OBJECTIVES

1. Describe the epigenetic mechanisms that involve protein, DNA, and RNA.

2. Recognize the effect of abnormal histone modifications on gene expression in the cancer cell.

3. Understand the relationship between DNA hypermethylation and tumor suppressor gene expression.

4. Compare and contrast the role of noncoding RNA with that of coding RNA in promoting the malignant cell phenotype.

5. Assess the potential applications of histone deacetylase inhibitors, DNA hypomethylating agents, and noncoding RNA in cancer therapy.

INTRODUCTION

Cancer is caused by structural and functional gene abnormalities affecting cell proliferation and survival. Deoxyribonucleic acid (DNA) sequence changes (mutations) are not the sole source of these abnormalities. Gene availability and expression are subject to DNA sequence-independent levels of control called epigenetics. Epigenetics is the study of sequence-independent yet heritable traits [1]. Epigenetic mechanisms encompass a variety of activities, from pretranscriptional to post-translational events. The most well studied of these involve control of transcription and translation, including histone modification, DNA methylation, and noncoding ribonucleic acids (ncRNA).

HISTONE MODIFICATION

Histones are small, basic nuclear proteins that associate with DNA in the nucleus. There are four core histones, designated H2A, H2B, H3, and H4 (Figure 4A.1). Each core histone is composed of a structured domain or histone fold and unstructured N- and C-terminal tails, accounting for about 30% of the protein. Histones associate with DNA to form nucleosomes. A core nucleosome is comprised of eight histone proteins, two each of H2A, H2B, H3, and H4, wrapped in 146 base pairs of DNA. The arginine- and lysine-rich tails of the histones further stabilize the coiled structure. An additional histone, H1, facilitates the compaction of DNA into interphase chromatin (30 micron fibers). Histone tails are required for nucleosome—nucleosome interaction and for establishing transcriptionally repressive heterochromatin versus transcriptionally active euchromatin [2,3].

Originally thought to be structural proteins maintaining DNA organization, histones are now known to also have functional roles in controlling gene transcription. Modification of histones affects their charge and ability to bind and position on DNA to carry out these processes and to interact with other nonhistone proteins. At least 11 different chemical modifications of histones have been described, including acetylation, methylation, phosphorylation, and ubiquitination of amino acids, mostly in the histone tails (Figure 4A.2). These modifications affect the availability of the DNA to transcription factors and RNA polymerase. For example, methylation of histones H3 and H4 regulates activity of origins of DNA replication [4], and acetylation of histones affects chromatin structure and gene transcription [5]. Modifications of specific histone amino acids, called marks, have been associated with specific biological effects [6], such as acetylation of lysine (K) at amino acid position 4 of histone H3, designated H3K4Ac. This mark modifies the positive charge on the histone, loosening its interaction with the negatively charged DNA and allowing entry of transcription factors and RNA polymerase. Trimethylation of lysine at amino acid position 9 on histone H3, designated H3K9Me3, attracts DNA methylase enzymes to DNA, thus repressing transcription. Histone marks are established and removed by specific enzymes.

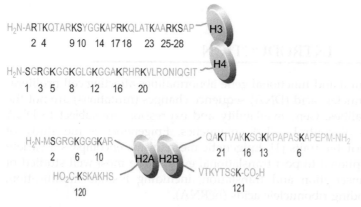

FIGURE 4A.1 Post-translational modifications of histone tails include acetylation, methylation, phosphorylation, and ubiquitination. These modifications occur mostly on lysine (K) amino acids, but may also occur on arginine (R) and serine (S) residues (dark blue). C-terminal lysines of histone 2A and histone 2B are sites of ubiquitination. Amino acid position numbers are indicated below the histone tails.

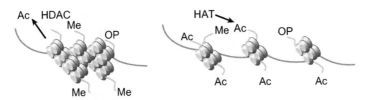

FIGURE 4A.2 Histone tails extend from nucleosomes and are modified by modifying enzymes including histone deacetylases (HDAC) and histone acetyl transferases (HAT). Closed chromatin (left) is transcriptionally inactive and rich in deacetylated and methylated histones and methylated DNA. Open chromatin (right) is transcriptionally active and has more acetylated histones and fewer methylated histones. Me: methyl group, Ac: acetyl group, OP: phosphate group.

Acetylation and Deacetylation

Acetylation and deacetylation of histones are among the most common post-translational protein modifications. Acetylation and deacetylation of lysine amino acids in histone proteins are catalyzed by histone acetyl transferase (HAT) and histone deacetylase (HDAC) enzyme activity, respectively. In humans, HAT1 catalyzes the transfer of an acetyl group from acetyl CoA to the epsilon nitrogen atom in lysine, resulting in an acetylated histone. The addition of the acetyl group lowers the positive charge on the histone and loosens its binding to the negatively charged DNA. Acetylation of histones therefore results in activation of gene transcription. Conversely, HDAC removes these acetyl groups, tightening the interaction between histones and DNA, rendering the DNA inaccessible for gene expression. These enzymes often function in a complex with transcription factors that may guide them to specific DNA sites.

Methylation

Histone methylation occurs on lysine and arginine residues in the H3 and H4 histone tails by histone methyltransferases (HMT) [7]. More than 29 of these enzymes have been identified in humans. Protein arginine N-methyltransferases (PRMTs) catalyze the sequential transfer of a methyl group from S-adenosyl-L-methionine (SAM) to the side-chain nitrogens of arginine residues in histones. SUV39H1, a human homologue of the *Drosophila* protein Su(var)3-9, is a histone H3-specific methyltransferase that selectively methylates lysine-9 of the N terminus of histone H3 [8]. Euchromatic histone-lysine n-methyltransferase 2 (EHMT2, G9a) is a major methyltransferase for histone 3 lysine-4 and lysine-9. Three histone 3 lysine-4 methylases—MLL1, MLL2, and hSet1—have been shown to be differentially expressed in the malignant cells as compared to the normal cells [9].

Methylation of histones has several effects. Histone methylation recruits factors such as heterochromatin protein 1 (HP1) and DNA methyltransferases that silence DNA by limiting accessibility or silencing transcription. Methylation of lysine-9 and lysine-27 in histone H3 (H3K9, H3K27) are marks of transcriptional silencing, and methylation of lysine-4 in histone H3 (H3K4) is a generally conserved mark for transcriptionally active regions. Histone methylation is naturally reversed by (1) enzymes that antagonize methyltransferase activity by histone replacement, (2) deiminase enzymes that antagonize histone arginine methylation, and (3) amine oxidase (AO) and hydroxylase enzymes that directly remove histone lysine methylation [10]. Histone demethylases (HDMs) have also been identified. HDMs, such as lysine-specific demethylase 1 (LSD1), which catalyzes removal of H3K4 methy groups through amine oxidation, are flavin-dependent amine oxidases. Alpha-ketoglutarate-Fe(II)-dependent dioxygenases, such as the JARIDI1 family of HDMs, share a conserved Jumanji enzymatic domain. LSD1 was the first enzyme identified to specifically demethylate lysine-4 of histone H3 (H3K4). Studies on cancer cell lines have revealed that HMT and HDM activities may be interdependent with regard to control of gene expression [11].

Histone Modification and Histone Deacetylase Inhibitors in Cancer

Abnormal histone modifications have been observed in cancer cells, where they interfere with gene expression and destabilize the genome. Deacetylation or methylation of histones associated with DNA containing tumor suppressor genes results in loss of their expression. Acetylation of histones associated with oncogenes increases cell proliferation and survival. Several agents have been designed to counteract aberrant histone acetylation. These include HDAC inhibitors, HATs, and inhibitors of HMT.

Histone deacetylase inhibitors are evaluated in clinical trials as anticancer agents as both monotherapy and in combination with radiation or other agents. *In vitro*, they induce differentiation, cell-cycle arrest, and apoptosis [12]. They can also inhibit migration, invasion, and angiogenesis and show antitumor activity. In 2006, vorinostat was the first HDAC inhibitor approved by the FDA, indicated for cutaneous T-cell lymphoma. A second HDAC inhibitor, the bicyclic depsipeptide romidepsin, was approved for cutaneous T-cell lymphoma in 2009. Alkylating agents, such as temozolomide, are currently in clinical trials as both single agents and in combination other chemotherapeutic agents or radiation therapy for hematologic and solid tumors. Temozolomide with valproic acid (a weak HDAC inhibitor) and radiation may show efficacy against glioblastoma [13]. Another hydroxamic acid compound, panobinostat, is a pan-deacetylase inhibitor that is being studied in many hematologic and solid malignancies, including lymphomas, multiple myeloma, acute myeloid leukemia, and myelodysplastic syndromes.

Inhibition of HDAC shows significant effect in some but not all patients [14,15]. Efforts to find biological factors predicting response are complicated by the incomplete understanding of the mechanism of action of the HDAC enzymes. The 11 known HDAC enzymes act on both histones and nonhistone proteins. The identity and contribution of these multiple substrates to the effect of aberrant HDAC activity will have to be better defined for clarification of the role of HDAC inhibitors in the clinic [16].

Although HMT activity is also associated with the cancer cell phenotype, HMT inhibitors are not as well developed as HDAC inhibitors [17]. SAM analogues effectively inhibit histone methylation; however, these agents inhibit other methyltransferase enzymes in addition to HMTs. To address this problem, more specific HMT inhibitors have been developed, and some have shown effects *in vitro* [18]. An example of such is an antisense (complementary) nucleic acid made of degradation-resistant substituted phosphorothioate DNA targeted against a component of the HMT complex. This compound causes selective and differential apoptosis in cell lines [9]. Another approach of using lysine mimics has also shown some promise *in vitro* [19]. Certain natural products, such as sinefungin and AdoHcy, act as competitive inhibitors of lysine methyltransferases *in vitro*, but their effects are not specific [20].

Similar to HMT, inhibition of HDM also affects gene expression patterns in tumor cells. The HDM LSD1, mentioned previously, is inhibited by long chain polyamine analogues (oligoamines). Oligoamines were observed to increase H3K4 methylation and alter transcription patterns in breast cancer cells [21]. Synergistic effects of specific oligoamines in combination with 2-difluoromethylornithine (DFMO), an inhibitor of ornithine decarboxylase on human colorectal cancer cells, has also been reported [22]. Screening of a large library of compounds revealed that 8-hydroxyquinolines effectively inhibited HDM in cell lines. These compounds modulated demethylation at the H3K9 locus [23].

Overall, the complex nature of histone methylation/demethylation and shared enzyme functions will make drug design difficult. Many normal functions are also dependent on the balance of modified histones. Counteracting aberrant histone modifications will require a highly specific compound, both in structure and in mechanism of action.

DNA METHYLATION

Addition of methyl groups to DNA by DNA methyltransferases is a frequent and normal mechanism to control gene expression. The enzymes that catalyze DNA methylation in humans are DNA methyltransferase 1 (DNMT1) and DNA methyltransferases 3a and 3b (DNMT3a and DNMT3b) [24]. DNMT1 reestablishes methylation patterns on hemimethylated DNA after replication. DNMT3a and 3b methylate hemimethylated and unmethylated DNA with equal efficiencies, suggesting that they are responsible for *de novo* methylation. These enzymes transfer methyl groups from a donor molecule (SAM) to the carbon at position 5 on the pyrimidine ring of nucleotides. 5 methyl-cytosine is the most frequent product of these reactions in humans.

DNA methylation generally silences transcription through interaction with histones and components of the transcriptional apparatus. Methylation of DNA occurs globally, throughout the genome in noncoding regions and at gene-specific sites (Figure 4A.3). Global methylation is thought to maintain chromosome stability by silencing retrotransposons that make up a large percentage of the genome and move from one location to another in the genome through transcriptional intermediates. Methylation at gene-specific sites occurs mainly on the cytosine residues of cytosine-guanine dinucleotides, or CpGs, in a region of DNA containing a cluster of these nucleotides called CpG islands.

FIGURE 4A.3 DNA methylation is a normal event in cell growth and development. In cancer cells, DNA methylation can be dysregulated in two ways. Hypermethylation occurs at individual tumor suppressor gene promoters (top). In addition, global hypomethylation in intergenic regions can result in activation of transposable elements (bottom) leading to chromosomal structural abnormalities and genomic instability.

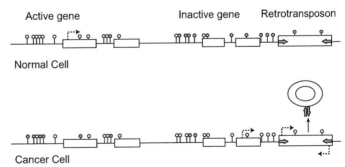

CpG islands are located close to the genes that they control. Normally, DNA methylation can be transient or permanent in particular cells. For some genes, methylation epigenetically regulates gene expression to provide gene product at the appropriate time in cell metabolism. DNA methylation is permanent for genes in cell types where those particular gene products are not or no longer required. Some genes have multiple promoters (start sites of transcription) with different methylation patterns. These promoters may be methylated in some cell types and unmethylated in others. For example, the protein-tyrosine phosphatase, nonreceptor-type, 6 (*PTPN6*) gene is expressed from two promoters, one of which is active in hematologic cells and one in other tissue types [25]. The hematologic promoter is unmethylated in blood but highly methylated in epithelial cells. Methylation of the tissue promoter is decreased in tissue but increased in blood. Methylation is also associated with maternal or paternal chromosome inheritance, resulting in imprinting that silences one of two inherited copies of genes. Imprinting is thought to be responsible for diseases such as Prader-Willi/ Angelman Syndrome, in which loss of the same site on chromosome 15 results in markedly different disease phenotypes, depending on whether the loss is from the maternal or paternal chromosome where different genes are paternally or maternally methylated [26].

Both aberrant hypermethylation and hypomethylation have been reported in cancer cells. Hypermethylation of CpG islands associated with tumor suppressor genes shuts down their expression, contributing to the malignant cell phenotype. Profiles of methylation patterns have been linked to cancer from tumorigenesis to patient outcome [27,28,29]. Global hypomethylation results in loss of chromosome stability due to activation of retrotransposons that move and duplicate themselves through RNA intermediates. Hypomethylation of the long interspersed nuclear element 1 (LINE-1) transposon, a marker for global hypomethylation, has been linked to outcome in lung cancer [30] and tumor progression in colon cancer [31]. Hypomethylation can also lead to changes in chromatin structure to forms that repress transcription of genes through recruiting of repressive methyl-domain binding (MDB) proteins, such as the methyl-CpG binding protein 1 (MeCP1). Silencing of genes through this mechanism has been reported in breast cancer [32].

Observations of detailed methylation profiles (sites of genomic methylation) are made using microarrays [33,34]. Methods are generally based on methylation-specific enzymatic digestion of DNA and/or sequence-specific DNA polymeration and analysis. Test samples are measured relative to a reference control. These profiles have been studied on a variety of tissue sources and in blood [35]. For example, arrays have been used to define methylation profiles correlated to estrogen receptor status in breast cancer [36] and tumorigenesis in lymphoma [37]. Methylation status may also predict response to therapy, as demonstrated in chronic myeloid leukemia [38]. In a phase II trial on imatinib-resistant CML, low-dose decitabine, which inhibits DNA methyltransferase activity, induced response in 46% of patients [39].

Drugs Targeting DNA Methylation

Because of the dual effects carried out by the methylase enzymes, development of drugs to counteract aberrant methylation is a difficult balancing act. Yet drugs targeting abnormal methylation have been devised and are clinically effective, mostly in hematological diseases [40]. These include the cytosine analogues, 5-aza-2′-deoxycytidine (5-aza-dC, decitabine) and 5-azacytidine (azacytidine), and the non-nucleoside analogue, DNMTi (RG 108,

N-phthalyl-L-tryptophan; Figure 4A.4). These agents are used in combination with HDAC inhibitors, such as trichostatin A (TSA, from *Streptomyces platensis*) and the benzamide compound entinostat. For example, the combination of either azacytidine with entinostat or decitabine with sodium phenylbutyrate or valproic acid showed activity for treatment of myelodysplastic syndrome [41]. Hypomethylating agents and histone deacetylase inhibitors are proposed or currently in clinical trials for graft-versus-host disease in allogeneic stem cell transplant patients [42]. There are ongoing clinical trials with decitabine in combination with more traditional anticancer agents, such as vincristine, prednisone, and doxorubicin, with the drug doses reduced to avoid toxicities. With some success of agents targeting abnormal methylation in hematological malignancies, agents directed at solid tumors are also being developed. Genomic studies on solid tumor cell lines have identified methylation-prone targets for therapeutic intervention [43]. The mechanism of action of these drugs *in vivo* is complex. For example, in colon cancer cell lines, 5-aza-dC reactivates gene expression without lowering cytosine methylation in gene-associated CpG islands [44]. Low-dose azacytidine with entinostat increased median survival from 6.4 months to 8.6 months in an intent-to-treat analysis of lung cancer patients [45].

FIGURE 4A.4 Histone deacetylase inhibitors include hydroxamic acids, such as vorinostat and panobinostat; fungal compounds, such as trichostatin A; and benzamide compounds, such as entinostat. They may be used alone or in combination with the DNA methylation inhibitors azacytidine, decitabine, or RG 108 for cancer treatment.

The use of DNA methylation inhibitors alone or in combination with other anticancer drugs presents a complicated set of options for patient care. Future use of methylation profiles may optimize drug selection for patients using the paradigm of agents targeting patient-specific molecular targets such as B-lymphocyte antigen CD20 expression for rituximab in B cell lymphoma, HER2 overexpression for trastuzumab in breast cancer, and epidermal growth factor receptor (EGFR) activation mutations for erlotinib in lung cancer. Promoter hypermethylation of tumor suppressor genes is found in over 70% of early-stage lung cancer patients, providing a potential for their use as early detection markers [46]. But just as the currently used biomarkers are not clear cut, so will there be multiple contributing factors to prediction of treatment response and outcome with DNA methylation.

NONCODING RNA (ncRNA)

Even though only 2% of the human genome codes for proteins, as much as 70% of the genome is transcribed into RNA. Recent studies have shown that these noncoding RNAs (ncRNAs)—that is, those that are not translated into protein products—have functional roles. Noncoding RNAs are classified into two types: small ncRNA, including microRNA (miRNA), and long ncRNA (lncRNA or lincRNA). A variety of other ncRNAs have been described, such as piRNA. However, their role in cancer therapeutics has not been explored to a great extent, and they will not be discussed here.

MicroRNA

MicroRNAs are short, single-stranded RNA molecules, approximately 20 to 30 nucleotides in length, that share homology with 3′ regions of their respective target messenger RNA (Figure 4A.5). They are transcribed in the nucleus as long precursor molecules (pri-miRNA), which form hairpin structures through intramolecular folding. The hairpins are metabolized into 70-nucleotide-long precursor miRNA (pre-miRNA) by a complex including the RNase enzyme Drosha and the double-stranded RNA binding protein DGCR8. The pre-miRNA

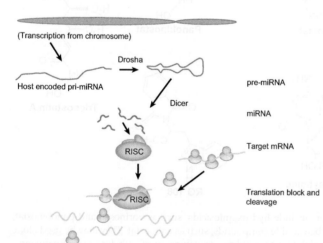

(Transcription from chromosome)

Drosha

Host encoded pri-miRNA

pre-miRNA

Dicer

miRNA

RISC

Target mRNA

RISC

Translation block and cleavage

FIGURE 4A.5 Micro RNAs are transcribed as long pre-miRNA transcripts that are metabolized by Drosha and Dicer enzymes into short single-stranded RNA species with imperfect complementarity to the 3′ untranslated regions of messenger RNA they control. Hybridization to the target mRNA in association with the miRNA induced silencing complex (RISC) results in disruption of translation and degradation of the mRNA.

then moves to the cytoplasm, where it is further processed by another enzyme, Dicer, and one strand is incorporated into the miRNA-induced silencing complex (RISC). The mature miRNA carry imperfect complementarity to messenger RNA that they control. Mismatches in base pairing between the miRNA and mRNA are important for miRNA function. Hybridization to target mRNA interrupts translation and also stimulates degradation of target mRNA, ultimately preventing gene expression.

The miRBase 18 microRNA database lists 2,042 unique mature human miRNAs [47]. MicroRNA are important regulators of developmental processes, cell proliferation, and survival. Their presence and levels are tissue specific and dependent on developmental stage. A cell may contain 800 to 50,000 miRNAs. Their expression is dysregulated in a variety of diseases including cancer, where abnormally expressed miRNA have been termed oncomirs.

Circulating cell-free miRNA can be found in normal human plasma [48]. MiR-15b, miRNA-16, and miR-24 have been measured at 8,910-133,970 copies/uL plasma [49], although the accurate measurement of circulating miRNA is affected by blood components and anticoagulants used for blood collection [50]. Circulation of specific miRNA has been associated with solid tumors, such as miR-92 and miR-17-3p in colon cancer [51].

The importance of miRNA in cancer became apparent through the early observation that miRNA genes are frequently located at fragile sites (sites of translocation breakpoints and other chromosomal abnormalities) in cancer cells [52]. Certain miRNAs act as oncogenes and are aberrantly up-regulated in malignant cells (e.g., miR-155 in breast cancer and miR-17-92 in leukemias and lymphomas) [53,54,55]. Some miRNA act as tumor suppressors and are aberrantly down-regulated in cancer cells (e.g., miR-29a/b/c family or miR15a/miR 16-1 in leukemia) [56,57].

Many studies have addressed the use of differential miRNA expression for stratifying tumor subtypes and predicting response to traditional, targeted, or epigenetic therapy or other prognoses [58,59,60]. The potential for miRNA in these roles or even as tumor targets is limited by the complexity of the miRNA system. MiRNA have a role of fine-tuning, rather than complete regulation, of the genes they control. There is also a network of miRNA/mRNA interactions, in that some miRNA control more than one mRNA, and a single mRNA may be affected by multiple miRNA. One way to address these factors is to combine miRNA and mRNA expression profiling [61].

Antisense molecules complementary to miRNA oncomirs have been investigated as therapeutic agents (antagomirs) for a variety of disease states, including cancer [62]. Use of miRNAs themselves as therapeutic agents have also been proposed—for example, miR-146 expression in breast cancer [63] and miR15a/miR 16-1 expression in leukemia [65].

Long Noncoding RNA

Long noncoding RNA (lncRNA) are defined by their length of more than 200 nucleotides. There may be more than 15,000 lncRNA expressed in the human genome. LncRNA are transcribed by RNA polymerase II, polyadenylated, and spliced similar to messenger RNA. They are distinguished from random products of "leaky" transcription throughout the genome because, similar to messenger RNA, their expression is regulated by transcription factors and histone marks.

Several classes of lncRNA have been described. Among these, large intergenic RNA, transcribed ultraconserved regions (T-UCRs), pseudogenes, antisense RNA that overlap genes they control, and long stress-induced ncRNA are reportedly aberrantly expressed in cancer [66]. Unlike miRNAs, most lncRNA manifest control at the transcriptional rather than the post-transcriptional level [67]. This control can occur through recruitment of histone modification factors to specific genes. Chromatin reorganization, such as X chromosome inactivation by the long noncoding RNA *XIST*, has also been described [68].

Genomic screening technologies such as tiling arrays, serial analysis of gene expression (SAGE), and RNA-Sequencing (RNA-Seq) have revealed subsets of lncRNA whose expression patterns are associated with malignancy in several types of cancer [69,70,71]. Such methods have revealed over 1,800 lncRNA in prostate tissue, 121 of which are dysregulated in prostate cancer [72].

The lncRNA, HOTAIRM1 (HOX antisense intergenic RNA myeloid 1), and HOTTIP (HOXA transcript at the distal tip) stimulate expression of developmental regulatory genes (HOX genes) by recruiting histone modification factors [73]. These lncRNAs may also have roles in carcinogenesis, as the HOX genes are implicated in oncogenesis in some leukemias. Another lncRNA, HOTAIR (HOX transcript antisense RNA), also affects chromatin structure through the polycomb repressive complex protein PRC2 [74]. HOTAIR expression is up-regulated in breast and liver cancers. LncRNA may be transcribed from loci they control. For example, ANRIL, an lncRNA encoded within the multigenic *p16/CDKN2A-p15/CDKN2B-p14/ARF* locus, recruits chromatin remodeling proteins to down-regulate *p16/CDKN2A* (Figure 4A.6). The effect of lncRNA on chromatin structure has also been associated with cancer metastasis [75].

The ability of lncRNA to affect chromatin through histone modifications led to the idea of lncRNA-associated therapy, where depleting specific oncogenic lncRNAs or replacing

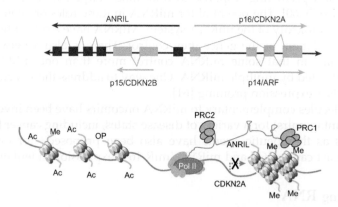

FIGURE 4A.6 Long noncoding RNAs are transcribed by RNA polymerase II similar to mRNA and may be associated with other genes. For example, the multigenic *p16/CDKN2A-p15/CDKN2B-p14/ARF* locus on chromosome 9 includes three protein coding genes and the lncRNA, ANRIL (antisense noncoding RNA in the INK4 locus). The *p16/CDKN2A, p15/CDKN2B, p14/ARF* genes are negative regulators of the cell division cycle and cell proliferation. The example in this figure shows ANRIL recruiting protein complexes, such as PRC2, resulting in reorganization of chromatin and decrease of *p16/CDKN2A* gene expression. In cancer cells, this decrease may result in overproliferation of cells.

lncRNA with tumor suppressor activity could epigenetically reprogram cancer cells, thus eliminating the malignant phenotype [76].

Challenges with Noncoding RNA Therapy

Although the generation of ncRNA (miRNA and lncRNA) therapy is a promising idea, there are several challenges. Half-life (degradation) is a concern for use of ncRNA as therapeutic agents *in vivo*. Use of alternate nucleic acid structures or modified nucleotides in synthetic RNA has shown some success *in vitro*, but targeting *in vivo* has potential drawbacks.

Another concern is potential toxicity, as ncRNA may have a variety of roles in the cell, not all of which may be known or defined. Such toxicity could be avoided by targeting specific tissues; however, this in turn presents the problem of delivery. Viral and nonviral vectors and physical methods such as electric shock or hydrodynamic force have been proposed as delivery methods [77]. Nonviral vectors, such as liposomes, synthetic and natural polymers, and nanoparticles are efficient *in vitro* but are still to be proven effective *in vivo* [78]. Physical methods, such as electroporation, are efficient as well but extra- and intracellular delivery may be nonspecific.

There has been a long history of use of viral vectors for gene therapy, for a variety of diseases. Viruses can be engineered to target specific cells or tissues and are relatively easy to produce. Historically, delivery systems using viruses have not been as successful as predicted, due to immune reactions, untargeted gene disruption, and other side effects, but newer methods, including engineered and oncolytic viruses, offer better results [79]. Oncolytic viruses target cells with altered proliferation and survival pathways. They may be engineered to target specific cells using ncRNA. For example, oncolytic viruses can be targeted to tumor cells not expressing the tumor suppressor miRNA, let-7, by introducing let-7 binding sites to essential genes in the viral genome. The oncolytic virus will affect only the tumor cells not expressing let-7 miRNA [80]. Such viral engineering will optimize use of these agents, as will recognition of their genetic and epigenetic interactions with cells [81]. To that end, in a phase I clinical trial for advanced cancer patients, RNA-based therapy has been successfully localized to tumors to reduce specific gene expression [82].

CONCLUSION

The field of epigenetics has opened an exciting new area for study and treatment of cancer. New ideas will bring new challenges. The complexity of the epigenetic lesions in cancer cells is yet to be fully characterized, so the optimal approach is still to be found. The variety of biological factors that affect how genes are expressed and how this expression is dysregulated in cancer has become increasingly important in diagnosis and treatment, as have the genetic abnormalities now well described in various cancer types. Epigenetics, including histone biology, DNA modification, and ncRNA, invites multiple fields of expertise to provide novel ideas and methods for these challenges. Multiple approaches will be required to overcome mutations and other mechanisms of drug resistance that have been observed in treatments from antibiotics to gene-targeted agents. Just as therapies targeted to genetic lesions have revolutionized cancer treatment, epigenetic therapies promise to offer novel, nontoxic, effective, easily administered, and cost-effective treatment strategies for cancer therapeutics.

QUESTIONS FOR DISCUSSION

1. Which epigenetic mechanisms have implications for cancer treatment?
2. What are examples of existing cancer therapies that target epigenetic mechanisms?
3. What are some challenges with applying epigenetic research to finding novel cancer therapies?

References

[1] Feinberg A. Phenotypic plasticity and the epigenetics of human disease. Nature 2007;447. 443−440.

[2] Luger K, Mader AW, Richmond RK, Sargent DF, Richmond TJ. Crystal structure of the nucleosome core particle at 2.8 Å resolution. Nature 1997;389:251−60.

[3] Grunstein M. Yeast heterochromatin: regulation of its assembly and inheritance by histones. Cell 1998;93:325−8.

[4] Dorn E, Cook JG. Nucleosomes in the neighborhood: new roles for chromatin modifications in replication origin control. Epigenetics 2011;6:552−9.

[5] Jayani R, Ramanujam PL, Galande S. Studying histone modifications and their genomic functions by employing chromatin immunoprecipitation and immunoblotting. Methods Cell Biol 2010;98:35−6.

[6] Tan M, Luo H, Lee S, Jin F, Yang JS, Montellier E, et al. Identification of 67 histone marks and histone lysine crotonylation as a new type of histone modification. Cell 2011;146:1016−28.

[7] Zhang Y, Reinberg D. Transcription regulation by histone methylation: interplay between different covalent modifications of the core histone tails. Genes Dev 2001;15:2343−60.

[8] Rea S, Eisenhaber F, O'Carroll D, Strahl BD, Sun Z-W, Schmid M, et al. Regulation of chromatin structure by site-specific histone H3 methyltransferases. Nature 2000;406:593−9.

[9] Yadav S, Singhal J, Singhal SS, Awasthi S. hSET1: a novel approach for colon cancer therapy. Biochem Pharmacol 2009;77:1635−41.

[10] Klose R, Zhang Y. Regulation of histone methylation by demethylimination and demethylation. Nat Rev Mol Cell Biol 2007;8:307−18.

[11] Islam A, Richter WF, Jacobs LA, Lopez-Bigas N, Benevolenskaya EV. Selective targeting of histone methylation. Cell Cycle 2011;10:413−24.

[12] Moreira J, Scheipers P, Sørensen P. The histone deacetylase inhibitor Trichostatin A modulates CD4+ T cell responses. BMC Cancer 2003;3:30−48.

[13] Mrugala M, Kesari S, Ramakrishna N, Wen PY. Therapy for recurrent malignant glioma in adults. Expert Rev Anticancer Ther 2004;4:759−82.

[14] Ma X, Ezzeldin HH, Diasio RB. Histone deacetylase inhibitors: current status and overview of recent clinical trials. Drugs 2009;69:1911−34.

[15] Shabason J, Tofilon PJ, Camphausen K. HDAC inhibitors in cancer care. Oncology 2010;24:1−8.

[16] Marks P. HDAC Inhibitors: much to learn about effective therapy. Oncology 2010;24:1−2.

[17] Spannhoff A, Sippl W, Jung M. Cancer treatment of the future: inhibitors of histone methyltransferases. Int J Biochem Cell Biol 2009;41:4−11.

[18] Yao Y, Chen P, Diao J, Cheng G, Deng L, Anglin JL, Prasad BV, et al. Selective inhibitors of histone methyltransferase DOT1L: design, synthesis, and crystallographic studies. J Am Chem Soc 2011;133:16746−9.

[19] Chang Y, Ganesh T, Horton JR, Spannhoff A, Liu J, Sun A, et al. Adding a lysine mimic in the design of potent inhibitors of histone lysine methyltransferases. J Mol Biol 2010;400:1−7.

[20] Krishnan S, Horowitz S, Trievel RC. Structure and function of histone H3 lysine 9 methyltransferases and demethylases. Chembiochem 2011;12:254−63.

[21] Zhu Q, Huang Y, Marton LJ, Woster PM, Davidson NE, Casero RA. Polyamine analogs modulate gene expression by inhibiting lysine-specific demethylase 1 (LSD1) and altering chromatin structure in human breast cancer cells. Amino Acids 2012;42:887−98.

[22] Wu Y, Steinbergs N, Murray-Stewart T, Marton LJ, Casero RA. Oligoamine analogues in combination with 2-difluoromethylornithine (DFMO) synergistically induce re-expression of aberrantly silenced tumor suppressor genes. Biochem J 2012;442:693−701.

[23] King O, Li XS, Sakurai M, Kawamura A, Rose NR, Ng SS, et al. Quantitative high-throughput screening identifies 8-hydroxyquinolines as cell-active histone demethylase inhibitors. PLoS One 2010;5:e15535.

[24] Robertson K, Uzolgyi E, Liang G, Talmadge C, Sumegi J, Gonzales FA, et al. The human DNA methyl-tranferases (DNMTs) 1, 3a and 3b: coordinate mRNA expression in normal tissues and overexpression in tumors. Nucleic Acids Res 1999;27:2291—8.

[25] Banville D, Stocco R, Shen S- H. Human protein tyrosine phosphatase 1C (PTPN6) gene structure: alternate promoter usage and exon skipping generate multiple transcripts. Genomics 1995;27:165—73.

[26] Buiting K. Prader-Willi syndrome and Angelman syndrome. Am J Med Genet 2010;154:365—76.

[27] Toyooka S, Tokumo M, Shigematsu H, Matsuo K, Asano H, Tomii K, et al. Mutational and epigenetic evidence for independent pathways for lung adenocarcinomas arising in smokers and never smokers. Cancer Res 2006;66:1371—5.

[28] Christensen B, Houseman EA, Godleski JJ, Marsit CJ, Longacker JL, Roelofs CR, et al. Epigenetic profiles distinguish pleural mesothelioma from normal pleura and predict lung asbestos burden and clinical outcome. Cancer Res 2009;69:227—34.

[29] Buckingham L, Faber LP, Kim A, Liptay M, Barger C, Basu S, et al. PTEN, RASSF1 and DAPK site-specific hypermethylation and outcome in surgically treated stage I and II nonsmall cell lung cancer patients. Int J Cancer 2010;126:1630—9.

[30] Saito K, Kawakami K, Matsumoto I, Oda M, Watanabe G, Minamoto T. Long interspersed nuclear element 1 hypomethylation is a marker of poor prognosis in stage IA non-small cell lung cancer. Clin Cancer Res 2010;16:2418—26.

[31] Sunami E, de Maat M, Vu A, Turner RR, Hoon DS. LINE-1 hypomethylation during primary colon cancer progression. PLoS One 2011;6:18884.

[32] Hon G, Hawkins RD, Caballero OL, Lo C, Lister R, Pelizzola M, et al. Global DNA hypomethylation coupled to repressive chromatin domain formation and gene silencing in breast cancer. Genome Res 2012;22:246—58.

[33] Bibikova M, Barnes B, Tsan C, Ho V, Klotzle B, Le JM, et al. High density DNA methylation array with single CpG site resolution. Genomics 2011;98:288—95.

[34] Cheung H, Lee TL, Rennert OM, Chan WY. Methylation profiling using methylated DNA immunoprecipitation and tiling array hybridization. Methods Mol Biol 2012;825:115—26.

[35] Cassinotti E, Melson J, Liggett T, Melnikov A, Yi Q, Replogle C, et al. DNA methylation patterns in blood of patients with colorectal cancer and adenomatous colorectal polyps. Int J Cancer 2012;131: 1153—7.

[36] Fackler M, Umbricht CB, Williams D, Argani P, Cruz LA, Merino VF, et al. Genome-wide methylation analysis identifies genes specific to breast cancer hormone receptor status and risk of recurrence. Cancer Res 2011;71:6195—207.

[37] Ammerpohl O, Haake A, Pellissery S, Giefing M, Richter J, Balint B, et al. Array-based DNA methylation analysis in classical Hodgkin lymphoma reveals new insights into the mechanisms underlying silencing of B cell-specific genes. Leukemia 2012;26:185—8.

[38] Jelinek J, Gharibyan V, Estecio MR, Kondo K, He R, Chung W, et al. Aberrant DNA methylation is associated with disease progression, resistance to imatinib and shortened survival in chronic myelogenous leukemia. PLoS One 2011;6:e2110.

[39] Issa J, Gharibyan V, Cortes J, Jelinek J, Morris G, Verstovsek S, et al. Phase II study of low-dose decitabine in patients with chronic myelogenous leukemia resistant to imatinib mesylate. J Clin Oncol 2005;23:3948—56.

[40] Song S, Han SW, Bang YJ. Epigenetic-based therapies in cancer: progress to date. Drugs 2011;71:2391—403.

[41] Lübbert M. Targets of epigenetic therapy—gene reactivation as a novel approach in MDS treatment. Cancer Treat Rev 2007;33:S47—52.

[42] Blazar B, Murphy WJ, Abedi M. Advances in graft-versus-host disease biology and therapy. Nat Rev Immunol 2012;12:443—58.

[43] McCabe M, Lee EK, Vertina PM. A multifactorial signature of DNA sequence and polycomb binding predicts aberrant CpG island methylation. Cancer Res 2009;69:282—91.

[44] Mossman D, Kim KT, Scott RJ. Demethylation by 5-aza-2'-deoxycytidine in colorectal cancer cells targets genomic DNA whilst promoter CpG island methylation persists. BMC Cancer 2010;10:366.

[45] Juergens R, Wrangle J, Vendetti FP, Murphy SC, Zhao M, Coleman B, et al. Combination epigenetic therapy has efficacy in patients with refractory advanced non-small cell lung cancer. Cancer Discov 2011:598—608.

[46] Begum S, Brait M, Dasgupta S, Ostro KL, Zahurak M, Carvalho AL, et al. An epigenetic marker panel for detection of lung cancer using cell-free serum DNA. Clin Cancer Res 2011;17:4494–503.

[47] Kozomara A, Griffiths-Jones S. miRBase: integrating microRNA annotation and deep-sequencing data. Nucleic Acids Res 2011;30(Database Issue):D152–7.

[48] Schöler N, Langer C, Kuchenbauer F. Circulating microRNAs as biomarkers—true Blood? Genome Med 2011;3:72.

[49] Mitchell P, Parkin RK, Kroh EM, Fritz BR, Wyman SK, Pogosova-Agadjanyan EL, et al. Circulating microRNAs as stable blood-based markers for cancer detection. Proc Natl Acad Sci USA 2008;105:10513–8.

[50] Kim D-J, Linnstaedt S, Palma J, Park JC, Ntrivalas E, Kwak-Kim JY-H, et al. Plasma components affect accuracy of circulating cancer-related microRNA quantitation. J Mol Diagn 2012;14:71–80.

[51] Schetter A, Harris CC. Plasma microRNAs: a potential biomarker for colorectal cancer? Gut 2009;58:1318–9.

[52] Calin GASC, Dumitru CD, Hyslop T, Noch E, Yendamuri S, Shimizu M, et al. Human microRNA genes are frequently located at fragile sites and genomic regions involved in cancers. Proc Natl Acad Sci USA 2004;101:2999–3004.

[53] Jiang S, Zhang HW, Lu MH, He XH, Li Y, Gu H, et al. MicroRNA-155 functions as an OncomiR in breast cancer by targeting the suppressor of cytokine signaling 1 gene. Cancer Res 2010;70:3119–27.

[54] Landais S, Landry S, Legault P, Rassart E. Oncogenic potential of the miR-106-363 cluster and its implication in human T-cell leukemia. Cancer Res 2007;67:5699–707.

[55] Bonauer A, Carmona G, Iwasaki M, Mione M, Koyanagi M, Fischer A, et al. MicroRNA-92a controls angio-genesis and functional recovery of ischemic tissues in mice. Science 2009;324:1710–3.

[56] Eyholzer M, Schmid S, Wilkens L, Mueller BU, Pabst T. The tumour-suppressive miR-29a/b1 cluster is regulated by CEBPA and blocked in human AML. Br J Cancer 2010;103:275–84.

[57] Pekarskya Y, Palamarchuka A, Maximova V, Efanova A, Nazaryana N, Santanama U, et al. Tcl1 functions as a transcriptional regulator and is directly involved in the pathogenesis of CLL. Cancer Res 2007;105:19643–8.

[58] Blower P, Verducci JS, Lin S, Zhou J, Chung J-H, Dai Z, et al. MicroRNA expression profiles for the NCI-60 cancer cell panel. Mol Cancer Ther 2007;6:1483–91.

[59] Blower P, Chung J-H, Verducci JS, Lin S, Park J-K, Dai Z, et al. MicroRNAs modulate the chemosensitivity of tumor cells. Mol Cancer Ther 2008;7:1–9.

[60] Weinstein J. MicroRNAs in cancer pharmacology and therapeutics: exploiting a natural synergy between "-omic" and hypothesis-driven research. Mol Cancer Ther 2008;7:2021.

[61] Buffa F, Camps C, Winchester L, Snell CE, Gee HE, Sheldon H, et al. microRNA-associated progression pathways and potential therapeutic targets identified by integrated mRNA and microRNA expression profiling in breast cancer. Cancer Res 2011;71:5635–45.

[62] Liu C, Kelnar K, Liu B, Chen X, Calhoun-Davis T, Li H, et al. The microRNA miR-34a inhibits prostate cancer stem cells and metastasis by directly repressing CD44. Nat Med 2011;17:211–5.

[63] Hurst D, Edmonds MD, Scott GK, Benz CC, Vaidya KS, Welch DR. Breast cancer metastasis suppressor 1 up-regulates miR-146, which suppresses breast cancer metastasis. Cancer Res 2009;69:1279–83.

[64] Xu B, Wang N, Wang X, Tong N, Shao N, Tao J, Li P, et al. MiR-146a suppresses tumor growth and progression by targeting EGFR pathway and in a p-ERK-dependent manner in castration-resistant prostate cancer. Prostate 2012;72:1171–8.

[65] Cimmino A, Calin GA, Fabbri M, Iorio MV, Ferracin M, Shimizu M, et al. miR-15 and miR-16 induce apoptosis by targeting BCL2. Proc Natl Acad Sci U S A 2005;102:13944–9.

[66] Prensner J, Chinnaiyan AM. The emergence of lncRNAs in cancer biology. Cancer Discov 2011;1:391–407.

[67] Wang K, Chang HY. Molecular mechanisms of long noncoding RNAs. Mol Cell 2011;43:904–14.

[68] Chow J, Hall LL, Baldry SEL, Thorogood NP, Lawrence JB, Brown CJ. Inducible XIST-dependent X-chromosome inactivation in human somatic cells is reversible. Proc Natl Acad Sci USA 2007;104:10104–9.

[69] Gibb E, Vucic EA, Enfield KS, Stewart GL, Lonergan KM, Kennett JY, et al. Human cancer long noncoding RNA transcriptomes. PLoS One 2011;6. e25915 (1–10).

[70] Tahira A, Kubrusly MS, Faria MF, Dazzani B, Fonseca RS, Maracaja-Coutinho V, et al. Long noncoding intronic RNAs are differentially expressed in primary and metastatic pancreatic cancer 2011;10. 141 (1–19).

[71] Huang R, Jaritz M, Guenzl P, Vlatkovic I, Sommer A, Tamir IM, et al. An RNA-Seq strategy to detect the complete coding and non-coding transcriptome including full-length imprinted macro ncRNAs. PLoS One 2011;6. e27288 (1–13).

[72] Prensner J, Iyer MK, Balbin A, Dhanasekaran SM, Cao Q, Brenner JC, et al. Transcriptome sequencing across a prostate cancer cohort identifies PCAT-1, an unannotated lincRNA implicated in disease progression. Nat Biotechnol 2011;29:742—9.

[73] Wang K, Yang YW, Liu B, Sanyal A, Corces-Zimmerman R, Chen Y, et al. A long noncoding RNA maintains active chromatin to coordinate homeotic gene expression. Nature 2011;472:120—4.

[74] Tsai M-C, Manor O, Wan Y, Mosammaparast N, Wang JK, Lan F, et al. Long noncoding RNA as modular scaffold of histone modification complexes. Science 2010;329:689—93.

[75] Gupta R, Shah N, Wang KC, Kim J, Horlings HM, Wong DJ, et al. Long noncoding RNA HOTAIR reprograms chromatin state to promote cancer metastasis. Nature 2010;464:1071—6.

[76] Moskalev E, Schubert M, Hoheisel JD. RNA-directed epigenomic reprogramming—an emerging principle of a more targeted cancer therapy? Genes Chromosomes Cancer 2012;51:105—10.

[77] Kamimura K, Suda T, Zhang G, Liu D. Advances in gene delivery systems. Pharmaceut Med 2011;25:293—306.

[78] Sun X, Liu C, Liu D, Li P, Zhang N. Novel biomimetic vectors with endosomal-escape agent enhancing gene transfection efficiency. Int J Pharm 2012;425:62—72.

[79] Tedcastle A, Cawood R, Di Y, Fisher K, Seymour L. Virotherapy—cancer targeted pharmacology. Drug Discov Today 2012;17:215—20.

[80] Jin H, Lv S, Yang J, Wang X, Hu H, Su C, et al. Use of microRNA Let-7 to control the replication specificity of oncolytic adenovirus in hepatocellular carcinoma cells. PLoS One 2011;6:e21307.

[81] Li Q, Tainsky MA. Epigenetic silencing of IRF7 and/or IRF5 in lung cancer cells leads to increased sensitivity to oncolytic viruses. PLoS One 2011;6:e28683.

[82] Davis M, Zuckerman JE, Choi CHJ, Seligson D, Tolcher A, Alabi CA, et al. Evidence of RNAi in humans from systemically administered siRNA via targeted nanoparticles. Nature 2010;464:1067—70.

Pharmacogenetics in Cardiovascular Diseases

Larisa H. Cavallari*, *Kathryn Momary*†

*Department of Pharmacy Practice, University of Illinois at Chicago,
Chicago, Illinois, USA
†Department of Pharmacy Practice, Mercer University,
College of Pharmacy and Health Sciences Center, Atlanta, Georgia, USA

OBJECTIVES

1. Describe potential approaches to implement warfarin and clopidogrel pharmacogenomics into clinical practice.

2. Provide examples of pharmacogenomic labeling for drugs used to manage cardiovascular disease.

3. Discuss guidelines for use of genetic information to guide therapy with cardiovascular agents.

4. Describe the potential applications of pharmacogenomics in the management of hypertension and heart failure.

INTRODUCTION

Cardiovascular disease is the most common cause of death in middle- and higher-income countries and among the top ten leading causes of death in lower-income countries [1]. Cardiovascular disease is associated with significant productivity loss and health care costs [2]. Cardiovascular drugs, including statins, clopidogrel, and angiotensin-converting enzyme (ACE) inhibitors, consistently rank among the top ten most commonly prescribed drugs in the United States [3].

Guidelines from expert consensus panels are available to guide the treatment for most cardiovascular diseases, including hypertension, heart failure, dyslipidemia, and ischemic heart disease [4–9]. These guidelines are based on data from large, randomized,

133

placebo-controlled clinical trials demonstrating significant improvements in clinical outcomes with certain medications in clinical trial populations. As illustrated in Figure 5.1, for some cardiovascular diseases, the same drug or drug combination is recommended for all affected persons, regardless of individual characteristics. Such is the case with ACE inhibitors and β-blockers, which are recommended for all patients with left ventricular dysfunction in the absence of a contraindication [4]. However, although these treatments were efficacious in clinical trial populations as a whole, there is no guarantee that they will be safe or effective in an individual patient. In fact, there is significant interpatient variability in response to ACE inhibitors and β-blockers, with some patients deriving no benefit and other patients experiencing intolerable adverse effects with these agents. Currently, it is difficult if not impossible to predict how a patient will respond to a cardiovascular agent based on clinical factors alone.

It is now well recognized that an individual's genotype impacts his or her response to cardiovascular drugs. As of November 2012, genetic information is included in the Food and Drug Administration (FDA)—approved labeling for 12 drugs used to treat cardiac and vascular disorders (Table 5.1). Genotype primarily influences cardiovascular drug response by affecting drug disposition in the body (pharmacokinetics) or a patient's sensitivity to a drug (pharmacodynamics), as described in detail in chapter 1.

This chapter reviews the pharmacogenomics of various cardiovascular agents. The strongest evidence exists for clopidogrel and warfarin, and thus the most in-depth discussion is devoted to these two drugs. This chapter also provides an overview of the pharmacogenomics of antihypertensive agents, statins, heart failure medications, and drugs that influence cardiac conduction. Challenges and opportunities with bringing cardiovascular pharmacogenomics to the clinical arena are also highlighted.

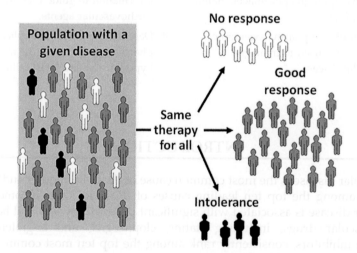

FIGURE 5.1 Current empiric approach of treating patients with cardiovascular disease. As shown, patients with a given cardiovascular disease are generally treated with similar therapy. The majority of patients will have a good response to such therapy. However, the problem with this approach is that a subset of patients will have little to no therapeutic response and another subset will develop intolerable adverse effects.

TABLE 5.1 Cardiovascular drugs with genetic labeling

Drug class/Drug	Biomarker	Location of label information	Context
Statins			
Atorvastatin	LDL receptor	Indications/Usage Dosage/Administration Warnings/Precautions Clinical Pharmacology Clinical Studies	Among other indications, atorvastatin is indicated in patients with familial hypercholesterolemia that is due to mutations in the LDL receptor gene.
Pravastatin	Genotype APOE E2/E2 and Fredrickson Type III dysbetalipoproteinemia	Clinical Studies	Response to pravastatin in patients with genotype E2/E2 and Fredrickson Type III dysbetalipoproteinemia is shown.
Beta-blocker			
Carvedilol	CYP2D6	Drug Interactions Clinical Pharmacology	Reduced carvedilol metabolism in poor metabolizers.
Metoprolol	CYP2D6	Precautions Clinical Pharmacology	Reduced metoprolol metabolism in poor metabolizers.
Propranolol	CYP2D6	Precautions Drug Interactions Clinical Pharmacology	Reduced propranolol metabolism in poor metabolizers.
Antiplatelet			
Clopidogrel	CYP2C19	Boxed Warning Dosage/Administration Warnings/Precautions Drug Interactions Clinical Pharmacology	Reduced clopidogrel efficacy in poor metabolizers.
Prasugrel	CYP2C19	Specific Populations Clinical Pharmacology Clinical Studies	No effect of CYP2C19 genotype on prasugrel efficacy.
Ticagrelor	CYP2C19	Clinical Studies	No effect of CYP2C19 genotype on ticagrelor efficacy.
Anticoagulant			
Warfarin	CYP2C9 VKORC1	Dosage/Administration Precautions Clinical Pharmacology	Lower warfarin doses needed with the *CYP2C9*2*, *CYP2C9*3*, and *VKORC1* -1639A alleles.
Antiarrhythmic			
Propafenone	CYP2D6	Clinical Pharmacology	The recommended dose is the same in slow and extensive metabolizers.
Quinidine	CYP2D6	Precautions	Quinidine can convert extensive metabolizers to poor metabolizers of CYP2D6 substrates.
Miscellaneous			
Isosorbide and hydralazine	NAT1, NAT2	Clinical Pharmacology	Fast acetylators have lower hydralazine exposure.

PHARMACOGENOMICS OF ANTIPLATELET AGENTS

Background on Antiplatelet Agents

Antiplatelet therapy plays a major role in cardiovascular risk reduction. Antiplatelet therapy began with aspirin monotherapy and has advanced to include multiple oral antiplatelet drugs affecting different mechanisms of platelet function [10]. The currently approved, oral antiplatelet drugs include ticlopidine, clopidogrel, prasugrel, and ticagrelor. Ticlopidine is rarely used because it increases the risk for neutropenia and thrombotic thrombocytopenic purpura; discussion will thus be limited to the other agents.

Clopidogrel has long been available; ticagrelor and prasugrel were more recently approved by the FDA. Although these agents have different pharmacokinetic and pharmacodynamic properties and indications, they all share the common mechanism of blocking the platelet $P2Y_{12}$ receptor, resulting in attenuation of adenosine diphosphate (ADP)—mediated platelet activation and aggregation. Thus, they are all classified as $P2Y_{12}$ receptor inhibitors.

Overview of Clopidogrel Metabolism and Pharmacodynamics

Clopidogrel is indicated in combination with aspirin for patients with an acute coronary syndrome (ACS) who are medically managed or undergo percutaneous coronary intervention (PCI) based on data that it reduces morbidity and mortality in these patient populations [11–13]. The combination of clopidogrel and aspirin also reduces the risk for coronary stent thrombosis following PCI [14]. There is significant interpatient variability in clopidogrel pharmacokinetics and pharmacodynamics [15]. Clopidogrel is a prodrug requiring bioactivation by multiple CYP450 enzymes. As shown in Figure 5.2, clopidogrel is a p-glycoprotein substrate, and once absorbed, the majority of clopidogrel is eliminated via esterases. The remaining drug requires conversion via a two-step process to its active form. Genetic variation in pathways involved in clopidogrel absorption and bioactivation can lead to lower levels of the active clopidogrel metabolite and reduced clopidogrel effectiveness.

Clopidogrel responsiveness can be characterized via drug effects on either platelet aggregation or clinical outcomes. Platelet aggregation tests involve *ex vivo* exposure of platelets to aggregating agents, including ADP. Decreased response to clopidogrel, as demonstrated by insufficient attenuation of platelet aggregation, has been linked to an increased risk of adverse cardiovascular events [15,16]. Investigators have also used clinical events, such as recurrent myocardial infarction (MI), stroke, or coronary artery stent thrombosis, as measures of clopidogrel response [16,17]. Genetic determinants of both measures of clopidogrel response will be discussed in this section.

ABCB1 Genotype and Clopidogrel Responsiveness

P-glycoprotein is encoded by the ATP-binding cassette, subfamily B, member 1 (*ABCB1*) gene. The most commonly studied *ABCB1* variant is the synonymous c.3435C>T polymorphism, located in a region that encodes for a cytoplasmic loop in the transporter [18]. A lower peak plasma concentration (Cmax) and total area under the plasma concentration-time curve (AUC) of clopidogrel and its active metabolite were noted after single 300 mg and 600 mg

FIGURE 5.2 Proteins involved in the absorption and metabolic activation of clopidogrel. Genes for proteins shown in bold contain polymorphisms linked to clopidogrel responsiveness. CYP, cytochrome P450.

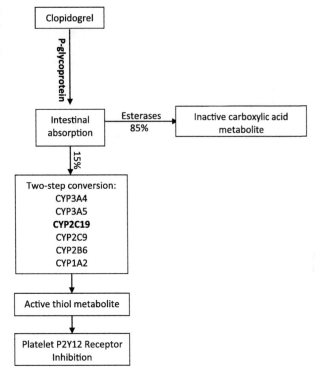

doses in subjects who were homozygous for the variant *ABCB1* 3435T allele [19]. Of note, increasing the clopidogrel dose to 900 mg overcame the effect of genotype on drug concentrations. Several studies have also assessed the association between *ABCB1* genotype and clinical response to clopidogrel with varying results [17,20—24]. The inconsistent results from these studies render it difficult to apply *ABCB1* testing to patients starting clopidogrel.

CYP450 Genotypes and Clopidogrel Responsiveness

Various isoenzymes of the CYP450 system, including CYP3A4, CYP3A5, and CYP2C19, are involved in clopidogrel metabolism. Of the *CYP3A4* polymorphisms assessed, only the IVS10+12A variant has been associated with response to clopidogrel, but with conflicting results [16,25]. Contradictory results have also been found with *CYP3A5* polymorphisms [16,23,26,27]. Because of the contradictory data, *CYP3A4* and *CYP3A5* genotyping is not currently recommended to guide clopidogrel therapy.

Polymorphisms within the gene for CYP2C19, which serves a major role in converting clopidogrel to its active form, have the greatest implications for clopidogrel response. The *CYP2C19* gene is located on chromosome 10q23.33. The *CYP2C19*2, *3, *4, *5, *6, *7*, and *8* alleles are loss-of-function alleles associated with decreased CYP2C19 function compared to the *CYP2C19*1* allele [28,29]. In contrast, the *CYP2C19*17* allele is associated with increased CYP2C19 function and deemed a gain-of-function allele. The *CYP2C19*2* allele is

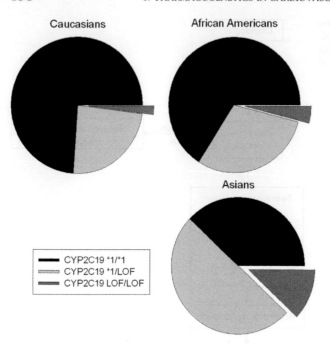

FIGURE 5.3 Cytochrome P450 2C19 (*CYP2C19*) allele frequencies among ethnic groups. LOF, loss-of-function.

by far the most common *CYP2C19* loss-of-function variant; however, its frequency differs by ancestral origin (Figure 5.3), with a higher frequency in Asians (approximately 30%) compared to Caucasians (13%) and African Americans (18%) [28,29]. The *CYP2C19*3* allele also occurs commonly in Asian populations (allele frequency of approximately 10%) but is rare in individuals of other ancestral backgrounds (allele frequency <1%). Approximately 14% of Asians, 2% of Caucasians, and 4% of African Americans are CYP2C19 poor metabolizers (with two loss-of-function alleles), and 50%, 25%, and 30%, respectively, are intermediate metabolizers (with one loss-of-function allele).

Numerous studies have demonstrated that possession of one or two *CYP2C19* loss-of-function alleles increases the risk of cardiovascular events with clopidogrel (Figure 5.4) [16,20,22,30−33]. In a meta-analysis of 9 studies and 9,685 total patients, the majority of whom underwent PCI (91%) and had an ACS (54%), carriers of at least one *CYP2C19* loss-of-function allele had a higher risk of adverse cardiovascular events, with a hazard ratio of 1.57 (95% confidence interval [CI]: 1.13−2.16) compared to noncarriers (i.e., the risk for adverse cardiovascular events was approximately 1.5-fold greater in loss-of-function allele carriers) [34]. The hazard ratio for stent thrombosis was 2.81 (95% CI: 1.81−4.37) for loss-of-function allele carriers compared to noncarriers. Similarly, another meta-analysis of nearly 12,000 patients reported that carriers of the *CYP2C19*2* allele had an increased risk for major adverse cardiovascular events (odds ratio [OR]: 1.29; 95% CI: 1.12−1.49) and stent thrombosis (OR: 3.45; 95% CI: 2.14 to 5.57) compared to noncarriers (i.e., the odds of adverse events were 1.29 times greater, and the odds for stent thrombosis was more than 3 times greater for carriers versus noncarriers) [35]. In contrast, two meta-analyses including more varied patient populations found no association between *CYP2C19* genotype and adverse events with clopidogrel

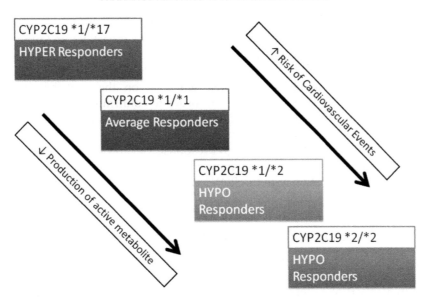

FIGURE 5.4 Effect of cytochrome P450 2C19 (*CYP2C19*) genotype on clopidogrel pharmacokinetics and efficacy.

[36,37]. However, these latter analyses have been criticized for including studies of lower-risk patients, such as those with atrial fibrillation or patients with an ACS managed medically (versus with PCI). The majority of data demonstrate that *CYP2C19* loss-of-function genotype significantly impacts formation of the active clopidogrel metabolite, *ex vivo* inhibition of platelet aggregation with clopidogrel, and clopidogrel's effectiveness in preventing adverse cardiovascular events, particularly for patients undergoing coronary artery stent placement.

The effect of the *CYP2C19*17* gain-of-function allele on clopidogrel responsiveness has also been examined; however, results are inconsistent. This allele has been associated with increased production of the clopidogrel active metabolite and greater inhibition of platelet aggregation with clopidogrel [38,39]. There is some evidence that carriers of a *CYP2C19*17* allele may be at greater bleeding risk [34]. However, the *CYP2C19*17* and *2 alleles are in linkage disequilibrium (LD) such that the *CYP2C19*17* single-nucleotide polymorphism (SNP) is not known to occur on the same allele as *2, thus complicating interpretation of effects observed in *CYP2C19*17* allele carriers. Until the consequences of the *CYP2C19*17* allele are better defined, for clinical purposes, it should probably be considered as a normal functioning allele.

Paraoxonase-1 (PON1) Genotype and Clopidogrel Responsiveness

Paraoxonase-1 (PON1) is an esterase that has been shown to facilitate the activation of clopidogrel *in vitro* [40]. A nonsynonymous polymorphism in the coding region of *PON1*, p.Q192R, has been evaluated for its role in clopidogrel responsiveness. The 192Q allele was associated with increased clopidogrel activation *in vitro* in one study [40]. The same study showed that possession of a 192Q allele was associated with decreased risk of stent thrombosis. However, in contrast to most previous data, the investigators found no association between *CYP2C19* genotype and stent thrombosis risk. As an explanation for their negative findings with *CYP2C19*, the authors argue that CYP2C19 may not have a major role in clopidogrel metabolism. There are no consistent data on the contribution of various CYP450 enzymes to clopidogrel metabolism in order to confirm or refute this explanation [41,42]. However, given that the majority of other studies show an increased risk of coronary events in carriers of a *CYP2C19* loss-of-function allele, especially in the setting of PCI, it is clear that *CYP2C19* contributes to clopidogrel response. Several studies have since demonstrated no association between *PON1* genotype and clopidogrel responsiveness [43—45]. Because of the lack of replication with the *PON1* genotype, *PON1* genotyping is not currently recommended.

Genome-Wide Association Study (GWAS) of Clopidogrel Responsiveness

Investigators for the Pharmacogenomics of Antiplatelet Intervention-1 (PAPI-1) study conducted a genome-wide association study (GWAS) of *ex vivo* platelet aggregation with clopidogrel in a cohort of generally healthy subjects from the Old Order Amish population ($n = 429$) [23]. Each subject was given a 300 mg clopidogrel loading dose, followed by a dose of 75 mg/day for six days, and platelet aggregation was measured before and after clopidogrel administration. Between 500,000 and 1 million SNPs were assessed for each subject to identify genetic associations with clopidogrel responsiveness based on *ex vivo* platelet aggregation. A cluster of 13 highly correlated SNPs on chromosome 10 in the genetic region encoding CYP2C18, CYP2C19, CYP2C9, and CYP2C8 were associated with clopidogrel response. These SNPs were in strong LD with the *CYP2C19*2* polymorphism and explained 12% of the inter-individual variation in platelet aggregation. Of note, no association was seen with the *CYP2C19*17* allele or with polymorphisms in the genes encoding CYP3A, ABCB1, or PON1. In a replication cohort of 227 patients undergoing nonemergent PCI and treated with clopidogrel, the investigators found that, similar to most previous data, the *CYP2C19*2* variant was associated with residual platelet aggregation and an increased risk for cardiovascular events or death at one year, with a hazard ratio of 2.4 (95% CI 1.18 to 4.99) indicating a nearly 2.5-fold greater risk for events or death with the *CYP2C19*2* allele [23].

Clopidogrel Labeling Revisions

Based on substantial data supporting an association between *CYP2C19* loss-of-function alleles and reduced clopidogrel responsiveness, the FDA approved the addition of genetic information to the clopidogrel labeling in March 2010. These label changes include a boxed warning about diminished antiplatelet response to clopidogrel in CYP2C19 poor metabolizers (with 2 loss-of-function alleles). The labeling further states that genetic testing is available and advises consideration of alternative therapy in CYP2C19 poor metabolizers. While these labeling updates highlight the importance of *CYP2C19* genotype in clopidogrel responsiveness, they provide no guidance on when or whom to genotype and little guidance on how to manage poor metabolizers. Further, they do not address CYP2C19 intermediate metabolizers.

Completed and Ongoing Prospective Pharmacogenomic Studies

Several studies have addressed alternative antiplatelet strategies in carriers of a *CYP2C19* loss-of-function allele. The primary strategy investigated to date is an increase in the clopidogrel dose. In a multicenter, double-blind clinical trial, patients with cardiovascular disease and the *CYP2C19*1/*2* or *2/*2* genotype were randomized to receive clopidogrel at varying doses (75 mg, 150 mg, 225 mg, and 300 mg), each for a 14-day period. Platelet function testing was conducted with each dose, and results were compared with those from noncarriers of the *CYP2C19*2* allele receiving clopidogrel 75 mg [46]. For carriers of a single *CYP2C19*2* allele, a clopidogrel dose of 225 mg/day resulted in levels of platelet inhibition similar to that attained with a 75 mg/day dose in noncarriers. However, in *CYP2C19*2* homozygotes, not even the 300 mg/day dose resulted in platelet inhibition comparable to the 75 mg/day dose in noncarriers.

A similar study in healthy volunteers found that poor metabolizers (*CYP2C19*2/*2* or *2/*3* genotypes) receiving a clopidogrel loading dose of 600 mg followed by a maintenance dose of 150 mg/day for five days had similar inhibition of platelet aggregation compared to extensive metabolizers (*CYP2C19*1/*1* genotype) receiving a 300 mg loading dose and 75 mg/day dosing [47]. Intermediate metabolizers (*CYP2C19*1/*2* or *1/*3* genotype) had a similar response as extensive metabolizers with all clopidogrel doses tested. In contrast, a study of patients undergoing PCI after an ACS found that doubling the maintenance dose of clopidogrel in *CYP2C19*2* carriers was not effective in overcoming reduced inhibition of platelet aggregation [48]. Another study of ACS patients undergoing PCI found that providing up to three additional 600 mg clopidogrel loading doses to *CYP2C19*2* carriers, according to the degree of platelet reactivity, was successful in overcoming reduced response with standard 600 mg dosing in some patients [49]. However, 12% of these patients never reached the desired level of inhibition of platelet aggregation. This latter study demonstrates that although titrating clopidogrel dosing based on platelet aggregation testing in carriers of a *CYP2C19* loss-of-function allele may be a viable approach to optimizing the clopidogrel loading dose for some patients, it is not an effective approach for all patients. The inconsistent results among studies are most likely due to differences in study populations (i.e., healthy subjects versus patients with an acute cardiac event undergoing PCI).

Perhaps a more feasible and effective approach to antiplatelet therapy based on *CYP2C19* genotype is to treat *CYP2C19*2* carriers with an alternative antiplatelet agent, namely prasugrel or ticagrelor. Like clopidogrel, prasugrel is a thienopyridine that binds covalently and irreversibly to the $P2Y_{12}$ receptor and is a prodrug requiring bioactivation [50]. In contrast, ticagrelor is administered in its active form and more reversibly binds to the $P2Y_{12}$ receptor to change its conformation.

Prasugrel is currently FDA-approved for use in patients with ACS undergoing PCI. Like clopidogrel, prasugrel is a p-glycoprotein substrate that is converted to the active metabolite via the CYP3A, CYP2B6, CYP2C9, and CYP2C19 enzymes. However, unlike clopidogrel, esterases convert prasugrel to an intermediate metabolite (rather than inactivate metabolite), and the CYP450 bioactivation occurs in a single step (rather than two steps). Likely because of prasugrel's unique bioactivation pathway, common genetic variants in CYP450 enzymes do not affect the pharmacokinetics or clinical efficacy of prasugrel [51]. There is also no association between *ABCB1* genotype and prasugrel pharmacokinetics, possibly because prasugrel is more rapidly metabolized compared to clopidogrel [17]. Ticagrelor is indicated for ACS, regardless of whether patients undergo PCI. The CYP3A4 enzyme is the primary enzyme responsible for ticagrelor metabolism. Similar to prasugrel, there is no evidence that common genetic variation affects ticagrelor pharmacokinetics or efficacy [24]. The data demonstrate that prasugrel and ticagrelor provide greater inhibition of platelet aggregation and greater protection against cardiovascular events compared to clopidogrel in *CYP2C19* loss-of-function allele carriers [17,24,52,53]. However, it is important to note that prasugrel use is contraindicated in patients with a history of stroke or transient ischemic attack, and its use is not recommended in patients 75 years of age or older because of increased bleeding risk.

Guidelines for the Clinical Use of *CYP2C19* Genotyping with Clopidogrel

Several statements by expert consensus panels address *CYP2C19* genotyping to determine clopidogrel responsiveness [29,54]. In 2010, the American College of Cardiology and the American Heart Association Foundation issued a joint clinical alert in response to the addition of genetic information to the clopidogrel labeling [54]. They surmised that the available data were insufficient to recommend genotyping for all patients prescribed clopidogrel. They specifically referred to the lack of outcomes data with routine genetic testing in large cohorts of patients. However, they further stated that *CYP2C19* genotyping may be considered in patients who are at moderate to high risk for poor cardiovascular outcomes, such as those undergoing elective high-risk PCI for extensive and/or very complex disease and others at the clinician's discretion. In these patients, alternative therapies (e.g., prasugrel or ticagrelor) can be considered. In addition, the guidelines state that platelet function testing may also be used in moderate- to high-risk patients to assess clopidogrel responsiveness. The guidelines further suggest that large, prospective, controlled trials will be required for widespread *CYP2C19* genotyping with clopidogrel therapy. However, it will likely be several years before these data are available.

In 2011, the Clinical Pharmacogenetics Implementation Consortium (CPIC) published guidelines regarding the pharmacogenetics of clopidogrel [29]. The guidelines suggest two potential approaches to genotyping for patients requiring dual antiplatelet therapy. One approach is to genotype all patients undergoing PCI, and the second approach is to target moderate- to high-risk patients, such as those with a history of stent thrombosis, diabetes, renal insufficiency, or high-risk coronary angiographic features. The authors state that the clinical data are sufficient to recommend genetic testing for *CYP2C19*2, but not for other variant alleles. Although the CPIC guidelines recommend prescribing prasugrel or ticagrelor in *CYP2C19*2 homozygotes, they state that the evidence is only "moderate" in patients with a single *CYP2C19*2 allele (Table 5.2). The guidelines state that clinicians should consider other factors, such as diabetes mellitus, age, body size, and drug interactions (especially with proton pump inhibitors), when assessing clopidogrel responsiveness. The authors also summarize the challenges related to clinical genetic testing, namely the need for rapid genotyping. In the setting of ACS or emergent PCI, genetic test results are unlikely to be available at the time of antiplatelet initiation.

Information on *CYP2C19* genotyping with clopidogrel has been incorporated into the recent guidelines for the management of patients with ACS or undergoing PCI [55,56]. The PCI guidelines suggest considering genetic testing to determine risk for inadequate platelet

TABLE 5.2   Phenotype classification and therapeutic recommendations from the CPIC based on *CYP2C19* genotype for patients requiring dual antiplatelet therapy[29]

CYP2C19 genotype	Phenotype classification	Therapeutic recommendation	Classification of recommendation
*1/*1	Extensive metabolizer	Clopidogrel label-recommended dosage and administration	Strong
*1/*17 or *17/*17	Ultra-rapid metabolizer	Clopidogrel label-recommended dosage and administration	Strong
*1/*2	Intermediate metabolizer	Prasugrel or ticagrelor if no contraindication	Moderate
*2/*2	Poor metabolizer	Prasugrel or ticagrelor if no contraindication	Strong

inhibition with clopidogrel, but only in patients at high risk for poor clinical outcomes, such as those undergoing high-risk PCI procedures (i.e., elective PCI in patients with unprotected left main, bifurcating left main, or last patent coronary artery). Alternative antiplatelet therapy is recommended for patients found to have the *CYP2C19* loss-of-function genotype. These recommendations are designated as Class IIb based on Level C evidence, meaning that the benefit of genotyping may be slightly greater than or equivalent to not genotyping, reflecting the lack of prospective studies evaluating the clinical utility of *CYP2C19* genotyping with clopidogrel (note to the reader: definitions of evidence levels are provided in the guidelines) [55,56]. The PCI guidelines also state that there is no benefit for the routine use of genetic testing; however, this is classified as the lowest level of recommendation based on the lowest level of evidence (Class III, Level C) [55].

The guidelines for the management of patients with non-ST segment elevation myocardial infarction (NSTEMI) or unstable angina (UA) provide similar recommendations [8]. Specifically, they suggest considering *CYP2C19* genotyping if results may alter management (Class IIb, Level C). Interestingly, these guidelines discuss the potential burden implementing genetic testing might have on clinicians, insurers, and society [8]. The guidelines state that the level of evidence does not warrant mandating genetic testing for the large number of patients with UA and NSTEMI, nor does it warrant the cost that mandatory testing would have on society. However, as additional data become available, the guideline recommendations may change.

Opportunities and Challenges of Pharmacogenomics for Antiplatelet Therapy

Pharmacogenomics may allow for personalized antiplatelet therapy for patients undergoing PCI, thus potentially improving overall patient outcomes. Specifically, in patients found to carry a *CYP2C19* loss-of-function allele who may not attain therapeutic clopidogrel metabolite levels, alternative antiplatelet therapy with prasugrel or ticagrelor may be preferred [55]. On the other hand, CYP2C19 extensive metabolizers could be treated with clopidogrel, rather than ticagrelor or prasugrel, thus avoiding the potential increase in bleeding risk with the newer agents and providing more cost-effective therapy, especially now that clopidogrel is available generically [57].

There are several challenges to clinical implementation of antiplatelet pharmacogenomics. First, some clinicians argue that the risk of adverse cardiovascular events in patients taking clopidogrel is low overall, regardless of genotype [55]. Thus, some have argued that it is not feasible to burden everyone with genetic testing to prevent only a few events [58]. Patients who would potentially benefit the most from CYP2C19 genotyping are those undergoing PCI. Although genotyping prior to a scheduled PCI is feasible, another barrier is the turn-around time for testing in the setting of emergent PCI. A novel point-of-care test for the CYP2C19*2 genotype was recently described, and implementation of such technology would essentially eliminate concerns about turnaround time [59]. In the absence of rapid genotyping capabilities, in which case genotyping results would not be available prior to the antiplatelet loading dose, one approach—not addressed in the guidelines—is to initiate therapy with prasugrel or ticagrelor, then transition to clopidogrel in patients later found to be CYP2C19 extensive or ultra-rapid metabolizers.

Another barrier is that genetic testing for clopidogrel efficacy may not be reimbursed by third-party insurers. Importantly, prasugrel and ticagrelor are both associated with a greater risk for noncoronary artery bypass surgery—related bleeding compared to clopidogrel. Thus, although routine substitution of these agents for clopidogrel may be done in lieu of genetic testing, such an approach may lead to increased bleeding events. Now that clopidogrel is available generically, resulting in a significant cost differential among agents, genotyping to identify CYP2C19 extensive and ultra-rapid metabolizers, who would be candidates for clopidogrel while reserving the newer, more expensive agents for poor metabolizers and possibly intermediate metabolizers may be a cost effective and safe approach to antiplatelet therapy. The PAPI-2 study is a multisite randomized clinical trial that is designed to evaluate the effects of such an approach on clinical outcomes and costs among 7,200 patients undergoing PCI. Patients are randomized to standard of care, with antiplatelet therapy chosen at the discretion of the treating physician, or genotype-directed therapy (clopidogrel for extensive or ultra-rapid metabolizers and prasugrel for poor or intermediate metabolizers). The trial is expected to be completed in late 2014 (clinicaltrials.org, NCT 01452152).

WARFARIN PHARMACOGENOMICS

Challenges with Warfarin

Warfarin is an oral anticoagulant that is widely prescribed for the prevention of thromboembolism in individuals with a history of venous thromboembolism or recent orthopedic surgery and for stroke prevention in patients with atrial fibrillation or heart valve replacement. Newer oral anticoagulants, including dabigatran and rivaroxaban, were recently approved by the FDA for prevention of thromboembolism in select indications. Newer agents have an advantage over warfarin in that they do not require frequent (e.g., at least once every 4 to 6 weeks) monitoring to assess their anticoagulant activity. However, even with expanding indications for these agents, warfarin will likely remain the treatment of choice for many patients requiring chronic anticoagulation, particularly for those who fail therapy with newer agents (i.e., develop thrombosis despite therapy), develop significant bleeding or other intolerable adverse effects (e.g., gastrointestinal effects) with newer agents, or have significant renal impairment and are unable to safely take newer agents.

Although in use for nearly 60 years, warfarin remains a difficult drug to manage primarily because of its narrow therapeutic index and the wide interpatient variability in the dose required to obtain optimal anticoagulation. For most indications, warfarin is dosed to achieve an international normalized ratio (INR, a measure of its anticoagulant activity) of 2 to 3. Failure to achieve optimal anticoagulation significantly increases the risk for adverse sequelae. Specifically, subtherapeutic anticoagulation increases the risk for thromboembolism, and supratherapeutic anticoagulation (particularly when the INR exceeds 4) increases the risk for bleeding [60,61]. Because of the difficulty in achieving therapeutic anticoagulation with warfarin, warfarin consistently ranks among the leading causes of serious drug-related adverse events, prompting a boxed warning in its FDA-approved labeling [62]. Achieving therapeutic anticoagulation in an efficient manner is therefore a priority for clinicians managing warfarin therapy.

The warfarin dose required to achieve an INR within the therapeutic INR range varies by as much as 20-fold among patients [63]. There are also significant differences in warfarin dose requirements by ancestral origin, with African Americans generally requiring higher doses and Asians requiring lower doses compared to Caucasians [64]. Thus, a major challenge with initiating warfarin therapy is predicting the dose that will produce therapeutic anticoagulation for a particular patient. Traditionally, warfarin is initiated at a similar dose for all patients, typically 5 mg/day, with the dose adjusted according to INR results. The problem with this trial-and-error dosing approach is that it often leads to over-anticoagulation during the initial months of therapy when the risk for bleeding is greatest [65]. Clinical factors, including age, body size, diet, and medications that interfere with warfarin metabolism, influence warfarin dose requirements [66]. More recently, severe renal dysfunction (estimated glomerular filtration rate <30 ml/min/1.73 m^2) was associated with 19% lower warfarin dose requirements and a twofold greater risk for warfarin-related major bleeding compared with mild to no renal dysfunction [67,68]. The mechanism underlying this association is unknown. Clinical factors alone account for only 15 to 20% of the overall variability in warfarin dose, and considering these factors alone is often insufficient to predict the dose of warfarin a patient will require [63,69,70].

Genes Affecting Warfarin Pharmacokinetics and Pharmacodynamics

Based on research over the past decade, it is now widely recognized that genotype significantly influences the pharmacokinetics and pharmacodynamics of warfarin and contributes to the interpatient variability in warfarin dose requirements [71,72]. The major genes influencing warfarin pharmacokinetics and pharmacodynamics are *CYP2C9* and *VKORC1*, respectively. As shown in Figure 5.5, CYP2C9 metabolizes the *S*-enantiomer of warfarin to the inactive 7-hydroxy warfarin protein. The *S*-enantiomers possess approximately three to five times the anticoagulant effects of *R*-warfarin [65]. The *VKORC1* gene encodes for the target site of warfarin. Specifically, warfarin inhibits vitamin K epoxide reductase complex 1 (VKORC1), thus preventing formation of vitamin K hydroquinone, a necessary cofactor for the gamma-carboxylation and activation of clotting factors II, VII, IX, and X.

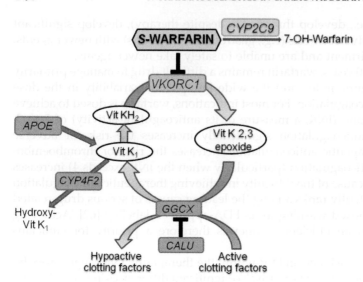

FIGURE 5.5 Genes involved in warfarin pharmacokinetics and pharmacodynamics. The *CYP2C9* gene influences the drug's pharmacokinetics; other genes affect warfarin's pharmacodynamics. *CYP2C9*, cytochrome P450 2C9; *VKORC1*, vitamin K epoxide reductase complex subunit 1; *GGCX*, gamma-glutamyl carboxylase; *CALU*, calumenin; *APOE*, apolipoprotein E; *CYP4F2*, cytochrome P450 4F2; vit, vitamin.

There are now substantial and convincing data from numerous candidate genes studies in addition to GWASes showing that the *CYP2C9* and *VKORC1* genotypes affect warfarin dose requirements [63,69,70,72—77]. Other genes, including *CYP4F2*, calumenin, and gamma glutamyl carboxylase (GGCX), produce lesser effects on warfarin pharmacodynamics and provide minor contributions to the variability in warfarin dose requirements [78—81]. These genes are described in detail in the following sections. The major goal of warfarin pharmacogenomics is to improve the accuracy of warfarin dosing and, consequently, to reduce the risk for adverse sequelae with warfarin therapy.

CYP2C9 Genotype and Warfarin Response

The *CYP2C9* gene is located on chromosome 10q24.1, and more than 35 *CYP2C9* alleles have been described, as detailed in chapter 1 [82]. The *CYP2C9*2* and *3* alleles are the most extensively studied and result from SNPs in the coding regions of the gene, as shown in Table 5.3. The *CYP2C9*2* amino acid substitution occurs on the outer surface of the enzyme, and the *3* substitution occurs internally [83,84]. Neither substitution appears to affect substrate binding. Rather, evidence suggests that the *CYP2C9*2* and *3* alleles disrupt formation of intermediate compounds in the CYP2C9 catalytic cycle leading to significant reductions in enzyme activity [85]. As a result, the clearance of *S*-warfarin is reduced approximately 40% with the *CYP2C9*1/*2* genotype, up to 75% with the *1/*3* genotype, and up to 90% with the *3/*3* genotype [71,86—88]. Accordingly, individuals with the *CYP2C9*1/*2* or *1/*3* genotypes required dose reductions of 30 to 47%, respectively, compared to those with the *CYP2C9*1/*1* (wild-type) genotype [71]. Individuals with the *CYP2C9*3/*3* genotype need up to 80% lower warfarin doses than *CYP2C9*1* homozygotes [87,88].

The *CYP2C9*2* and *3* alleles are the most common variant *CYP2C9* alleles in Caucasians but much less prevalent among Asians and African Americans, as shown in Table 5.3. The

TABLE 5.3 Nucleotide base pair or amino acid substitution, location, and minor allele frequencies of variants associated with warfarin dose response in various populations[64,73,80,96,98,225,226]

Polymorphism	Base pair or amino acid substitution	Location	Minor allele frequency			
			Caucasian	African American	Asian	Egyptian
CYP2C9						
*2 (rs1799853)	p.R144C	Exon 3	0.13-0.14	0.01-0.02	<0.01	0.12
*3 (rs1057910)	p.I359L	Exon 7	0.06-0.11	0.01	0.02-0.04	0.09
*5 (rs28371686)	p.D360E	Exon 7	<0.01	0.01	<0.01	0.01
*6 (rs9332131)	10601delA	Exon 5	<0.01	0.01	<0.01	NR
*8 (rs7900194)	p.R150H	Exon 3	<0.01	0.05-0.07	0.01	<0.01
*11 (rs28371685)	p.R335W	Exon 7	<0.01	0.01-0.04	<0.01	NR
VKORC1						
rs9923231	c.-1639G>A	Promoter	0.39	0.11	0.91	0.46
rs9934438	c.1173C>T	Intronic	0.40	0.10	0.90	NR
CYP4F2						
rs2108622	p.V433M	Exon 11	0.24	0.07	0.23	0.42
CALU						
rs339097	c.A>G	Intronic	<0.01	0.14	0.02	0.02
GGCX						
rs11676382	c.C>G	Intronic	0.15	0.01	<0.01	NR

CYP, cytochrome P450; VKORC1, vitamin K epoxide reductase complex subunit 1; APOE, apolipoprotein E; CALU, calumenin; GGCX, gamma-glutamyl carboxylase

CYP2C9*8 allele is one of the most common variant CYP2C9 alleles in African Americans but is virtually absent in other populations [89]. The CYP2C9 *5, *6, and *11 alleles also occur almost exclusively in African Americans but at much lower frequencies than the *8 allele. The CYP2C9*5, *8, and *11 alleles result from nonsynonymous SNPs in gene coding regions, whereas CYP2C9*6 results from a nucleotide deletion (Table 5.3). Decreased enzyme activity and clearance of CYP2C9 substrates have been reported with the CYP2C9*5, *6, *8 and *11 alleles [90–92]. However, allele effects appear to be somewhat substrate specific. For example, CYP2C9*8 decreases enzyme activity toward warfarin and phenytoin, increases enzyme activity toward tolbutamide, and has no effect on losartan metabolism [91–94]. The CYP2C9*8 allele decreases clearance of S-warfarin by 25 to 30% [95]. This decrease coincides with about a 20% reduction in warfarin dose requirements with the CYP2C9*8 allele [73]. Similarly, lower warfarin dose requirements have been reported in individuals with a CYP2C9*5, *6, or *11 allele [73,74,96].

A novel CYP2C9 variant located in intron 3 has also been associated with higher warfarin dose requirements in African Americans [96]. Specifically, the CYP2C9 c.18786A>T variant, which occurs in approximately 40% of African Americans, is associated with 3.7 mg/week higher warfarin dose requirements per each 18786T allele [96].

In addition to affecting warfarin dose requirements, the *CYP2C9* genotype is associated with the risk of over-anticoagulation and bleeding during warfarin therapy [88,97,98]. Specifically, warfarin-treated patients with a *CYP2C9* variant allele have about a twofold greater risk for bleeding compared to *CYP2C9*1* homozygotes [98,99]. Although the risk for bleeding with a *CYP2C9* variant allele is highest during the initial months of warfarin therapy, there is evidence that it persists during chronic therapy. Thus, patients with a *CYP2C9* variant allele should be monitored closely for signs and symptoms of bleeding throughout warfarin therapy [98].

VKORC1 Genotype and Warfarin Dose Requirements

The *VKORC1* gene is located on chromosome 16p11.2 and was initially discovered in the context of warfarin resistance, where exceptionally high doses of warfarin (e.g., >20 mg/day) are needed to achieve therapeutic anticoagulation [100]. Warfarin resistance is due to nonsynonymous (or missense) mutations in the *VKORC1* coding region. Variants contributing to warfarin resistance are deemed mutations (versus polymorphisms) because they are rare in most populations. An exception is in the Ashkenazi Jewish population, where individuals have a relatively high prevalence (8%) of the *VKORC1* p.D36Y variant, leading to a higher prevalence of warfarin resistance in this population [101].

In 2005, investigators described more common *VKORC1* variants that contribute to warfarin dose variability across the general population [72,102]. Common *VKORC1* polymorphisms occur in the gene's regulatory region rather than coding regions that house the warfarin resistance mutations. Seven common *VKORC1* variants were initially associated with warfarin dose requirements in Caucasians [72]. These SNPs comprise two major haplotypes, commonly designated as *VKORC1* haplotypes A and B [72]. Haplotype A is associated with lower mRNA expression and warfarin maintenance dose compared to haplotype B. Mean warfarin dose requirements reported with the AA, AB, and BB haplotype combinations are 2.7 mg/day, 4.9 mg/day, and 6.2 mg/day, respectively [72].

Of the SNPs comprising *VKORC1* haplotypes A and B, only the c.-1639G>A (rs9923231) and possibly the c.1173C>T (rs9934438) SNP appear to be functional. Specifically, *in vitro* studies in liver tissue showed that the -1639G>A and 1173C>T variants were associated with twofold allelic mRNA expression imbalance (e.g., twofold lower gene expression) [103]. The -1639A and 1173T SNPs are in near complete linkage disequilibrium in individuals of European, Asian, and African descent [64]. Numerous studies have consistently demonstrated that the -1639A and 1173T alleles are associated with significantly lower warfarin dose requirements in these populations [63,64,69,70,73,74,104–106]. On average, the -1639 AA, AG, and GG genotypes predict warfarin maintenance doses of 3 mg/day, 5 mg/day, and 6 mg/day, respectively, which coincide with dose predictions according to the AA, AB, and BB haplotype combinations. The -1639G>A and 1173C>T SNP are equally predictive of dose requirements [64]. Thus, only one of these SNPs needs to be considered for warfarin dosing decisions. This greatly simplifies genotype-guided warfarin dosing compared to dosing based on *VKORC1* haplotype because only one SNP needs to be genotyped.

As shown in Table 5.3, the frequency of the *VKORC1* -1639A allele differs significantly by ancestry, with a greater frequency in Asians and lower frequency in African Americans compared to Caucasians. Approximately 50% of Caucasians have the -1639AG genotype, associated with intermediate VKORC1 sensitivity and usual (i.e., 5 mg/day) warfarin dose requirements. The -1639 AA genotype is the most common genotype in Asians and is associated with high VKORC1 sensitivity and low warfarin dose requirements. The most common genotype in African Americans is -1639GG, which is associated with lower VKORC1 sensitivity and high dose requirements. The difference in *VKORC1* genotype distribution among ancestral groups contributes to the higher mean warfarin maintenance dose in African Americans and lower mean dose in Asians, compared to Caucasians, independent of the effects associated with *CYP2C9* genotype [64].

Investigators recently resequenced the *VKORC1* gene in African Americans and discovered a novel variant associated with higher warfarin dose requirements [96]. Specifically, the *VKORC1* c.-8191A>G SNP was predictive of a 5.2 mg/week dose increase for each *VKORC1* -8191G allele. The -8191A>G allele is located upstream from the gene transcriptional start site and occurs in approximately 70% of individuals of African ancestry. A linear regression model including the novel *VKORC1* -8191A>G and *CYP2C9* 18786A>T variants plus the *VKORC1* 1173C>T; *CYP2C9*2, *3, *5, *6, *8,* and *11* alleles; and clinical factors explained 40% of the overall variability in warfarin dose in African Americans [96]. In contrast, the model without the novel variants explains only 26% of the variability in warfarin dose in African Americans [70]. These data demonstrate the importance of accounting for variants that are common in African Americans and associated with warfarin dose requirements when making warfarin dosing decisions.

CYP4F2 Genotype and Warfarin Dose Requirements

The CYP4F2 enzyme is responsible for metabolizing vitamin K_1 to hydroxyvitamin K_1, as shown in Figure 5.5 [107]. This process results in less vitamin K_1 being available for reduction to vitamin KH_2, which is necessary for clotting factor activation. Thus, increased CYP4F2 activity leads to reduced clotting factor activation. The *CYP4F2* p.V433M SNP in exon 11 leads to lower CYP4F2 protein concentration and consequently to greater vitamin K availability [107].

In an initial study of three independent Caucasian cohorts, the *CYP4F2* 433M/M genotype was associated with approximately 1 mg/day higher warfarin dose requirements compared to the V/V genotype, with heterozygotes requiring intermediate doses [78]. Subsequent studies in Caucasians and Asians confirmed the association between V433M genotype and warfarin dose requirements [76,77,108–110]. The *CYP4F2* V433M genotype explains approximately 1 to 3% of the overall variability in warfarin dose in these populations [76,110]. Interestingly, the association between *CYP4F2* genotype and warfarin dose requirements was not observed in African Americans, Indonesians, Egyptians, or children [73,111–113]. The lack of association in African Americans is probably because of the low frequency of the 433M allele in individuals of African ancestry. Body size provides a greater contribution to warfarin dose variability in children versus adults, potentially explaining the negative findings with the *CYP4F2* genotype in a pediatric population. However, the 433M allele is common in Indonesians and Egyptians, and the explanation for the negative association in these groups is unclear.

OTHER GENETIC CONTRIBUTIONS TO WARFARIN DOSE VARIABILITY AND RESPONSE

Calumenin

Calumenin inhibits gamma-carboxylation of vitamin K–dependent proteins, suggesting that *CALU* may influence warfarin dose requirements [114]. The *CALU* variant, rs339097 A>G, was associated with warfarin maintenance dose in a diverse patient cohort [79]. Specifically, the minor rs339097G allele was significantly overrepresented among patients requiring high (mean dose of 13 mg/day) versus low (mean dose of 2.6 mg/day) warfarin doses. The association between the rs339097 variant and warfarin dose requirements was validated in a separate diverse cohort and in a cohort of African Americans [79]. In a pooled analysis of 241 African Americans, the G allele was associated with an 11% higher warfarin dose than predicted based on clinical factors, *CYP2C9*, and *VKORC1*. The correlation of the rs339097G allele with higher warfarin doses was confirmed in a separate study of Egyptian patients, in whom the variant allele was associated with 14 mg/week higher dose requirements [112]. The rs339097G allele is common among African Americans but rare in other populations, as shown in Table 5.3.

GGCX

The gamma-glutamyl carboxylase (GGCX) enzyme plays an essential role in biosynthesis of vitamin K–dependent clotting factors by carboxylating protein-bound glutamate residues. Thus the *GGCX* gene is a candidate for affecting warfarin pharmacodynamics. Rare *GGCX* mutations cause deficiencies in vitamin K–dependent clotting factors [115]. There are also common *GGCX* variants that have been reported to influence warfarin dose in the general population [116–119]. In particular, the *GGCX* rs11676382 SNP was associated with a 6% reduction in warfarin dose per minor G allele. The rs11676382 variant is common in Caucasians but rare in African Americans [119,120]. On the other hand, the *GGCX* rs10654848 (CAA) 16/17 repeat polymorphism occurs almost exclusively in African Americans and is associated with warfarin dose requirements of greater than 7.5 mg/day in this population [121].

Genome-Wide Association Studies

Three GWASes with warfarin have been completed and confirm that the *CYP2C9* and *VKORC1* genes are the primary contributors to warfarin dose requirements in Caucasian and Asian populations [75,76,77]. In an initial GWAS, investigators surveyed over 538,000 SNPs in a discovery cohort of 181 Caucasians and 2 independent replication cohorts consisting of 374 Caucasians taking warfarin [75]. An SNP in complete linkage disequilibrium with the *VKORC1* -1639G>A variant had the most significant effect on warfarin dose in the index population and explained approximately 25% of the overall variance in dose requirements. The *CYP2C9*2* and *CYP2C9*3* alleles provided modest contributions to warfarin dose and explained an additional 9% of the variability. These associations were validated in the replication cohort. The combination of *VKORC1*, *CYP2C9*, and clinical factors (age, sex, weight, amiodarone use, and losartan use) explained 47% of total variance in warfarin maintenance dose [75].

In a second GWAS, over 325,000 SNPs were tested for their association with warfarin dose in 1,053 Swedish patients [76]. Similar to the first GWAS, the *VKORC1* locus had the strongest association with warfarin dose, followed by SNPs clustered around *CYP2C9*. After adjustment for *VKORC1*, *CYP2C9*, age, and gender, the only other SNP reaching genome-wide significance with warfarin dose was *CYP4F2* V433M, which explained an additional 1 to 2% of the variability. Results were confirmed in a replication cohort of 588 Swedish patients. A third GWAS was conducted in Japanese patients [77]. Similar to the studies in Caucasians, *VKORC1* was found to provide the greatest contribution to warfarin maintenance dose, with *CYP2C9* and *CYP4F2* providing lesser contribution.

A GWAS in African Americans was recently completed. Initial results suggest that the *VKORC1* -1639G>A variant is also the major contributor to warfarin dose in those of African descent [122]. However, investigators also identified a novel association between the rs12777823 variant, located near the *CYP2C18* gene, and warfarin dose requirements in African Americans. This association was replicated in a separate cohort of African American patients. Neither the *CYP4F2* nor *CALU* genotypes were associated with warfarin dose in the GWAS, but this result may be reflective of their low minor allele frequencies in African Americans (e.g., insufficient power to detect an association). Neither the *GGCX* repeat polymorphism nor novel *VKORC1* -8191A>G variant were included on the GWAS platform, and thus their association with warfarin dose was not assessed. This situation illustrates the limitation of GWASes in that not all variants, particularly those of interest in minority groups, are captured [123].

International Warfarin Pharmacogenetics Consortium (IWPC)

There are a number of published algorithms to assist clinicians with warfarin dosing when genotype is known [69,105,124–128]. Most contain the *VKORC1* -1639G>A or 1173C>T SNP, *CYP2C9* *2 and *3 alleles, and clinical factors, including age, body size, and amiodarone use [69,70,128–132]. Warfarin pharmacogenomic dosing algorithms explain between 30% and 60% of the variability in warfarin dose requirements in Caucasians but less of the variability in African Americans and Asians [64,128]. A limitation of these algorithms is that most were derived from geographically limited populations of relatively small size.

The International Warfarin Pharmacogenetics Consortium (IWPC) was formed by members of the Pharmacogenomics Knowledge Base (PharmGKB) in collaboration with investigators from the international community with the initial purpose of creating a dosing equation that would have global clinical utility (see chapter 3 for further information about the PharmGKB) [70,133]. Researchers from 21 groups representing 11 countries and 4 continents pooled genotype and phenotype data for over 5,700 chronic warfarin-treated patients [64,70]. The resulting equation contains clinical factors and *CYP2C9* and *VKORC1* genotypes. The algorithm is freely available through the PharmGKB and www.warfarindosing.org websites and explains 40% of the variability in warfarin dose among Caucasians and approximately 25% among Asians and African Americans.

More recently, IWPC investigators assessed the influence of common *VKORC1* polymorphisms on warfarin dose among Asian, African American, and Caucasian patients and found that none of the seven originally described *VKORC1* SNPs or haplotypes contributed to warfarin dose requirements beyond the -1639G>A or 1173C>T variant [64,72]. The -1639G>A variant explained greater variability in warfarin dose among Caucasians compared to African Americans or Asians. However, the differences in -1639A allele frequency among ancestral groups (e.g., significantly lower frequency in African Americans) largely accounted for the ancestral difference in the contribution of *VKORC1* genotype to dose variability.

Warfarin Labeling Revisions

In August 2007, the FDA approved the addition of pharmacogenomic data to the warfarin labeling. The pharmacogenomic content of the label was further revised in January 2010, with the addition of a dosing table based on *CYP2C9* and *VKORC1* genotypes (Table 5.4). The table may be used to determine initial warfarin dose when genotype is known, with subsequent dose adjustment based on INR results. An advantage of the table over dosing algorithms is its ease of use. However, the revised warfarin labeling does not require pharmacogenetic testing for warfarin dosing. Thus, genetic testing and use of the dosing table is at the discretion of the clinician.

Comparison of Pharmacogenomic Dosing Tools

Of the available warfarin dosing algorithms, the three derived from the largest populations and most commonly cited are the IWPC algorithm,[70] the dosing table in the warfarin labeling, and an algorithm by Gage and colleagues, [69] which is freely available through the www.warfarindosing.org website. The www.warfarindosing.org algorithm can account for previous warfarin doses and INR values and thus may be preferred over the use of other algorithms when genotype results are not immediately available.

The IWPC and www.warfarindosing.org algorithms are consistently shown to be the most accurate of the warfarin dosing methods [134–136]. The pharmacogenetic table in the warfarin labeling is generally less accurate at predicting warfarin dose, possibly because

TABLE 5.4   Warfarin dose by *CYP2C9* and *VKORC1* genotypes, as reproduced from the FDA-approved warfarin label

VKORC1 -1639G>A	CYP2C9					
	*1/*1	*1/*2	*1/*3	*2/*2	*2/*3	*3/*3
GG						
GA						
AA						

5–7 mg/day

3–4 mg/day

0.5–2 mg/day

the table predicts doses within the confined range of 0.5 to 7 mg/day, whereas dosing algorithms allow for dose prediction above 7 mg/day [137]. In addition, pharmacogenomic algorithms may better account for the clinical factors influencing warfarin dose variability. Pharmacogenomic algorithms are especially superior compared to other dosing methods for patients requiring low (≤3 mg/day) or high (≥7 mg/day) warfarin doses [70,138].

Warfarin pharmacogenomic algorithms have several limitations. First, they estimate doses within 20% of the actual dose about 50% of the time [134–136]. Pharmacogenomic algorithms do not include all of the factors known to affect warfarin dose variability, such as vitamin K intake and many of the drugs known to interact with warfarin. In addition, most algorithms, including the www.warfarindosing.org and IWPC algorithms, do not contain genetic variables that are specific to African Americans (e.g., *CYP2C9*8*), likely contributing to lesser accuracy in individuals of African ancestry [135]. Also, many algorithms do not include genetic variants associated with warfarin resistance and are thus less accurate at predicting higher than usual doses [139]. Finally, pharmacogenomic algorithms may overestimate doses in elderly patients (>65 years) who often require warfarin doses of less than 2 mg/day [140]. As such, pharmacogenomic algorithms are useful to reduce uncertainty about initial warfarin doses. However, they should not replace routine INR monitoring and clinical judgment.

Prospective Studies of Genotype-Guided Warfarin Dosing

Two small prospective trials and a comparative effectiveness study provided initial evidence of benefit with genotype-guided warfarin dosing [126,141]. Specifically, in a study of 191 patients, warfarin dosing based on clinical factors plus *CYP2C9* (but not *VKORC1*) genotype resulted in more rapid attainment of stable anticoagulation, more time spent within the therapeutic range, and a lower incidence of minor bleeding than dosing according to clinical factors alone [126]. Similarly, another small trial showed that warfarin dosing based on both *CYP2C9* and *VKORC1* genotypes improved the time to achieve stable dosing [105]. A comparative effectiveness study further showed that patients who were offered free *CYP2C9* and *VKORC1* genotyping, with results provided to their physician with an interpretive report, had fewer hospitalizations for any cause and fewer hospitalizations for bleeding or thromboembolism during the initial six months of warfarin therapy compared to historical controls [141].

In contrast to data supporting genotype-guided warfarin therapy, two small, randomized trials showed no benefit with such an approach over traditional dosing [139,142]. In particular, both trials showed that the percent of time spent within the therapeutic range, which is often used as a marker of bleeding or thrombotic risk, was similar between patients dosed based on genotype plus clinical factors or clinical factors alone. However, these trials were small in size, including only 206 to 230 patients. In addition, an exploratory analysis of one trial, called the CoumaGen-I trial, showed a benefit with pharmacogenomic dosing for two groups of patients: those with more than one variant allele and those with the wild-type genotype (*VKORC1* -1639 CC and *CYP2C9 *1/*1*) [139]. In contrast, single-variant allele carriers appeared to have no benefit from genotype-guided dosing, likely because patients with a single variant usually require a warfarin dose of about 5 mg/day, which is the dose commonly started in patients new to warfarin. In contrast, those with multiple variant alleles usually require lower doses (e.g., 3 to 4 mg/day), and those with the wild-type genotype usually require higher doses

(6 to 7 mg/day). Thus, starting a dose of 5 mg/day in individuals with multiple or no variant allele would probably result in over- and under-coagulation, respectively.

The results from CoumaGen-II were published more recently [143]. CoumaGen-II involved (1) a blinded, randomized comparison of two pharmacogenomic dosing algorithms and (2) a clinical effectiveness comparison of genotype-guided warfarin dosing ($n = 504$) versus standard dosing ($n = 1911$). For the comparison of dosing algorithms, a modified version of the IWPC algorithm (taking into account smoking status and different INR targets) was compared to a three-step algorithm in which the *CYP2C9* genotype was not taken into account until day 3, and a dose-revision algorithm was used starting on day 4, taking into account warfarin dosing history and INR. The three-step algorithm was found to be noninferior, but not superior, to the modified IWPC algorithm in terms of the percent of out-of-range INR values at one and three months. Thus, the two pharmacogenomic dosing approaches were combined for comparison with standard dosing. Genotype-guided therapy (using either algorithm) was superior to standard warfarin dosing in reducing the percent of out-of-range INRs, the percent of INRs greater than or equal to 4 or less than or equal to 1.5, and serious adverse events at three months.

Ongoing Clinical Trials of Genotype-Guided Warfarin Therapy

There are at least four multicenter, randomized, clinical trials underway in the United States and Europe to assess the clinical utility of warfarin pharmacogenomic testing [144]. These include the Clarification of Optimal Anticoagulation through Genetics (COAG); the Genetics Informatics Trial (GIFT) of Warfarin to Prevent Deep Venous Thrombosis; the Clinical, Economic Implication of Genetic Testing for Warfarin Management trial; and the European Pharmacogenetics of Anticoagulation Trial (EU-PACT). The details of these trials are shown in Table 5.5. The COAG trial, in particular, is powered to account for the potential lack of benefit with genotype-guided therapy in patients with a single variant. Most trials include the percent of time spent within the therapeutic INR range during the initial weeks of therapy as their primary outcome. Results from these trials will likely have a significant impact on the future of warfarin pharmacogenomics in that they will provide evidence for or against the clinical utility of genotype-guided warfarin therapy. These trials will also provide data on the cost-effectiveness of warfarin pharmacogenomics.

Opportunities and Challenges for Warfarin Pharmacogenomics

Genotype-guided warfarin dosing has the potential to improve time to reach therapeutic anticoagulation. In addition, information about genotype would allow for starting lower-than-usual warfarin doses in patients known to carry a *CYP2C9* or *VKORC1* variant, thus potentially avoiding over-anticoagulation and bleeding. Knowledge of *CYP2C9* genotype may also better inform warfarin dose adjustments, with slower dose titration in individuals known to have a *CYP2C9* variant allele. However, despite substantial data supporting genetic determinants of warfarin response, warfarin labeling revisions, the availability of decision support tools (dosing algorithms), and the FDA-clearance of at least four warfarin genotyping platforms, pharmacogenomics is not routinely integrated into the management of warfarin-treated patients.

There are several challenges to clinical implementation of warfarin pharmacogenomics that must first be addressed. First, guidelines from expert consensus groups, including the

TABLE 5.5 Details of ongoing clinical trials to assess the clinical utility of genotype-guided warfarin dosing

Trial name or acronym	Intervention	Outcomes	Estimated enrollment	Estimated completion date
Clarification of Optimal Anticoagulation Through Genetics (COAG)	Genotype-guided versus clinical-guided warfarin dosing algorithm	Primary: Time spent within the therapeutic INR range in the first 4 weeks Secondary: Occurrence of an INR >4 or serious clinical event in the first 4 weeks	1238	3/2013
Clinical and Economic Implications of Genetic Testing for Warfarin Management	Prescribers are provided with genetic data and dosing recommendations versus standard of care	Primary: Clinical outcomes and costs associated with use of genetic testing Secondary: Clinical outcomes: inpatient length of stay, supratherapeutic dosing; time in range, incidence of DVT, stroke, PE, GI, and intracranial hemorrhage in the first year Costs: hospital, genetic, medication, laboratory, and total costs	268	7/2011
European Pharmacogenetics of AntiCoagulant Therapy (EU-PACT)	Genotype versus nongenotype guided warfarin, acenocoumarol, or phenprocouman dosing	Primary: Percent of time in an INR range at 12 weeks No. of patients with an INR ≥4 Secondary: Time to an INR ≥4 Percent of time with an INR ≥4 Percent of time with an INR ≤2 Time to reach therapeutic INR Others	970	11/2012
Genetics Informatics Trial (GIFT) of Warfarin to Prevent DVT	Pharmacogenetic versus clinical warfarin initiation	Primary: Nonfatal venous thromboembolism; Nonfatal hemorrhage; Death from any cause; INR ≥ 4 (all during the initial 4-6 weeks) Secondary: Percent time in range Time to first INR>1.5 and target INR	1600	8/2015

DVT, deep vein thrombosis; PE, pulmonary embolism; GI, gastrointestinal; INR, international normalized ratio

American College of Chest Physicians and the American College of Medical Genetics, recommended against routine genetic testing to guide warfarin therapy, [65,145,146] predominantly because these groups traditionally rely on evidence from randomized, controlled clinical trials to guide therapy. The ongoing trials, discussed above, should provide evidence for or against the clinical utility of genotype-guided warfarin therapy and will likely influence future guidelines [147].

In the meantime, the CPIC has provided guidelines on how to interpret and apply genetic test results to warfarin dosing when such results are available [148]. The CPIC does not however address when or whom to genotype, leaving these determinations to the discretion of the clinician. The CPIC guidelines were written in recognition that the available data strongly support a genetic influence on dose requirements and that the dose should be adjusted when genotype is known. These guidelines provide a rating of A (strong) for warfarin dosing based on genotype when genotyping results are available [148].

Another important challenge for genotype-guided warfarin therapy is the lack of reimbursement for genetic testing from some insurers. Although some private companies, such as Medco, pay for warfarin genetic testing, the Centers for Medicare and Medicaid Services has announced that coverage for genetic testing to guide warfarin therapy would be denied unless testing is provided in the context of a controlled clinical study. Thus, even if clinicians are in favor of genotype-guided warfarin dosing, the lack of reimbursement may prevent them from going forward with genetic testing.

A third important challenge is the need to establish clinical utility, an often-cited reason for hesitancy to embrace warfarin pharmacogenomics. The CoumaGen-II study (discussed above), other small, prospective, randomized trials, and a comparative effectiveness study provide some evidence of the clinical utility of genotype-guided therapy. Larger, randomized, clinical trials are ongoing to further address the clinical utility of a genotype-guided approach to warfarin dosing.

Another challenge is obtaining timely genotype results. Clinical laboratories that lack the personnel or equipment for rapid genotyping may need to send samples to an outside facility, which may delay the availability of results. Nonetheless, there are data showing that even with a four- to seven-day delay to obtain genotype data, a pharmacogenomic algorithm incorporating previous doses and INR results provides better prediction of warfarin maintenance dose than clinical factors alone [132,149]. In addition, with continuing technological advances, time to genotyping results will continue to decrease. Further, some institutions, such as the University of Florida, are exploring *a priori* genotyping for patients who provide consent. Genotype results may be placed in the medical records so that they are available in the event that a drug such as warfarin is needed.

TRIALS AND TRIBULATIONS OF PHARMACOGENOMICS OF AGENTS USED TO TREAT DYSLIPIDEMIA

Overview of Statin Pharmacokinetics and Pharmacodynamics

The 3-hydroxy-3-methylglutaryl coenzyme A (HMG-CoA) reductase inhibitors, also known as statins, are commonly prescribed to reduce low-density lipoprotein (LDL) cholesterol. Multiple randomized, placebo-controlled, clinical trials have demonstrated that statins reduce the relative risk of major coronary events [150]. However, there is substantial variability in LDL cholesterol lowering and clinical outcomes with statin therapy [151,152]. Genes associated with the pharmacokinetics and pharmacodynamics of statins have been studied for their contribution to this variability.

Table 5.6 shows the various enzymes involved in statin transport and metabolism. The CYP3A4 enzyme plays an important role in the metabolism of lovastatin, simvastatin, and atorvastatin; fluvastatin and rosuvastatin are metabolized primarily by CYP2C9 [153]. Pravastatin is primarily eliminated unchanged in the feces and urine, and pitavastatin is a substrate for UGT1A3 and UGT2B7. Most statins are transported by OATP1B1 into hepatocytes, [154] where they are competitive inhibitors of HMG-CoA reductase, the rate-limiting enzyme involved in cholesterol synthesis. All statins share this uniform mechanism of action.

Genetic Contributors to Plasma Lipid Levels

Plasma lipid levels are highly heritable traits, with over 50% of the interindividual variation in LDL cholesterol levels attributed to genetic factors [155]. Mutations in single genes with severe functional consequences contribute to Mendelian lipid disorders (also referred to as familial hypercholesterolemia); polymorphisms in multiple genes, each with fairly weak to moderate effects, contribute to variation in lipid levels across the general population. Among the most notable discoveries from Mendelian studies were genetic mutations in the LDL receptor that cause significantly elevated LDL cholesterol and premature coronary heart disease [156]. Information about variants in the LDL receptor gene and other variants associated with Mendelian lipid disorders are included in the product labeling for some statins.

As evidence of multigenic contributions to cholesterol levels across the population, a recent, large GWAS examining approximately 2.6 million SNPs in over 100,000 individuals identified variants at 95 loci associated with lipid levels [157]. In addition to genotype, lipid levels are also affected by lifestyle, diet, and other environmental factors, thus underscoring

TABLE 5.6 Drug metabolizing enzymes and transporter proteins for various statins

Statin	Metabolizing CYP450 enzymes	Active metabolite	Transporter proteins
Atorvastatin	3A4, 3A5	Yes	OATP1B1, ABCB1
Fluvastatin	2C9	No	OATP1B1, ABCB1
Lovastatin	3A4, 3A5	Yes	OATP1B1, ABCB1
Pitavastatin	None$^\infty$		
Pravastatin	None	No	OATP1B1, ABCB1
Rosuvastatin	None*	Yes	OATP1B1
Simvastatin	3A4, 3A5, 2C8	Yes	OATP1B1, ABCB1

* Approximately 10% of rosuvastatin is metabolized by CYP2C.
$^\infty$ UGT substrate.

the complexity of dyslipidemia [9]. This complexity renders it difficult to identify the genetic factors that influence statin response.

Pharmacogenomics of Statin Efficacy

Given the important role of statins in reducing cardiovascular disease risk, pharmacogenomic studies of statins are plentiful. The majority of data are related to statin efficacy. There are two major outcomes in these studies: LDL cholesterol lowering or clinical event risk lowering with statin therapy. The efficacy-related studies follow either a candidate gene approach (single or multiple genes) or GWASes. There are several plausible candidate genes that have been well studied for their role in statin response. These include genes encoding for HMG-CoA reductase (HMGCR), the target of statin therapy; apolipoprotein E (ApoE), which transports cholesterol through the bloodstream; and organic anion transporting polypeptide 1B1 (OATP1B1), which transports statins to the liver [154,158–163]. However, the data with these genotypes are inconsistent. In addition, a meta-analysis of three GWASes, including approximately 3,900 subjects, did not yield any significant SNP associated with differential LDL cholesterol reduction [164]. Although GWASes may be limited by their incomplete coverage of genetic variation across the genome, these results further underscore that variation in LCL cholesterol reduction from statin therapy is genetically complex.

Another gene that has been well studied for its role in statin response is the gene for the kinesin-like protein 6 (KIF6), a kinesin involved in intracellular transport. The data for *KIF6* genotype and its association with statin efficacy have also been inconsistent. Initial studies showed an association between the *KIF6* p.W719R polymorphism and clinical outcomes with statins. Specifically, the 719R allele was associated with both an increased risk for coronary events and a greater reduction in coronary event risk with pravastatin in the retrospective analyses of the secondary prevention study, the Cholesterol and Recurrent Events (CARE) trial, and the primary prevention study, West of Scotland Coronary Prevention Study (WOSCOPS) [165]. Similar findings were observed among patients with prior vascular disease in the Prospective Study of Pravastatin in the Elderly at Risk (PROSPER) trial [166]. In the Pravastatin or Atorvastatin Evaluation and Infection Therapy: Thrombolysis in Myocardial Infarction 22 (PROVE IT–TIMI22) trial, intensive statin therapy (atorvastatin 80 mg/day) resulted in significantly greater reduction in coronary events compared to moderate therapy (pravastatin 40 mg/day) in carriers of the 719R allele but not in noncarriers [167]. Based on these data, a commercially available assay for *KIF6* genotyping was developed and marketed to predict risk for cardiac event and response to statins. However, subsequent studies with rosuvastatin and simvastatin revealed no association between *KIF6* genotype and either risk for coronary events or response to statins [168,169]. Potential contributors to the inconsistencies in the data with *KIF6* include variation among studies in the statin used, study population, and trial design, in addition to statistical anomalies in the studies. Nonetheless, the clinical utility of using *KIF6* genotype as a marker of vascular risk for statin response is questionable given the inconsistencies in the data.

Despite the many studies assessing the pharmacogenomics of statin responsiveness, no concrete genotype associations with statin efficacy have been made. There are several reasons that genetic association studies with statins are difficult. First, each statin has its own specific metabolic process. Therefore, genetic variation in a particular metabolizing enzyme will not affect response to all statins. In addition, baseline lipid levels are affected by many factors beyond genetics. Thus, the effect of statin therapy on lipid levels is laid over the backdrop of an already complex physiology. Because each study assesses a different statin and a different patient population, with varying underlying pathophysiologies, it is difficult to find genotypes that consistently affect statin response. The *KIF6* studies described thus far demonstrate these complexities. Although the *KIF6* genotype was consistently associated with response to pravastatin and, to a lesser extent, atorvastatin, this association was not replicated in studies with other statins. In addition, as the patient population of the different studies expanded, the association with *KIF6* genotype was not maintained. A composite of SNPs from several genes and clinical factors, each explaining some small portion of statin response, will likely be necessary to truly predict statin response.

Pharmacogenomics of Statin Safety

Statins are generally well tolerated but can sometimes produce myopathies, with symptoms ranging from mild myalgias to life-threatening rhabdomyolysis. In clinical trials, the reported incidence of statin-associated myalgias is 3 to 5%, with greater risk with the use of high-dose statin therapy [170]. Fatal rhabdomyolysis is rare, occurring in an estimated 1.5 patients per 10 million prescriptions [170].

The mechanism underlying statin-associated myopathies is unknown but appears to be related to increased statin concentrations [170]. Statin concentrations are affected by extensive first-pass uptake into hepatocytes and the rate of metabolism by hepatic CYP450 enzymes. Hepatic uptake appears to be necessary for statin clearance. Genetic variants for hepatic uptake and statin metabolism have been associated with altered statin concentrations and risk for myopathy [171].

The strongest genetic association with statin-induced myopathy has been documented with genes affecting statin hepatic uptake. Statins are transported into hepatocytes by OATP1B1, which is encoded by the *SLCO1B1* gene. Organic anion transporting polypeptides or solute carrier organic (SLCO) anion transporters are vital for drug uptake into tissues and organ systems. These transporters are found in the liver, intestine, and central nervous system. All statins, except for fluvastatin, are transported by this mechanism into hepatocytes.

The first study to demonstrate an association between *SLCO1B1* genotype and myopathy risk with statin therapy was a genome-wide analysis in participants of the Study of the Effectiveness of Additional Reductions in Cholesterol and Homocysteine (SEARCH) study [172]. More than 300,000 variants were genotyped in 85 patients who developed confirmed myopathy (cases) and 90 patients who did not develop myopathy (controls) during treatment with simvastatin 80 mg/day. The only variant reaching genome-wide significance for association with statin-induced myopathy was rs4363657, a noncoding SNP located within the *SLCO1B1* gene on chromosome 12. The rs4363657 SNP was in near complete LD with the nonsynonymous rs4149056 (c.521T>C, p.V174A) SNP. The odds ratio for myopathy was 4.5 (95% CI: 2.6 to 7.7) with a single rs4149056 C allele and nearly 17 (95% CI: 4.7 to 61) with the CC versus TT genotype. In a replication cohort of patients who received simvastatin 40 mg/day as part of the Heart Protection Study, the rs4149056 SNP remained associated with statin-induced myopathy (OR: 2.6, 95% CI: 1.3 to 5.0).

The haplotype containing the *SLCO1B1* 521C allele is termed *SLCO1B1**5. This allele is associated with low OATP1B1 activity and increased plasma concentrations of relevant substrates [173]. Consistent with previous data, in a study of patients receiving atorvastatin, simvastatin, or pravastatin, the *SLCO1B1**5 allele was associated with increased adverse effects from statins, defined as statin discontinuation for any side effect, myalgia, or creatinine kinase greater than three times the upper limit of normal [174]. The association between the *SLCO1B1**5 allele and statin-induced myopathy was further validated in two additional studies [175,176]. However, data from one of these studies suggest the association may be stronger for simvastatin than atorvastatin [176]. Unlike genetic studies of statin efficacy, this association persisted in studies of multiple statins and patient populations.

The CPIC recently published guidelines related to statin dosing when *SLCO1B1* genotype results are available [177]. They do not make recommendation for when or whom to genotype. Their recommendations are limited to simvastatin, for which the most data exist. Regardless of genotype, the simvastatin 80 mg dose should be avoided. For heterozygotes (CT genotype), the guidelines recommend using a lower simvastatin dose (<40 mg/day). For homozygous variant carriers (CC genotype), either a low simvastatin dose or alternative therapy is recommended.

Ezetimibe Pharmacogenomics and *NPC1L1* Genotype

Ezetimibe lowers LDL cholesterol by blocking the Niemann-Pick C1-like 1 (NPC1L1) intestinal cholesterol transporter. The first genetic association reported with ezetimibe was in a treatment-resistant patient who was found to have rare nonsynonymous *NPC1L1* gene mutation [178]. The gene was subsequently sequenced in additional patients, and 140 SNPs and 5 insertion/deletion polymorphisms were identified.

Two studies have assessed the association between *NPC1L1* genotype and LDL cholesterol response to ezetimibe. The first study found a haplotype, consisting of three SNPs (1735C, 25342A, and 27677T), associated with the percent of LDL cholesterol reduction from baseline [179]. Specifically, subjects possessing at least one copy of the *NPC1L1* haplotype had smaller LDL cholesterol reduction from baseline with ezetimibe ($-23.6 \pm 1.6\%$ versus $-35.9 \pm 4.0\%$, $p<0.01$). The second study also used three *NPC1L1* SNPs to create haplotype groups, albeit different SNPs from the previous study [180]. They found that possession of the haplotype -133A/-18A/1679G was associated with greater ezetimibe-induced LDL cholesterol lowering. However, because each study found different *NPC1L1* SNPs and haplotypes to be associated with ezetimibe response, it is as of yet unclear which polymorphism(s) is actually underlying altered LDL cholesterol response. In addition, there were impressive differences in the allele frequencies for studied SNPs by ancestral origin. Thus, whether ancestral differences exist in the genotype—ezetimibe response association is unclear. At this time, because of these issues, regular genotyping for *NPC1L1* polymorphisms to predict ezetimibe response cannot be recommended. In addition, as discussed with statins, lipid homeostasis involves several pathways with many different genes. Therefore, a polygenetic approach will likely be necessary to assess ezetimibe response.

Genetic variability may also affect the pharmacokinetics of ezetimibe. Ezetimibe is primarily metabolized by UDP glucuronosyltransferase 1 family, polypeptide A1 (UGT1A1) [181]. Both ezetimibe and ezetimibe glucuronide are substrates for the efflux pumps, ABCB1 and ATP-binding cassette, subfamily C, member 2 (ABCC2). These efflux pumps and UGT1A1 are susceptible to drug interactions, and these interactions have been shown to modulate the effects of ezetimibe. When subjects were given the ABCB1, ABCC2, and UGT1A1 inducer rifampin, clearance of ezetimibe and ezetimibe glucuronide was significantly increased, and the sterol-lowering effects of ezetimibe were abolished [182]. *UGT1A1*, *ABCB1*, and *ABCC2* also have functionally characterized genetic variation. The role of such genetic variation in ezetimibe response is unknown; however, theoretically, patients with variants associated with decreased functionality could have increased response to ezetimibe, and patients with increased functionality could have decreased ezetimibe response. Genotyping for polymorphisms associated with ezetimibe pharmacokinetics cannot be recommended at this time. However, pharmacists should consider the implications of drug interactions on ezetimibe response.

Opportunity in Pharmacogenomics: Potential to Improve Management of Dyslipidemia

In clinical trials, statins have been shown to reduce the risk for adverse cardiovascular events in patients with established cardiovascular disease as well as those at high risk for cardiovascular disease [9]. However, not all patients derive protection against cardiovascular events with statins, and some patients experience intolerable (e.g., myopathy) and potentially life-threatening (e.g., rhabdomyolysis) adverse effects. Pharmacogenomics offers the potential to identify patients who will either not benefit from statin therapy or who are at high risk for experiencing adverse statin-induced effects, in whom a statin may be avoided. At present, the evidence most strongly supports a genetic determinant of adverse statin-induced effects (e.g., *SLCO1B1* genotype and statin-induced myopathy).

Pharmacogenomics may also have a potential role in predicting patients who will respond to combination statin–fibrate therapy. Specifically, the Action to Control Cardiovascular Risk in Diabetes (ACCORD) trial investigated whether adding a fibrate to statin therapy would reduce the risk for cardiovascular disease in patients with type 2 diabetes mellitus [183]. In the overall clinical trial population of over 5,000 patients, the combination failed to reduce the risk for cardiovascular events compared to statin therapy alone. However, there was significant variability in outcomes among trial participants. Genetic samples were collected as part of the trial design, and investigators are examining whether certain genotypes may be predictive of benefit with combination therapy.

Pharmacogenomics of Antihypertensives

Hypertension is the most common chronic disease in this country, affecting more than 76 million Americans [2]. Thus, agents to treat hypertension are among the most commonly prescribed drugs in the United States and other countries. There is significant interpatient variability in response to antihypertensive agents, and factors underlying this variability are not well understood [184]. Clinicians currently treat hypertension with a largely trial-and-error approach. It is often necessary to try several agents or combinations of agents before achieving adequate blood pressure control with acceptable tolerability for a given patient. The ability to predict antihypertensive response may allow for earlier initiation of effective antihypertensive therapy, thus reducing the time to adequate blood pressure control and potentially reducing the risk for adverse sequelae from prolonged untreated hypertension. In addition, it may also help to decrease adverse event risk with antihypertensive therapy. With this idea in mind, a number of investigators are searching for genetic determinants of antihypertensive responses. However, in contrast to pharmacogenetic data with warfarin and clopidogrel, pharmacogenetic data with antihypertensives are often inconsistent and even conflicting, which is particularly true with agents that antagonize the renin–angiotensin system. Thus, the potential for improving blood pressure control with pharmacogenetics is largely unrealized. The following section discusses only the most consistently replicated genetic associations with blood-pressure-lowering effects with antihypertensive agents. In addition, emerging data on genetic determinants of clinical outcomes and adverse drug effects with antihypertensive agents will be discussed.

Genetic Determinants of β-Blocker Response

β-blockers are indicated for the treatment of a number of cardiovascular disorders, including hypertension, coronary artery disease, heart failure, and cardiac arrhythmias. Many β-blockers are metabolized to some degree by CYP2D6 to inactive metabolites, and all β-blockers exert their therapeutic effects by primarily antagonizing the ADRB1. Both the *CYP2D6* and *ADRB1* genes are highly polymorphic and can have significant effects on β-blocker plasma concentration and therapeutic effects, respectively [185,186].

Metoprolol is the β-blocker most extensively metabolized by the CYP2D6 enzyme. A description of the *CYP2D6* genetic variants is provided in chapter 1. The clinical relevance of alterations in β-blocker plasma concentration due to *CYP2D6* polymorphism is questionable, given that β-blockers have a wide therapeutic index. Nonetheless, investigators have reported a higher risk of adverse effects with β-blocker therapy among CYP2D6 PMs compared to EMs, with normal CYP2D6 function [187]. Specifically, in a cohort of more than 700 metoprolol users, the PM phenotype was associated with a significantly lower heart rate and diastolic blood pressure and a nearly fourfold higher risk of bradycardia compared to the EM phenotype [187].

Common variants in the *ADRB1* gene, p.S49G and p.R389G, have been correlated with blood pressure lowering effects of β-blocker therapy. These variants are in strong LD, as described in detail in chapter 1. Greater blood pressure reduction has been observed with the homozygous RR389 genotype and the S49-R389 haplotype [188–190]. Given that blood pressure is a surrogate marker and that the ultimate goal of antihypertensive therapy is to reduce hypertension-related morbidity and mortality, genetic associations with clinical outcomes have particular relevance. The influence of *ADRB1* genotype on the incidence of death, nonfatal myocardial infarction, or nonfatal stroke was examined in participants in the International VErapail SR/Trandolapril (INVEST) study [191]. Patients in this trial had both hypertension and coronary heart disease and were assigned to either atenolol- or verapamil sustained release (SR)-based treatment. The *ADRB1* S49-R389 haplotype was associated with an increased risk for death among patients randomized to verapamil but not those randomized to atenolol. These data suggest that atenolol exerts a protective effect in individuals with hypertension, coronary heart disease, and the S49-R389 genotype. Further evaluations of genetic data from the INVEST study revealed that the gene for the alpha(2C)-adrenergic receptor (ADRA2C) modulated the risk for adverse outcomes with atenolol and that the β2 subunit of the voltage-gated calcium channel (*CACNB2*) and large-conductance calcium and voltage-dependent potassium channel β1 subunit (*KCNMB1*) genotypes influenced responsiveness and risk for adverse outcomes with verapamil [191–193].

The ability to predict response to β-blockade based on genotype could have important clinical implications. Specifically, in the absence of compelling indications for β-blocker therapy, β-blockers could be reserved for hypertension management in individuals expected to have a good blood pressure response to this drug class based on *ADRB1* genotype. Other antihypertensive agents could be used in those expected to have little to no blood pressure reduction with β-blockade. β-blockers could also be used as first-line therapy for hypertensive patients with coronary heart disease and the *ADRB1* genotype predictive of poor survival. However, further confirmatory data are necessary before genotype will be used clinically for antihypertensive therapy.

Pharmacogenomics in Antihypertensive Drug Development

Rostafuroxin is an example of a novel antihypertensive agent developed based on an understanding of genetic mechanisms underlying hypertension. Specifically, rostafuroxin was developed based on associations between the *ADD1* genotype and renal sodium pump function. The *ADD1* gene encodes for α-adducin, a cytoskeletal protein involved in signal transduction and renal sodium transport. The *ADD1* p.G460W variant is associated with greater renal sodium-potassium pump activity, renal sodium retention, and salt-sensitive hypertension [194,195]. Rostafuroxin selectively interferes with the effects of the variant ADD1 460W protein on renal sodium handling, without affecting the activity of the wild-type 460G protein [196]. In a recent study, response to rostafuroxin was predicted based on a genetic signature, composed of variants in the *ADD1* gene as well as genes for proteins involved in synthesis and transport of oubain, an endogenous ligand for the renal sodium-potassium pump [196]. The genetic signature predicted a large and clinically significant (e.g., 14 mm Hg) reduction in blood pressure with rostafuroxin, whereas it was not predictive of responses to other antihypertensive drugs (losartan, hydrochlorothiazide). Rostafuroxin thus serves as a model for developing unique genetic signatures to predict drug response.

Genetic Determinants of Glycemic Response to Thiazide Diuretics

Although thiazide diuretics are recommended as first-line therapy for most patients with hypertension, their use is not without risk. In particular, thiazide diuretics are associated with metabolic disturbances, most notably, increased blood glucose levels and new-onset diabetes. In the Genetics of Hypertension Associated Treatment (GenHAT) study, several candidate genes were interrogated for their association with thiazide diuretic-induced elevation in blood glucose [197]. The GenHAT study was an ancillary study of the Antihypertensive and Lipid-Lowering Treatment to Prevent Heart Attack Trial (ALLHAT), in which over 42,000 patients with hypertension and at least one other cardiovascular risk factor were randomized to one of four treatments (chlorthalidone, amlodipine, lisinopril, or doxazosin) [198]. The primary outcome of ALLHAT was the incidence of fatal coronary heart disease or nonfatal myocardial infarction. Chlorthalidone was shown to be superior to other agents in preventing cardiovascular disease; however, chlorthalidone was also associated with a higher incidence of new-onset diabetes compared to other agents. Of 11 genes examined in the Gen-HAT study, only the *SCNN1A* gene, encoding for the alpha subunit of the epithelial sodium channel, was associated with chlorthalidone-induced glucose elevation [197]. Specifically, the 633TT genotype was associated with higher fasting glucose levels with chlorthalidone versus amlodipine. Blockade of renal sodium transport leading to excess potassium excretion is believed to be important in glucose dysregulation with thiazide diuretics, providing biological plausibility for the association [197,199]. Understanding the risk factors for adverse drug effects, such as glucose intolerance with thiazide diuretics, may ultimately allow for measures to prevent or attenuate such effects and improve the safety of antihypertensive therapy.

Opportunity in Pharmacogenomics: A Look to the Future of Hypertension Management

There are a number of antihypertensive agents available, and it is often difficult to choose which agent to prescribe for a particular patient. Even when following guideline recommendations, there is no guarantee that the prescribed drug will effectively lower blood pressure and prevent adverse outcomes in a particular patient. The ability to predict response to antihypertensive therapy could revolutionize hypertension management. Though we are still likely years away from being able to use genotype-guided antihypertensive therapy, the ability to eliminate the trial-and-error approach to blood pressure management based on genotype is quite attractive. Looking forward, antihypertensive therapy could be potentially individualized based on unique genetic signatures, such as those described with rostafuroxin. Specifically, an improved understanding of genetic contributions to the mechanisms underlying hypertension could lead to the development of novel strategies to combat the disease. This development might pave the way for safer and more effective antihypertensive therapy. However, unlike oncology, in which there are a number of examples of drugs whose response is dependent on specific genetic defects (e.g., trastuzumab and Her2 gene over expression), there are not yet any similar examples in the field of cardiology.

PHARMACOGENOMIC POTENTIAL IN HEART FAILURE

Current Approach to Heart Failure Management

As shown in Figure 5.6, standard therapy for heart failure generally consists of an ACE inhibitor and a β-blocker for morbidity and mortality reduction, with the addition of a diuretic with or without digoxin for symptom control. Other agents, including aldosterone antagonists, angiotensin receptor blockers, and the combination of isosorbide dinitrate (ISDN) and hydralazine have been shown to further improve outcomes when added to the standard heart failure drug regimen in select patients [4,200,201]. Thus, patients may require four to six medications for their heart failure alone, in addition to therapy needed to treat any concomitant diseases.

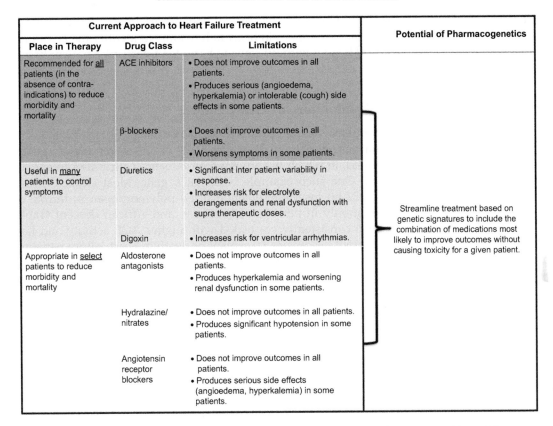

Current Approach to Heart Failure Treatment			Potential of Pharmacogenetics
Place in Therapy	**Drug Class**	**Limitations**	
Recommended for <u>all</u> patients (in the absence of contra-indications) to reduce morbidity and mortality	ACE inhibitors	• Does not improve outcomes in all patients. • Produces serious (angioedema, hyperkalemia) or intolerable (cough) side effects in some patients.	
	β-blockers	• Does not improve outcomes in all patients. • Worsens symptoms in some patients.	
Useful in <u>many</u> patients to control symptoms	Diuretics	• Significant inter patient variability in response. • Increases risk for electrolyte derangements and renal dysfunction with supra therapeutic doses.	Streamline treatment based on genetic signatures to include the combination of medications most likely to improve outcomes without causing toxicity for a given patient.
	Digoxin	• Increases risk for ventricular arrhythmias.	
Appropriate in <u>select</u> patients to reduce morbidity and mortality	Aldosterone antagonists	• Does not improve outcomes in all patients. • Produces hyperkalemia and worsening renal dysfunction in some patients.	
	Hydralazine/ nitrates	• Does not improve outcomes in all patients. • Produces significant hypotension in some patients.	
	Angiotensin receptor blockers	• Does not improve outcomes in all patients. • Produces serious side effects (angioedema, hyperkalemia) in some patients.	

FIGURE 5.6 Current approach and potential of pharmacogenetics in the treatment of heart failure.

There are several limitations with our current approach to heart failure treatment. First, patients often have difficulty adhering to the multidrug regimens that have become the norm in heart failure. Second, many patients cannot safely take target doses of all recommended heart failure therapies because of low blood pressure. Thus, clinicians must decide which drug to up-titrate and which drug to continue at suboptimal doses or abandon all together. Third, although data from multiple randomized trials demonstrate reductions in morbidity and mortality with ACE inhibitors and β-blockers in overall heart failure study populations, not all study participants derived benefits from these agents, and some experienced serious adverse effects requiring drug discontinuation. For example, approximately 13% of enalapril-treated subjects in the Studies of Left Ventricular Dysfunction (SOLVD) discontinued the drug because of worsening heart failure or adverse drug effects [202]. Similarly, 14% of patients in the β-blocker arm of the Metoprolol CR/XL Randomised Intervention Trial in Congestive Heart Failure (MERIT-HF) discontinued the drug prematurely because of poor tolerability [203]. Thus, although ACE inhibitors and β-blockers improved outcomes in clinical trial populations as a whole, there is no guarantee that they will improve outcomes without causing harm in an individual patient. Currently, there is no reliable method of predicting response to heart failure medications, and all patients are treated with a similar "cocktail" of medications. Pharmacogenomics in heart failure aims to identify the combination of drugs most likely to be of benefit without causing harm for a particular patient based on genotype.

Pharmacogenomics of ACE Inhibitors in Heart Failure

The genes discussed thus far in this chapter primarily influence drug response by altering drug pharmacokinetics or pharmacodynamics. However, there are also examples of genes associated with disease prognosis, in which the adverse consequences attributed to a gene are modified by drug therapy. One such example is the *ACE* gene. Most studies of the *ACE* gene have focused on a 287-bp insertion/deletion (I/D) polymorphism in intron 16 of the gene, which occurs commonly in persons of European and African descent (Table 5.7). The *ACE* D allele has been consistently correlated with higher ACE activity and has been shown to confer increased risk for cardiac transplant or death in heart failure patients, likely because of the deleterious effects of the renin—angiotensin system on heart failure progression [204—208]. Inhibition of the renin—angiotensin system appears to attenuate the detrimental effects of the *ACE* D allele. For example, a study of patients with systolic heart failure showed that the adverse effect of the *ACE* D allele on transplant-free survival was greatest among patients who were not taking β-blockers or were taking less than or equal to 50% of the recommended target ACE inhibitor dose (dose associated with mortality reduction in clinical trials) [208]. Both ACE inhibitors and β-blockers attenuate the renin—angiotensin system. Use of β-blockers and higher ACE inhibitor doses, defined as doses greater than 50% of the target dose, attenuated the detrimental effects of the *ACE* D allele. A subsequent study in diastolic heart failure revealed similar findings [209].

In contrast to candidate gene studies linking the *ACE* I/D genotype to adverse outcomes in heart failure, a recent GWAS examining over 2.4 million SNPs in nearly 21,000 Caucasians and 3,000 African Americans found no association between the *ACE* gene and heart failure prognosis. However, the investigators did not account for heart failure treatment, which has been shown to modify the effect of *ACE* genotype on outcomes. Nonetheless, it is certainly premature to suggest that ACE inhibitors may not be necessary in individuals without an *ACE* D allele. However, if a patient is known to carry the *ACE* D allele, it may

TABLE 5.7 Minor allele frequencies for genes associated with responses to hypertension and heart failure therapies[215,228,229]

Gene	Variant	Caucasians	African Americans
ACE	I/D	0.42	0.56
ADRB1	S49G	0.15	0.13
	R389G	0.27	0.42
ADRA2C	Del322-325	0.04	0.40
NOS3	E298D	0.37	0.14

be particularly beneficial to use ACE inhibitors at recommended target doses to potentially ameliorate adverse consequences of this genotype [4].

ADRB1 Genotype: A Case for Targeted β-Blocker Therapy?

β-blockers are well recognized to reduce morbidity and mortality in heart failure by inhibiting the excessive sympathetic nervous system activity that propagates heart failure progression [4]. However, not all patients benefit from β-blocker therapy. In addition, because β-blockers inhibit cardiac contractility, they must be started in very small doses with careful up-titration to help prevent worsening heart failure. Nonetheless, some patients still suffer cardiac decompensation during β-blocker initiation.

The *ADRB1* gene has been extensively studied for its effects on β-blocker response. Although the data are not always consistent, the *ADRB1* R389G genotype is associated with the degree of improvement in left ventricular ejection fraction with either metoprolol or carvedilol, with the greatest improvement observed with the RR389 genotype [210,211]. The RR389 genotype has also been associated with greater survival benefits with the β-blocker, bucindolol. Unlike metoprolol, carvedilol, and bisoprolol, which significantly improved survival in heart failure clinical trials, bucindolol was shown to have a neutral effect on survival [4,212]. However, unlike clinical trials with the other β-blockers, the trial with bucindolol included a larger number of African Americans, and a subgroup analysis revealed improved survival with bucindolol in Caucasians but not African Americans [212]. A subsequent genetic analysis showed that response to bucindolol was dependent on ADBR1 genotype, with a reduced risk for hospitalization and death with bucindolol in RR389 homozygotes, but not G389 allele carriers [213]. The G389 allele is more common among African Americans than Caucasians (Table 5.7), potentially accounting for the negative effects of bucindolol in persons of African descent.

In contrast to data with bucindolol, the *ADRB1* genotype was not associated with clinical outcomes with metoprolol or carvedilol [214]. However, there are important pharmacological differences among β-blockers, including a sympatholytic effect with bucindolol, which may contribute to differential genotype interactions with response to various drugs.

Based on the pharmacogenetic data with bucindolol, ARCA Biopharma sought FDA approval of bucindolol in patients with the *ADRB1* RR389 genotype. Their initial request was denied. The company now has plans to pursue a clinical trial of bucindolol compared to metoprolol in *ADRB1* R389 homozygotes, as requested by the FDA.

Pharmacogenomics of Nitrates/Hydralazine

In the African-American Heart Failure Trial (AHeFT), the addition of isosorbide dinitrate (ISDN)/hydralazine to standard therapy with an ACE inhibitor plus/minus a β-blocker significantly improved the primary composite end point of death, hospitalizations for heart failure, and quality of life compared to placebo [201]. Based on these data, the ISDN/hydralazine combination was FDA-approved for the treatment of heart failure in self-identified African Americans. Because the effects of adjunctive ISDN/hydralazine therapy has been examined only in African Americans, the benefits of the combination in individuals of other descents are unknown. Consistent with the FDA-approved labeling, current joint guidelines from the American College of Cardiology and American Heart Association recommend ISDN/hydralazine for African Americans with continued symptoms despite optimal treatment with ACE inhibitors, β-blockers, and diuretics [4].

Ancestral origin is a poor and controversial marker of drug response. Because any difference in drug response may be attributable, at least in part, to genotype, investigators have attempted to identify a genetic marker for response to ISDN/hydralazine in the A-HeFT population. ISDN/hydralazine is believed to exert its beneficial effects by increasing nitric oxide variability, and as such, several variants in the endothelial nitric oxide synthase (eNOS) gene have been examined for their effects on ISDN/hydralazine response. Of these, only the p.E298D variant in exon 7 was found to influence response to ISDN/hydralazine [215]. The EE298 genotype is more common in African Americans than Caucasians and was associated with a greater improvement in the study's composite end point with ISDN/hydralazine, an association largely driven by the improvement in the quality of life score with the EE298 genotype. These data suggest that eNOS genotype, rather than ancestral background, may be useful as a predictor of response to ISDN/hydralazine therapy.

Opportunity in Pharmacogenomics: Potential to Streamline Heart Failure Therapy

At this point, there are insufficient data to support withholding any heart failure therapy because of a potential lack of benefit based on genotype. However, future prospective studies evaluating the benefits of pharmacogenomic-based prescribing compared to traditional prescribing of heart failure medications are conceivable. Ultimately, results of pharmacogenomic research efforts could lead to genotype-guided prescribing of heart failure therapy. Specifically, rather than initiating the same "cocktail" of drugs for all patients, regimens might be tailored according to each individual's genetic predisposition for obtaining benefit or experiencing harm from a particular drug. Drugs predicted to be of minimal to no benefit could be avoided, thus simplifying drug regimens and reducing the associated costs, while potentially improving overall patient outcomes.

GENETIC INFLUENCES OF DRUG-INDUCED ARRHYTHMIA

Cardiac arrhythmias are potentially fatal if not treated appropriately. However, the drugs used to treat arrhythmias are themselves arrhythmogenic. Thus, there has been significant study of the genetics of cardiac arrhythmia and the pharmacogenomics of antiarrhythmic therapy in order to improve drug effectiveness and limit proarrhythmic effects.

Antiarrhythmic Medications

Antiarrhythmic agents in general have a narrow therapeutic index. Thus, they are often susceptible to drug–drug interactions and can cause significant adverse events. Therefore, therapeutic monitoring and pharmacogenomics have been well studied in efforts to improve outcomes and decrease adverse events with these medications.

Polymorphisms in the genes encoding drug metabolizing enzymes have been examined for their role in antiarrhythmic efficacy and toxicity. The highly polymorphic CYP2D6 enzyme metabolizes propafenone and quinidine. Propafenone is a class Ic antiarrhythmic that exerts its effects by blocking the fast inward sodium current in addition to having some β-receptor blocking properties at higher concentrations. Propafenone is primarily metabolized by CYP2D6, though CYP1A2 and CYP3A4 also contribute to its metabolism. Genetic classification of CYP2D6 activity is complex and can be determined via genotyping or phenotyping. Patients are generally classified as poor, extensive, or ultra-rapid metabolizers. Patients who are classified as CYP2D6 poor metabolizers have decreased propafenone clearance, which leads to an increase in propafenone serum concentrations. Propafenone has more β-blocking properties than its active metabolites, which is more pronounced at low doses compared to high doses. However, because high doses result in only a slight increase in propafenone serum concentrations and this increase is balanced by a decrease in production of an active metabolite, the recommended dosing regimen is the same regardless of CYP2D6 phenotype.

Although it is well documented that *CYP2D6* genotype, and thus metabolizer status, affects the pharmacokinetics of propafenone, it is unclear whether this effect translates into differences in clinical outcomes. The greater variability in plasma concentrations of propafenone and its metabolites in CYP2D6 poor metabolizers does require that propafenone be titrated carefully and the echocardiogram (ECG) be monitored for evidence of toxicity [216]. Importantly, the additional β-blockade seen in CYP2D6 poor metabolizers can potentially lead to adverse events in asthmatic patients. In addition, there is evidence that subjects with paroxysmal atrial fibrillation classified as CYP2D6 poor metabolizers are more likely to maintain normal sinus rhythm with propafenone compared to extensive metabolizers [217]. However, the data are inconsistent. For this reason, routine *CYP2D6* genotyping cannot be recommended for patients receiving propafenone.

In contrast, quinidine is not a CYP2D6 substrate. However, the prescribing information for quinidine contains information on *CYP2D6* pharmacogenomics because quinidine is a CYP2D6 inhibitor [218]. Quinidine can convert patients who are extensive metabolizers to poor metabolizers. Therefore, it is important to monitor for adverse events when quinidine is coadministered with CYP2D6 substrates. Although pharmacogenomic studies have been done for several other antiarrhythmic medications, no strong and reliable associations have been observed. Therefore, genotyping to predict response to antiarrhythmic therapy is not indicated at this time.

Pharmacogenomics of Drug-Induced Long QT Syndrome

The proarrhythmic effects of medications (both those used to treat arrhythmia and those used for other indications) have been well studied. Interest in these rare adverse events is due to the fact that they can be life threatening and require a significant amount of patient monitoring. Proarrhythmia is generally defined as the worsening of the arrhythmia being treated or generation of a new arrhythmia with drug therapy [219]. Genetic studies have focused on drug-induced increases in the QT interval on the ECG and drug-induced torsades de pointes. The knowledge gained from many years of studying and evaluating congenital long QT syndromes has aided the study of genetic factors associated with drug-induced prolonged

QT intervals. In addition, there is a well-documented association between a prolonged QT interval and the life-threatening arrhythmia, torsades de pointes.

The QT interval on an ECG represents the action potential of ventricular myocytes. The ventricular action potential is made up of several currents produced by different ion channels. The action potential is prolonged when there is either increased inward current or decreased outward current. The heart has significant built-in redundancy, as several ion channels participate in the ventricular action potential. This redundancy is termed "repolarization reserve." Thus, variation in one ion channel will not necessarily lead to an increase in the QT interval. A combination of factors is generally necessary for patients to exhibit both congenital or drug-induced long QT syndromes. Several clinical factors, beyond genetic variation in ion channels, have been associated with an increased risk of drug-induced long QT syndromes, and it is thought that these factors affect the heart's "repolarization reserve."[219] These factors include but are not limited to hypokalemia, recent conversion of atrial fibrillation, advanced heart disease, and female gender.

Factors influencing the pharmacokinetics of medications may also affect the risk of drug-induced long QT syndrome. The risk of drug-induced long QT syndrome may be increased if the clearance of a drug is decreased via either a drug interaction or genetic variants in hepatic enzyme systems. Clinicians should be particularly vigilant in monitoring for drug interactions with medications known to prolong the QT interval. Drug interactions contributing to long QT syndrome are frequently ignored despite the well-documented risk [220]. In addition, if genetic variability in hepatic enzyme systems for a patient is known, this should be considered as well. An ECG should be checked, with the corrected QT interval measured for all patients with increased risk of drug-induced long QT syndrome.

Genetic variation in ion channels associated with ventricular action potential has been well studied in congenital long QT syndromes because of the possible effect on "repolarization reserve."[221] Mutations found in genes encoding potassium (KCNQ1, KCNH2, KCNE1, and KCNE2) and sodium (SCN5A) voltage-gated channels have been associated with risk for congenital long QT syndrome. In addition, medications can prolong the QT interval by blocking the ion channel pore, inducing conformational changes in the ion channel pore, and/or decreasing production of the proteins encoding the ion channels. The amino acid structure of KCNH2 appears to make this ion channel pore particularly susceptible to drug blockade. Polymorphisms in the gene encoding KCNH2 may affect its susceptibility to drug binding [221]. Polymorphisms in *SCN5A* may also contribute to the risk of drug-induced long QT syndrome. However, although polymorphisms in several genes encoding ion channels have been found to be associated with risk of drug-induced long QT syndrome, the data are inconsistent [221].

GWAS have identified novel genes associated with congenital long QT syndrome. *NOS1AP*, which encodes for an accessory protein for the nitric oxide synthase type 1 gene, was initially linked to variability of QT interval across the normal population [222]. Variants in *NOS1AP* have also been linked to arrhythmia risk, sudden cardiac death, congenital long QT syndrome, and the QT interval prolonging effects of verapamil [222]. The mechanism by which this gene affects the QT interval is unclear. Another gene associated with prolongation of the QT interval in GWAS is *GINS3*. The role of *GINS3* polymorphisms in QT interval prolongation was validated in a zebra fish model assessing the effect of dofetilide (a QT-prolonging drug) on the action potential duration [222]. However, the mechanism underlying this effect is unclear. Prior to utilizing pharmacogenomic information to predict proarrhythmic risk with medications, a genetic variant must be validated in separate, distinct populations. In addition, an understanding of the biology and physiology of the polymorphism is necessary.

Currently, using genetic information to predict drug-induced long QT syndrome cannot be recommended. The volume of knowledge on this topic is growing rapidly, and with validated genetic markers, genotyping may in the future be clinically useful. However, it is unlikely that polymorphisms in a single gene or a single clinical risk factor will be sufficient to predict risk because of the redundancy in the system. Predicting drug-induced long QT syndrome will likely require a complex combination of multiple polymorphisms and clinical and environmental information. In the FDA guidance for industry for screening non-antiarrhythmic drug potential to cause delay in cardiac repolarization, they recommend considering genotyping for cardiac ion channel mutations for patients who experience marked QT interval prolongation or torsades de pointes in early clinical trials [223].

Opportunities in Pharmacogenomics: Potential to Resurrect Old Drugs

One of the primary reasons that drugs in development do not succeed or that approved drugs are withdrawn from the market is because of proarrhythmic effects. The ability to predict risk for proarrhythmia with a drug could potentially revive some agents, particularly if few other treatment options are available. In this case, genetic testing for the "at-risk" variant(s) would likely be required prior to drug use. The drug could then be avoided in patients genetically at risk for drug-induced proarrhythmia.

CONCLUSION

To date, the majority of cardiovascular pharmacogenomic research has focused on candidate genes involved in drug metabolism or mechanism of action. However, we are beginning to see more genome-wide approaches to identifying novel genes affecting cardiovascular drug response. Genome-wide association data are already available with clopidogrel and warfarin, and in both cases, these data confirm that previously identified candidate genes are the most predictive of drug response. In the case of statins, a GWAS led to the discovery of a novel *SLCO1B1* SNP as the primary determinant of statin-induced myopathy. Genome-wide studies to date have predominantly focused on associations in Caucasian patients. Future GWASes are likely to include more diverse ethnic groups in order to determine which variants universally affect drug response.

Clinical implementation of cardiovascular pharmacogenomics lags behind that of cancer pharmacogenomics, which has the most examples of clinical application of genetic information. Nonetheless, several institutions are beginning to embrace cardiovascular pharmacogenomics. Clopidogrel pharmacogenomics is the farthest along in this regard. In 2011, Vanderbilt University announced plans to genotype all persons at risk for ACS for genes related to clopidogrel's ability to inhibit platelet aggregation [224]. For individuals later requiring antiplatelet therapy, this genetic information will be available to guide the choice of therapy. Vanderbilt also has a decision support algorithm for simvastatin dosing that includes *SLCO1B1* genotyping [177]. Such applications of genetic information are paving the way for clinical implementation of genotype-guided therapy for other drugs. Warfarin is probably the next in line for widespread clinical implementation. The major barrier to genotype-guided warfarin therapy at present is the refusal of many third-party payers to cover genotyping for warfarin therapy in the absence of clinical trial participation. This situation may change, however, pending the outcome of ongoing trials.

A number of governmental groups have taken an active role in promoting pharmacogenomic research and clinical implementation. One noteworthy example is the NIH-funded Pharmacogenomics Research Network (PGRN), which consists of scientists across the United States focusing on understanding genetic determinants of response to various medications, including medications used to treat hypertension and cardiac arrhythmias. The PharmGKB is part of the PGRN and serves as a central repository for pharmacogenomic data. The PharmGKB hosts the CPIC, which, as described in this chapter, has already published recommendations for clopidogrel, warfarin, and simvastatin pharmacogenomics. See chapter 3 for further information on efforts by governmental groups in the area of pharmacogenomics.

One approach to pharmacogenomic implementation is to genotype variants related to a specific drug at the time of drug prescribing, as discussed above. Another approach is to genotype a panel of variants influencing responses to numerous drugs preemptively. This way, genetic information is available at the time various drugs are prescribed. Such an approach would likely begin with variants involved in drug metabolism and transport, as these variants are not usually involved in disease susceptibility or progression and have minimal ethical concerns. Drug metabolizing enzyme gene panels are commercially available or may be custom built for this type of approach.

DISCUSSION POINTS

1. Contrast the effect of a poor drug metabolizer phenotype on response to warfarin versus clopidogrel.
2. Describe feasible approaches to implement warfarin and clopidogrel pharmacogenetics into clinical practice.

DISCUSSION QUESTIONS

1. What are examples of novel pharmacogenetic findings from genome-wide association studies?
2. What are barriers to the clinical implementation of pharmacogenomics to manage cardiovascular disease?
3. How might pharmacogenetic findings lead to new drug development for cardiovascular disorders?

References

[1] The top 10 causes of death. (accessed 20.11.2011, at http://www.who.int/mediacentre/factsheets/fs310/en/index.html.)

[2] Roger VL, Go AS, Lloyd-Jones DM, et al. Heart disease and stroke statistics—2011 update: a report from the American Heart Association. Circulation 2011;123. e18—e209.

[3] Bartholow M. Top 200 Drugs of 2010. Pharmacy Times 2011. Available at http://www.pharmacytimes.com/publications/issue/2011/May2011/Top-200-Drugs-of-2010 (accessed 12.22.12).

[4] Hunt SA, Abraham WT, Chin MH, et al. 2009 focused update incorporated into the ACC/AHA 2005 Guidelines for the Diagnosis and Management of Heart Failure in Adults: a report of the American College of Cardiology Foundation/American Heart Association Task Force on Practice Guidelines, developed in collaboration with the International Society for Heart and Lung Transplantation. J Am College Cardiol 2009;53:e1–e90.

[5] Chobanian AV, Bakris GL, Black HR, et al. The Seventh Report of the Joint National Committee on Prevention, Detection, Evaluation, and Treatment of High Blood Pressure: the JNC 7 report. J Am Med Assoc 2003;289:2560–72.

[6] Mancia G, De Backer G, Dominiczak A, et al. 2007 guidelines for the management of arterial hypertension: the Task Force for the Management of Arterial Hypertension of the European Society of Hypertension (ESH) and of the European Society of Cardiology (ESC). European Heart J 2007;28:1462–536.

[7] Dickstein K, Cohen-Solal A, Filippatos G, et al. ESC guidelines for the diagnosis and treatment of acute and chronic heart failure 2008: the Task Force for the Diagnosis and Treatment of Acute and Chronic Heart Failure 2008 of the European Society of Cardiology. Developed in collaboration with the Heart Failure Association of the ESC (HFA) and endorsed by the European Society of Intensive Care Medicine (ESICM). Eur J Heart Fail 2008;10:933–89.

[8] Anderson JL, Adams CD, Antman EM, et al. 2011 ACCF/AHA focused update incorporated into the ACC/AHA 2007 Guidelines for the Management of Patients with Unstable Angina/Non-ST-Elevation Myocardial Infarction: a report of the American College of Cardiology Foundation/American Heart Association Task Force on Practice Guidelines. Circulation 2011;123. e426–e579.

[9] Executive summary of the Third Report of the National Cholesterol Education Program (NCEP) Expert Panel on Detection, Evaluation, and Treatment of High Blood Cholesterol In Adults (Adult Treatment Panel III). J Am Med Assoc 2001;285:2486–97.

[10] Born GV, Gorog P, Begent NA. The biologic background to some therapeutic uses of aspirin. Am J Med 1983;74:2–9.

[11] Patrono C, Baigent C, Hirsh J, Roth G. Antiplatelet drugs: American College of Chest Physicians Evidence-Based Clinical Practice Guidelines (8th Edition). Chest 2008;133:199S–233S.

[12] Chen ZM, Jiang LX, Chen YP, et al. Addition of clopidogrel to aspirin in 45,852 patients with acute myocardial infarction: randomised placebo-controlled trial. Lancet 2005;366:1607–21.

[13] Yusuf S, Zhao F, Mehta SR, Chrolavicius S, Tognoni G, Fox KK. Effects of clopidogrel in addition to aspirin in patients with acute coronary syndromes without ST-segment elevation. N Eng J Med 2001;345:494–502.

[14] Steinhubl SR, Berger PB, Mann JT, 3rd, et al. Early and sustained dual oral antiplatelet therapy following percutaneous coronary intervention: a randomized controlled trial. J Am Med Assoc 2002;288:2411–20.

[15] Combescure C, Fontana P, Mallouk N, et al. Clinical implications of clopidogrel non-response in cardiovascular patients: a systematic review and meta-analysis. J Thromb Haemost 2010;8:923–33.

[16] Mega JL, Close SL, Wiviott SD, et al. Cytochrome p-450 polymorphisms and response to clopidogrel. N Eng J Med 2009;360:354–62.

[17] Mega JL, Close SL, Wiviott SD, et al. Genetic variants in ABCB1 and CYP2C19 and cardiovascular outcomes after treatment with clopidogrel and prasugrel in the TRITON-TIMI 38 trial: a pharmacogenetic analysis. Lancet 2010;376:1312–9.

[18] Hodges LM, Markova SM, Chinn LW, et al. Very important pharmacogene summary: ABCB1 (MDR1, P-glycoprotein). Pharmacogenet Genomics 2011;21:152–61.

[19] Taubert D, von Beckerath N, Grimberg G, et al. Impact of P-glycoprotein on clopidogrel absorption. Clin Pharmacol Ther 2006;80:486–501.

[20] Cayla G, Hulot JS, O'Connor SA, et al. Clinical, angiographic, and genetic factors associated with early coronary stent thrombosis. J Am Med Assoc 2011;306:1765–74.

[21] Simon T, Steg PG, Becquemont L, et al. Effect of paraoxonase-1 polymorphism on clinical outcomes in patients treated with clopidogrel after an acute myocardial infarction. Clin Pharmacol Ther 2011;90:561–7.

[22] Simon T, Verstuyft C, Mary-Krause M, et al. Genetic determinants of response to clopidogrel and cardiovascular events. N Eng J Med 2009;360:363–75.

[23] Shuldiner AR, O'Connell JR, Bliden KP, et al. Association of cytochrome P450 2C19 genotype with the anti-platelet effect and clinical efficacy of clopidogrel therapy. J Am Med Assoc 2009;302:849—57.

[24] Wallentin L, James S, Storey RF, et al. Effect of CYP2C19 and ABCB1 single nucleotide polymorphisms on outcomes of treatment with ticagrelor versus clopidogrel for acute coronary syndromes: a genetic substudy of the PLATO trial. Lancet 2010;376:1320—8.

[25] Angiolillo DJ, Fernandez-Ortiz A, Bernardo E, et al. Contribution of gene sequence variations of the hepatic cytochrome P450 3A4 enzyme to variability in individual responsiveness to clopidogrel. Arterioscler Thromb Vasc Biol 2006;26:1895—900.

[26] Smith SM, Judge HM, Peters G, et al. Common sequence variations in the P2Y12 and CYP3A5 genes do not explain the variability in the inhibitory effects of clopidogrel therapy. Platelets 2006;17:250—8.

[27] Suh JW, Koo BK, Zhang SY, et al. Increased risk of atherothrombotic events associated with cytochrome P450 3A5 polymorphism in patients taking clopidogrel. Can Med Assoc J 2006;174:1715—22.

[28] Desta Z, Zhao X, Shin JG, Flockhart DA. Clinical significance of the cytochrome P450 2C19 genetic poly-morphism. Clin Pharmacokinet 2002;41:913—58.

[29] Scott SA, Sangkuhl K, Gardner EE, et al. Clinical Pharmacogenetics Implementation Consortium guidelines for cytochrome P450-2C19 (CYP2C19) genotype and clopidogrel therapy. Clin Pharmacol Ther 2011;90:328—32.

[30] Collet JP, Hulot JS, Pena A, et al. Cytochrome P450 2C19 polymorphism in young patients treated with clo-pidogrel after myocardial infarction: a cohort study. Lancet 2009;373:309—17.

[31] Trenk D, Hochholzer W, Fromm MF, et al. Cytochrome P450 2C19 681G>A polymorphism and high on-clopidogrel platelet reactivity associated with adverse 1-year clinical outcome of elective percutaneous coronary intervention with drug-eluting or bare-metal stents. J Am Coll Cardiol 2008;51:1925—34.

[32] Sibbing D, Stegherr J, Latz W, et al. Cytochrome P450 2C19 loss-of-function polymorphism and stent thrombosis following percutaneous coronary intervention. Eur Heart J 2009;30:916—22.

[33] Giusti B, Gori AM, Marcucci R, et al. Relation of cytochrome P450 2C19 loss-of-function polymorphism to occurrence of drug-eluting coronary stent thrombosis. Am J Cardiol 2009;103:806—11.

[34] Mega JL, Simon T, Collet JP, et al. Reduced-function CYP2C19 genotype and risk of adverse clinical outcomes among patients treated with clopidogrel predominantly for PCI: a meta-analysis. J Am Med Assoc 2010;304:1821—30.

[35] Hulot JS, Collet JP, Silvain J, et al. Cardiovascular risk in clopidogrel-treated patients according to cytochrome P450 2C19*2 loss-of-function allele or proton pump inhibitor coadministration: a systematic meta-analysis. J Am Coll Cardiol 2010;56:134—43.

[36] Bauer T, Bouman HJ, van Werkum JW, Ford NF, ten Berg JM, Taubert D. Impact of CYP2C19 variant geno-types on clinical efficacy of antiplatelet treatment with clopidogrel: systematic review and meta-analysis. Br Med J (Clinical research ed.) 2011;343:d4588.

[37] Holmes MV, Perel P, Shah T, Hingorani AD, Casas JP. CYP2C19 genotype, clopidogrel metabolism, platelet function, and cardiovascular events: a systematic review and meta-analysis. J Am Med Assoc 2011;306:2704—14.

[38] Frere C, Cuisset T, Gaborit B, Alessi MC, Hulot JS. The CYP2C19*17 allele is associated with better platelet response to clopidogrel in patients admitted for non-ST acute coronary syndrome. J Thromb Haemost 2009;7:1409—11.

[39] Sibbing D, Koch W, Gebhard D, et al. Cytochrome 2C19*17 allelic variant, platelet aggregation, bleeding events, and stent thrombosis in clopidogrel-treated patients with coronary stent placement. Circulation 2010;121:512—8.

[40] Bouman HJ, Schomig E, van Werkum JW, et al. Paraoxonase-1 is a major determinant of clopidogrel efficacy. Nat Med 2011;17:110—6.

[41] Brandt JT, Close SL, Iturria SJ, et al. Common polymorphisms of CYP2C19 and CYP2C9 affect the pharma-cokinetic and pharmacodynamic response to clopidogrel but not prasugrel. J Thromb Haemost 2007;5:2429—36.

[42] Clarke TA, Waskell LA. The metabolism of clopidogrel is catalyzed by human cytochrome P450 3A and is inhibited by atorvastatin. Drug Metab Dispos 2003;31:53—9.

[43] Trenk D, Hochholzer W, Fromm MF, et al. Paraoxonase-1 Q192R polymorphism and antiplatelet effects of clopidogrel in patients undergoing elective coronary stent placement. Circ Cardiovasc Genet 2011;4:429—36.

[44] Lewis JP, Fisch AS, Ryan K, et al. Paraoxonase 1 (PON1) gene variants are not associated with clopidogrel response. Clin Pharmacol Ther 2011;90:568—74.

[45] Hulot JS, Collet JP, Cayla G, et al. CYP2C19 but not PON1 genetic variants influence clopidogrel pharmacokinetics, pharmacodynamics, and clinical efficacy in post-myocardial infarction patients. Circ Cardiovasc Interv 2011;4:422—8.

[46] Mega JL, Hochholzer W, Frelinger AL, 3rd, et al. Dosing clopidogrel based on CYP2C19 genotype and the effect on platelet reactivity in patients with stable cardiovascular disease. J Am Med Assoc 2011;306:2221—8.

[47] Simon T, Bhatt DL, Bergougnan L, et al. Genetic polymorphisms and the impact of a higher clopidogrel dose regimen on active metabolite exposure and antiplatelet response in healthy subjects. Clin Pharmacol Ther 2011;90:287—95.

[48] Cuisset T, Quilici J, Cohen W, et al. Usefulness of high clopidogrel maintenance dose according to CYP2C19 genotypes in clopidogrel low responders undergoing coronary stenting for non ST elevation acute coronary syndrome. Am J Cardiol 2011;108:760—5.

[49] Bonello L, Armero S, Ait Mokhtar O, et al. Clopidogrel loading dose adjustment according to platelet reactivity monitoring in patients carrying the 2C19*2 loss of function polymorphism. J Am Coll Cardiol 2010;56:1630—6.

[50] Desager JP. Clinical pharmacokinetics of ticlopidine. Clin Pharmacokinet 1994;26:347—55.

[51] Mega JL, Close SL, Wiviott SD, et al. Cytochrome P450 genetic polymorphisms and the response to prasugrel: relationship to pharmacokinetic, pharmacodynamic, and clinical outcomes. Circulation 2009;119:2553—60.

[52] Alexopoulos D, Dimitropoulos G, Davlouros P, et al. Prasugrel overcomes high on-clopidogrel platelet reactivity post-stenting more effectively than high-dose (150-mg) clopidogrel: the importance of CYP2C19*2 genotyping. J Am Coll Cardiol 2011;4:403—10.

[53] Sorich MJ, Vitry A, Ward MB, Horowitz JD, McKinnon RA. Prasugrel vs. clopidogrel for cytochrome P450 2C19-genotyped subgroups: integration of the TRITON-TIMI 38 trial data. J Thromb Haemost 2010;8:1678—84.

[54] Holmes Jr DR, Dehmer GJ, Kaul S, Leifer D, O'Gara PT, Stein CM. ACCF/AHA Clopidogrel clinical alert: approaches to the FDA "boxed warning": a report of the American College of Cardiology Foundation Task Force on Clinical Expert Consensus Documents and the American Heart Association. Circulation 2010;122:537—57.

[55] Levine GN, Bates ER, Blankenship JC, et al. 2011 ACCF/AHA/SCAI Guideline for Percutaneous Coronary Intervention: a report of the American College of Cardiology Foundation/American Heart Association Task Force on Practice Guidelines and the Society for Cardiovascular Angiography and Interventions. Circulation 2011;124:e574—651.

[56] Wright RS, Anderson JL, Adams CD, et al. 2011 ACCF/AHA focused update of the Guidelines for the Management of Patients with Unstable Angina/Non-ST-Elevation Myocardial Infarction (updating the 2007 guideline): a report of the American College of Cardiology Foundation/American Heart Association Task Force on Practice Guidelines developed in collaboration with the American College of Emergency Physicians, Society for Cardiovascular Angiography and Interventions, and Society of Thoracic Surgeons. J Am Coll Cardiol 2011;57:1920—59.

[57] Reese ES, Daniel Mullins C, Beitelshees AL, Onukwugha E. Cost-effectiveness of cytochrome P450 2C19 genotype screening for selection of antiplatelet therapy with clopidogrel or prasugrel. Pharmacotherapy 2012;32:323—32.

[58] Pare G, Eikelboom JW, Sibbing D, Bernlochner I, Kastrati A. Testing should not be done in all patients treated with clopidogrel who are undergoing percutaneous coronary intervention. Circ Cardiovasc Interv 2011;4:514—21. discussion 21.

[59] Roberts JD, Wells GA, Le May MR, et al. Point-of-care genetic testing for personalisation of antiplatelet treatment (RAPID GENE): a prospective, randomised, proof-of-concept trial. Lancet 2012;379(9827):1705—11.

[60] Hylek EM, Singer DE. Risk factors for intracranial hemorrhage in outpatients taking warfarin. Ann Intern Med 1994;120:897—902.

[61] Hylek EM, Go AS, Chang Y, et al. Effect of intensity of oral anticoagulation on stroke severity and mortality in atrial fibrillation. N Eng J Med 2003;349:1019—26.

[62] Coumadin (warfarin sodium) package insert. Princeton, NJ: Bristol-Myers Squibb; 2010 January.

[63] Wadelius M, Chen LY, Lindh JD, et al. The largest prospective warfarin-treated cohort supports genetic forecasting. Blood 2009;113:784—92.

[64] Limdi NA, Wadelius M, Cavallari L, et al. Warfarin pharmacogenetics: a single VKORC1 polymorphism is predictive of dose across 3 racial groups. Blood 2010;115:3827—34.

[65] Ansell J, Hirsh J, Hylek E, Jacobson A, Crowther M, Palareti G. Pharmacology and management of the vitamin K antagonists: American College of Chest Physicians Evidence-Based Clinical Practice Guidelines (8th Edition). Chest 2008;133:160S—98S.

[66] Absher RK, Moore ME, Parker MH. Patient-specific factors predictive of warfarin dosage requirements. Ann Pharmacother 2002;36:1512—7.

[67] Limdi NA, Limdi MA, Cavallari L, et al. Warfarin dosing in patients with impaired kidney function. Am J Kidney Dis 2010;56:823—31.

[68] Limdi NA, Beasley TM, Baird MF, et al. Kidney function influences warfarin responsiveness and hemorrhagic complications. J Am Soc Nephrol 2009;20:912—21.

[69] Gage BF, Eby C, Johnson JA, et al. Use of pharmacogenetic and clinical factors to predict the therapeutic dose of warfarin. Clin Pharmacol Ther 2008;84:326—31.

[70] Klein TE, Altman RB, Eriksson N, et al. Estimation of the warfarin dose with clinical and pharmacogenetic data. N Engl J Med 2009;360:753—64.

[71] Scordo MG, Pengo V, Spina E, Dahl ML, Gusella M, Padrini R. Influence of CYP2C9 and CYP2C19 genetic polymorphisms on warfarin maintenance dose and metabolic clearance. Clin Pharmacol Ther 2002;72:702—10.

[72] Rieder MJ, Reiner AP, Gage BF, et al. Effect of VKORC1 haplotypes on transcriptional regulation and warfarin dose. N Eng J Med 2005;352:2285—93.

[73] Cavallari LH, Langaee TY, Momary KM, et al. Genetic and clinical predictors of warfarin dose requirements in African Americans. Clin Pharmacol Ther 2010;87:459—64.

[74] Limdi NA, Arnett DK, Goldstein JA, et al. Influence of CYP2C9 and VKORC1 on warfarin dose, anticoagulation attainment and maintenance among European-Americans and African-Americans. Pharmacogenomics 2008;9:511—26.

[75] Cooper GM, Johnson JA, Langaee TY, et al. A genome-wide scan for common genetic variants with a large influence on warfarin maintenance dose. Blood 2008;112:1022—7.

[76] Takeuchi F, McGinnis R, Bourgeois S, et al. A genome-wide association study confirms VKORC1, CYP2C9, and CYP4F2 as principal genetic determinants of warfarin dose. PLoS Genet 2009;5:e1000433.

[77] Cha PC, Mushiroda T, Takahashi A, et al. Genome-wide association study identifies genetic determinants of warfarin responsiveness for Japanese. Hum Mol Genet 2011;19:4735—44.

[78] Caldwell MD, Awad T, Johnson JA, et al. CYP4F2 genetic variant alters required warfarin dose. Blood 2008;111:4106—12.

[79] Voora D, Koboldt DC, King CR, et al. A polymorphism in the VKORC1 regulator calumenin predicts higher warfarin dose requirements in African Americans. Clin Pharmacol Ther 2010;87:445—51.

[80] Kimmel SE, Christie J, Kealey C, et al. Apolipoprotein E genotype and warfarin dosing among Caucasians and African Americans. Pharmacogenomics J 2007.

[81] King CR, Deych E, Milligan P, et al. Gamma-glutamyl carboxylase and its influence on warfarin dose. Thromb Haemostasis 2011;104:750—4.

[82] CYP2CP allele nomenclature. Home page of the Human Cytochrome P450 (CYP) Allele Nomenclature Committee, <http://www.cypalleles.ki.se/cyp2c9.htm>; 2011 (accessed 28.09.2011).

[83] Rettie AE, Wienkers LC, Gonzalez FJ, Trager WF, Korzekwa KR. Impaired (S)-warfarin metabolism catalysed by the R144C allelic variant of CYP2C9. Pharmacogenetics 1994;4:39—42.

[84] Sullivan-Klose TH, Ghanayem BI, Bell DA, et al. The role of the CYP2C9-Leu359 allelic variant in the tolbutamide polymorphism. Pharmacogenetics 1996;6:341—9.

[85] Wei L, Locuson CW, Tracy TS. Polymorphic variants of CYP2C9: mechanisms involved in reduced catalytic activity. Mol Pharmacol 2007;72:1280—8.

[86] Takahashi H, Kashima T, Nomizo Y, et al. Metabolism of warfarin enantiomers in Japanese patients with heart disease having different CYP2C9 and CYP2C19 genotypes. Clin Pharmacol Ther 1998;63:519—28.

[87] Lindh JD, Holm L, Andersson ML, Rane A. Influence of CYP2C9 genotype on warfarin dose requirements—a systematic review and meta-analysis. Eur J Clin Pharmacol 2009;65:365—75.

[88] Higashi MK, Veenstra DL, Kondo LM, et al. Association between CYP2C9 genetic variants and anticoagulation-related outcomes during warfarin therapy. J Am Med Assoc 2002;287:1690—8.

[89] Scott SA, Khasawneh R, Peter I, Kornreich R, Desnick RJ. Combined CYP2C9, VKORC1 and CYP4F2 frequencies among racial and ethnic groups. Pharmacogenomics 2010;11:781−91.

[90] Dickmann LJ, Rettie AE, Kneller MB, et al. Identification and functional characterization of a new CYP2C9 variant (CYP2C9*5) expressed among African Americans. Mol Pharmacol 2001;60:382−7.

[91] Allabi AC, Gala JL, Horsmans Y. CYP2C9, CYP2C19, ABCB1 (MDR1) genetic polymorphisms and phenytoin metabolism in a Black Beninese population. Pharmacogenetics Genomics 2005;15:779−86.

[92] Allabi AC, Gala JL, Horsmans Y, et al. Functional impact of CYP2C95, CYP2C96, CYP2C98, and CYP2C911 in vivo among black Africans. Clin Pharmacol Ther 2004;76:113−8.

[93] Liu Y, Jeong HY, Takahashi H, et al. Decreased warfarin clearance with the CYP2C9 R150H (*8) polymorphism. Clin Pharmacol Ther 2012;91(4):660−5.

[94] Blaisdell J, Jorge-Nebert LF, Coulter S, et al. Discovery of new potentially defective alleles of human CYP2C9. Pharmacogenetics 2004;14:527−37.

[95] Liu Y, Jeong H, Takahashi H, et al. Decreased warfarin clearance associated with the CYP2C9 R150H (*8) polymorphism. Clin Pharmacol Ther 2012;91:660−5.

[96] Perera MA, Gamazon E, Cavallari LH, et al. The missing association: sequencing-based discovery of novel SNPs in VKORC1 and CYP2C9 that affect warfarin dose in African Americans. Clin Pharmacol Ther 2011;89: 408−15.

[97] Margaglione M, Colaizzo D, D'Andrea G, et al. Genetic modulation of oral anticoagulation with warfarin. Thromb Haemostasis 2000;84:775−8.

[98] Limdi NA, McGwin G, Goldstein JA, et al. Influence of CYP2C9 and VKORC1 1173C/T genotype on the risk of hemorrhagic complications in African-American and European-American patients on warfarin. Clin Pharmacol Ther 2008;83:312−21.

[99] Sanderson S, Emery J, Higgins J. CYP2C9 gene variants, drug dose, and bleeding risk in warfarin-treated patients: a HuGEnet systematic review and meta-analysis. Genet Med 2005;7:97−104.

[100] Rost S, Fregin A, Ivaskevicius V, et al. Mutations in VKORC1 cause warfarin resistance and multiple coagulation factor deficiency type 2. Nature 2004;427:537−41.

[101] Scott SA, Edelmann L, Kornreich R, Desnick RJ. Warfarin pharmacogenetics: CYP2C9 and VKORC1 genotypes predict different sensitivity and resistance frequencies in the Ashkenazi and Sephardi Jewish populations. Am J Hum Genet 2008;82:495−500.

[102] D'Andrea G, D'Ambrosio RL, Di Perna P, et al. A polymorphism in the VKORC1 gene is associated with an interindividual variability in the dose-anticoagulant effect of warfarin. Blood 2005;105:645−9.

[103] Wang D, Chen H, Momary KM, Cavallari LH, Johnson JA, Sadee W. Regulatory polymorphism in vitamin K epoxide reductase complex subunit 1 (VKORC1) affects gene expression and warfarin dose requirement. Blood 2008;112:1013−21.

[104] Geisen C, Watzka M, Sittinger K, et al. VKORC1 haplotypes and their impact on the interindividual and interethnic variability of oral anticoagulation. Thromb Haemostasis 2005;94:773−9.

[105] Huang SW, Chen HS, Wang XQ, et al. Validation of VKORC1 and CYP2C9 genotypes on interindividual warfarin maintenance dose: a prospective study in Chinese patients. Pharmacogenet Genomics 2009;19:226−34.

[106] Lee MT, Chen CH, Chou CH, et al. Genetic determinants of warfarin dosing in the Han-Chinese population. Pharmacogenomics 2009;10:1905−13.

[107] McDonald MG, Rieder MJ, Nakano M, Hsia CH, Rettie AE. CYP4F2 is a Vitamin K1 oxidase: an explanation for altered warfarin dose in carriers of the V433M variant. Mol Pharmacol 2009;75:1337−46.

[108] Pautas E, Moreau C, Gouin-Thibault I, et al. Genetic factors (VKORC1, CYP2C9, EPHX1, and CYP4F2) are predictor variables for warfarin response in very elderly, frail inpatients. Clin Pharmacol Ther 2011;87:57−64.

[109] Choi JR, Kim JO, Kang DR, et al. Proposal of pharmacogenetics-based warfarin dosing algorithm in Korean patients. J Hum Genet 2011;56:290−5.

[110] Chan SL, Suo C, Lee SC, Goh BC, Chia KS, Teo YY. Translational aspects of genetic factors in the prediction of drug response variability: a case study of warfarin pharmacogenomics in a multi-ethnic cohort from Asia. Pharmacogenomics J 2011.

[111] Suriapranata IM, Tjong WY, Wang T, et al. Genetic factors associated with patient-specific warfarin dose in ethnic Indonesians. BMC Med Genet 2011;12:80.

[112] Shahin MH, Khalifa SI, Gong Y, et al. Genetic and nongenetic factors associated with warfarin dose requirements in Egyptian patients. Pharmacogenet Genomics 2011;21:130−5.

[113] Biss TT, Avery PJ, Brandao LR, et al. VKORC1 and CYP2C9 genotype and patient characteristics explain a large proportion of the variability in warfarin dose requirement among children. Blood 2012;119(3):868–73.

[114] Wallin R, Hutson SM, Cain D, Sweatt A, Sane DC. A molecular mechanism for genetic warfarin resistance in the rat. Faseb J 2001;15:2542–4.

[115] Rost S, Fregin A, Koch D, Compes M, Muller CR, Oldenburg J. Compound heterozygous mutations in the gamma-glutamyl carboxylase gene cause combined deficiency of all vitamin K-dependent blood coagulation factors. Br J Haematol 2004;126:546–9.

[116] Kimura R, Miyashita K, Kokubo Y, et al. Genotypes of vitamin K epoxide reductase, gamma-glutamyl carboxylase, and cytochrome P450 2C9 as determinants of daily warfarin dose in Japanese patients. Thromb Res 2007;120:181–6.

[117] Chen LY, Eriksson N, Gwilliam R, Bentley D, Deloukas P, Wadelius M. Gamma-glutamyl carboxylase (GGCX) microsatellite and warfarin dosing. Blood 2005;106:3673–4.

[118] Wadelius M, Chen LY, Downes K, et al. Common VKORC1 and GGCX polymorphisms associated with warfarin dose. Pharmacogenomics J 2005;5:262–70.

[119] King CR, Deych E, Milligan P, et al. Gamma-glutamyl carboxylase and its influence on warfarin dose. Thromb Haemostasis 2010;104:750–4.

[120] Rieder MJ, Reiner AP, Rettie AE. Gamma-glutamyl carboxylase (GGCX) tagSNPs have limited utility for predicting warfarin maintenance dose. J Thromb Haemost 2007;5(11):2227–34.

[121] Cavallari LH, Perera M, Wadelius M, et al. Association of the GGCX (CAA)16/17 repeat polymorphism with higher warfarin dose requirements in African Americans. Pharmacogenet Genomics 2012;22(2):152–8.

[122] Perera MA, Limdi NA, Cavallari L, et al. Novel SNPs associated with warfarin dose in a large multicenter cohort of African Americans: genome wide association study and replication results. Circulation 2011:124.

[123] Gamazon ER, Skol AD, Perera MA. The limits of genome-wide methods for pharmacogenomic testing. Pharmacogenet Genom;22:261–72.

[124] Sconce EA, Khan TI, Wynne HA, et al. The impact of CYP2C9 and VKORC1 genetic polymorphism and patient characteristics upon warfarin dose requirements: proposal for a new dosing regimen. Blood 2005;106:2329–33.

[125] Caldwell MD, Berg RL, Zhang KQ, et al. Evaluation of genetic factors for warfarin dose prediction. Clin Med Res 2007;5:8–16.

[126] Caraco Y, Blotnick S, Muszkat M. CYP2C9 genotype-guided warfarin prescribing enhances the efficacy and safety of anticoagulation: a prospective randomized controlled study. Clin Pharmacol Ther 2008;83:460–70.

[127] Sagreiya H, Berube C, Wen A, et al. Extending and evaluating a warfarin dosing algorithm that includes CYP4F2 and pooled rare variants of CYP2C9. Pharmacogenet Genomics 2010;20:407–13.

[128] Schelleman H, Chen J, Chen Z, et al. Dosing algorithms to predict warfarin maintenance dose in Caucasians and African Americans. Clin Pharmacol Ther 2008;84:332–9.

[129] Herman D, Peternel P, Stegnar M, Breskvar K, Dolzan V. The influence of sequence variations in factor VII, gamma-glutamyl carboxylase and vitamin K epoxide reductase complex genes on warfarin dose requirement. Thromb Haemost 2006;95:782–7.

[130] Kim HS, Lee SS, Oh M, et al. Effect of CYP2C9 and VKORC1 genotypes on early-phase and steady-state warfarin dosing in Korean patients with mechanical heart valve replacement. Pharmacogenet Genomics 2009;19:103–12.

[131] Takahashi H, Wilkinson GR, Nutescu EA, et al. Different contributions of polymorphisms in VKORC1 and CYP2C9 to intra- and inter-population differences in maintenance dose of warfarin in Japanese, Caucasians and African-Americans. Pharmacogenet Genomics 2006;16:101–10.

[132] Lenzini P, Wadelius M, Kimmel S, et al. Integration of genetic, clinical, and INR data to refine warfarin dosing. Clin Pharmacol Ther 2010;87:572–8.

[133] Owen RP, Altman RB, Klein TE. PharmGKB and the International Warfarin Pharmacogenetics Consortium: the changing role for pharmacogenomic databases and single-drug pharmacogenetics. Human Mutation 2008;29:456–60.

[134] Roper N, Storer B, Bona R, Fang M. Validation and comparison of pharmacogenetics-based warfarin dosing algorithms for application of pharmacogenetic testing. J Mol Diagn;12:283–91.

[135] Shin J, Cao D. Comparison of warfarin pharmacogenetic dosing algorithms in a racially diverse large cohort. Pharmacogenomics 2011;12:125–34.

[136] Langley MR, Booker JK, Evans JP, McLeod HL, Weck KE. Validation of clinical testing for warfarin sensitivity: comparison of CYP2C9-VKORC1 genotyping assays and warfarin-dosing algorithms. J Mol Diagn 2009;11:216−25.

[137] Company. B-MS. Coumadin® package insert. In: Company. B-MS, ed. Princeton, NJ: Bristol Myers Squibb Company; 2010.

[138] Finkelman BS, Gage BF, Johnson JA, Brensinger CM, Kimmel SE. Genetic warfarin dosing: tables versus algorithms. J Am Coll Cardiol 2011;57:612−8.

[139] Anderson JL, Horne BD, Stevens SM, et al. Randomized trial of genotype-guided versus standard warfarin dosing in patients initiating oral anticoagulation. Circulation 2007;116:2563−70.

[140] Schwartz JB, Kane L, Moore K, Wu AH. Failure of pharmacogenetic-based dosing algorithms to identify older patients requiring low daily doses of warfarin. J Am Med Dir Assoc 2011;2:633−8.

[141] Epstein RS, Moyer TP, Aubert RE, et al. Warfarin genotyping reduces hospitalization rates results from the MM-WES (Medco-Mayo Warfarin Effectiveness study). J Am Coll Cardiol 2010;55:2804−12.

[142] Burmester JK, Berg RL, Yale SH, et al. A randomized controlled trial of genotype-based Coumadin initiation. Genet Med 2011;13:509−18.

[143] Anderson JL, Horne BD, Stevens SM, et al. Randomized and clinical effectiveness trial comparing two pharmacogenetic algorithms and standard care for individualizing warfarin dosing: CoumaGen-II. Circulation 2012;125:1997−2005.

[144] ClinicalTrials.gov. Accessed 3/14/2011.

[145] Flockhart DA, O'Kane D, Williams MS, et al. Pharmacogenetic testing of CYP2C9 and VKORC1 alleles for warfarin. Genet Med 2008;10:139−50.

[146] Guyatt GH, Akl EA, Crowther M, Gutterman DD, Schunemann HJ. Executive summary: antithrombotic therapy and prevention of thrombosis, 9th ed: American College of Chest Physicians Evidence-Based Clinical Practice Guidelines. Chest 2012;141:7S−47S.

[147] French B, Joo J, Geller NL, et al. Statistical design of personalized medicine interventions: the Clarification of Optimal Anticoagulation through Genetics (COAG) trial. Trials 2010;11:108.

[148] Johnson JA, Gong L, Whirl-Carrillo M, et al. Clinical Pharmacogenetics Implementation Consortium Guidelines for CYP2C9 and VKORC1 genotypes and warfarin dosing. Clin Pharmacol Ther 2011;90:625−9.

[149] Horne BD, Lenzini PA, Wadelius M, et al. Pharmacogenetic warfarin dose refinements remain significantly influenced by genetic factors after one week of therapy. Thromb Haemostasis 2012;107:232−40.

[150] Vrecer M, Turk S, Drinovec J, Mrhar A. Use of statins in primary and secondary prevention of coronary heart disease and ischemic stroke. Meta-analysis of randomized trials. Int J Clin Pharmacol Therapeut 2003;41:567−77.

[151] Libby P. The forgotten majority: unfinished business in cardiovascular risk reduction. J Am Coll Cardiol 2005;46:1225−8.

[152] Zineh I, Johnson JA. Pharmacogenetics of chronic cardiovascular drugs: applications and implications. Expert Opin Pharmacother 2006;7:1417−27.

[153] Neuvonen PJ, Niemi M, Backman JT. Drug interactions with lipid-lowering drugs: mechanisms and clinical relevance. Clin Pharmacol Ther 2006;80:565−81.

[154] Romaine SP, Bailey KM, Hall AS, Balmforth AJ. The influence of SLCO1B1 (OATP1B1) gene polymorphisms on response to statin therapy. Pharmacogenomics J 2010;10:1−11.

[155] Heller DA, de Faire U, Pedersen NL, Dahlen G, McClearn GE. Genetic and environmental influences on serum lipid levels in twins. N Eng J Med 1993;328:1150−6.

[156] Hobbs HH, Brown MS, Goldstein JL. Molecular genetics of the LDL receptor gene in familial hypercholesterolemia. Hum Mutat 1992;1:445−66.

[157] Teslovich TM, Musunuru K, Smith AV, et al. Biological, clinical and population relevance of 95 loci for blood lipids. Nature 2010;466:707−13.

[158] Krauss RM, Mangravite LM, Smith JD, et al. Variation in the 3-hydroxyl-3-methylglutaryl coenzyme A reductase gene is associated with racial differences in low-density lipoprotein cholesterol response to simvastatin treatment. Circulation 2008;117:1537−44.

[159] Donnelly LA, Doney AS, Dannfald J, et al. A paucimorphic variant in the HMG-CoA reductase gene is associated with lipid-lowering response to statin treatment in diabetes: a GoDARTS study. Pharmacogenet Genomics 2008;18:1021−6.

[160] Thompson JF, Man M, Johnson KJ, et al. An association study of 43 SNPs in 16 candidate genes with atorvastatin response. Pharmacogenomics J 2005;5:352−8.

[161] Chasman DI, Posada D, Subrahmanyan L, Cook NR, Stanton Jr VP, Ridker PM. Pharmacogenetic study of statin therapy and cholesterol reduction. J Am Med Assoc 2004;291:2821−7.

[162] Tirona RG, Leake BF, Merino G, Kim RB. Polymorphisms in OATP-C: identification of multiple allelic variants associated with altered transport activity among European- and African-Americans. J Biol Chem 2001;276:35669−75.

[163] Pasanen MK, Fredrikson H, Neuvonen PJ, Niemi M. Different effects of SLCO1B1 polymorphism on the pharmacokinetics of atorvastatin and rosuvastatin. Clin Pharmacol Ther 2007;82:726−33.

[164] Barber MJ, Mangravite LM, Hyde CL, et al. Genome-wide association of lipid-lowering response to statins in combined study populations. PloS One 2010;5:e9763.

[165] Iakoubova OA, Tong CH, Rowland CM, et al. Association of the Trp719Arg polymorphism in kinesin-like protein 6 with myocardial infarction and coronary heart disease in 2 prospective trials: the CARE and WOSCOPS trials. J Am Coll Cardiol 2008;51:435−43.

[166] Iakoubova OA, Robertson M, Tong CH, et al. KIF6 Trp719Arg polymorphism and the effect of statin therapy in elderly patients: results from the PROSPER study. Eur J Cardiovasc Prev Rehabil 2010;17:455−61.

[167] Iakoubova OA, Sabatine MS, Rowland CM, et al. Polymorphism in KIF6 gene and benefit from statins after acute coronary syndromes: results from the PROVE IT-TIMI 22 study. J Am Coll Cardiol 2008;51: 449−55.

[168] Ridker PM, MacFadyen JG, Glynn RJ, Chasman DI. Kinesin-like protein 6 (KIF6) polymorphism and the efficacy of rosuvastatin in primary prevention. Circ Cardiovasc Genet 2011;4:312−7.

[169] Hopewell JC, Parish S, Clarke R, et al. No impact of KIF6 genotype on vascular risk and statin response among 18,348 randomized patients in the heart protection study. J Am Coll Cardiol 2011;57:2000−7.

[170] Ghatak A, Faheem O, Thompson PD. The genetics of statin-induced myopathy. Atherosclerosis 2010;210:337−43.

[171] Maggo SD, Kennedy MA, Clark DW. Clinical implications of pharmacogenetic variation on the effects of statins. Drug Saf 2011;34:1−19.

[172] Link E, Parish S, Armitage J, et al. SLCO1B1 variants and statin-induced myopathy—a genomewide study. N Eng J Med 2008;359:789−99.

[173] Niemi M. Transporter pharmacogenetics and statin toxicity. Clin Pharmacol Ther 2010;87:130−3.

[174] Voora D, Shah SH, Spasojevic I, et al. The SLCO1B1*5 genetic variant is associated with statin-induced side effects. J Am Coll Cardiol 2009;54:1609−16.

[175] Donnelly LA, Doney AS, Tavendale R, et al. Common nonsynonymous substitutions in SLCO1B1 predispose to statin intolerance in routinely treated individuals with type 2 diabetes: a go-DARTS study. Clin Pharmacol Ther 2011;89:210−6. .

[176] Brunham LR, Lansberg PJ, Zhang L, et al. Differential effect of the rs4149056 variant in SLCO1B1 on myopathy associated with simvastatin and atorvastatin. Pharmacogenomics J 2011.

[177] Wilke RA, Ramsey LB, Johnson SG, et al. The Clinical Pharmacogenomics Implementation Consortium: CPIC Guideline for SLCO1B1 and simvastatin-induced myopathy. Clin Pharmacol Ther 2012;92:112−7.

[178] Wang J, Williams CM, Hegele RA. Compound heterozygosity for two non-synonymous polymorphisms in NPC1L1 in a non-responder to ezetimibe. Clin Genet 2005;67:175−7.

[179] Hegele RA, Guy J, Ban MR, Wang J. NPC1L1 haplotype is associated with inter-individual variation in plasma low-density lipoprotein response to ezetimibe. Lipids Health Disease 2005;4:16.

[180] Simon JS, Karnoub MC, Devlin DJ, et al. Sequence variation in NPC1L1 and association with improved LDL-cholesterol lowering in response to ezetimibe treatment. Genomics 2005;86:648−56.

[181] Schmitz G, Schmitz-Madry A, Ugocsai P. Pharmacogenetics and pharmacogenomics of cholesterol-lowering therapy. Curr Opin Lipidol 2007;18:164−73.

[182] Oswald S, Haenisch S, Fricke C, et al. Intestinal expression of P-glycoprotein (ABCB1), multidrug resistance associated protein 2 (ABCC2), and uridine diphosphate-glucuronosyltransferase 1A1 predicts the disposition and modulates the effects of the cholesterol absorption inhibitor ezetimibe in humans. Clin Pharmacol Ther 2006;79:206−17.

[183] Ginsberg HN, Elam MB, Lovato LC, et al. Effects of combination lipid therapy in type 2 diabetes mellitus. N Eng J Med 2010;362:1563−74.

[184] Materson BJ, Reda DJ, Cushman WC. Department of Veterans Affairs single-drug therapy of hypertension study. Revised figures and new data. Department of Veterans Affairs Cooperative Study Group on Antihypertensive Agents. Am J Hypertens 1995;8:189−92.

[185] Rau T, Wuttke H, Michels LM, et al. Impact of the CYP2D6 genotype on the clinical effects of metoprolol: a prospective longitudinal study. Clin Pharmacol Ther 2009;85:269—72.

[186] Terra SG, Pauly DF, Lee CR, et al. beta-Adrenergic receptor polymorphisms and responses during titration of metoprolol controlled release/extended release in heart failure. Clin Pharmacol Ther 2005;77:127—37.

[187] Bijl MJ, Visser LE, van Schaik RH, et al. Genetic variation in the CYP2D6 gene is associated with a lower heart rate and blood pressure in beta-blocker users. Clin Pharmacol Ther 2009;85:45—50.

[188] Johnson JA, Zineh I, Puckett BJ, McGorray SP, Yarandi HN, Pauly DF. Beta 1-adrenergic receptor polymorphisms and antihypertensive response to metoprolol. Clin Pharmacol Ther 2003;74:44—52.

[189] Liu J, Liu ZQ, Tan ZR, et al. Gly389Arg polymorphism of beta1-adrenergic receptor is associated with the cardiovascular response to metoprolol. Clin Pharmacol Ther 2003;74:372—9.

[190] Shin J, Johnson JA. Pharmacogenetics of beta-blockers. Pharmacotherapy 2007;27:874—87.

[191] Pacanowski MA, Gong Y, Cooper-Dehoff RM, et al. beta-Adrenergic receptor gene polymorphisms and beta-blocker treatment outcomes in hypertension. Clin Pharmacol Ther 2008;84:715—21.

[192] Niu Y, Gong Y, Langaee TY, et al. Genetic variation in the beta2 subunit of the voltage-gated calcium channel and pharmacogenetic association with adverse cardiovascular outcomes in the INternational VErapamil SR-Trandolapril STudy GENEtic Substudy (INVEST-GENES). Circ Cardiovasc Genet 2010;3:548—55.

[193] Beitelshees AL, Gong Y, Wang D, et al. KCNMB1 genotype influences response to verapamil SR and adverse outcomes in the INternational VErapamil SR/Trandolapril STudy (INVEST). Pharmacogenet Genomics 2007;17:719—29.

[194] Manunta P, Lavery G, Lanzani C, et al. Physiological interaction between alpha-adducin and WNK1-NEDD4L pathways on sodium-related blood pressure regulation. Hypertension 2008;52:366—72.

[195] Cusi D, Barlassina C, Azzani T, et al. Polymorphisms of alpha-adducin and salt sensitivity in patients with essential hypertension. Lancet 1997;349:1353—7.

[196] Lanzani C, Citterio L, Glorioso N, et al. Adducin- and ouabain-related gene variants predict the antihypertensive activity of rostafuroxin, part 2: clinical studies. Sci Transl Med 2010;2:59. ra87.

[197] Irvin MR, Lynch AI, Kabagambe EK, et al. Pharmacogenetic association of hypertension candidate genes with fasting glucose in the GenHAT Study. J Hypertens 2010;28:2076—83.

[198] Major outcomes in high-risk hypertensive patients randomized to angiotensin-converting enzyme inhibitor or calcium channel blocker vs diuretic: The Antihypertensive and Lipid-Lowering Treatment to Prevent Heart Attack Trial (ALLHAT). J Am Med Assoc 2002;288:2981—97.

[199] Bozkurt O, de Boer A, Grobbee DE, et al. Variation in renin-angiotensin system and salt-sensitivity genes and the risk of diabetes mellitus associated with the use of thiazide diuretics. Am J Hypertens 2009;22:545—51.

[200] Pitt B, Zannad F, Remme WJ, et al. The effect of spironolactone on morbidity and mortality in patients with severe heart failure. Randomized Aldactone Evaluation Study Investigators. N Eng J Med 1999;341:709—17.

[201] Taylor AL, Ziesche S, Yancy C, et al. Combination of isosorbide dinitrate and hydralazine in blacks with heart failure. N Eng J Med 2004;351:2049—57.

[202] Effect of enalapril on survival in patients with reduced left ventricular ejection fractions and congestive heart failure. The SOLVD Investigators. N Eng J Med 1991;325:293—302.

[203] Effect of metoprolol CR/XL in chronic heart failure: Metoprolol CR/XL Randomised Intervention Trial in Congestive Heart Failure (MERIT-HF). Lancet 1999;353:2001—7.

[204] Tiret L, Rigat B, Visvikis S, et al. Evidence, from combined segregation and linkage analysis, that a variant of the angiotensin I-converting enzyme (ACE) gene controls plasma ACE levels. Am J Hum Genet 1992;51:197—205.

[205] Danser AH, Derkx FH, Hense HW, Jeunemaitre X, Riegger GA, Schunkert H. Angiotensinogen (M235T) and angiotensin-converting enzyme (I/D) polymorphisms in association with plasma renin and prorenin levels. J Hypertens 1998;16:1879—83.

[206] Winkelmann BR, Nauck M, Klein B, et al. Deletion polymorphism of the angiotensin I-converting enzyme gene is associated with increased plasma angiotensin-converting enzyme activity but not with increased risk for myocardial infarction and coronary artery disease. Ann Intern Med 1996;125:19—25.

[207] Andersson B, Sylven C. The DD genotype of the angiotensin-converting enzyme gene is associated with increased mortality in idiopathic heart failure. J Am Coll Cardiol 1996;28:162—7.

[208] McNamara DM, Holubkov R, Postava L, et al. Pharmacogenetic interactions between angiotensin-converting enzyme inhibitor therapy and the angiotensin-converting enzyme deletion polymorphism in patients with congestive heart failure. J Am Coll Cardiol 2004;44:2019–26.

[209] Wu CK, Luo JL, Tsai CT, et al. Demonstrating the pharmacogenetic effects of angiotensin-converting enzyme inhibitors on long-term prognosis of diastolic heart failure. Pharmacogenomics J 2010;10:46–53.

[210] Terra SG, Hamilton KK, Pauly DF, et al. Beta1-adrenergic receptor polymorphisms and left ventricular remodeling changes in response to beta-blocker therapy. Pharmacogenet Genomics 2005;15:227–34.

[211] Chen L, Meyers D, Javorsky G, et al. Arg389Gly-beta1-adrenergic receptors determine improvement in left ventricular systolic function in nonischemic cardiomyopathy patients with heart failure after chronic treatment with carvedilol. Pharmacogenet Genomics 2007;17:941–9.

[212] A trial of the beta-blocker bucindolol in patients with advanced chronic heart failure. N Eng J Med 2001;344:1659–67.

[213] Liggett SB, Mialet-Perez J, Thaneemit-Chen S, et al. A polymorphism within a conserved beta(1)-adrenergic receptor motif alters cardiac function and beta-blocker response in human heart failure. Proc Natl Acad Sci U S A 2006;103:11288–93.

[214] Sehnert AJ, Daniels SE, Elashoff M, et al. Lack of association between adrenergic receptor genotypes and survival in heart failure patients treated with carvedilol or metoprolol. J Am Coll Cardiol 2008;52:644–51.

[215] McNamara DM, Tam SW, Sabolinski ML, et al. Endothelial nitric oxide synthase (NOS3) polymorphisms in African Americans with heart failure: results from the A-HeFT trial. J Card Fail 2009;15:191–8.

[216] RYTHMOL® (propafenone hydrochloride) prescribing information. Triange Park, NC: GlaxoSmithKline. Research; June 2011.

[217] Jazwinska-Tarnawska E, Orzechowska-Juzwenko K, Niewinski P, et al. The influence of CYP2D6 polymorphism on the antiarrhythmic efficacy of propafenone in patients with paroxysmal atrial fibrillation during 3 months propafenone prophylactic treatment. Int J Clin Pharm Th 2001;39:288–92.

[218] Quinidine sulfate prescribing information. Corona, CA: Watson Laboratories, Inc; June 2005.

[219] Roden DM. Proarrhythmia as a pharmacogenomic entity: a critical review and formulation of a unifying hypothesis. Cardiovasc Res 2005;67:419–25.

[220] van der Sijs H, Kowlesar R, Klootwijk AP, Nelwan SP, Vulto AG, van Gelder T. Clinically relevant QTc prolongation due to overridden drug-drug interaction alerts: a retrospective cohort study. Br J Clin Pharmacol 2009;67:347–54.

[221] Kannankeril P, Roden DM, Darbar D. Drug-induced long QT syndrome. Pharmacol Rev 2010;62:760–81.

[222] Roden DM. Personalized medicine and the genotype-phenotype dilemma. J Interv Card Electrophysiol 2011;31:17–23.

[223] Food and Drug Administration, guidance for industry. E14 Clinical Evaluation of QT/QTc Interval Prolongation and Proarrhythmic Potential for Non-Antiarrhythmic Drugs. (Accessed 1.07.2012, at http://www.fda.gov/downloads/RegulatoryInformation/Guidances/ucm129357.pdf.)

[224] Huges S. Vanderbilt now also routinely gene testing for clopidogrel metabolizer status [Clinical Conditions> Interventional/Surgery> Interventional/Surgery]. Accessed at, <http://www.theheart.org/article/1139495.do>; Oct 21, 2010. on Aug 11, 2011.

[225] Takahashi H, Wilkinson GR, Nutescu EA, et al. Different contributions of polymorphisms in VKORC1 and CYP2C9 to intra- and inter-population differences in maintenance dose of warfarin in Japanese, Caucasians and African-Americans. Pharmacogenet Genomics 2006;16:101–10.

[226] Lal S, Sandanaraj E, Jada SR, et al. Influence of APOE genotypes and VKORC1 haplotypes on warfarin dose requirements in Asian patients. Br J Clin Pharmacol 2008;65:260–4.

[227] Shyamala G, Sowmya P, Madhavan HN, Malathi J. Relative efficiency of polymerase chain reaction and enzyme-linked immunosorbant assay in determination of viral etiology in congenital cataract in infants. J Postgrad Med 2008;54:17–20.

[228] Johnson JA, Liggett SB. Cardiovascular pharmacogenomics of adrenergic receptor signaling: clinical implications and future directions. Clin Pharmacol Ther 2011;89:366–78.

[229] Maliarik MJ, Rybicki BA, Malvitz E, et al. Angiotensin-converting enzyme gene polymorphism and risk of sarcoidosis. Am J Respir Crit Care Med 1998;158:1566–70.

5A

A Look to the Future: Cardiovascular Pharmacoepigenetics

Julio D. Duarte

Department of Pharmacy Practice, University of Illinois at Chicago, Chicago, Illinois, USA

LEARNING OBJECTIVES

1. Describe how pharmacoepigenetics can further explain variation in drug response beyond pharmacogenomics.

2. Provide examples of how investigating drug effects on epigenetic status can improve our understanding of drug mechanisms and lead to discovery of novel drug targets.

3. Explain additional challenges that exist in pharmacoepigenetic research beyond those found in pharmacogenomics.

INTRODUCTION

Pharmacogenomic research has helped us better understand interindividual differences in cardiovascular drug response, but a considerable amount of observed variability has yet to be explained. Epigenetics can help further elucidate these unexplained differences. Epigenetics describes heritable alterations of gene expression that do not involve DNA sequence variation and are changeable throughout an organism's lifetime. Pharmacoepigenetic research investigates both how epigenetic status affects drug response and how drugs affect epigenetic status. These areas of study can provide valuable insight into cardiovascular drug response and lead to new strategies for cardiovascular disease management.

Epigenetic modification of gene expression can include:

- DNA methylation
- regulation by noncoding RNA
- histone protein modification

Pharmacogenomics
http://dx.doi.org/10.1016/B978-0-12-391918-2.00015-9

DNA methylation, or the covalent addition of a methyl group by methyltransferase proteins, occurs at cytosines in the genome that are immediately followed by a guanine (referred to as CpG sites). DNA methylation at these sites, particularly around the gene promoter, is associated with decreased gene transcription.

The currently most-studied and best-understood noncoding regulatory RNA is micro-RNA (miRNA). MiRNAs are approximately 22 nucleotides long and negatively regulate protein expression by binding to mRNA and marking it for degradation before it can be translated into protein [1].

Histone modifications can include acetylation, methylation, phosphorylation, ubiquitylation, sumoylation, ADP-ribosylation, deimination, and proline isomerization [2]. Acetylation and methylation of histone proteins are the most-common and the currently best-understood modifications. Histone acetylation is associated with a more open chromatin configuration, thus allowing increased gene transcription and expression. Histone methylation is often associated with a more closed chromatin configuration, which usually decreases transcription, but is sometimes associated with increases in transcription.

Refer to chapter 3 for a more detailed explanation of epigenetic modification.

The study of how epigenetic pathways are affected by cardiovascular drugs is just beginning, [3,4,5] and although the effect of variations in one's epigenetic status on drug response in cancer has been studied for some time, [6,7,8] new opportunities for study exist in cardiovascular disease. As with any new research area, opportunities come with challenges that must be overcome before adoption of research findings into clinical practice can occur.

OPPORTUNITIES

Variability in Drug Responses

Much of the interindividual variation in drug metabolism has been attributed to differences in cytochrome P450 (CYP) enzyme expression. Many, but not all, of these differences in expression are explained by genetic polymorphisms. DNA methylation, especially at the gene promoter, can also regulate expression of CYP enzymes. For drugs that are metabolized by CYP enzymes, hypermethylation, leading to decreased enzyme expression, may increase the concentration of drug at the site of action and/or increase the duration that the drug occupies the site. This epigenetic control of CYP enzyme expression presents a possible opportunity to explain more of the interindividual variation in drug response. The CYP enzymes CYP1A1, CYP1A2, CYP1B1, CYP2C19, CYP2D6, CYP3A4, and CYP3A5 are thought to be regulated, at least in part, by DNA methylation [9,10,11]. CYP enzyme expression can also be regulated by other epigenetic mechanisms. For example, expression of CYP3A4, which metabolizes the cardiovascular drugs simvastatin, sildenafil, and quinidine, is negatively modulated by miRNA-27 (miR-27) at both the transcriptional and post-transcriptional levels [12].

As with CYP enzymes, expression of the multidrug efflux transporter P-glycoprotein (P-gp), encoded by the ATP binding cassette, subfamily B (MDR/TAP) member 1 (*ABCB1*) gene, can be regulated by methylation status at the gene promoter [6,13]. By hypermethylating the *ABCB1* promoter, less protein would theoretically be available to efflux cardiovascular drugs that are P-gp substrates (e.g., digoxin, diltiazem, and verapamil) from their sites of action. *ABCB1* expression is also modulated by miR-27a and miR-451, with these miR-NAs increasing expression of *ABCB1* in cancer cell lines, likely by inhibiting transcriptional factors that suppress *ABCB1* [14]. Antagonizing these miRNAs increased susceptibility of cancer cells to the cytotoxic P-gp substrate vinblastine [14]. If also true in normal cells, such as intestinal enterocytes, decreased miR-27a and miR-451 levels could decrease the amount of P-gp available to efflux P-gp substrates, leading to increased drug bioavailability and risk for supratherapeutic drug concentrations and potential toxicity.

As described above, epigenetic regulation of cardiovascular drug transport and metabolism could affect drug disposition in patients, contributing to the response variability observed in clinical practice. However, little research has been done to quantify the effect of epigenetic regulation of these enzymes on drug transport and metabolism *in vivo*, providing significant research opportunity.

Epigenetics not only has the ability to influence pharmacokinetic properties but may also affect drug pharmacodynamics. However, investigators have just begun to explore epigenetic effects on pharmacodynamic properties, and few examples of such exist. In one recent report, DNA methylation influenced expression of the alpha-1d adrenergic receptor, with DNA hypermethylation at the promoter decreasing gene expression *in vitro* [15]. In a second study, promoter hypermethylation decreased expression of the alpha-1b subtype in a gastric cancer model [16]. Importantly, whether other alpha adrenergic receptor subtypes are regulated by DNA methylation remains unknown. The alpha-1 adrenergic receptor family plays an important role in cardiovascular homeostasis, and thus epigenetically controlled variations in alpha-adrenergic expression could affect responses to drugs with alpha-adrenergic modulating effects. Examples of such drugs include midodrine, terazosin, labetalol, and carvedilol. However, as with pharmacokinetic variations, pharmacologic research into epigenetic influences at the drug target level is needed *in vivo* to assess the extent to which epigenetic variation may affect drug response in humans.

Drug-Induced Alterations of Epigenetic Status

Unlike the genome, an organism's epigenome can be modified by environmental factors, including drug therapy. Thus, pharmacoepigenetic research can further our understanding of the mechanisms of drug action and potentially reveal novel targets for drug development. In cardiovascular disease, the best described example of drug-induced epigenome alteration is that of histone deacetylase (HDAC) inhibition. Inhibiting HDACs increases histone acetylation, which in turn increases the expression of a wide range of genes. Evidence shows that HDAC inhibition may provide protection against proliferative vascular diseases by regulating multiple cyclin-dependent kinase inhibitors and inducing cellular growth arrest at the G_1 stage [17]. Inhibition of HDACs can also reduce ischemia-reperfusion injury in mice by reducing acetylation of histones H3 and H4 [18]. This reduction in acetylation prevents up-regulation of ischemia-induced genes, such as hypoxia-inducible factor 1 alpha [18]. Interestingly, HDAC inhibitors can also repress hypoxia-inducible factors by preventing their direct acetylation, [19] providing a potential post-translational mechanism for cardiovascular protection. Thus, HDAC inhibitors have implications for the management of many cardiovascular diseases, including heart failure, atherosclerosis, and intracoronary artery stent thrombosis [20,21].

Research shows that HMG-CoA reductase inhibitors, or statins, may also act as HDAC inhibitors. By protecting against histone deacetylation, simvastatin and fluvastatin were shown to reduce endothelial expression of the proinflammatory cytokines IL-8 and MCP-1, which are associated with atherosclerosis [22]. Statin-induced reduction in HDAC activity was also shown *in vitro* to increase expression of p21, a cyclin-dependent kinase inhibitor known to reduce proliferation [5]. This antiproliferative effect is thought to be one of the pleiotropic mechanisms by which statins provide benefit in atherosclerosis beyond cholesterol reduction [23,24].

In addition to HDAC inhibition, a recent report indicates that some statins, such as atorvastatin (but not pravastatin) can decrease miR-221 and miR-222 levels in endothelial progenitor cells from patients with coronary artery disease [25]. These miRNAs have been shown to decrease angiogenesis [26] and such reduction could be detrimental in ischemic heart disease, where increased blood vessel formation would be beneficial.

Heparin is another example of a drug affecting angiogenesis via miRNA modulation. In endothelial cells, heparin down-regulates miR-10b, an miRNA associated with induction of angiogenesis [4]. By inhibiting angiogenesis, heparin could theoretically slow recovery from ischemic disease. On the contrary, heparin is a well-established treatment for acute coronary syndromes, including myocardial infarction. Thus, any anti-angiogenic effect of heparin in this setting is likely outweighed by the drug's beneficial anticoagulative properties. An improved understanding of the consequences of miR-10b, miR- 221, and miR-222 modulation could lead to exciting opportunities for developing new cardiovascular treatment strategies, as angiogenesis is thought to have therapeutic potential in the recovery of ischemic cardiovascular disease [27,28].

β-blockers also appear to affect epigenetic regulatory pathways. For example, out of 102 miRNAs measured in rats, expression of nearly 60 was altered by the β-blocker propranolol, suggesting a widespread epigenetic effect of β-blockers [3]. In addition, 31 of these 102 miRNAs were found to be dysregulated in a rat myocardial infarction model, and propranolol reversed 18 of the 31 dysregulated miRNAs [3]. In a separate study, propranolol protected against an increase in myocardial miR-1, an miRNA associated with myocardial ischemia [29]. Interestingly, miR-1 has also been associated with cardiac arrhythmias by repressing the potassium channel subunit Kir2.1 (encoded by *KCNJ2*) and connexin 43 (encoded by *GJA1*) [30]. This modulation of miR-1 could potentially contribute to the antiarrhythmic properties of β-blockers.

Pharmacoepigenetic research has also identified dietary compounds with epigenetic-modifying properties that could assist in future drug discovery. For example, butyrate, a short chain fatty acid derived from the intestinal microbial fermentation of dietary fiber, was recently shown to act as an HDAC inhibitor. By causing chromatin remodeling through increased acetylation, butyrate reduced vascular smooth muscle proliferation, a well-known contributor to angiogenesis [31]. Inhibiting smooth muscle proliferation could be crucial in preventing arterial and coronary in-stent thrombosis [31]. Butyrate and its derivatives are already being developed for the treatment of various cancers and could also have a role in the treatment of cardiovascular disease.

The dietary compound curcumin, a substance found in the spice turmeric, showed the ability to prevent histone acetylation in cardiomyocytes [32]. Inhibiting acetylation helped prevent hypertension-induced heart failure in rats [32]. These findings are important because a more complete understanding of such cardiovascular effects could lead to novel targets for drug development.

CHALLENGES

As with other burgeoning fields of research, challenges exist in pharmacoepigenetic research and translating research findings into practice. Unlike one's genome, which usually remains unaltered over the patient's lifetime, the epigenomic profile changes over time. Thus, measurements of epigenetic status may need to be assessed more than once, increasing the cost and effort required to accurately use epigenetic compared to genetic information. Also, unlike genetic data, epigenetic profiles are tissue specific. In the cancer field, biopsies of suspected tumors are done routinely, and tumors are often surgically removed, thus providing ample tissue for research. In contrast, cardiovascular biopsies are seldom done, making cardiac or endothelial tissue acquisition difficult. Thus, in order to routinely make epigenetic measurements in cardiovascular-related tissue types, less invasive methods of measuring epigenetic status will need to be developed. Because little clinical research has been done in the cardiovascular pharmacoepigenetic field, these techniques do not yet exist.

CONCLUSION

Although epigenetic information has the potential to enhance decisions about cardiovascular drug therapy, many challenges must be overcome in order to study and to one day use this information in clinical practice. Currently, variability in epigenetic status and the epigenetic effects of cardiovascular drugs are in the early stages of research. In the future, epigenetic status will likely be used in clinical practice along with other commonly used biomarkers. The consideration of epigenetic data along with genetic information and other patient characteristics may allow for more accurate prediction of response to cardiovascular drug treatment. However, before this prediction can occur, advances in the field of pharmacoepigenetics must take place.

QUESTIONS FOR DISCUSSION

1. What are the mechanisms by which epigenetics can affect cardiovascular drug response?
2. What are some examples of drug-induced alterations of the epigenome?
3. What are some challenges specific to the field of cardiovascular epigenetics?

References

[1] Esteller M. Non-coding RNAs in human disease. Nat Rev Genet 2011;12:861—74.
[2] Kouzarides T. Chromatin modifications and their function. Cell 2007;128:693—705.
[3] Zhu W, Yang L, Shan H, Zhang Y, Zhou R, Su Z, et al. MicroRNA expression analysis: clinical advantage of propranolol reveals key microRNAs in myocardial infarction. PLoS One 2011;6:e14736.

[4] Shen X, Fang J, Lv X, Pei Z, Wang Y, Jiang S, et al. Heparin impairs angiogenesis through inhibition of microRNA-10b. J Biol Chem 2011;286:26616−27.

[5] Lin YC, Lin JH, Chou CW, Chang YF, Yeh SH, Chen CC. Statins increase p21 through inhibition of histone deacetylase activity and release of promoter-associated HDAC1/2. Cancer Res 2008;68:2375−83.

[6] Kusaba H, Nakayama M, Harada T, Nomoto M, Kohno K, Kuwano M, et al. Association of 5′ CpG deme-thylation and altered chromatin structure in the promoter region with transcriptional activation of the multidrug resistance 1 gene in human cancer cells. Eur J Biochem 1999;262:924−32.

[7] Dejeux E, Ronneberg JA, Solvang H, Bukholm I, Geisler S, Aas T, et al. DNA methylation profiling in doxo-rubicin treated primary locally advanced breast tumours identifies novel genes associated with survival and treatment response. Mol Cancer 2010;9:68.

[8] Toyota M, Suzuki H, Yamashita T, Hirata K, Imai K, Tokino T, et al. Cancer epigenomics: implications of DNA methylation in personalized cancer therapy. Cancer Sci 2009;100:787−91.

[9] Nakajima M, Iwanari M, Yokoi T. Effects of histone deacetylation and DNA methylation on the constitutive and TCDD-inducible expressions of the human CYP1 family in MCF-7 and HeLa cells. Toxicol Lett 2003;144:247−56.

[10] Ingelman-Sundberg M, Sim SC, Gomez A, Rodriguez-Antona C. Influence of cytochrome P450 polymorphisms on drug therapies: pharmacogenetic, pharmacoepigenetic and clinical aspects. Pharmacol Ther 2007;116:496−526.

[11] Li Y, Cui Y, Hart SN, Klaassen CD, Zhong XB. Dynamic patterns of histone methylation are associated with ontogenic expression of the Cyp3a genes during mouse liver maturation. Mol Pharmacol 2009;75:1171−9.

[12] Pan YZ, Gao W, Yu AM. MicroRNAs regulate CYP3A4 expression via direct and indirect targeting. Drug Metab Dispos 2009;37:2112−7.

[13] David GL, Yegnasubramanian S, Kumar A, Marchi VL, De Marzo AM, et al. MDR1 promoter hyper-methylation in MCF-7 human breast cancer cells: changes in chromatin structure induced by treatment with 5-Aza-cytidine. Cancer Biol Ther 2004;3:540−8.

[14] Zhu H, Wu H, Liu X, Evans BR, Medina DJ, Liu CG, et al. Role of MicroRNA miR-27a and miR-451 in the regulation of MDR1/P-glycoprotein expression in human cancer cells. Biochem Pharmacol 2008;76:582−8.

[15] Michelotti GA, Brinkley DM, Morris DP, Smith MP, Louie RJ, Schwinn DA. Epigenetic regulation of human alpha1d-adrenergic receptor gene expression: a role for DNA methylation in Sp1-dependent regulation. FASEB J 2007;21:1979−93.

[16] Noda H, Miyaji Y, Nakanishi A, Konishi F, Miki Y. Frequent reduced expression of alpha-1B-adrenergic receptor caused by aberrant promoter methylation in gastric cancers. Br J Cancer 2007;96:383−90.

[17] Findeisen HM, Gizard F, Zhao Y, Qing H, Heywood EB, Jones KL, et al. Epigenetic regulation of vascular smooth muscle cell proliferation and neointima formation by histone deacetylase inhibition. Arterioscler Thromb Vasc Biol 2011;31:851−60.

[18] Granger A, Abdullah I, Huebner F, Stout A, Wang T, Huebner T, et al. Histone deacetylase inhibition reduces myocardial ischemia-reperfusion injury in mice. FASEB J 2008;22:3549−60.

[19] Chen S, Sang N. Histone deacetylase inhibitors: the epigenetic therapeutics that repress hypoxia-inducible factors. J Biomed Biotechnol 2011;2011:197946.

[20] Natarajan R. Drugs targeting epigenetic histone acetylation in vascular smooth muscle cells for restenosis and atherosclerosis. Arterioscler Thromb Vasc Biol 2011;31:725−7.

[21] Bush EW, McKinsey TA. Protein acetylation in the cardiorenal axis: the promise of histone deacetylase inhibitors. Circ Res 2010;106:272−84.

[22] Dje N'Guessan P, Riediger F, Vardarova K, Scharf S, Eitel J, Opitz B, et al. Statins control oxidized LDL-mediated histone modifications and gene expression in cultured human endothelial cells. Arterioscler Thromb Vasc Biol 2009;29:380−6.

[23] Vaughan CJ, Murphy MB, Buckley BM. Statins do more than just lower cholesterol. Lancet 1996;348:1079−82.

[24] Liao JK, Laufs U. Pleiotropic effects of statins. Annu Rev Pharmacol Toxicol 2005;45:89−118.

[25] Minami Y, Satoh M, Maesawa C, Takahashi Y, Tabuchi T, Itoh T, et al. Effect of atorvastatin on microRNA 221/222 expression in endothelial progenitor cells obtained from patients with coronary artery disease. Eur J Clin Invest 2009;39:359−67.

[26] Poliseno L, Tuccoli A, Mariani L, Evangelista M, Citti L, Woods K, et al. MicroRNAs modulate the angiogenic properties of HUVECs. Blood 2006;108:3068−71.

[27] Beohar N, Rapp J, Pandya S, Losordo DW. Rebuilding the damaged heart: the potential of cytokines and growth factors in the treatment of ischemic heart disease. J Am Coll Cardiol 2010;56:1287—97.

[28] Ahn A, Frishman WH, Gutwein A, Passeri J, Nelson M. Therapeutic angiogenesis: a new treatment approach for ischemic heart disease—part I. Cardiol Rev 2008;16:163—71.

[29] Lu Y, Zhang Y, Shan H, Pan Z, Li X, Li B, et al. MicroRNA-1 downregulation by propranolol in a rat model of myocardial infarction: a new mechanism for ischaemic cardioprotection. Cardiovasc Res 2009;84:434—41.

[30] Yang B, Lin H, Xiao J, Lu Y, Luo X, Li B, et al. The muscle-specific microRNA miR-1 regulates cardiac arrhythmogenic potential by targeting GJA1 and KCNJ2. Nat Med 2007;13:486—91.

[31] Mathew OP, Ranganna K, Yatsu FM. Butyrate, an HDAC inhibitor, stimulates interplay between different posttranslational modifications of histone H3 and differently alters G1-specific cell cycle proteins in vascular smooth muscle cells. Biomed Pharmacother 2010;64:733—40.

[32] Morimoto T, Sunagawa Y, Kawamura T, Takaya T, Wada H, Nagasawa A, et al. The dietary compound curcumin inhibits p300 histone acetyltransferase activity and prevents heart failure in rats. J Clin Invest 2008;118:868—78.

6

Pharmacogenomics in Psychiatric Disorders

Y.W. Francis Lam[*], *Naoki Fukui*[†],
Takuro Sugai[†], *Junzo Watanabe*[†], *Yuichiro Watanabe*[†],
Yutato Suzuki[†], *Toshiyuki Someya*[†]

[*]Department of Pharmacology, University of Texas Health Science Center
at San Antonio, San Antonio, Texas, USA
[†]Department of Psychiatry, Niigata University Graduate School of Medical
and Dental Sciences, Chuo-ku, Niigata, Japan

OBJECTIVES

1. Discuss the utility of *CYP* genotyping in psychopharmacology.

2. List and define the rationale of different drug targets for psychopharmacogenomic investigations.

3. Discuss alternative approaches to pharmacogenomic studies in psychopharmacology.

4. Describe how pharmacogenomics may play a role in minimizing the adverse effects of antipsychotics.

INTRODUCTION

Evaluations and predictions of treatment response and potentials of adverse drug reactions in psychiatric patients have in the past been partially limited by the patients' subjective reports and the subjective elements in clinicians' assessments. Despite the availability of different clinical rating scales, there remains no reliable biological marker of response. Since the completion of the Human Genome Project a decade ago, the implications of pharmacogenomics in psychiatry have been increasingly demonstrated. Many candidate genes have been identified, with the hope that they can be utilized to improve patient outcome. However, so far applications of pharmacogenomics have been primarily more successful in predicting adverse drug reactions than treatment response. This chapter will review the pharmacogenetic findings, discuss the current evidence and challenges of genotyping biomarkers in psychopharmacotherapeutics, and address the future potentials of applying pharmacogenomics in psychopharmacology. Because a comprehensive review of all research in psychiatric pharmacogenomics is beyond the scope of this chapter, affective disorder and schizophrenia will be the focus to highlight the principles and issues in this emerging field.

POLYMORPHISMS IN PROTEINS THAT AFFECT DRUG CONCENTRATIONS

Genes Encoding Drug-Metabolizing Enzymes

Antidepressants

Several polymorphic cytochrome P-450 isoenzymes, notably CYP2D6 and CYP2C19, are involved in metabolism and elimination of psychotropics. The lack of therapeutic response even with standard dosage regimens of the tricyclic antidepressant (TCA) nortriptyline provided one of the earliest clinical examples of how altered expression of CYP2D6 can affect drug response in patients who have multiple copies of the CYP2D6*2 allele. The original clinical observation [1] was followed up with additional studies that elucidated the molecular basis [2] and the gene—dose relationship in nortriptyline pharmacokinetics [3] in patients with the ultra-rapid metabolizer (UM) phenotype for CYP2D6. In the single-dose pharmacokinetic study of nortriptyline, Dalen et al. reported that in subjects with either 0, 1, 2, 3, or 13 copies of CYP2D6*2, the increases in nortriptyline clearance and formation of the 10-hydroxy-nortriptyline metabolite were proportional to the number of gene copies [3]. Nevertheless, the effect of genetics on CYP2D6 activity or the pharmacokinetics of CYP2D6 substrates can be modulated by drug dose [4,5], patient-specific factors such as presence of concurrent CYP2D6 inhibitors [6], or smoking status [7]. The effect of CYP2D6 variants (*5 and *10) on fluvoxamine and paroxetine pharmacokinetics are shown only in patients treated with the lower doses of 50 mg/day and 10 mg/day, respectively. This finding is most likely a result of CYP2D6 being a low-capacity enzyme, with saturation of its metabolic capacity occurring with higher-dosage regimens: 100 to 200 mg/day of fluvoxamine and 20 to 40 mg/day of paroxetine, respectively [4,5], thus effectively diminishing the impact of the genetic polymorphism at higher dosages. Because fluvoxamine is a CYP1A2 substrate, the effect of the CYP2D6 genotype on fluvoxamine pharmacokinetics is further modulated by smoking, which together accounted for 23% of the variance in fluvoxamine concentrations for patients treated with the low-dose regimen of 50 mg/day [7].

In their report of antidepressant dose recommendations based on pharmacokinetics and pharmacogenetics relationships, Kirchheiner et al. [8] suggested increased dose requirement in UMs receiving nortriptyline (up to 230% of the usual dose), despiramine (up to 260%), and mianserin (up to 300%). In two separate retrospective data analyses, Rau et al. [9] and

Kawanishi et al. [10] showed a preponderance of UMs in, respectively, 16 and 81 nonresponders who received at least four weeks of standard recommended dosage regimens of TCAs and selective serotonin reuptake inhibitors (SSRIs) that are CYP2D6 substrates. A more recent study reported that the CYP metabolic genotypes have no correlation with either response to antidepressants or remission of depression, although most of the 197 nonresponders received antidepressants that depend on multiple CYP enzymes for metabolism [11]. The number of literature reports of lower efficacy in UMs is significantly less for the SSRIs, which is not unexpected, given their flatter dose-response curve. The lack of data supporting CYP450 genotyping to improve outcome in patients receiving SSRIs is also confirmed by the reports of the Evaluation of Genomic Applications in Practice and Prevention (EGAPP) Working Group and the Sequenced Treatment Alternatives to Relieve Depression (STAR*D) trial [12,13]. In analyzing the DNA samples from 1,953 patients enrolled in the STAR*D trial, the investigators did not find an association of CYP genes (*CYP2C19, CYP2D6, CYP3A4, CYP3A5*) with response to citalopram [13]. The patient population in the STAR*D trial [13] included primarily Caucasians (78.1%) and African Americans (16.1%). In another study of Chinese subjects, Tsai et al. reported polymorphisms in *CYP2D6* and *CYP2C19* impact on the therapeutic outcome and serum concentrations of S-citalopram [14]. Because multiple *CYP2D6* and *CYP2C19* alleles occur at variable frequencies among different ethnic groups, the conflicting study results [13,14] underscore the importance of heterogeneity of study populations and the need of defining ethnicity in pharmacogenomic research. This issue of ethnicity is further discussed in chapter 10 of this book.

Current evidence suggests that the utility of pharmacogenetic testing for antidepressants lies more in anticipating adverse drug effects than in predicting their therapeutic efficacy. In a study of 1,198 elderly patients treated with antidepressants, poor metabolizers (PMs) of CYP2D6 were five times more likely to show significant adverse effects with the use of CYP2D6-dependent tricyclic antidepressants [15]. Rau et al. also reported a predominance of PMs in 28 patients who were treated with CYP2D6-dependent tricyclic antidepressants and SSRIs [9], and patients with the intermediate metabolizer (IM) phenotype for CYP2D6 were found not able to tolerate venlafaxine doses larger than 75 mg [16]. When compared to CYP2D6, the contribution of *CYP2C19* polymorphism to antidepressant pharmacokinetics and response is less, as is the value of genotyping for this isoenzyme.

Antipsychotics

Although literature data provide good evidence that differences in antipsychotic pharmacokinetics are related to polymorphic metabolism mediated by CYP2D6 and to a lesser extent CYP2C19 [8], little evidence exists for a role of *CYP2D6* and *CYP2C19* genotypes in determining antipsychotic efficacy. In a prospective study, Pollock et al. reported no significant differences in improvement of psychotic symptoms between five CYP2D6 PMs and 40 extensive metabolizers (EMs) treated with perphenazine for 17 days [17]. In addition, very few published studies separated UMs from EMs, which likely would negate any possible difference in efficacy between the UM and other CYP2D6 genotypes.

Even though a trend of lower haloperidol efficacy in UMs and higher efficacy in the PMs was shown in the study of Brockmoller et al. [18], there were significant overlaps in the haloperidol daily doses among the four metabolic groups, with 14 ±10 mg in UMs versus 13 ±9 mg in the PMs. These overlaps preclude the possibility of any useful genotype-based dose recommendation. In 235 patients with schizophrenia or schizoaffective disorder who failed to respond to typical antipsychotics, subsequent *CYP2D6* genotyping showed the presence of the UM phenotype in less than 1 percent of the patients, suggesting that the UM genotype was not a major contributing factor to the therapeutic failure [19]. In the Clinical Antipsychotics Trials of Intervention Effectiveness (CATIE), Grossman et al. reported little evidence of difference in efficacy of either perphenazine or risperidone in patients with different CYP2D6 genotypes, although there were no UMs included in the study [20].

In their report of no difference in perphenazine efficacy between 40 EMs and 5 PMs, Pollock et al. also found that PMs experienced more severe adverse effects, including oversedation and parkinsonism, than EMs during the first 10 days of treatment, although there were no drug concentration measurements performed [17]. *CYP2D6*10*, a predominant allele in Asian IMs, had been reported to be associated with weight gain in risperidone-treated Chinese patients [21], although it is not known whether plasma concentration correlates with weight gain. Several studies have also shown PMs to have a higher incidence of adverse drug reactions, including extrapyramidal side effects and drug discontinuance associated with the use of antipsychotic agents [18,22-28], whereas the evidence of a role of CYP2D6 in the etiology of tardive dyskinesia in PMs was less clear [29]. Without good data to suggest a concentration-dependent relationship, it is not surprising that tardive dyskinesia might not be related to the CYP2D6 genotype.

Although retrospective analysis psychotropic use in psychiatric patients who are CYP2D6 UMs and PMs would "cost" US$4,000 to $6,000 more than EMs and IMs, there has been no prospective controlled trial to further validate the therapeutic and/or economic benefit of *CYP2D6* genotyping in dose recommendation and/or avoidance of adverse drug reactions [25,30]. In addition, given the known genetically determined interethnic differences in activities of CYP2D6 and CYP2C19 (multiple alleles existing at variable frequencies), pharmacogenetic-guided pharmacokinetic bridging-studies would be needed in different ethnic populations for rational use of psychotropics dependent on these polymorphisms for elimination.

Mood Stabilizers

Despite the common clinical practice of monitoring plasma concentration and an inadequate response rate of <50% in lithium-treated patients, there are no pharmacogenomic studies on its pharmacokinetics. Published studies have mainly focused on the genes involved in the signaling and biochemical pathways involved in the mechanism of action of lithium, which will be described in a later section.

Genes Encoding Drug Transporters

The lack of data supporting a primary role for CYP gene polymorphisms in determining psychotropic drug response might be due to the presence of the drug efflux ATP-binding cassette (ABC, and formerly known as multidrug resistance [MDR]) superfamily of transporters residing at the blood—brain barrier (BBB). P-glycoprotein (Pgp) was the first recognized and the most studied ABC transporter and together with other more recently discovered ABC transporters such as multidrug resistance—associated proteins (MRPs) and breast cancer resistance protein (BCRP), plays a significant role in limiting the amount of drug crossing the BBB and reaching the cerebral circulation. *In vitro* measurement of either Pgp-mediated ATP-ase activity [31] or Pgp-mediated efflux across human colon adenocarcinoma (Caco-2) cell monolayer [32] has been utilized to assess which psychotropics are Pgp substrates. However, evaluation of potential functional significance would require *in vivo* studies of psychotropic drug distribution across the BBB, such as studies comparing brain concentrations of antipsychotics in wild-type mice and Pgp knockout (KO) mice, specifically the abcb1a $(-/-)$ KO mice [32,33]. Amitriptyline [34], nortriptyline [35], paroxetine, venlafaxine [36], and risperidone [32] are Pgp substrates, whereas the literature data for citalopram [37], fluoxetine [34], clozapine [38], haloperidol [31,32,38], olanzapine [38], and quetiapine [31,32,38] are less convincing. The *in vivo* results in general also showed a more modest effect of Pgp on brain penetration (up to 2.5 times higher brain concentrations for antidepressants, and 13 times and 30 times higher for risperidone and 9-hydroxyrisperidone, respectively [39]), as opposed to other substrates such as amiodarone, nelfinavir, and verapamil.

Pgp is encoded by the *ABCB1* gene (also known as the *MDR1* gene) in humans. Over the years, several polymorphisms have been identified in the promoter and exon regions of the *ABCB1* gene. The most common single-nucleotide polymorphisms (SNPs) are the c.C1236T (rs1128503) polymorphism in exon 12, the c.G2677T (rs2032583) polymorphism in exon 21, and the c.C3435T (rs1045642) polymorphism in exon 26. In a randomized study of the effect of C3435T polymorphism in 54 nortriptyline-treated patients and 72 fluoxetine-treated patients, Roberts et al. found no difference in nortriptyline serum concentrations among the three genotypes (C/C, C/T, and T/T) but observed a higher incidence of postural hypotension for homozygous carriers of the T allele [35]. Fukui et al. [40] showed that the effect of the C3435T polymorphism on fluvoxamine pharmacokinetics is dose dependent, with the TT homozygote showing a significantly higher concentration to dose ratio than the CC homozygote only at the highest dose of 200 mg per day. Therefore, the effect of Pgp polymorphism on drug concentrations could be similar to the dose-dependency effect shown with the *CYP2D6* polymorphism.

Although each of the three *ABCB1* SNPs is associated with altered Pgp expression, studies investigating their effect on antidepressant response have been conflicting with both positive [35,40-42] and negative [13,43-45] associations. This discrepancy might be due to the presence of strong linkage disequilibrium (LD) between several of these *ABCB1* polymorphisms, although conflicting results have also been reported for haplotype association studies [13,41]. In addition, some negative studies evaluated the association with C3435T polymorphism for too many drugs ($n = 9$) in too few patients ($n = 55$) [46], which poses a problem for statistical power. The choice of drug to be evaluated would also be important, as better remission rate was demonstrated only for patients carrying the C allele for the rs2032583 polymorphism *and* receiving a Pgp substrate (amitriptyline, citalopram, paroxetine, or venlafaxine), whereas the response prediction associated with *ABCB1* polymorphism disappeared when data from all patients or from patients receiving non-Pgp substrates were analyzed [47]. Although currently significant differences in adverse effects or extent of BBB penetration have not been consistently demonstrated for different Pgp polymorphisms and expressions, whether mutation resulting in loss of function, change in substrate specificity, or functionality [48] would have a bigger impact is not known and awaits further studies for clarification.

POLYMORPHISMS IN PROTEINS THAT MEDIATE DRUG RESPONSE

Recent work has revealed that genes encoding drug targets such as receptors, ion channels, and intracellular signaling proteins also play a significant role in determining drug efficacy and safety in patients. Multiple targets for the psychotropics exist for the neurotransmitter systems, including those that affect synthesis, degradation, or uptake of neurotransmitters, as well as their binding to pre- and postsynaptic receptors, and the cascade of downstream signal transduction proteins within the postsynaptic synapse (Figure 6.1). Abundant pharmacogenomic data on target polymorphisms exist for the psychotropics, in particular the antidepressants and the antipsychotics.

Antidepressants

Serotonin Transporter (SERT or 5-HTT)

The gene polymorphism that has received the most attention for predicting antidepressant response is that affecting the human serotonin transporter gene (Table 6.1). The serotonin

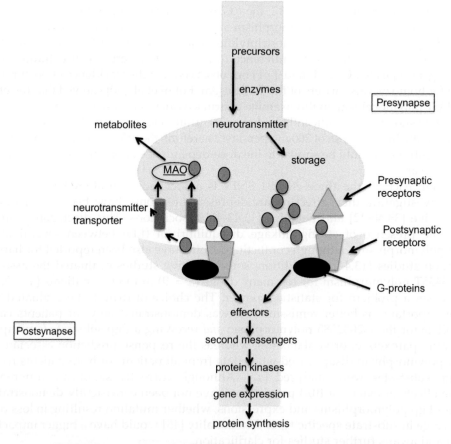

FIGURE 6.1   Schematic representation of psychotropic target proteins.

TABLE 6.1 Overview of selected pharmacogenomic studies with *SLC6A4* polymorphisms

SNP	Antidepressant study duration	Demographics of subjects	Results	References
5-HTTLPR	Sertraline 8 weeks	Mostly Caucasians $n = 106$ Major depressive disorder (MDD)	Homozygous carriers of L allele showed better response at weeks 1 and 2	[54]
	Paroxetine or nortriptyline 12 weeks	Caucasians $n = 57$ MDD	Homozygosity for L allele associated with faster response to paroxetine within two weeks	[55]
	Fluvoxamine and placebo, fluvoxamine and pindolol 6 weeks	Caucasians $n = 155$ MDD, Bipolar disorder	Homozygous carriers of S allele showed slower response. Addition of pindolol to fluvoxamine improved treatment response in homozygous carriers of S allele	[56]
	Fluoxetine or paroxetine 6 weeks	Koreans $n = 120$ MDD, Bipolar disorder	Homozygous carriers of S allele showed better response	[60]
	Fluvoxamine 6 weeks	Japanese $n = 54$ MDD	Homozygous carriers of S allele showed better response	[61]
	Citalopram 12 weeks	Caucasians $n = 130$ MDD	Homozygosity for S allele and G alleles of *HTR1A* gene associated with nonresponse	[91]
5-HTTLPR, rs25531	Citalopram 12 weeks	Non-Hispanic white, blacks, mixed race $n = 1,655$ MDD	No association between treatment outcome and 5HTTLPR alleles or haplotypes. However, lower incidence of adverse effects were reported with the L_A allele of rs25531	[65]
STin2	Fluvoxamine 6 weeks	Asians $n = 54$ MDD	No association with treatment outcome	[67]
5-HTTLPR, rs25531, STin2	Citalopram 14 weeks	Non-Hispanic white, blacks $n = 1,074$ MDD	Association between remission and a haplotype consisting of S allele of 5-HTTLPR, L_A allele of rs25531, and 12-repeat allele of STin2	[68]

transporter, alternatively known as solute carrier family 6 [neurotransmitter transporter, serotonin], member 4 (SLC6A4); transports serotonin within the synapse back to the presynaptic neurons and in this role represents the major pharmacological target for SSRIs and serotonin norepinephrine reuptake inhibitors (SNRIs). A functional polymorphism in the promoter region (rs4795541, serotonin transporter linked promoter region polymorphism, or 5HTTLPR) of the *SLC6A4* gene, resulting in the insertion/deletion of a 44-base pair repeat, was reported with the long (L) allele of the gene having higher transcriptional activity of the *SLC6A4* gene promoter and hence higher 5-HTT basal expression and serotonin uptake than the short (S) allele [49], although *in vivo* neuroimaging studies reported the contradictory effects of 5HTTLPR on brain 5-HTT availability [50–52]. Nevertheless, given the ability of the SSRIs to down-regulate the SERT function, investigators hypothesized that SSRI efficacy could be affected by 5HTTLPR polymorphism. Since then, many studies have shown an association between homozygosity for the S allele and inferior response to the SSRIs, in contrast to homo- or heterozygosity for the L allele of the gene, which predict beneficial outcome with SSRI treatment. Using positron emission tomography (PET) to evaluate the influence of genetic factors on 5-HT$_{1A}$ receptor expression in a living human brain, David et al. observed that the S allele was associated with a reduction in availability of the postsynaptic 5-HT$_{1A}$ receptors in man [53], which might provide a possible biological basis of the decreased response to SSRIs in carriers of the S allele.

Not only was the therapeutic outcome reported to be different among patients with different *SLC6A4* genotypes, but in addition, elderly patients with the L/L genotype treated with paroxetine or sertraline had a more rapid response, as early as after one week of treatment, than those with the L/S and S/S genotypes [54,55]. In addition, the lack of similar change in the onset of response in elderly patients treated with nortriptyline suggests that the difference in response is relevant only to antidepressants with a selective effect on serotonin [55]. Based on these findings, a case could be made for a preferential use of SSRI in patients with the L/L and L/S genotypes versus a TCA or a noradrenergic agent in patients with the S/S genotype. In addition, in patients with the S/S genotype, augmentation strategy of combining fluvoxamine and pindolol (being a 5-HT$_{1A}$ antagonist as well and accelerating the antidepressant effects of SSRIs) has been shown to reduce the difference in response between carriers of the S and the L allele, resulting in comparable treatment outcomes in all three genotypes [56]. Based on a decision–analytic model of pretreatment genetic testing for *SLC6A4* genotypes, Smits et al. concluded that the testing might result in a greater number of patients achieving remission earlier in the course of the treatment [57].

Serretti et al. performed a meta-analysis of the literature through 2006 and included 15 studies with 1,435 SSRI-treated patients who had been genotyped for the L/S *SLC6A4* variants. Patients with L/L and L/S genotypes were found to have increased remission rate (OR = 2.21) and better response rate within four weeks of treatment (OR = 1.72) with SSRIs. The investigators also evaluated ethnicity as a confounding variable in the association reported in the literature, and reported that the strength of the association was more notable in Caucasians than in Asians [58]. Whether the more heterogeneous results within Asian populations could partially account for the lack of association between 5HTTLPR polymorphism and SSRI response reported in a more recent meta-analysis that included more studies ($n = 28$) and patients ($n = 5,408$) is not known, as the investigators did not analyze the result

TABLE 6.2 Ethnic differences in allele frequencies (%) of 5HTTLPR and STin2 VNTR

Ethnicity	L allele of 5HTTLPR	S allele of 5HTTLPR	9-repeat allele of VNTR	10-repeat allele of VNTR	12-repeat allele of VNTR
Caucasians	60	40	1	47	52
African Americans	83	17	1	26	73
Chinese	26	74	<0.1	8	92
Japanese	20	80	<0.1	2	98
Koreans	23	77	<0.1	10	90

separately in Caucasians and Asians [59]. This finding is especially important because opposite yet comparable associations (S allele conferring good therapeutic response) have been reported in Korean and Japanese populations, possibly at least partially related to either ethnicity-based differences in the 5HTTLPR S allele frequency, being higher in Asians (74 to 80%) than in Caucasians (40%) (Table 6.2) [60,61], interaction with other functional gene variants, or gene–environment interaction. In addition, a study showed that 5HTTLPR is not a simple insertion/deletion of a 44-base pair repeat but a complex and highly polymorphic structure consisting of 14 kinds of alleles in different populations, including the Japanese and Caucasian, with variable distribution frequency [62]. These findings should be taken into consideration for current and future pharmacogenomic studies of the 5HTTLPR.

Perhaps more intriguing is the discovery of rs25531, an SNP located just upstream of the 5HTTLPR, from genetic analysis of the STAR*D sample. A functional A>G variation in the L allele (but not S allele) of 5HTTLPR, known as the L_G allele, reduces mRNA expression of the *SLC6A4* to a level comparable to that of the S allele and therefore changes the functional significance of the L allele of 5HTTLPR. On the other hand, the L_A allele increases *SLC6A4* mRNA expression, resulting in a "higher function" phenotype [63,64]. Therefore, by changing the expression of the L allele, this previously unrecognized L_G allele would further modulate the SSRI response predictive value of the *SLC6A4* L/S and L/L genotypes. In essence, within *SLC6A4*, there would be two promoter polymorphisms and three alleles of functional importance: the high-expression L_A allele and the low expression S and L_G alleles. Because the S and L_G alleles are very comparable in SERT expression, the possible genotypes based on this L_A and L_G difference would be genotype with no L_A alleles (either S/S, S/L_G, or L_G/L_G), one L_A allele (S/L_A or L_G/L_A), and genotype with two L_A alleles (L_A/L_A). Among the haplotypes constructed, only LL_A is associated with high *SLC6A4* transcription [63].

Using a two-stage genotyping to differentiate the S and L alleles as well as the L_A and L_G alleles in a second study of the STAR*D samples, Hu et al. reported an association between the L_A allele and citalopram adverse effects in all 1,655 subjects (Caucasians, blacks, and mixed race), with lower adverse effects associated with L_A/L_A genotype ($p = 0.004$) and L_A allele ($p < 0.001$) in all subjects, and a lesser association in a subset of 1,131 Caucasian subjects ($p = 0.03$ and $p = 0.007$, respectively). The adverse effect association was also evident for the entire study population even when the L_A and L_G alleles were combined in the analysis. On the other hand, association for the Caucasian subjects was present only with differentiation of the L allele into L_A and L_G alleles. There was no association between treatment outcome and 5HTTLPR alleles or genotypes in the Caucasian subjects [65].

In addition to 5HTTLPR, additional polymorphisms of the SERT gene have been identified with potential roles in modulating the response to SSRIs. Ogilvie et al. discovered a 17-base pair, variable number of tandem repeats (VNTR) polymorphism within intron 2 (STin2) of *SLC6A4*, resulting in three alleles containing 9, 10, and 12 copies of the VNTR element [66]. However, similar to 5HTTLPR polymorphism, both positive [60] and negative [67] associations have been reported. What is interesting is that despite the lack of association between 5HTTLPR alleles or haplotypes and citalopram response [65], reanalysis of the same data set from the STAR*D study revealed an association between remission and a haplotype that consists of 5HTTLPR, rs25531, and STin2 (haplotype S-L_A-12) [68].

Therefore, even for *SLC6A4*, a candidate gene with an obvious relevance to the therapeutic effect of antidepressants, especially the SSRIs, it is clear that predicting response in a patient solely with any one SNP would likely yield misleading and conflicting results. Despite genotyping multiple SNP markers that span the coding region of *SLC6A4*, no association with outcome was found in the primary outcome candidate gene study of STAR*D. It remains to be seen whether additional haplotypes would provide better predictive value. Together with the usual heterogeneous study limitations (ethnicity [Table 6.2], outcome assessment, study design, sample characteristics, and sample size), significant work remains in much-needed recommendations for appropriate pharmacogenomic studies before any findings can be meaningfully translated from the bench to the bedside.

In this regard, a recent study reported an entirely different approach to search for SSRI response biomarkers. Based on similar genome-wide expression profiling in human lymphoblastoid cell lines (LCLs) demonstrated for anticancer drugs [69-71], investigators first identified and demonstrated the existence of LCLs with high or low sensitivities to different SSRIs. Each of the cell lines functionally expresses the SERT, and based on the *in vitro* growth inhibition phenotypes, appears to reflect downstream cellular pathways [72]. The investigators then screened 80 LCLs for growth inhibition by paroxetine. For genome-wide expression profiling in search of response biomarkers, seven LCLs with high sensitivity and seven LCLs with low sensitivity to paroxetine were chosen for comparison of basal gene-expression difference by microarray analysis. Genes with basal expression difference and also implicated in CNS function were then validated with real-time polymerase chain reaction (PCR) [73].

A 6.4-fold difference in expression between the two paroxetine-sensitivity phenotypes was demonstrated for the cell adhesion molecule with homology to L1CAM gene (*CHL1*), which encodes a neuronal cell adhesion protein that is implicated in correct brain circuitry, as well as schizophrenia and autism. The microarray result was confirmed by real-time PCR demonstrating a 36-fold difference in *CHL1* expression level between the two phenotypes. In addition to *CHL1*, 12 additional genes implicated in brain function or psychiatric disorders also showed more than a 1.5-fold difference in expression between the two phenotypic groups. Although one can argue that the discovery of yet another set of genes for predicting SSRI response does not necessarily translate to definitive and clinically relevant biomarkers, the investigators also discussed the functional significance of *CHL1* in *CHL1* KO mice. Because the cell lines were derived from healthy females, these preliminary study results provide the basis for additional investigations—for example, identification of the same genes in cell lines from male subjects, as well as comparison of gene expression levels from patients with major depression.

5-Hydroxytryptamine Receptors

5-Hydroxytryptamine 2A (5-HT$_{2A}$) Receptor

The postsynaptic 5-HT$_{2A}$ receptor represents another serotonin-related target for psychotropics. Antidepressants and typical and atypical antipsychotics all act as antagonists and down-regulate the receptor [74], reportedly overexpressed in patients with depression [75]. In humans, the polymorphic 5-hydroxytryptamine receptor 2A (HTR2A) gene encodes the 5-HT$_{2A}$ receptor, and several polymorphisms had been investigated, including a c.-1438 A/G (rs6311) promoter polymorphism and two coding region polymorphisms: the silent c.T102C (rs6313) polymorphism in exon 1 and the c.C1354T (rs6314) polymorphism resulting in a p.His452Tyr amino acid substitution. Two of these (rs6311 and rs6313) are in LD and their SNP had been associated with antidepressant response (Table 6.3). [76–78] Although specific alleles such as the C variant of the C102T polymorphism [78] and the G allele of the -1438 A/G polymorphism [76,77] were reported to be associated with antidepressant response, the findings are conflicting and not supported by the large-scale association study of 68 candidate genes in the STAR*D sample. In the STAR*D study, a single synonymous variant of HTR2A, IVS2 A/G (rs7997012) within intron 2, emerged as the only SNP with sufficient predictive value for response to citalopram in a Caucasian population. Homozygous carriers of the A allele have better response (18% reduction in absolute risk of treatment failure) than homozygous carriers of the G allele. In addition, analysis of the genetic data showed that blacks had a higher frequency of the "nonresponding" allele [79]. Whether this might partially account for the findings of poorer response among citalopram-treated black patients in the clinical STAR*D study [80] is not known.

Lucae et al. provided the first replicated confirmation of the role of rs7997012 shown in the genetic STAR*D study. In evaluating 637 German Caucasian patients with a major depressive episode, the SNP rs7997012 was significantly associated with remission of depression after five weeks treatment with a variety of antidepressants. However, the association (A allele conferred impaired treatment response) was inverse to that of the genetic STAR*D study [81]. Ethnic differences in patient samples (Caucasians versus a more heterogeneous population [Caucasian, black, and mixed race] in the genetic STAR*D study), time of evaluation of treatment response (after five-week treatment versus at study exit versus regardless of length of duration since study entry for the genetic STAR*D study), and a smaller sample size (Table 6.3) in the study of Lucae et al. could confound the result and limit the comparability and require additional studies to ascertain the direction of the association.

5-Hydroxytryptamine 1A (5-HT$_{1A}$) Receptor

The 5-HT$_{1A}$ receptor is encoded by the 5-hydroxytryptamine receptor 1A (HTR1A) gene. Desensitation (or down-regulation) of the somatodendritic 5-HT$_{1A}$ receptor by chronic SSRI treatment results in enhanced serotonergic neurotransmission [82,83]. In addition, antagonism of the 5-HT$_{1A}$ receptor has also been suggested to be associated with antidepressant effects [84,85]. Therefore, genetic variation of the HTR1A might change the functional properties of the 5-HT$_{1A}$ receptor, resulting in differences in antidepressant response.

TABLE 6.3   Overview of selected pharmacogenomic studies with serotonin receptor polymorphisms

Genes	SNP	Antidepressants and study duration	Subject demographics	Main findings	References
HTR2A	rs6311 and rs6313	Citalopram, 4 weeks, Paroxetine or fluvoxamine, 6 weeks	Koreans, $n = 71$ Major depressive disorder (MDD) Japanese, $n = 100$ MDD	G allele and GG genotype of rs6311 associated with better response	[76,77]
		TCAs, SSRIs, or mirtazapine, 4 weeks	Caucasians, $n = 173$ MDD	C allele and CC genotype of rs6313 associated with better response	[78]
		Citalopram, 12 weeks	Non-Hispanic white, blacks $n = 1,248$, MDD	No association with response	[79]
	rs7997012	Citalopram, 12 weeks	Non-Hispanic white, blacks $n = 1,248$, MDD	A allele associated with better response	[79,95]
		Citalopram, 5 weeks	Caucasians $n = 637$, MDD	G allele associated with better response	[81]
HTR1A	rs6295	Fluoxetine, nefazodone, 4 weeks Citalopram, other SSRIs, mirtazapine, MAOI, TCAs, 6 weeks	Caucasians $n = 118$, MDD Caucasians $n = 340$, MDD	CC genotype associated with better response. Better response seen in patients with melancholic depression	[87–89,91]
		SSRIs, 4 weeks	Caucasians, African Americans, $n = 133$, depression	No association with SSRI response	[90]
GRIK4	rs1954787	Citalopram, 12 weeks	Non-Hispanic white, blacks $n = 1,704$, MDD	C allele associated with better response	[95]
		Duloxetine, 6 weeks	Caucasians $n = 200$, MDD	No association	[96]

Of the ten polymorphisms identified in the *HTR1A* gene, the most investigated ones are c.-1019C/G (rs6295) located in the promoter region, p.Gly22Ser (rs1799920), and p.Ile28Val (rs1799921) (Table 6.3). The G allele of the rs6295 polymorphism has been associated with up-regulation of 5-HT$_{1A}$ receptor expression [86] and response prediction with antidepressant treatment [87]. In 118 patients treated with fluoxetine or nefazodone augmented with pindolol, or monotherapy with flibanserin (a 5-HT$_{1A}$ agonist), the homozygous GG genotype was more prevalent in nonresponders than the homozygous CC genotype ($p = 0.0497$ for the augmentation group and $p = 0.039$ for the monotherapy group) [87]. However, other investigators reported positive association being evident only for females [88] or in patients with specific depressive manifestation [89]. In a retrospective study, Levin et al. found no association between seven *HTR1A* polymorphisms, including rs6295, and SSRI response in 100 responders and 33 nonresponders [90]. As additional evidence that response to antidepressants likely is influenced by more than one gene, Arias et al. reported in 130 subjects treated with citalopram that homozygosity for both the G allele of the *HTR1A* polymorphism and S allele of the *SLC6A4* polymorphism predict nonresponse to SSRI treatment ($p = 0.009$) [91]. Differences in ethnic and allele distributions in study subjects could partially account for the conflicting results in replication studies. As an example, with very low frequencies of the Gly22Ser and Ile28Val polymorphisms in Japanese populations, the effect of the more common Gly272Asp polymorphism of the *HTR1A* on clinical response to fluvoxamine was studied in 65 depressed Japanese patients. Subjects with the Asp allele had a significantly higher percentage reduction in score of the 17-item Hamilton Rating Scale for Depression than homozygous carrier of the Gly allele at week 2 ($p = 0.009$), week 6 ($p = 0.036$), and at week 12 ($p = 0.031$) [92].

Glutamate Receptor

Glutamate is the primary excitatory neurotransmitter in the brain and increased levels have been observed in patients with depression [93]. Glutamate receptors selectively bind to glutamate to modulate excitatory neurotransmission. Chronic use of an SSRI such as citalopram was shown to attenuate glutaminergic transmission and reduce excitatory glutamate activity [94]. Within the STAR*D samples, the second marker shown to have significant association with antidepressant response was rs1954787, a C/T SNP residing in the first intron of the glutamate receptor, ionotropic, kainite 4 gene (*GRIK4*) that encodes a kainic-acid type glutamate receptor (Table 6.3) [95]. The C allele was associated with better outcome and suggested that the glutamate system could have a significant role in antidepressant response. In addition, homozygous carriers of both the A allele of *HTR2A* and the C allele of *GRIK4* were twice as likely to be associated with better response to citalopram than patients who did not carry either of these two outcome-related alleles. However, Perlis et al. could not replicate the rs1954787 association in 250 Caucasian patients with nonpsychotic major depressive disorder and treated with daily regimens of duloxetine 60 mg per day for six weeks. In addition to smaller sample size and difference in study population, one additional reason for the discrepancy could be related to the differential mechanisms of action of duloxetine (a serotonin—norepinephrine reuptake inhibitor) versus SSRIs (selectively inhibiting the reuptake of serotonin). Interestingly, the investigators also reported the failure to replicate previously reported associations with rs25531, 5HTTLPR, and the 17-base pair VNTR polymorphism in intron 2 (STin2) for *SLC6A4*. Additional negative associations were also shown for four SNPs for *ABCB1*, six SNPs for four genes coding for phosophodiesterases, and a single SNP for *OPRM1* coding for the opioid receptor mu 1 [96].

In summary, despite significant progress in antidepressant pharmacogenomic research over the years, the lack of consistent findings among all studies of different neurotransmitter receptors and transporters, including single candidate gene association studies, genome-wide association studies, and meta-analyses, make it difficult to identify definitive associations that can be used to predict antidepressant response. Differences in study design, disease phenotypes, patient population, response definition and assessment, and sample size all contribute to the conflicting results.

Antipsychotics

Dopamine Receptors

All antipsychotic agents, especially the first-generation antipsychotics, are dopamine D_2 receptor blockers [97]. Functional brain imaging studies suggest, and pooled and meta-analyses confirm, that a threshold level (60 to 65%) of D_2 receptor binding by antipsychotic agents is needed for sustained therapeutic effect, and excessive blockade (≥ 78 to 80%) is associated with extrapyramidal side effects [98-101]. Of the five subtypes of dopamine receptors, D_2, D_3, and D_4 receptors are the most studied for pharmacogenetic evaluation of antipsychotic efficacy. Several polymorphisms of the D_2 receptor gene (*DRD2*) have been identified: the -141-C Ins/Del polymorphism (rs1799732) in the promoter region, the Taq1A polymorphism (rs1800497), and the p.Ser311Cys coding polymorphism (rs1801028). The Del allele of the -141-C Ins/Del polymorphism is associated with not only lower expression of the D_2 receptor *in vitro*[102] but also higher striatal D_2 receptor density *in vivo* [103]. Several studies have also shown that the Del allele predicts less beneficial response from antipsychotics (Table 6.4).[104-106] In a recent study, 60 drug-naïve schizophrenic patients received 6 mg of risperidone for four weeks. The investigators confirmed the role of the Del allele in lesser improvement in total Brief Psychiatric Rating Scale (BPRS) and positive symptoms (51.0% and 51.1%, respectively) in carriers of the Del variant compared to 66.8% and 74.8%, respectively, in noncarriers. Interestingly, the investigators also reported a correlation between plasma concentration of the risperidone active moiety (risperidone and 9-hydroxy-risperidone) and improvement in total BPRS scores ($p < 0.01$) that is independent of the -141-C Ins/Del polymorphism, although there was no determination of the *CYP2D6* genotype of the patients [107].

The Taq1A polymorphism is located downstream of *DRD2* and has two variants: A1 and A2, with lower striatal D_2 receptor density reported in carriers of the A1 allele [108]. Although the functional effect of Taq1A polymorphism is not clear, the A1 allele and A1/A1 genotype had been reported to be associated with better response (greater improvement in positive symptoms) to haloperidol and nemonapride [109,110]. However, in 117 Chinese patients with first-episode schizophrenia who were previously untreated, the Taq1A polymorphism was not associated with change in total Positive and Negative Syndrome Scale (PANSS) score after treatment with antipsychotics, primarily risperidone ($n = 43$) and chlorpromazine ($n = 66$) [111]. Although ethnicity might play a role in the discrepancy, the population of drug-naïve patients in this study suggested prior drug treatment would not be a confounding variable to assess the association, even though different durations of drug washout prior to baseline assessment using clinical scales will influence the extent of change in the scores. The recent study of Yasui-Furukori et al. also reported a lack of association with response in their 60 drug-naïve schizophrenic patients treated with risperidone [107].

In 123 Chinese patients treated with risperidone for up to 42 days, patients with the p.Ser/Cys genotype of *DRD2* polymorphism showed greater absolute score reduction and greater percent change in negative symptoms than patients with the Ser/Ser genotype. However, the

TABLE 6.4 Overview of selected genetic association studies for antipsychotic response and toxicity

Genes	SNP	Antipsychotics	Main findings	References
DRD1	-48 A>G (rs4532), rs5326, rs265975	Haloperidol, chlorpromazine, sulpride, flupenthixol, zuclopenthixol	CGC haplotype of the three SNPs associated with tardive dyskinesia (TD) risk	[133]
	rs4532	Typical and atypical antipsychotics	No association with TD	[134,135]
DRD2	-141-C del/ins	Chlorpormazine, bromperidol, nemonapride, risperidone	Del allele associated with less response	[104–107]
	Taq1A	Haloperidol, nemonapride	A1 allele and A1/A1 genotype associated with better response	[109,110]
		Risperidone, chlorpromazine	No association with response	[107,111]
		Antipsychotics	A2 carriers at risk for TD	[131]
		Haloperidol, perphenazine, levomepromazine, fluphenazine, chlorpromazine, thioridazine, zyclopenthixol	A1 carriers associated with EPS	[129]
		Bromperidol, nemonapride	No association of A1 allele with EPS	[130]
		Risperidone, quetiapine, olanzapine	A1 carriers associated with increased prolactin level	[147]
	Ser311Cys	Risperidone	Ser/Cys genotype showed better response	[112]
DRD3	Ser9Gly	Risperidone, chlorpromazine	Gly allele and Gly/Gly genotype associated with less response	[111,114]
		Antipsychotics	Gly allele associated with TD	[137–139]
DRD4	VNTR	Clozapine	No association with response	[116,117,122]
SLC6A4	44 bp del/ins	Typical and atypical antipsychotics	No association with response	[118]

(Continued)

TABLE 6.4 Overview of selected genetic association studies for antipsychotic response and toxicity—cont'd

Genes	SNP	Antipsychotics	Main findings	References
HTR2A	102-T/C	Clozapine	T allele associated with better response	[120,122]
		Risperidone	CC genotype associated with better response	[127]
		Antipsychotics	C allele associated with TD risk	[140]
	His425Tyr	Clozapine	His allele associated with better response	[120,122]
		Antipsychotics	No association with TD	[140]
HTR2C	-759C/T (rs3813929)	Risperidone, chlorpromazine	C allele and C/C genotype associated with less response	[111]
		Atypical antipsychotics	T allele associated with risk of weight gain	[141,142]
RSG4	rs951439	Perphenazine, ziprasidone, olanzapine, quetiapine, risperidone	CC genotype associated with response to perphenazine in patients with African descent. TT genotype associated with response to risperidone in patients with European descent	[125]
	rs2842030	Perphenazine, ziprasidone, olanzapine, quetiapine, risperidone	TT genotype associated with response to perphenazine in patients of African descent	[125]
			GG genotype associated with response to risperidone in patients of European descent	
	rs2661319	Risperidone	AA genotype associated with better response in patients of Chinese descent	[126]
LEP	-2548 A/G (rs7799039)	Clozapine	G allele, A/G and G/G genotypes associated with risk for weight gain	[146]
		Clozapine, olanzapine, risperidone	No association	[145]
LEPR	223 Gln/Arg (rs1137101) [also known as Q223R in the literature]	Clozapine, olanzapine, risperidone	QQ genotype associated with risk for weight gain in females but not in males	[145]

result should be considered preliminary because there were only 12 subjects with the Ser/Cys genotype and no patient had the homozygous Cys/Cys genotype [112].

Dopamine binds to the D_2 receptor and inhibits prolactin secretion. Therefore, Fukui et al. hypothesized that basal prolactin level accurately reflects DRD2 function and investigated the association of the basal prolactin levels of 140 healthy Japanese subjects with *DRD2* "tagging" SNPs that covered the *DRD2* gene, as well as with the Taq1 A, Ser311Cys, and -141C Ins/Del polymorphisms. Significant associations were found between two DRD2 polymorphisms (rs7131056 and rs4648317) in intron 1 and serum prolactin levels, but only in the female subjects, which is consistent with the known gender difference in prolactin concentration [113]. These preliminary data suggest that the two new polymorphisms can be considered as candidate functional *DRD2* polymorphisms and should be further investigated in future studies.

The *DRD3* gene contains an SNP that results in a serine–glycine substitution. Although the p.Ser9Gly polymorphism had been implicated with lesser antipsychotic response in carriers of the Gly allele [111,114], literature data mostly evaluated its association with development of tardive dyskinesia, which will be discussed in latter sections of this chapter. The tenfold higher affinity of the atypical antipsychotic agent clozapine for the D_4 receptor than for the D_2 and D_3 receptors results in a lower risk of inducing extrapyramidal side effects. Despite the earlier report of a common variable number of tandem repeat polymorphism of the *DRD4* gene linked to clozapine response [115], subsequent studies were not able to detect significant association [116,117].

Serotonergic System

Although no differences in response to typical antipsychotic agents were reported in 684 patients with different *SLC6A4* genotypes [118], the pharmacological action of the atypical antipsychotic agents partially involves the serotonergic system, with single-photon emission computed tomography (SPECT) evidence of high occupancy of the 5-HT$_{2C}$ receptor by clozapine and risperidone [119], making it a logical candidate gene for evaluation of response association. Based on clozapine's high affinities for the 5-HT$_{2A}$ and 5-HT$_{2C}$ receptors, several polymorphisms of the *HTR2A* gene (c.-1438-G/A and c.102-T/C in the promoter region and p.His425Tyr in the coding region) and the *HTR2C* gene (c.-759-T/C [rs3813929] in the promoter region and p.Cys23Ser [rs6318] in the coding region) have been extensively investigated in the literature for response prediction. Meta-analyses of literature data reported association between 102-T/C and His425Tyr polymorphisms and response [120]. Although a significant association was found between the Ser allele of the Cys23Ser polymorphism of the *HTR2C* gene [121], subsequent studies were not able to replicate the results.

Arranz et al. evaluated 19 genetic polymorphisms that affect the different pharmacological targets of clozapine. Based on association studies of these polymorphisms in 133 responders and 67 nonresponders, the investigators reported that a combination of six different polymorphisms (the -1438-G/A and 102-T/C polymorphisms that are in LD, as well as the His425Tyr polymorphism of the *HTR2A* gene; the Cys23Ser and -330-GT/-244-CT polymorphisms of the *HTR2C* gene; the 5HTTLPR polymorphism of the *SLC6A4* gene; and the -1018-G/A polymorphism for the histamine-2 receptor) resulted in a 76.7% success in predicting response to clozapine. In addition, about 50% of the patients are homozygous carriers of the T allele of the 102-T/C polymorphism and the His allele of the His425Tyr polymorphism of the *HTR2A* gene, and good response was evident in 80% of this patient subgroup. Interestingly, despite the high affinity of clozapine for the D_4 receptor, no association was found with response [122]. Nevertheless, despite the appeal of this polymorphism combination approach to more accurately predict clozapine response, to date there have not been any studies that replicate the primary findings of this study.

G-Protein-Coupled Receptors

The dopamine and serotonin receptors targeted by the antipsychotics are G-protein-coupled receptors (GPCRs) and signal to effector proteins through intracellular G-protein subunits. Regulators of G-protein signaling shorten the time period of neurotransmitter signaling through the GPCRs. The regulator of G-protein signaling 4 (RGS4) is one such regulator, and it regulates the activity of the GPCRs. The gene that encodes RSG4 had been identified as a vulnerability gene for schizophrenia [123,124], and variants of *RSG4* have been studied as predictors for antipsychotic treatment response.

Three SNPs of *RSG4* have been reported to confer differential treatment responses in three ethnic groups. In patients of African descent, those with the CC genotype of the rs951439 SNP had longer (391 days) and better (21% improvement based on the PANSS) response to perphenazine than ziprasidone (124 days and 5% worsening, respectively). On the other hand, the same patient population with the TT genotype of the rs2842030 SNP responded better to perphenazine (24% improvement in the PANSS) than to quetiapine, risperidone, and ziprasidone. A sharp contrast in association was shown in patients with European descent, in which risperidone treatment resulted in better response with the TT genotype of the rs951439 SNP and GG genotype of the rs2842030 SNP [125]. In 120 schizophrenic patients of Chinese descent, rs2661319 of *RSG4* was found to predict response to risperidone treatment [126]. These data with *RSG4* polymorphisms underscore the importance of stratification of patient population by ethnicity in pharmacogenomic investigations.

It is noteworthy that the investigators of the Chinese study also had reported in other studies that polymorphisms affecting the D_2 receptor (Ser311Cys), D_3 receptor (Ser9Gly), and 5-HT2A receptor (102-T/C) predict treatment response to risperidone [112,114,127]. Whether a combination of polymorphism approach similar to that for clozapine could result in better response prediction remains to be investigated.

In summary, although some association studies with individual candidate genes encoding their respective targets showed positive findings with response prediction to antipsychotic drugs, the data are far from convincing, and there are just about as many negative associations reported in the literature. Although it is obvious that a combination of different genes would account for a greater portion of the response variance than an individual gene, analysis of how genetic variants influence improvement in positive or negative symptoms as well as in cognitive function would likely yield more useful insight than improvement in overall symptomatology.

Antipsychotic use is associated with a variety of adverse effects, with extrapyramidal symptoms (EPS) and weight gain being the most commonly reported and also the focus of many of the pharmacogenetic studies of psychotropic-induced adverse drug reactions. Among the different EPS, tardive dyskinesia (TD) is a debilitating and irreversible movement disorder that develops in up to 30% of patients after long-term antipsychotic treatment. As indicated earlier, excessive blockade of D_2 receptor is associated with extrapyramidal side effects, although this issue is primarily a problem for the typical antipsychotics [99,128].

Both positive [129] and negative [130] associations with EPS had been reported in A1 carriers of the Taq1A polymorphism of the *DRD2* gene, whereas a meta-analysis found a risk-increasing effect for TD in carriers of the A2 allele [131]. Because an imbalance between D_1 and D_2 receptors had been suggested to result in TD [132], the conflicting results reported for association between *DRD2* polymorphism and EPS as well as the risk of TD [129-131] could be related to genetic variants in *DRD1* as well. In a recent study involving 220 Chinese patients with TD and 162 Chinese patients without TD treated with stable dosage regimens of typical antipsychotics for at least 6 months, the SNP rs4532 (also known as -48 A>G) in *DRD1* was significantly associated with TD risk in the schizophrenic patients. The positive association was also evident in haplotype analyses involving two additional SNPs: rs5326 and rs265975—specifically, the haplotype CGC (rs5326-rs4532-rs265975) [133]. The study result contrasted with the negative association reported by two studies [134,135], which could be related to ethnic differences in allele frequency of rs4532 (18% frequency for the G allele in Chinese versus 39% in Caucasians [134]) and contribution of *DRD1* to TD, as well as the inclusion of patients treated with atypical antipsychotics in the two negative studies.

In a transfected Chinese hamster ovary cell line, the Gly allele for the Ser9Gly polymorphism of *DRD3* gene was shown to have a higher affinity for the dopamine receptor [136]. Brain imaging studies also showed that haloperidol-treated patients with the Gly/Gly genotype had greater fluorodeoxyglucose metabolism in the anterior striatum than patients who were either heterozygous or homozygous carriers of the Ser allele. The increased brain activity observed in the patients correlated with the presence of the most severe TD symptoms [137]. In a meta-analysis of data from 317 patients with TD and 463 patients without TD, patients with the Gly allele were found to experience a higher incidence of TD ($p = 0.04$). In addition, patients who were homozygous carrier for the Gly allele had higher abnormal involuntary movement scores than heterozygotes ($p = 0.006$) and homozygotes for the Ser allele ($p < 0.0001$). The effect of the Gly allele, though significant, was modest with an odds ratio of 1.33 [138]. Nevertheless, the role of the Ser9Gly polymorphism in TD was confirmed in another meta-analysis, which also suggested that the association was related to ethnicity, with a stronger association in non-Asians versus Asians [139]. Finally, a pooled analysis of 256 patients with TD and 379 patients without TD showed a positive association for the C allele of the 102-T/C polymorphism of *HTR2A*, especially in the elderly. This suggests that 5-HT receptors can also be involved in etiology of TD [140].

Although the atypical antipsychotic agents have lower propensity to produce extrapyramidal side effects, their use is associated with a higher incidence of weight gain than the typical antipsychotics. Given the deleterious effects of weight gain on the cardiovascular system as well as lipid and glucose metabolism, identification of potential markers for weight gain in at-risk patients treated with psychotropics would be beneficial. Among the various neurotransmitters involved in etiology of schizophrenia and/or mechanism of antipsychotic drug action, the involvement of the 5-hydroxytryptamine 2C (5-HT_{2C}) receptor is the most convincing, with evidence converging on the -759C/T (rs3813929) polymorphism in the promoter region of the *HTR2C* gene as a predictor of risk of weight gain associated with atypical antipsychotic use, [141,142] despite conflicting reporting of the functional significance of the C versus the T allele [143,144]. In contrast to studies that showed positive association of weight gain with the T allele, atypical antipsychotic treatment duration (less than three months) and ethnicity (European Americans and not African Americans or Asians) are variables that are found to be more prominent in studies with positive association of the C allele as a risk for weight gain.

More recently, research has also focused on the leptin—melanocortin system. Leptin is a peptide hormone secreted by the adipose tissue, with a proportional correlation between the adipose tissue amount and leptin level. Leptin activates secondary signals associated with food intake inhibition and increased energy expenditure, and high serum leptin level results in appetite suppression and energy storage. A functional -2548 A/G (rs7799039) polymorphism in the promoter region of the gene coding the leptin protein (*LEP*) and a 223 Gln/Arg (rs1137101) polymorphism of the gene coding the leptin receptor (*LEPR*) have been reported as risk predictors for weight gain [145,146].

Dopamine binds to the D_2 receptor and inhibits prolactin secretion; therefore, use of antipsychotic agents results in increased prolactin level, although the effect is less with the atypical antipsychotics. A recent study showed that hyperprolactinemia is related to the Taq1A polymorphism, with the A1 allele associated with elevated prolactin level, as well as being drug specific, with the effect being more prominent with risperidone and olanzapine than with quetiapine [147].

Mood Stabilizers

Even though therapeutic efficacy of lithium as a mood stabilizer has been shown to be associated with *SLC6A4* genotypes, with better outcomes for patients with the L/L or L/S genotypes [148], most of the published pharmacogenomic studies of lithium primarily focus on the inositol turnover signaling pathway and the inhibition of glycogen synthase kinase 3-ß (GSK3B). Patients with bipolar disorder are reported to have hyperactive signaling in the inositol turnover signaling pathway, and lithium use inhibits the activity of inositol polyphosphate-1-phosphatase (INPP1) and inositol monophosphotases (IMPA1 and IMPA2), resulting in reduced amount of free inositol available for signaling activity [149]. SNP (rs2067421) in the *INPPI* gene had been reported to be associated with lithium response [150], and Bremer et al. reported that the association is likely dependent on clinical subtype [151]. In the same study by Bremer et al., the SNP (rs2199503) for *GSK3B* also was shown to be associated with lithium response in patients with post-traumatic stress disorder [151]. Failure to differentiate clinical comorbidity in past association studies might contribute to the conflicting results with *INPPI* and *GSK3B* polymorphisms in the literature.

One of the most useful applications of pharmacogenomics in psychiatry rests with the use of the anticonvulsant carbamazepine as a mood stabilizer. Despite its usefulness for patients with bipolar disorder, carbamazepine use is associated with severe adverse effects such as aplastic anemia and life-threatening cutaneous drug reactions such as Stevens—Johnson syndrome/toxic epidermal necrosis (SJS/TEN). Abundant literature data support findings that the major histocompatibility complex *HLA-B*1502* is a strong predictor of carbamazepine-induced Stevens—Johnson syndrome, primarily in patients of Asian descent. The presence of *HLA-B*1502* was documented in all 44 Taiwanese Chinese of Han descent with SJS/TEN. Another study reported a positive association with *HLA-B*1502* in 98% of 60 Han Chinese patients with the adverse drug reaction compared to 4% of patients who did not have the reaction [152,153]. A more recent study confirmed the positive association in 94% of Han Chinese patients with SJS/TEN compared to 9.5% of carbamazepine-tolerant patients and 9% of healthy control subjects [154]. Similar associations have been reported for other Asian populations, despite variability in the frequency of *HLA-B*1502* in those populations (Figure 6.2) [155-159]. A black-box warning regarding this association in specific populations of susceptible individuals carrying the *HLA-B*1502* allele was issued by the FDA in 2007, with a recommendation that regardless of their countries of origin, all patients of Asian descent should be screened for *HLA-B*1502* prior to initiation of carbamazepine therapy, and an alternative agent should be used in patients who test positive for the allele. However, it should be noted that (1) phenytoin also causes SJS/TEN [160] and is not a suitable alternative agent for carbamazepine in patients with the *HLA-B*1502* variant, and (2) *HLA-B*1502* is rare in both Japanese and Korean patients. Instead, another more common HLA allele, *HLA-B*1511*, is a risk factor for carbamazepine-induced SJS/TEN in these two populations [155,161,162].

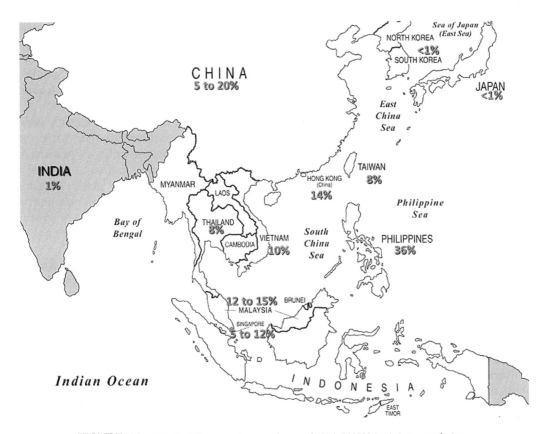

FIGURE 6.2 Ethnic differences in prevalence of *HLA-B*1502* in Asian populations.

APPLICATION OF PHARMACOGENOMICS IN PSYCHIATRY

Drug development in psychiatry has made little progress over the last several decades. Although there have been better safety profiles for newer psychotropics, the CATIE study showed that the atypical antipsychotics represent only a small improvement over the typical antipsychotics. There were no real advantages of any newer SSRIs over their older counterparts. Over the years, there have been many advances in pharmacogenomics and expectations of what it could bring to psychiatric practice. Arguments for utilizing genetic information to maximize effectiveness of current drugs have been made by many investigators within the field of psychopharmacogenomics.

Genetic differences in psychotropic metabolism, most of which are mediated by the cytochrome P-450 enzyme systems, are well established, and the frequency of the cytochrome P-450 enzyme polymorphisms also have been characterized in different ethnic groups. Screening of common *CYP2C19* and *CYP2D6* alleles could be achieved with the FDA-approved AmpliChip CYP genotyping test (Roche Diagnostics) and is the most common application of pharmacogenomic advances in clinical practice when abnormal metabolic capacity is suspected to contribute to unexpected response [163]. Current evidence demonstrates the genotype test possesses high analytic validity with excellent sensitivity and specificity in genotype prediction. However, despite documented substantial variability in psychotropic drug exposure among subjects of different genotypes or metabolic status and suggested dosage regimens for the different genotypes (Table 6.5) [8], there were only a few studies that provided evidence of association between genotypes and adverse effects, with even less documented clinical validity based on psychotropic response prediction.

Not surprisingly, with very few well-designed clinical trials using patient-specific genotypes to demonstrate the clinical relevance of pharmacogenomic-guided dosing to optimize response rates and/or minimize adverse drug reactions, the utility of pharmacogenomics in clinical practice to influence prescribing pattern and patient outcome is nonexistent. Unlike drug approval, the approval of the AmpliChip CYP genotyping test was therefore not based on evidence of clinical validity or utility, and demonstration of clinical benefit rests with the practitioners [163]. This not only leads to absence of specific dosing guidance from the regulatory agency for psychotropics, including atomoxetine, but also provides support against reimbursing CYP genotyping in psychopharmacotherapeutics. Therefore, given the current literature data and the emerging role of neurotransmitter receptors and transporters in psychotropics response association, the utility of CYP genotyping in improving drug treatment could ultimately be in reducing side effects and improving medication adherence. Nevertheless, current literature suggest that the serotonergic system, in particular the -759C/T polymorphism affecting the 5-HT$_{2C}$ receptor, is much more promising in predicting weight gain associated with the use of antipsychotics.

Even with the abundance of research with neurotransmitter receptors and transporters, predicting psychotropic drug response remains a significant challenge (Table 6.6). Similar to CYP genotyping, the same limitations of lack of large-scale, prospective clinical trials, sample size, and ethnic variability need to be overcome. In addition, unlike drug metabolism, drug response is more likely to be mediated by multiple genes, and haplotype analyses

TABLE 6.5 Dosage adjustment (percent of normal dose) of selected antidepressants and antipsychotics based on CYP2D6 genotypes

Drugs	UM	EM	IM	PM
Amitriptyline		120	90	50
Clomipramine		120	90	60
Desipramine	260	130	80	30
Imipramine		130	80	30
Nortriptyline	230	120	90	50
Trimipramine		130	80	40
Fluoxetine		110	90	70
Fluvoxamine		120	90	60
Paroxetine		110	90	70
Maprotiline		130	80	40
Mianserine	300	110	90	70
Venlafaxine		130	80	30
Haloperidol		110	100	80
S-Flupenthixol		120	90	70
Olanzapine		140	60	40
Risperidone		110	100	50
Zuclopenthixol		110	90	70

(Adapted from Kirchheiner et al [8].)

would be critical in identifying appropriate genotypes for prediction. In this regard, it is of note that most pharmacogenomic studies to date focus either on combination of variants of different CYP isoenzymes or combination of variants of different drug targets but seldom both. Based on PET studies, there is evidence of a threshold 76 to 85% serotonin transporter occupancy for therapeutic response from different SSRI treatment [164-168]. In a 2001 study that investigated the relationship between paroxetine concentration and serotonin transporter occupancy, Meyer et al. showed that plateau occupancy of about 85% occurred when serum concentration of paroxetine exceeded 28 ng/ml [169]. Because paroxetine is metabolized by the polymorphic CYP2D6, the threshold drug concentration in the range of 28 ng/ml would not be achieved in some patients administered the standard dosage regimens, especially the UM. One could argue that without sufficient drug exposure at the target site, the relevance of any target polymorphism might be less. Perhaps another approach to clinical psychopharmacogenomic investigations would be with a concentration-controlled trial to integrate relevant pharmacokinetic variants with important pharmacodynamic variants and complemented with PET evidence of drug target occupancy—for example, serotonin transporter occupancy for SSRI [164,169]. For practicing psychiatrists and clinicians,

TABLE 6.6   Challenges for psychopharmacogenomic evaluation and implementation

Genomic studies mostly based on *post hoc* analyses of DNA samples collected from clinical trials that are not initially designed for pharmacogenomic evaluations

Heterogeneity of study populations with respect to

- allele frequencies
- ethnicity
- patient-specific variables (gender, age, concurrent drug)
- disease phenotypes
- prior drug use

Study designs with differences in

- prospective versus retrospective versus naturalistic study
- gene(s) investigated
- selection of genetic biomarkers: single SNP versus haplotypes
- treatment duration
- response definition
- response assessment

Inadequate sample size

Few prospective trials of pharmacogenomic-based clinical practice versus standard of care

Small incremental value in current quality and evidence-based-driven clinical environment

this approach might provide a more pragmatic approach and ultimately a more likely solution than an endless list of potential and almost completely different sets of biomarkers of SSRI efficacy that have been identified by genome-wide association studies (GWASes) [73, 170-172].

The recently approved antidepressant vilazodone provides another example of pharmacogenomic application. Approved by the FDA in January 2011, vilazodone is the first of a new class of antidepressant (the indolalkylamines) with dual action of serotonin reuptake inhibition and partial agonist activity at the 5-HT_{1A} receptor [173]. In addition, development of vilazodone included a clinical trial with patient stratification according to genetic biomarker most likely to be associated with therapeutic response. A report described the association of haplotypes of biomarkers involved in neurotransmitter signaling and vilazodone metabolism with clinical response, although the identity of the biomarkers was not revealed. The result indicated that 75.5% of 49 vilazodone-treated patients who also possess one specific biomarker ($M1^+$) responded to therapy (defined as a decrease of at least 50% from the baseline Montgomery–Asberg Depression Rating Scale [MADRS] score after eight weeks of treatment), whereas only 35.2% of 108 "marker-negative" patients ($M1^-$) responded with vilazodone treatment. Remission (defined as final MADRS score of less than 10) was achieved in 44.9% of "marker-positive" patients and 20.4% of "marker-negative" patients. 57.1% of 14 vilazodone-treated patients with another biomarker ($M2^+$) were reported to have nausea and vomiting, compared to 15.5% of patients without the same biomarker [174]. Despite the small number of patients, the study represents an example of early use of biomarkers in drug development. Additional results from ongoing replication studies will provide insight as to whether these biomarkers could eventually be useful for identifying responders to vilazodone. If confirmed, the unique dual pharmacological action of vilazodone and availability of clinically relevant biomarkers could provide significant contribution to individualized clinical treatment.

CONCLUSION

Despite many findings within the field of psychopharmacogenomics, only a few of the results are ready for translation into clinical practice. Although CYP genotyping was previously recommended for incorporation into the therapeutic decision-making process, the current evidence-based approach significantly limits its application in clinical practice. Compared to other therapeutic areas such as cardiovascular disease and cancer, promising research findings to predict drug response in psychiatry is still in its infancy. Multiple genetic biomarkers have been identified by either candidate gene approach or GWASes and evaluated in clinical studies involving various designs and ethnic populations. However, lack of consistent results among the studies does not point to definitive associations for most biomarkers. Given the currently available psychotropics and the lack of novel compounds in the foreseeable future, pharmacogenomics holds significant promise in optimizing drug therapy for the mentally ill populations. Further advances in the field would require in-depth understanding of mental disease etiology, developing clear definitions of response phenotypes and outcome measurements, and refining current molecular approaches.

QUESTIONS FOR DISCUSSION

1. How do ethnic variabilities in allele frequencies affect interpretation of study results in psychopharmacogenomics?
2. Are there significant roles for *ABCB1* polymorphism in psychotropic disposition and response?
3. What is the significance of the STAR*D study with respect to 5HTTLPR polymorphism?
4. What are some of the factors that slow translation of pharmacogenomic findings to practice for psychopharmacology, in contrast to other therapeutic areas such as oncology and cardiovascular diseases?

References

[1] Bertilsson L, Aberg-Wistedt A, Gustafsson LL, Nordin C. Extremely rapid hydroxylation of debrisoquine: a case report with implication for treatment with nortriptyline and other tricyclic antidepressants. Ther Drug Monit 1985;7:478—80.
[2] Bertilsson L, Dahl ML, Sjoqvist F, Aberg-Wistedt A, Humble M, Johansson I, et al. Molecular basis for rational megaprescribing in ultrarapid hydroxylators of debrisoquine. Lancet 1993;341:63.
[3] Dalen P, Dahl ML, Bernal Ruiz ML, Nordin J, Bertilsson L. 10-Hydroxylation of nortriptyline in white persons with 0, 1, 2, 3, and 13 functional CYP2D6 genes. Clin Pharmacol Ther 1998;63:444—52.
[4] Sawamura K, Suzuki Y, Someya T. Effects of dosage and CYP2D6-mutated allele on plasma concentration of paroxetine. Eur J Clin Pharmacol 2004;60:553—7.
[5] Watanabe J, Suzuki Y, Fukui N, Sugai T, Ono S, Inoue Y, et al. Dose-dependent effect of the CYP2D6 genotype on the steady-state fluvoxamine concentration. Ther Drug Monit 2008;30:705—8.
[6] Alfaro CL, Lam YW, Simpson J, Ereshefsky L. CYP2D6 status of extensive metabolizers after multiple-dose fluoxetine, fluvoxamine, paroxetine, or sertraline. J Clin Psychopharmacol 1999;19:155—63.
[7] Suzuki Y, Sugai T, Fukui N, Watanabe J, Ono S, Inoue Y, et al. CYP2D6 genotype and smoking influence fluvoxamine steady-state concentration in Japanese psychiatric patients: lessons for genotype-phenotype association study design in translational pharmacogenetics. J Psychopharmacol 2011;25:908—14.

[8] Kirchheiner J, Brosen K, Dahl ML, Gram LF, Kasper S, Roots I, et al. CYP2D6 and CYP2C19 genotype-based dose recommendations for antidepressants: a first step towards subpopulation-specific dosages. Acta Psychiatr Scand 2001;104:173–92.

[9] Rau T, Wohlleben G, Wuttke H, Thuerauf N, Lunkenheimer J, Lanczik M, et al. CYP2D6 genotype: impact on adverse effects and nonresponse during treatment with antidepressants-a pilot study. Clin Pharmacol Ther 2004;75:386–93.

[10] Kawanishi C, Lundgren S, Agren H, Bertilsson L. Increased incidence of CYP2D6 gene duplication in patients with persistent mood disorders: ultrarapid metabolism of antidepressants as a cause of nonresponse. A pilot study. Eur J Clin Pharmacol 2004;59:803–7.

[11] Serretti A, Calati R, Massat I, Linotte S, Kasper S, Lecrubier Y, et al. Cytochrome P450 CYP1A2, CYP2C9, CYP2C19 and CYP2D6 genes are not associated with response and remission in a sample of depressive patients. Int Clin Psychopharmacol 2009;24:250–6.

[12] Recommendations from the EGAPP Working Group: testing for cytochrome P450 polymorphisms in adults with nonpsychotic depression treated with selective serotonin reuptake inhibitors. Genet Med 2007;9: 819–25.

[13] Peters EJ, Slager SL, Kraft JB, Jenkins GD, Reinalda MS, McGrath PJ, et al. Pharmacokinetic genes do not influence response or tolerance to citalopram in the STAR*D sample. PLoS One 2008;3:e1872.

[14] Tsai MH, Lin KM, Hsiao MC, Shen WW, Lu ML, Tang HS, et al. Genetic polymorphisms of cytochrome P450 enzymes influence metabolism of the antidepressant escitalopram and treatment response. Pharmacogenomics 2010;11:537–46.

[15] Bijl MJ, Visser LE, Hofman A, Vulto AG, van Gelder T, Stricker BH, et al. Influence of the CYP2D6*4 polymorphism on dose, switching and discontinuation of antidepressants. Br J Clin Pharmacol 2008;65:558–64.

[16] McAlpine DE, O'Kane DJ, Black JL, Mrazek DA. Cytochrome P450 2D6 genotype variation and venlafaxine dosage. Mayo Clin Proc 2007;82:1065–8.

[17] Pollock BG, Mulsant BH, Sweet RA, Rosen J, Altieri LP, Perel JM. Prospective cytochrome P450 phenotyping for neuroleptic treatment in dementia. Psychopharmacol Bull 1995;31:327–31.

[18] Brockmoller J, Kirchheiner J, Schmider J, Walter S, Sachse C, Muller-Oerlinghausen B, et al. The impact of the CYP2D6 polymorphism on haloperidol pharmacokinetics and on the outcome of haloperidol treatment. Clin Pharmacol Ther 2002;72:438–52.

[19] Aitchison KJ, Munro J, Wright P, Smith S, Makoff AJ, Sachse C, et al. Failure to respond to treatment with typical antipsychotics is not associated with CYP2D6 ultrarapid hydroxylation. Br J Clin Pharmacol 1999;48:388–94.

[20] Grossman I, Sullivan PF, Walley N, Liu Y, Dawson JR, Gumbs C, et al. Genetic determinants of variable metabolism have little impact on the clinical use of leading antipsychotics in the CATIE study. Genet Med 2008;10:720–9.

[21] Lane HY, Liu YC, Huang CL, Chang YC, Wu PL, Lu CT, et al. Risperidone-related weight gain: genetic and nongenetic predictors. J Clin Psychopharmacol 2006;26:128–34.

[22] Crescenti A, Mas S, Gasso P, Parellada E, Bernardo M, Lafuente A. Cyp2d6*3, *4, *5 and *6 polymorphisms and antipsychotic-induced extrapyramidal side-effects in patients receiving antipsychotic therapy. Clin Exp Pharmacol Physiol 2008;35:807–11.

[23] Mulder H, Wilmink FW, Belitser SV, Egberts AC. The association between cytochrome P450-2D6 genotype and prescription of antiparkinsonian drugs in hospitalized psychiatric patients using antipsychotics: a retrospective follow-up study. J Clin Psychopharmacol 2006;26:212–5.

[24] Schillevoort I, de Boer A, van der Weide J, Steijns LS, Roos RA, Jansen PA, et al. Antipsychotic-induced extrapyramidal syndromes and cytochrome P450 2D6 genotype: a case-control study. Pharmacogenetics 2002;12:235–40.

[25] Chen S, Chou WH, Blouin RA, Mao Z, Humphries LL, Meek QC, et al. The cytochrome P450 2D6 (CYP2D6) enzyme polymorphism: screening costs and influence on clinical outcomes in psychiatry. Clin Pharmacol Ther 1996;60:522–34.

[26] de Leon J, Barnhill J, Rogers T, Boyle J, Chou WH, Wedlund PJ. Pilot study of the cytochrome P450-2D6 genotype in a psychiatric state hospital. Am J Psychiatry 1998;155:1278–80.

[27] Vandel P, Haffen E, Vandel S, Bonin B, Nezelof S, Sechter D, et al. Drug extrapyramidal side effects. CYP2D6 genotypes and phenotypes. Eur J Clin Pharmacol 1999;55:659–65.

[28] de Leon J, Susce MT, Pan RM, Fairchild M, Koch WH, Wedlund PJ. The CYP2D6 poor metabolizer phenotype may be associated with risperidone adverse drug reactions and discontinuation. J Clin Psychiatry 2005;66:15—27.

[29] Patsopoulos NA, Ntzani EE, Zintzaras E, Ioannidis JP. CYP2D6 polymorphisms and the risk of tardive dyskinesia in schizophrenia: a meta-analysis. Pharmacogenet Genomics 2005;15:151—8.

[30] Chou WH, Yan FX, de Leon J, Barnhill J, Rogers T, Cronin M, et al. Extension of a pilot study: impact from the cytochrome P450 2D6 polymorphism on outcome and costs associated with severe mental illness. J Clin Psychopharmacol 2000;20:246—51.

[31] Boulton DW, DeVane CL, Liston HL, Markowitz JS. In vitro P-glycoprotein affinity for atypical and conventional antipsychotics. Life Sci 2002;71:163—9.

[32] Ejsing TB, Pedersen AD, Linnet K. P-glycoprotein interaction with risperidone and 9-OH-risperidone studied in vitro, in knock-out mice and in drug-drug interaction experiments. Hum Psychopharmacol 2005;20:493—500.

[33] Schinkel AH, Smit JJ, van Tellingen O, Beijnen JH, Wagenaar E, van Deemter L, et al. Disruption of the mouse mdr1a P-glycoprotein gene leads to a deficiency in the blood-brain barrier and to increased sensitivity to drugs. Cell 1994;77:491—502.

[34] Uhr M, Steckler T, Yassouridis A, Holsboer F. Penetration of amitriptyline, but not of fluoxetine, into brain is enhanced in mice with blood-brain barrier deficiency due to mdr1a P-glycoprotein gene disruption. Neuropsychopharmacology 2000;22:380—7.

[35] Roberts RL, Joyce PR, Mulder RT, Begg EJ, Kennedy MA. A common P-glycoprotein polymorphism is associated with nortriptyline-induced postural hypotension in patients treated for major depression. pharmacogenomics J 2002;2:191—6.

[36] Weiss J, Dormann SM, Martin-Facklam M, Kerpen CJ, Ketabi-Kiyanvash N, Haefeli WE. Inhibition of P-glycoprotein by newer antidepressants. J Pharmacol Exp Ther 2003;305:197—204.

[37] Rochat B, Baumann P, Audus KL. Transport mechanisms for the antidepressant citalopram in brain microvessel endothelium. Brain Res 1999;831:229—36.

[38] El Ela AA, Hartter S, Schmitt U, Hiemke C, Spahn-Langguth H, et al. Identification of P-glycoprotein substrates and inhibitors among psychoactive compounds—implications for pharmacokinetics of selected substrates. J Pharm Pharmacol 2004;56:967—75.

[39] Wang JS, Ruan Y, Taylor RM, Donovan JL, Markowitz JS, DeVane CL. The brain entry of risperidone and 9-hydroxyrisperidone is greatly limited by P-glycoprotein. Int J Neuropsychopharmacol 2004;7:415—9.

[40] Fukui N, Suzuki Y, Sawamura K, Sugai T, Watanabe J, Inoue Y, et al. Dose-dependent effects of the 3435 C>T genotype of ABCB1 gene on the steady-state plasma concentration of fluvoxamine in psychiatric patients. Ther Drug Monit 2007;29:185—9.

[41] Kato M, Fukuda T, Serretti A, Wakeno M, Okugawa G, Ikenaga Y, et al. ABCB1 (MDR1) gene polymorphisms are associated with the clinical response to paroxetine in patients with major depressive disorder. Prog Neuropsychopharmacol Biol Psychiatry 2008;32:398—404.

[42] Nikisch G, Eap CB, Baumann P. Citalopram enantiomers in plasma and cerebrospinal fluid of ABCB1 genotyped depressive patients and clinical response: a pilot study. Pharmacol Res 2008;58:344—7.

[43] Laika B, Leucht S, Steimer W. ABCB1 (P-glycoprotein/MDR1) gene G2677T/a sequence variation (polymorphism): lack of association with side effects and therapeutic response in depressed inpatients treated with amitriptyline. Clin Chem 2006;52:893—5.

[44] Gex-Fabry M, Eap CB, Oneda B, Gervasoni N, Aubry JM, Bondolfi G, et al. CYP2D6 and ABCB1 genetic variability: influence on paroxetine plasma level and therapeutic response. Ther Drug Monit 2008;30:474—82.

[45] Mihaljevic Peles A, Bozina N, Sagud M, Rojnic Kuzman M, Lovric M. MDR1 gene polymorphism: therapeutic response to paroxetine among patients with major depression. Prog Neuropsychopharmacol Biol Psychiatry 2008;32:1439—44.

[46] De Luca V, Mundo E, Trakalo J, Wong GW, Kennedy JL. Investigation of polymorphism in the MDR1 gene and antidepressant-induced mania. Pharmacogenomics J 2003;3:297—9.

[47] Uhr M, Tontsch A, Namendorf C, Ripke S, Lucae S, Ising M, et al. Polymorphisms in the drug transporter gene ABCB1 predict antidepressant treatment response in depression. Neuron 2008;57:203—9.

[48] Kimchi-Sarfaty C, Oh JM, Kim IW, Sauna ZE, Calcagno AM, Ambudkar SV, et al. A "silent" polymorphism in the MDR1 gene changes substrate specificity. Science 2007;315:525—8.

[49] Heils A, Teufel A, Petri S, Stober G, Riederer P, Bengel D, et al. Allelic variation of human serotonin transporter gene expression. J Neurochem 1996;66:2621–4.

[50] Heinz A, Jones DW, Mazzanti C, Goldman D, Ragan P, Hommer D, et al. A relationship between serotonin transporter genotype and in vivo protein expression and alcohol neurotoxicity. Biol Psychiatry 2000;47: 643–9.

[51] Shioe K, Ichimiya T, Suhara T, Takano A, Sudo Y, Yasuno F, et al. No association between genotype of the promoter region of serotonin transporter gene and serotonin transporter binding in human brain measured by PET. Synapse 2003;48:184–8.

[52] van Dyck CH, Malison RT, Staley JK, Jacobsen LK, Seibyl JP, et al. Central serotonin transporter availability measured with [123I]beta-CIT SPECT in relation to serotonin transporter genotype. Am J Psychiatry 2004;161:525–31.

[53] David SP, Murthy NV, Rabiner EA, Munafo MR, Johnstone EC, Jacob R, et al. A functional genetic variation of the serotonin (5-HT) transporter affects 5-HT1A receptor binding in humans. J Neurosci 2005;25:2586–90.

[54] Durham LK, Webb SM, Milos PM, Clary CM, Seymour AB. The serotonin transporter polymorphism, 5HTTLPR, is associated with a faster response time to sertraline in an elderly population with major depressive disorder. Psychopharmacology 2004;174:525–9.

[55] Pollock BG, Ferrell RE, Mulsant BH, Mazumdar S, Miller M, Sweet RA, et al. Allelic variation in the serotonin transporter promoter affects onset of paroxetine treatment response in late-life depression. Neuropsychopharmacology 2000;23:587–90.

[56] Zanardi R, Serretti A, Rossini D, Franchini L, Cusin C, Lattuada E, et al. Factors affecting fluvoxamine antidepressant activity: influence of pindolol and 5-HTTLPR in delusional and nondelusional depression. Biol Psychiatry 2001;50:323–30.

[57] Smits KM, Smits LJ, Schouten JS, Peeters FP, Prins MH. Does pretreatment testing for serotonin transporter polymorphisms lead to earlier effects of drug treatment in patients with major depression? A decision-analytic model. Clin Ther 2007;29:691–702.

[58] Serretti A, Kato M, De Ronchi D, Kinoshita T. Meta-analysis of serotonin transporter gene promoter polymorphism (5-HTTLPR) association with selective serotonin reuptake inhibitor efficacy in depressed patients. Mol Psychiatry 2007;12:247–57.

[59] Taylor MJ, Sen S, Bhagwagar Z. Antidepressant response and the serotonin transporter gene-linked polymorphic region. Biol Psychiatry 2010;68:536–43.

[60] Kim DK, Lim SW, Lee S, Sohn SE, Kim S, Hahn CG, et al. Serotonin transporter gene polymorphism and antidepressant response. Neuroreport 2000;11:215–9.

[61] Yoshida K, Ito K, Sato K, Takahashi H, Kamata M, Higuchi H, et al. Influence of the serotonin transporter gene-linked polymorphic region on the antidepressant response to fluvoxamine in Japanese depressed patients. Prog Neuropsychopharmacol Biol Psychiatry 2002;26:383–6.

[62] Nakamura M, Ueno S, Sano A, Tanabe H. The human serotonin transporter gene linked polymorphism (5-HTTLPR) shows ten novel allelic variants. Mol Psychiatry 2000;5:32–8.

[63] Hu XZ, Lipsky RH, Zhu G, Akhtar LA, Taubman J, Greenberg BD, et al. Serotonin transporter promoter gain-of-function genotypes are linked to obsessive-compulsive disorder. Am J Hum Genet 2006;78:815–26.

[64] Kraft JB, Slager SL, McGrath PJ, Hamilton SP. Sequence analysis of the serotonin transporter and associations with antidepressant response. Biol Psychiatry 2005;58:374–81.

[65] Hu XZ, Rush AJ, Charney D, Wilson AF, Sorant AJ, Papanicolaou GJ, et al. Association between a functional serotonin transporter promoter polymorphism and citalopram treatment in adult outpatients with major depression. Arch Gen Psychiatry 2007;64:783–92.

[66] Ogilvie AD, Battersby S, Bubb VJ, Fink G, Harmar AJ, Goodwim GM, et al. Polymorphism in serotonin transporter gene associated with susceptibility to major depression. Lancet 1996;347:731–3.

[67] Ito K, Yoshida K, Sato K, Takahashi H, Kamata M, Higuchi H, et al. A variable number of tandem repeats in the serotonin transporter gene does not affect the antidepressant response to fluvoxamine. Psychiatry Res 2002;111:235–9.

[68] Mrazek DA, Rush AJ, Biernacka JM, O'Kane DJ, Cunningham JM, Wieben ED, et al. SLC6A4 variation and citalopram response. Am J Med Genet B Neuropsychiatr Genet 2009;150B:341–51.

[69] Bleibel WK, Duan S, Huang RS, Kistner EO, Shukla SJ, Wu X, et al. Identification of genomic regions contributing to etoposide-induced cytotoxicity. Hum Genet 2009;125:173–80.

[70] Huang RS, Duan S, Bleibel WK, Kistner EO, Zhang W, Clark TA, et al. A genome-wide approach to identify genetic variants that contribute to etoposide-induced cytotoxicity. Proc Natl Acad Sci U S A 2007;104:9758–63.

[71] Li L, Fridley B, Kalari K, Jenkins G, Batzler A, Safgren S, et al. Gemcitabine and cytosine arabinoside cytotoxicity: association with lymphoblastoid cell expression. Cancer Res 2008;68:7050–8.

[72] Morag A, Kirchheiner J, Rehavi M, Gurwitz D. Human lymphoblastoid cell line panels: novel tools for assessing shared drug pathways. Pharmacogenomics 2010;11:327–40.

[73] Morag A, Pasmanik-Chor M, Oron-Karni V, Rehavi M, Stingl JC, Gurwitz D. Genome-wide expression profiling of human lymphoblastoid cell lines identifies CHL1 as a putative SSRI antidepressant response biomarker. Pharmacogenomics 2011;12:171–84.

[74] Meyer JH, Kapur S, Eisfeld B, Brown GM, Houle S, DaSilva J, et al. The effect of paroxetine on 5-HT(2A) receptors in depression: an [(18)F]setoperone PET imaging study. Am J Psychiatry 2001;158:78–85.

[75] Stanley M, Mann JJ. Increased serotonin-2 binding sites in frontal cortex of suicide victims. Lancet 1983;1:214–6.

[76] Choi MJ, Kang RH, Ham BJ, Jeong HY, Lee MS. Serotonin receptor 2A gene polymorphism (-1438A/G) and short-term treatment response to citalopram. Neuropsychobiology 2005;52:155–62.

[77] Kato M, Fukuda T, Wakeno M, Fukuda K, Okugawa G, Ikenaga Y, et al. Effects of the serotonin type 2A, 3A and 3B receptor and the serotonin transporter genes on paroxetine and fluvoxamine efficacy and adverse drug reactions in depressed Japanese patients. Neuropsychobiology 2006;53:186–95.

[78] Minov C, Baghai TC, Schule C, Zwanzger P, Schwarz MJ, Zill P, et al. Serotonin-2A-receptor and -transporter polymorphisms: lack of association in patients with major depression. Neurosci Lett 2001;303:119–22.

[79] McMahon FJ, Buervenich S, Charney D, Lipsky R, Rush AJ, Wilson AF, et al. Variation in the gene encoding the serotonin 2A receptor is associated with outcome of antidepressant treatment. Am J Hum Genet 2006;78:804–14.

[80] Lesser IM, Castro DB, Gaynes BN, Gonzalez J, Rush AJ, Alpert JE, et al. Ethnicity/race and outcome in the treatment of depression: results from STAR*D. Med Care 2007;45:1043–51.

[81] Lucae S, Ising M, Horstmann S, Baune BT, Arolt V, Muller-Myhsok B, et al. HTR2A gene variation is involved in antidepressant treatment response. Eur Neuropsychopharmacol 2010;20:65–8.

[82] Rotondo A, Nielsen DA, Nakhai B, Hulihan-Giblin B, Bolos A, Goldman D. Agonist-promoted downregulation and functional desensitization in two naturally occurring variants of the human serotonin1A receptor. Neuropsychopharmacology 1997;17:18–26.

[83] Artigas F, Romero L, de Montigny C, Blier P. Acceleration of the effect of selected antidepressant drugs in major depression by 5-HT1A antagonists. Trends Neurosci 1996;19:378–83.

[84] Perez V, Gilaberte I, Faries D, Alvarez E, Artigas F. Randomised, double-blind, placebo-controlled trial of pindolol in combination with fluoxetine antidepressant treatment. Lancet 1997;349:1594–7.

[85] Zanardi R, Artigas F, Franchini L, Sforzini L, Gasperini M, Smeraldi E, et al. How long should pindolol be associated with paroxetine to improve the antidepressant response? J Clin Psychopharmacol 1997;17:446–50.

[86] Lemonde S, Turecki G, Bakish D, Du L, Hrdina PD, Bown CD, et al. Impaired repression at a 5-hydroxytryptamine 1A receptor gene polymorphism associated with major depression and suicide. J Neurosci 2003;23:8788–99.

[87] Lemonde S, Du L, Bakish D, Hrdina P, Albert PR. Association of the C(-1019)G 5-HT1A functional promoter polymorphism with antidepressant response. Int J Neuropsychopharmacol 2004;7:501–6.

[88] Yu YW, Tsai SJ, Liou YJ, Hong CJ, Chen TJ. Association study of two serotonin 1A receptor gene polymorphisms and fluoxetine treatment response in Chinese major depressive disorders. Eur Neuropsychopharmacol 2006;16:498–503.

[89] Baune BT, Hohoff C, Roehrs T, Deckert J, Arolt V, Domschke K. Serotonin receptor 1A-1019C/G variant: impact on antidepressant pharmacoresponse in melancholic depression? Neurosci Lett 2008;436:111–5.

[90] Levin GM, Bowles TM, Ehret MJ, Langaee T, Tan JY, Johnson JA, et al. Assessment of human serotonin 1A receptor polymorphisms and SSRI responsiveness. Mol Diagn Ther 2007;11:155–60.

[91] Arias B, Catalan R, Gasto C, Gutierrez B, Fananas L. Evidence for a combined genetic effect of the 5-HT(1A) receptor and serotonin transporter genes in the clinical outcome of major depressive patients treated with citalopram. J Psychopharmacol 2005;19:166–72.

[92] Suzuki Y, Sawamura K, Someya T. The effects of a 5-hydroxytryptamine 1A receptor gene polymorphism on the clinical response to fluvoxamine in depressed patients. Pharmacogenomics J 2004;4:283–6.

[93] Hashimoto K, Sawa A, Iyo M. Increased levels of glutamate in brains from patients with mood disorders. Biol Psychiatry 2007;62:1310—6.

[94] Bobula B, Tokarski K, Hess G. Repeated administration of antidepressants decreases field potentials in rat frontal cortex. Neuroscience 2003;120:765—9.

[95] Paddock S, Laje G, Charney D, Rush AJ, Wilson AF, Sorant AJ, et al. Association of GRIK4 with outcome of antidepressant treatment in the STAR*D cohort. Am J Psychiatry 2007;164:1181—8.

[96] Perlis RH, Fijal B, Dharia S, Heinloth AN, Houston JP. Failure to replicate genetic associations with antidepressant treatment response in duloxetine-treated patients. Biol Psychiatry 2010;67:1110—3.

[97] Creese I, Burt DR, Snyder SH. Dopamine receptor binding predicts clinical and pharmacological potencies of antischizophrenic drugs. Science 1976;192:481—3.

[98] Kapur S, Zipursky R, Jones C, Remington G, Houle S. Relationship between dopamine D(2) occupancy, clinical response, and side effects: a double-blind PET study of first-episode schizophrenia. Am J Psychiatry 2000;157:514—20.

[99] Mamo D, Kapur S, Shammi CM, Papatheodorou G, Mann S, Therrien F, et al. A PET study of dopamine D2 and serotonin 5-HT2 receptor occupancy in patients with schizophrenia treated with therapeutic doses of ziprasidone. Am J Psychiatry 2004;161:818—25.

[100] Stone JM, Davis JM, Leucht S, Pilowsky LS. Cortical dopamine D2/D3 receptors are a common site of action for antipsychotic drugs—an original patient data meta-analysis of the SPECT and PET in vivo receptor imaging literature. Schizophr Bull 2009;35:789—97.

[101] Uchida H, Takeuchi H, Graff-Guerrero A, Suzuki T, Watanabe K, Mamo DC. Dopamine D2 receptor occupancy and clinical effects: a systematic review and pooled analysis. J Clin Psychopharmacol 2011;31:497—502.

[102] Arinami T, Gao M, Hamaguchi H, Toru M. A functional polymorphism in the promoter region of the dopamine D2 receptor gene is associated with schizophrenia. Hum Mol Genet 1997;6:577—82.

[103] Jonsson EG, Nothen MM, Grunhage F, Farde L, Nakashima Y, Propping P, et al. Polymorphisms in the dopamine D2 receptor gene and their relationships to striatal dopamine receptor density of healthy volunteers. Mol Psychiatry 1999;4:290—6.

[104] Lencz T, Robinson DG, Xu K, Ekholm J, Sevy S, Gunduz-Bruce H, et al. DRD2 promoter region variation as a predictor of sustained response to antipsychotic medication in first-episode schizophrenia patients. Am J Psychiatry 2006;163:529—31.

[105] Wu S, Xing Q, Gao R, Li X, Gu N, Feng G, et al. Response to chlorpromazine treatment may be associated with polymorphisms of the DRD2 gene in Chinese schizophrenic patients. Neurosci Lett 2005;376:1—4.

[106] Suzuki A, Kondo T, Mihara K, Yasui-Furukori N, Ishida M, Furukori H, et al. The -141C Ins/Del polymorphism in the dopamine D2 receptor gene promoter region is associated with anxiolytic and antidepressive effects during treatment with dopamine antagonists in schizophrenic patients. Pharmacogenetics 2001;11: 545—50.

[107] Yasui-Furukori N, Tsuchimine S, Saito M, Nakagami T, Sugawara N, Fujii A, et al. Comparing the influence of dopamine D polymorphisms and plasma drug concentrations on the clinical response to risperidone. J Clin Psychopharmacol 2011;31:633—7.

[108] Thompson J, Thomas N, Singleton A, Piggott M, Lloyd S, Perry EK, et al. D2 dopamine receptor gene (DRD2) Taq1 A polymorphism: reduced dopamine D2 receptor binding in the human striatum associated with the A1 allele. Pharmacogenetics 1997;7:479—84.

[109] Schafer M, Rujescu D, Giegling I, Guntermann A, Erfurth A, Bondy B, et al. Association of short-term response to haloperidol treatment with a polymorphism in the dopamine D(2) receptor gene. Am J Psychiatry 2001;158:802—4.

[110] Suzuki A, Mihara K, Kondo T, Tanaka O, Nagashima U, Otani K, et al. The relationship between dopamine D2 receptor polymorphism at the Taq1 A locus and therapeutic response to nemonapride, a selective dopamine antagonist, in schizophrenic patients. Pharmacogenetics 2000;10:335—41.

[111] Reynolds GP, Yao Z, Zhang X, Sun J, Zhang Z. Pharmacogenetics of treatment in first-episode schizophrenia: D3 and 5-HT2C receptor polymorphisms separately associate with positive and negative symptom response. Eur Neuropsychopharmacol 2005;15:143—51.

[112] Lane HY, Lee CC, Chang YC, Lu CT, Huang CH, Chang WH. Effects of dopamine D2 receptor Ser311Cys polymorphism and clinical factors on risperidone efficacy for positive and negative symptoms and social function. Int J Neuropsychopharmacol 2004;7:461—70.

[113] Fukui N, Suzuki Y, Sugai T, Watanabe J, Ono S, Tsuneyama N, et al. Exploring functional polymorphisms in the dopamine receptor D2 gene using prolactin concentration in healthy subjects. Mol Psychiatry 2011;16:356–8.

[114] Lane HY, Hsu SK, Liu YC, Chang YC, Huang CH, Chang WH. Dopamine D3 receptor Ser9Gly polymorphism and risperidone response. J Clin Psychopharmacol 2005;25:6–11.

[115] Van Tol HH, Wu CM, Guan HC, Ohara K, Bunzow JR, et al. Multiple dopamine D4 receptor variants in the human population. Nature 1992;358:149–52.

[116] Kohn Y, Ebstein RP, Heresco-Levy U, Shapira B, Nemanov L, Gritsenko I, et al. Dopamine D4 receptor gene polymorphisms: relation to ethnicity, no association with schizophrenia and response to clozapine in Israeli subjects. Eur Neuropsychopharmacol 1997;7:39–43.

[117] Rao PA, Pickar D, Gejman PV, Ram A, Gershon ES, Gelernter J. Allelic variation in the D4 dopamine receptor (DRD4) gene does not predict response to clozapine. Arch Gen Psychiatry 1994;51:912–7.

[118] Kaiser R, Tremblay PB, Schmider J, Henneken M, Dettling M, Muller-Oerlinghausen B, et al. Serotonin transporter polymorphisms: no association with response to antipsychotic treatment, but associations with the schizoparanoid and residual subtypes of schizophrenia. Mol Psychiatry 2001;6:179–85.

[119] Travis MJ, Busatto GF, Pilowsky LS, Mulligan R, Acton PD, Gacinovic S, et al. 5-HT2A receptor blockade in patients with schizophrenia treated with risperidone or clozapine. A SPET study using the novel 5-HT2A ligand 123I-5-I-R-91150. Br J Psychiatry 1998;173:236–41.

[120] Arranz MJ, Munro J, Sham P, Kirov G, Murray RM, Collier DA, et al. Meta-analysis of studies on genetic variation in 5-HT2A receptors and clozapine response. Schizophr Res 1998;32:93–9.

[121] Sodhi MS, Arranz MJ, Curtis D, Ball DM, Sham P, Roberts GW, et al. Association between clozapine response and allelic variation in the 5-HT2C receptor gene. Neuroreport 1995;7:169–72.

[122] Arranz MJ, Munro J, Birkett J, Bolonna A, Mancama D, Sodhi M, et al. Pharmacogenetic prediction of clozapine response. Lancet 2000;355:1615–6.

[123] Levitt P, Ebert P, Mirnics K, Nimgaonkar VL, Lewis DA. Making the case for a candidate vulnerability gene in schizophrenia: convergent evidence for regulator of G-protein signaling 4 (RGS4). Biol Psychiatry 2006;60:534–7.

[124] Talkowski ME, Seltman H, Bassett AS, Brzustowicz LM, Chen X, Chowdari KV, et al. Evaluation of a susceptibility gene for schizophrenia: genotype based meta-analysis of RGS4 polymorphisms from thirteen independent samples. Biol Psychiatry 2006;60:152–62.

[125] Campbell DB, Ebert PJ, Skelly T, Stroup TS, Lieberman J, Levitt P, et al. Ethnic stratification of the association of RGS4 variants with antipsychotic treatment response in schizophrenia. Biol Psychiatry 2008;63:32–41.

[126] Lane HY, Liu YC, Huang CL, Chang YC, Wu PL, Huang CH, et al. RGS4 polymorphisms predict clinical manifestations and responses to risperidone treatment in patients with schizophrenia. J Clin Psychopharmacol 2008;28:64–8.

[127] Lane HY, Chang YC, Chiu CC, Chen ML, Hsieh MH, Chang WH. Association of risperidone treatment response with a polymorphism in the 5-HT(2A) receptor gene. Am J Psychiatry 2002;159:1593–5.

[128] Kapur S, Seeman P. Does fast dissociation from the dopamine d(2) receptor explain the action of atypical antipsychotics? A new hypothesis. Am J Psychiatry 2001;158:360–9.

[129] Guzey C, Scordo MG, Spina E, Landsem VM, Spigset O. Antipsychotic-induced extrapyramidal symptoms in patients with schizophrenia: associations with dopamine and serotonin receptor and transporter polymorphisms. Eur J Clin Pharmacol 2007;63:233–41.

[130] Mihara K, Suzuki A, Kondo T, Nagashima U, Ono S, Otani K, et al. No relationship between Taq1 A polymorphism of dopamine D(2) receptor gene and extrapyramidal adverse effects of selective dopamine D(2) antagonists, bromperidol, and nemonapride in schizophrenia: a preliminary study. Am J Med Genet 2000;96:422–4.

[131] Bakker PR, van Harten PN, van Os J. Antipsychotic-induced tardive dyskinesia and polymorphic variations in COMT, DRD2, CYP1A2 and MnSOD genes: a meta-analysis of pharmacogenetic interactions. Mol Psychiatry 2008;13:544–56.

[132] Gerlach J, Casey DE. Tardive dyskinesia. Acta Psychiatr Scand 1988;77:369–78.

[133] Lai IC, Mo GH, Chen ML, Wang YC, Chen JY, Liao DL, et al. Analysis of genetic variations in the dopamine D1 receptor (DRD1) gene and antipsychotics-induced tardive dyskinesia in schizophrenia. Eur J Clin Pharmacol 2011;67:383–8.

[134] Dolzan V, Plesnicar BK, Serretti A, Mandelli L, Zalar B, Koprivsek J, et al. Polymorphisms in dopamine receptor DRD1 and DRD2 genes and psychopathological and extrapyramidal symptoms in patients on long-term antipsychotic treatment. Am J Med Genet B Neuropsychiatr Genet 2007;144B:809—15.

[135] Srivastava V, Varma PG, Prasad S, Semwal P, Nimgaonkar VL, Lerer B, et al. Genetic susceptibility to tardive dyskinesia among schizophrenia subjects: IV. Role of dopaminergic pathway gene polymorphisms. Pharmacogenet Genomics 2006;16:111—7.

[136] Lundstrom K, Turpin MP. Proposed schizophrenia-related gene polymorphism: expression of the Ser9Gly mutant human dopamine D3 receptor with the Semliki Forest virus system. Biochem Biophys Res Commun 1996;225:1068—72.

[137] Potkin SG, Kennedy JL, Basile VS. Combining brain imaging and pharmacogenetics in understanding clinical response in Alzheimer's disease and schizophrenia. In: Lerer B, editor. Pharmacogenetics of Psychotropics. New York: Cambridge University Press; 2002. p. 391—400.

[138] Lerer B, Segman RH, Fangerau H, Daly AK, Basile VS, Cavallaro R, et al. Pharmacogenetics of tardive dyskinesia: combined analysis of 780 patients supports association with dopamine D3 receptor gene Ser9Gly polymorphism. Neuropsychopharmacology 2002;27:105—19.

[139] Bakker PR, van Harten PN, van Os J. Antipsychotic-induced tardive dyskinesia and the Ser9Gly polymorphism in the DRD3 gene: a meta analysis. Schizophr Res 2006;83:185—92.

[140] Lerer B, Segman RH, Tan EC, Basile VS, Cavallaro R, Aschauer HN, et al. Combined analysis of 635 patients confirms an age-related association of the serotonin 2A receptor gene with tardive dyskinesia and specificity for the non-orofacial subtype. Int J Neuropsychopharmacol 2005;8:411—25.

[141] De Luca V, Mueller DJ, de Bartolomeis A, Kennedy JL. Association of the HTR2C gene and antipsychotic induced weight gain: a meta-analysis. Int J Neuropsychopharmacol 2007;10:697—704.

[142] Sicard MN, Zai CC, Tiwari AK, Souza RP, Meltzer HY, Lieberman JA, et al. Polymorphisms of the HTR2C gene and antipsychotic-induced weight gain: an update and meta-analysis. Pharmacogenomics 2010;11:1561—71.

[143] Buckland PR, Hoogendoorn B, Guy CA, Smith SK, Coleman SL, O'Donovan MC. Low gene expression conferred by association of an allele of the 5-HT2C receptor gene with antipsychotic-induced weight gain. Am J Psychiatry 2005;162:613—5.

[144] Hill MJ, Reynolds GP. 5-HT2C receptor gene polymorphisms associated with antipsychotic drug action alter promoter activity. Brain Res 2007;1149:14—7.

[145] Gregoor JG, van der Weide J, Mulder H, Cohen D, van Megen HJ, et al. Polymorphisms of the LEP- and LEPR gene and obesity in patients using antipsychotic medication. J Clin Psychopharmacol 2009;29:21—5.

[146] Zhang XY, Tan YL, Zhou DF, Haile CN, Cao LY, Xu Q, et al. Association of clozapine-induced weight gain with a polymorphism in the leptin promoter region in patients with chronic schizophrenia in a Chinese population. J Clin Psychopharmacol 2007;27:246—51.

[147] Lopez-Rodriguez R, Roman M, Novalbos J, Pelegrina ML, Ochoa D, Abad-Santos F. DRD2 Taq1A polymorphism modulates prolactin secretion induced by atypical antipsychotics in healthy volunteers. J Clin Psychopharmacol 2011;31:555—62.

[148] Serretti A, Artioli P. Predicting response to lithium in mood disorders: role of genetic polymorphisms. Am J Pharmacogenomics 2003;3:17—30.

[149] Harwood AJ. Lithium and bipolar mood disorder: the inositol-depletion hypothesis revisited. Mol Psychiatry 2005;10:117—26.

[150] Steen VM, Lovlie R, Osher Y, Belmaker RH, Berle JO, Gulbrandsen AK. The polymorphic inositol polyphosphate 1-phosphatase gene as a candidate for pharmacogenetic prediction of lithium-responsive manic-depressive illness. Pharmacogenetics 1998;8:259—68.

[151] Bremer T, Diamond C, McKinney R, Shehktman T, Barrett TB, Herold C, et al. The pharmacogenetics of lithium response depends upon clinical co-morbidity. Mol Diagn Ther 2007;11:161—70.

[152] Chung WH, Hung SI, Hong HS, Hsih MS, Yang LC, Ho HC, et al. Medical genetics: a marker for Stevens-Johnson syndrome. Nature 2004;428:486.

[153] Hung SI, Chung WH, Jee SH, Chen WC, Chang YT, Lee WR, et al. Genetic susceptibility to carbamazepine-induced cutaneous adverse drug reactions. Pharmacogenet Genomics 2006;16:297—306.

[154] Zhang Y, Wang J, Zhao LM, Peng W, Shen GQ, Xue L, et al. Strong association between HLA-B*1502 and carbamazepine-induced Stevens-Johnson syndrome and toxic epidermal necrolysis in mainland Han Chinese patients. Eur J Clin Pharmacol 2011;67:885—7.

[155] Kashiwagi M, Aihara M, Takahashi Y, Yamazaki E, Yamane Y, Song Y, et al. Human leukocyte antigen genotypes in carbamazepine-induced severe cutaneous adverse drug response in Japanese patients. J Dermatol 2008;35:683−5.

[156] Man CB, Kwan P, Baum L, Yu E, Lau KM, Cheng AS, et al. Association between HLA-B*1502 allele and antiepileptic drug-induced cutaneous reactions in Han Chinese. Epilepsia 2007;48:1015−8.

[157] Mehta TY, Prajapati LM, Mittal B, Joshi CG, Sheth JJ, Patel DB, et al. Association of HLA-B*1502 allele and carbamazepine-induced Stevens-Johnson syndrome among Indians. Indian J Dermatol Venereol Leprol 2009;75:579−82.

[158] Romphruk A, Phongaen K, Chotechai J, Puapairoj C, Leelayuwat C, Romphruk AV. HLA-B*15 subtypes in the population of north-eastern Thailand. Eur J Immunogenet 2003;30:153−8.

[159] Tassaneeyakul W, Tiamkao S, Jantararoungtong T, Chen P, Lin SY, Chen WH, et al. Association between HLA-B*1502 and carbamazepine-induced severe cutaneous adverse drug reactions in a Thai population. Epilepsia 2010;51:926−30.

[160] Locharernkul C, Loplumlert J, Limotai C, Korkij W, Desudchit T, Tongkobpetch S, et al. Carbamazepine and phenytoin induced Stevens-Johnson syndrome is associated with HLA-B*1502 allele in Thai population. Epilepsia 2008;49:2087−91.

[161] Kaniwa N, Saito Y, Aihara M, Matsunaga K, Tohkin M, Kurose K, et al. HLA-B*1511 is a risk factor for carbamazepine-induced Stevens-Johnson syndrome and toxic epidermal necrolysis in Japanese patients. Epilepsia 2010;51:2461−5.

[162] Kim YH, Kang HC, Kim DS, Kim SH, Shim KW, Kim HD, et al. Neuroimaging in identifying focal cortical dysplasia and prognostic factors in pediatric and adolescent epilepsy surgery. Epilepsia 2011;52: 722−7.

[163] Guzey C, Spigset O. Low serum concentrations of paroxetine in CYP2D6 ultrarapid metabolizers. J Clin Psychopharmacol 2006;26:211−2.

[164] Meyer JH, Wilson AA, Sagrati S, Hussey D, Carella A, Potter WZ, et al. Serotonin transporter occupancy of five selective serotonin reuptake inhibitors at different doses: an [11C]DASB positron emission tomography study. Am J Psychiatry 2004;161:826−35.

[165] Bymaster FP, Dreshfield-Ahmad LJ, Threlkeld PG, Shaw JL, Thompson L, Nelson DL, et al. Comparative affinity of duloxetine and venlafaxine for serotonin and norepinephrine transporters in vitro and in vivo, human serotonin receptor subtypes, and other neuronal receptors. Neuropsychopharmacology 2001;25:871−80.

[166] Owens MJ, Knight DL, Nemeroff CB. Second-generation SSRIs: human monoamine transporter binding profile of escitalopram and R-fluoxetine. Biol Psychiatry 2001;50:345−50.

[167] Takano A, Suzuki K, Kosaka J, Ota M, Nozaki S, Ikoma Y, et al. A dose-finding study of duloxetine based on serotonin transporter occupancy. Psychopharmacology 2006;185:395−9.

[168] Tatsumi M, Groshan K, Blakely RD, Richelson E. Pharmacological profile of antidepressants and related compounds at human monoamine transporters. Eur J Pharmacol 1997;340:249−58.

[169] Meyer JH, Wilson AA, Ginovart N, Goulding V, Hussey D, Hood K, et al. Occupancy of serotonin transporters by paroxetine and citalopram during treatment of depression: a [(11)C]DASB PET imaging study. Am J Psychiatry 2001;158:1843−9.

[170] Garriock HA, Kraft JB, Shyn SI, Peters EJ, Yokoyama JS, Jenkins GD, et al. A genomewide association study of citalopram response in major depressive disorder. Biol Psychiatry 2010;67:133−8.

[171] Ising M, Lucae S, Binder EB, Bettecken T, Uhr M, Ripke S, et al. A genomewide association study points to multiple loci that predict antidepressant drug treatment outcome in depression. Arch Gen Psychiatry 2009;66:966−75.

[172] Uher R, Perroud N, Ng MY, Hauser J, Henigsberg N, Maier W, et al. Genome-wide pharmacogenetics of antidepressant response in the GENDEP project. Am J Psychiatry 2010;167:555−64.

[173] Hughes ZA, Starr KR, Langmead CJ, Hill M, Bartoszyk GD, Hagan JJ, et al. Neurochemical evaluation of the novel 5-HT1A receptor partial agonist/serotonin reuptake inhibitor, vilazodone. Eur J Pharmacol 2005;510:49−57.

[174] Athanasiou M, Reed CR, Rickels K. Vilazodone, a novel, dual-acting antidepressant: current status, future promise and potential for individualized treatment of depression. Per Med 2009;6:217−24.

6A

A Look to the Future: Epigenetics in Psychiatric Disorders and Treatment

John A. Bostrom, Monsheel Sodhi

Department of Pharmacy Practice and Center for Pharmaceutical Biotechnology,
College of Pharmacy, University of Illinois at Chicago, Chicago, Illinois, USA

OBJECTIVES

1. Describe epigenetic processes that may contribute to the development of psychiatric illness and how environmental stressors can influence these epigenetic processes.

2. Provide examples of drugs that modulate epigenetic processes thought to be important in the pathophysiology of schizophrenia.

3. Describe challenges in the development of pharmacoepigenetic biomarkers for psychotropic treatment response.

NATURE AND NURTURE BOTH CONTRIBUTE TO THE ETIOLOGIES OF PSYCHIATRIC DISORDERS

The "nature versus nurture" debate for the etiology of psychiatric disorders is as old as the field of psychiatric genetics. The contribution of "nature" to these disorders is supported by a large body of genetic epidemiological evidence, including family, twin, and adoption studies, which indicate that psychiatric disorders have a large heritable component [1]. Path analyses predict that schizophrenia and other psychiatric disorders are probably caused by a number of interacting genes in combination with nongenetic factors. However, these nongenetic factors, which would be considered to be aberrant "nurture," have been more difficult to characterize. Even though epidemiological research revealed no single environmental risk factor as a strong predictor of psychiatric illness [2,3], studies have shown that accumulating stressful life events in combination with genetic risk factors can increase vulnerability to poor outcomes in psychiatric patients [4].

EPIGENETIC PROCESSES ARE LIKELY TO CONTRIBUTE TO THE BIOCHEMICAL BASIS OF "NURTURE"

Epigenetic processes, which regulate gene expression through the modification of chromatin, appear to be altered by environmental stress with lifelong consequences. These modifications can take place through direct methylation of DNA or through the methylation and/or acetylation of particular lysines on histone proteins. The speed at which these processes occur ranges from the order of minutes to hours [5–7] and may have lifelong implications [8]. Maternal behavior during development initiates lasting epigenetic modifications that have a powerful influence on stress and mood of offspring later in life. Meaney and colleagues demonstrated that high levels of maternal care during the first week of life produced lower glucocorticoid levels and lower hypothalamic pituitary adrenal (HPA) axis responses to stress in adult rats. By contrast, rats experiencing relatively low levels of maternal care early in life displayed high levels of response to stress as adults. The chemical basis of these lifelong behavioral changes is thought to be mediated through the methylation of the glucocorticoid receptor promoter. In adult rats, the long-term effects of "good" or "poor" maternal care could be reversed by infusion of L-methionine, whose metabolite (S-adenosyl methionine [SAM]) serves as a methyl group donor for DNA methylation [9] (Figure 6A.1). These observations indicate that epigenetic processes underlie an important biochemical basis for the "nurture" component of the etiology of psychiatric disorders [10–12]. These findings also correlate with data from human studies of maternal care, as described later in this chapter.

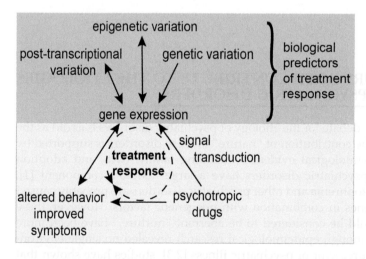

FIGURE 6A.1 Epigenetic processes are likely to alter behavior and treatment response in psychiatric disorders.

EPIGENETIC DYSFUNCTION HAS BEEN LINKED TO PSYCHIATRIC AND NEUROLOGICAL DISORDERS

A large body of data indicates that epigenetic processes are disrupted in disorders of the central nervous system. Dysfunction in epigenetic processing due to deleterious chromosomal or genetic variation is exemplified by Rett syndrome [13], which is a neurodevelopmental disorder characterized by symptoms including severe motor and cognitive impairment in addition to characteristics of autism. Most cases are accompanied by a mutation in the gene encoding the methyl CpG binding protein 2 (MeCP2). MeCP2 is predicted to bind to methylated CpG DNA sequences in the promoter regions of specific genes, leading to transcriptional silencing at those loci [14–18]. Therefore, Rett syndrome is an example of a disease caused by a genetic variation that disrupts epigenetic processes, leading to severe neurological dysfunction. In addition, a comprehensive screen of the DNA methylation covering more than 8,000 genes revealed subtle modifications of methylation patterns in schizophrenia and bipolar disorder [19], suggesting that entire physiological pathways may be impaired by altered epigenetic processes. Moreover, investigations in rodents have revealed the importance of DNA methylation in the induction of synaptic plasticity, learning, and the formation of associative memories [20–23], which are complex, essential brain functions shown to be disrupted in major psychiatric disorders.

In addition to genetic factors, environmental stressors can also influence epigenetic processes and thereby contribute to the pathophysiological mechanisms leading to psychiatric disorders. Adverse life events, including childhood abuse, have been shown to increase vulnerability to psychiatric disorders, particularly post-traumatic stress disorders [24,25], depression, and suicide [4]. Furthermore, schizophrenia-like symptoms can be precipitated after childhood abuse and neglect [26]. Removal of or reductions in environmental stressors seem to contribute to good prognosis and treatment outcomes in psychiatric patients, which have been shown in several studies demonstrating improved outcomes in children with psychiatric disorders when their mothers are treated for depression [27–30]. The impact of environmental stress on epigenetic processing of genes is likely to contribute to these adverse behavioral outcomes. Animal and human studies have shown that early-life stress and poor parenting cause altered methylation of key genes in the brain, which can have lifelong psychiatric consequences [31,32].

Animal models have been used to demonstrate the effects of environmental stress on epigenetic processes, leading to behavioral changes that correlate to symptoms of mental illness. The effects on glucocorticoid activity and brain-derived neurotrophic factor (BDNF) function of maltreatment early in life has been studied in both human subjects and animal models because both have been strongly associated with stress response and psychiatric illness. Childhood abuse has been associated with increased methylation of a neuron-specific glucocorticoid receptor gene [32]. Similar findings have been reported in rats [33,34]. Moreover, exposure of rat pups to negligent and abusive treatment by a "caregiver" produces a marked and prolonged increase in the level of methylation of the gene encoding BDNF, followed by decreased BDNF expression [35]. These methylation changes were reversed by the administration of zebularine [35], a DNA methyltransferase (DNMT) inhibitor [36], and this reversal of DNA methylation led to improvements in behavior. There are four mammalian DNMT enzymes divided into two families of functionally distinct proteins. *De novo*, initial CpG methylation patterns are established by the DNMT3 family, while DNMT1 proteins maintain this pattern during chromosome replication and repair [37]. A study of suicide victims showed increased DNMT-3B expression in the postmortem frontopolar cortex [38], and another study found hypermethylation of the promoter of a truncated gene encoding a BDNF receptor, tropomyosin-related kinase (TrkB), accompanied by decreased TrkB expression [39]. Therefore altered DNMT activity could be related to the reduced levels of BDNF expression reported in the dorsolateral prefrontal cortex of patients with schizophrenia and other psychiatric disorders [40].

Besides BDNF and glucocorticoid receptors, studies indicate dysfunction of many additional genes due to altered DNA methylation in psychiatric disorders. The major inhibitory neurotransmitter system in the brain, activated by gamma-aminobutyric acid (GABA), has also been shown to be altered in psychiatric disorders [41–44]. DNMT1 is coexpressed with reelin and glutamic acid decarboxylase-67 (GAD67) in cortical GABAergic neurons [45–47]. Several postmortem studies have shown that the expression levels of genes encoding reelin and GAD67 are down-regulated in brain regions shown to be critically impaired in schizophrenia [48–50] and may be an important pathophysiological mechanism of the disorder. Reelin is a glycoprotein synthesized and secreted primarily by GABAergic interneurons in the frontal cortex [51] and hippocampus [52] that is thought to play a role in the plasticity of dendritic spines and proper neural positioning during brain development [52,53]. GAD67 is responsible for synthesizing GABA from glutamate [54]. Studies of patients with schizophrenia and bipolar disorder reveal increased cortical DNMT1 activity, which is thought to directly contribute to the reduced expression of reelin and GAD67 after the hypermethylation of their gene promoters. This could be a critical pathophysiological mechanism in the development of psychosis [45,48,55–59].

EPIGENETIC PROCESSES MAY BE CRITICAL TO TREATMENT RESPONSE IN PSYCHIATRIC DISORDERS

Drugs that alter behavior have been shown to also alter epigenetic processing. These include the drugs that are commonly abused by psychiatric patients, in addition to drugs that are used to treat patients, including antipsychotic drugs and antidepressants. Conversely, drugs that are known to alter epigenetic processes seem to alter behavior and could be useful therapeutically. Examples of these findings are discussed later in this chapter.

Modulation of Epigenetic Processes May Contribute to the Behavioral Effects of Drugs Abused by Psychiatric Patients

The frequency of cigarette smoking is very high in patients with schizophrenia compared with the general public [60–66], and nicotine ingested during cigarette smoking induces epigenetic modifications that may be relevant to the pathophysiology of schizophrenia. Nicotine appears to reduce the expression of DNMT1 mRNA, which is related to reduced cortical methylation of the promoter of the gene encoding GAD67 [67], leading to increased cortical GAD67 expression. In contrast, GAD67 is down-regulated in cortical GABAergic neurons in schizophrenia. Therefore, the increased cigarette consumption observed in patients with schizophrenia could be a method by which these patients "correct" this GAD67 imbalance. Other substances that are commonly abused by psychiatric patients include alcohol (ethanol) and opiate drugs such as diamorphine (heroin). Ethanol and its metabolites (such as acetaldehyde and acetate) have been shown to specifically modify various histones via acetylation, such as the histone residue H3K9, or methylation in the case of H3K4. Ethanol has also been shown to increase histone acetyl transferase activity [68]. Heroin appears to play a role in the aberrant methylation of the promoter region of the mu-opioid receptor gene (OPRM1), leading to altered OPRM1 expression [69].

Modulation of Epigenetic Processes May Contribute to the Therapeutic Effects of Currently Prescribed Psychotropic Drugs

In addition to drugs abused by psychiatric patients, drugs administered for therapeutic purposes can also alter epigenetic processing. For example, long-term administration of antipsychotics in schizophrenia patients has been strongly correlated with methylation of a CpG island upstream of emitogen-activated protein kinase I gene (MEK1) in postmortem prefrontal cortex. This is interesting considering the critical role of mitogen-activated protein kinase (MAPK) in signaling pathways in intraneuronal signaling and the finding that clozapine, an atypical antipsychotic drug, activates MAPK pathways by modulating MEK1 [19]. Valproic acid, which is known to be an inhibitor of histone deacetylases (HDACs) and has many other pharmacological effects, is prescribed alone or with other medications to treat certain types of seizures and mania in bipolar disorder and in the prevention of migraine headaches. It is worth noting that valproic acid has many other pharmacological sites of action besides its inhibition of HDACs and that its efficacy in the treatment of CNS disorders may not be through epigenetic mechanisms, although accumulating data indicate that some psychotropic drugs may exert their efficacy by modulating epigenetic mechanisms [70,71], as described in the following sections.

Drugs That Alter Epigenetic Processing May Also Have Psychotropic Effects

Epigenetic processes may be a therapeutic target in neurological and psychiatric disorders (Figure 6A.2). It has been demonstrated that drug-induced increases in DNA methylation can alter behavior. In animal models, infusing methionine into the brain raises the level of SAM, a cofactor involved in DNA methylation at CpG islands, which in turn increases the methylation of the promoter regions of the GAD67 and reelin genes. This process results in reduced GAD67 and reelin gene expression [72], which is consistent with findings of reduced cortical expression of these genes in postmortem studies of schizophrenia [48,50,56]. Furthermore, methionine administration has been shown to reduce the expression of GAD67 and reelin in rodents, with resulting alterations in prepulse inhibition and startle response. These behaviors correlate with symptoms of schizophrenia and form the basis of some animal models of the disorder [59,72].

SAM has been the focus of clinical trials, with mixed results. Studies have indicated that SAM, through either oral or parenteral administration, may act as a relatively safe and fast-acting antidepressant [73–75]. In a trial including schizophrenia patients, administration

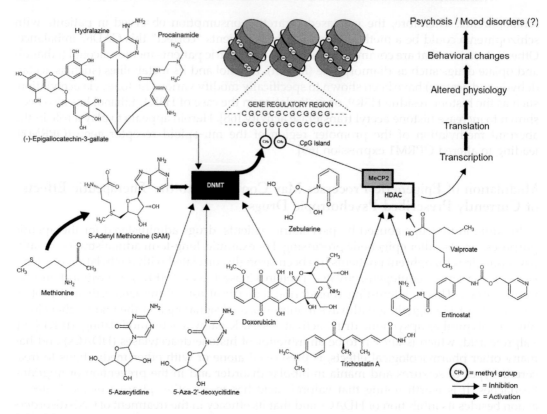

FIGURE 6A.2   Drugs directly alter epigenetic processing. The term "epigenetics" is used to describe stable but reversible chemical alterations to the gene that occur without changing the nucleotide composition of the gene sequence [25,53,82–84]. According to recent studies, these processes regulate learning and memory [20,147], chronic drug abuse [148–150], and perhaps mental illness [151–153]. Epigenetic modifications include DNA methylation and histone acetylation, which have the potential to inhibit or enhance transcription [14,154–156], thus regulating the development and differentiation of cells [157–161]. DNA methylation involves the covalent attachment of a methyl group to the 5-carbon position of cytosine in DNA [162–164] and is catalyzed by DNA methyl transferase enzymes (DNMTs) [165–168]. Increased susceptibility to DNA methylation occurs in regions that have high concentrations of cytosine residues connected to guanosine by phosphodiester bonds, which are known as CpG islands [169], indicated in the figure. These CpG islands are often found in gene promoters and are usually 0.5-3 kb, with greater than 55% GC content [170]. Increased methylation levels in these regions usually result in the suppression of gene transcription [170–172]. Once methylated, the cytosines recruit methyl-DNA binding proteins such as methyl CpG-binding protein 2 (MeCP2) [173] that attract histone deacetylase enzymes (HDACs) [14,15,165,174]. HDACs are generally associated with transcriptional repression through the removal of negatively charged acetyl groups from histone proteins, which inhibits their association with transcriptional proteins [127]. Removal of DNA methylation could be achieved by reducing histone deacetylase activity. Although it is very common for DNA methylation to be associated with the inhibition of transcription, DNA methylation can also be associated with transcriptional activation; for instance, MeCP2 has been shown to activate transcription in the hypothalamus (specifically the major transcriptional activator CREB1) [175,176], and site-specific methylation of histone H4 at arginine 3 has been shown to activate transcription [177]. Doxorubicin [178], 5-azacytidine, and 5-aza-2'-deoxycitidine have been shown to be effective DNMT inhibitors [179], and trichostatin A, valproate, and entinostat act as HDAC inhibitors [85]. The therapeutic effects of the three non-nucleoside DNMT inhibitors shown in the figure—(-)-Epigallocatechin-3-gallate, hydralazine, and procainamide [180–184]—are questionable.

of SAM was shown to reduce aggressive behavior and improve quality of life [76]. In contrast, several trials of SAM have associated the drug with significantly increased irritability and also mania [76–79].

Trichostatin A (TSA) has not been tested for efficacy as a psychotropic drug, although clinical trials of valproate indicated that the drug effectively reduces mania and acute mood episodes in bipolar disorder [80], and valproic acid has in fact been used safely for years as treatment for bipolar disorder, epilepsy, and mania without debilitating side effects [81,82]. However, given the lack of pharmacological specificity of sodium valproate, it is unclear whether this efficacy in mania and depression is due to its interaction with HDACs.

These epigenetic modifications can be reversed by the coadministration of other drugs. The methionine-induced changes in gene expression and behaviors are reversed after coadministration of methionine with valproic acid. Animal studies show that this drug combination prevents hypermethylation of critical gene promoters and histones through the inhibition of neuronal HDACs, which results in the increased expression of reelin and the GAD67 gene in mice administered methionine with valproic acid relative to mice that were treated with methionine alone [59,72]. The behavioral effects of methionine administration have also been inhibited by imidazenil, which is an allosteric modulator of the GABAα-5 subunit of the GABA$_A$ receptor [59]. These data indicate that epigenetic processing within the GABAergic system may be an important target for psychotropic drug development.

Reducing DNA methylation may be therapeutic and achieved indirectly through the inhibition of HDAC enzymes. HDAC inhibitors with greater specificity than valproic acid, such as entinostat and TSA (illustrated in Figure 6A.2), have been shown to reduce DNA methylation, thereby activating the expression of reelin and GAD67 [83–86]. As described earlier, the cortical expression of genes encoding reelin and GAD67 is reduced in schizophrenia [48–50], suggesting that restoration of reelin and GAD67 levels may be beneficial in modifying the disease pathophysiology. It is worth noting that valproic acid is currently almost exclusively used adjunctively with other antipsychotics to treat schizophrenia and mood disorders [8].

Studies have also been conducted on drugs with greater specificity for HDAC inhibition than valproic acid. In mice with "negligent" mothers, depressive symptoms in adulthood were alleviated by infusion of the HDAC inhibitor TSA, possibly due to changes in DNA methylation, histone acetylation, and adrenocortical stress response [31,87]. Extensive clinical trials of drugs influencing epigenetic processes for the treatment of psychiatric disorders in humans have yet to be conducted. The activities of HDAC inhibitors are likely to be complex, since they appear to cause severe side effects due to cell cycle arrest, apoptosis, altered angiogenesis, and immune modulation [88]. Trials of HDAC inhibitors for the treatment of cancers have revealed side effects including progressive fatigue and nausea [89], gastrointestinal disturbances, elevated liver function, and moderate tinnitus [90]. A trial of valproate in patients with acute myeloid leukemia revealed side effects including fatigue, petechiae, vasculitis, conjunctivitis, dyspnea, granulocytopenia, dizziness, confusion, and even sepsis associated with the lack of granulocytes [90]. Coadministration of valproate with antipsychotic drugs for the treatment of schizophrenia produce less severe side effects, such as moderate weight gain and gastrointestinal symptoms [91,92]. The occurrence of these side effects should be considered in the development of HDAC inhibitors as medication for psychiatric disorders.

HDACs are grouped into four classes according to structural similarities. Class I enzymes are most closely related to the yeast transcriptional regulator RPD3; class II enzymes share similarities with the yeast deacetylase enzyme HDA1. Class I and II are expressed in the brain, although Class I is thought to be ubiquitously expressed. Class III HDACs are sirtuin deacetylases (a class of nicotinamide adenine dinucleotide-dependent deacetylases with multiple protein substrates), and Class IV contains only HDAC11 [93]. Benzamide-based HDAC inhibitors are unique because they inhibit Class I, II, and III HDACs, while other HDAC inhibitors are more specific [83]. The benzamide compound MS-275 has been shown to be brain selective and has high potency for histone acetylation. MS-275 was shown to be approximately 100 times more potent in HDAC inhibition than valproic acid in the frontal cortex and therefore may have potential as a psychotropic compound [94]. Another small molecule antagonist that works to activate transcription, BIX-01294, has been shown to inhibit histone methyltransferases (HMTs) G9a and GLP [95,96], and although administration of the drug only shows a modest up-regulation of gene expression, it may have future therapeutic relevance.

EPIGENETIC VARIATION OF GENES REGULATING PHARMACOKINETIC PROCESSES IS LIKELY TO INFLUENCE TREATMENT RESPONSE

Biomarkers predicting drug response have yet to be validated in psychiatric disorders, most likely because the mechanisms for drug efficacy in these disorders are not clear and therefore pharmacodynamic targets are less certain. However, many psychotropic drugs are metabolized or otherwise influenced by pharmacokinetic processes, including metabolism by cytochrome P450 enzymes. Therefore, it is likely that epigenetic processing of proteins within pharmacokinetic pathways could provide pharmacoepigenetic biomarkers for psychotropic drug response, as described in the following section.

Cytochrome P450

The majority of drugs used therapeutically in psychiatry are metabolized by the cytochrome 450 (CYP) enzyme complex. This complex is a superfamily of proteins encoded by genes containing numerous genetic variants, some of which have been found to influence the activity of the enzyme. Typically, subjects can be classified as ultra-rapid, extensive, intermediate, and poor metabolizers of drugs targeted by a specific enzyme. Table 6A.1 summarizes the influence of these enzymes on the metabolism of psychotropic drugs [97].

Several members of the CYP family have been shown to be susceptible to DNA methylation [98]. The genes *CYP1A1*, *CYP1A2*, *CYP1B1*, *CYP2W1*, and *CYP2E1* have been shown to be methylated. A high density of CG sequences in several other CYP enzymes (e.g., *CYP2D6* and *CYP2C19*) suggests that they are also potential targets for DNA methylation [99]. The methylation status of *CYP1A1* is modified by smoking [100], and

TABLE 6A.1 Cytochrome P450 enzymes have specific metabolic activities that are modified by drugs and genetic and epigenetic factors. Drugs metabolized by each enzyme subtype are listed in addition to drugs that inhibit and induce the enzymes. Known genetic polymorphisms that alter enzyme activity are listed

Cytochrome P450 subtype:	1A2	2B6	2C8	2C9	2C19	2D6	2E1	3A4, 5, 7
Substrates	**Antidepressants:**	**Antidepressants:**		**Antiepileptics:**	**Antiepileptics:**	**Antidepressants:**	**Anesthetics:**	**Anxiolytics:**
	duloxetine	bupropion		phenytoin	diazepam	amitriptyline	enflurane	alprazolam
	fluvoxamine	**Opiates:**			phenytoin	clomipramine	halothane	diazepam
	imipramine	methadone			phenobarbitone	desipramine	isoflurane	midazolam
	Antipsychotics:				**Antidepressants:**	imipramine	methoxyflurane	triazolam
	clozapine				amitriptyline	nortriptyline	sevoflurane	buspirone
	olanzapine				clomipramine	mirtazapine	**Others:**	**Antidepressants:**
	haloperidol					paroxetine	ethanol	trazodone
	riluzole					duloxetine		**Antipsychotics:**
	tacrine					venlafaxine		aripiprazole
	tizanidine					citalopram		haloperidol
	zolmitriptan					escitalopram		pimozide
	cyclobenzaprine					**Antipsychotics:**		**Opiates:**
						haloperidol		methadone
						risperidone		
						thioridazine		
						flupentixol		
						aripiprazole		
						Opiates:		
						codeine		
						tramadol		

(Continued)

TABLE 6A.1 Cytochrome P450 enzymes have specific metabolic activities that are modified by drugs and genetic and epigenetic factors. Drugs metabolized by each enzyme subtype are listed in addition to drugs that inhibit and induce the enzymes. Known genetic polymorphisms that alter enzyme activity are listed—cont'd

Cytochrome P450 subtype:	1A2	2B6	2C8	2C9	2C19	2D6	2E1	3A4, 5, 7
Inhibitors	**Antidepressants:** fluvoxamine				**Antidepressants:** fluoxetine fluvoxamine	**Antidepressants:** fluoxetine paroxetine duloxetine bupropion clomipramine doxepin **Antipsychotics:** haloperidol **Opiates:** methadone	disulfiram	nefazodone grapefruit juice fluvoxamine
Inducers	tobacco	**Antiepileptics:** phenobarbital phenytoin		**Antiepileptics:** secobarbital			ethanol	**Antiepileptics:** carbamazepine phenobarbital phenytoin **Antidepressants:** St. John's wort

Genetics

*2 = PM	*2 = PM (99% in Asians)	*2 = EM (similar to WT)
*3 = PM	*3 = PM (87% in Whites)	*3 = PM (truncates protein)
		*4 = PM (nonfunctional)
		*5 = PM (whole gene del.)
		*6 = PM (truncates protein)
		*9 = IM/EM (reduces func.)
		*10 = IM (reduces func.)
		*17 = IM (reduces func.)
		*29 = IM (reduces func.)
		*41 = IM (reduces func.)

PM = Poor metabolizer, IM = intermediate metabolizer, EM = extensive metabolizer, UM = ultrarapid metabolizer. *2 = allele 2 of gene, *3 = allele 3 of gene, etc.
Shaded columns indicate which cytochrome subtypes are encoded by genes that are epigenetically modified.
http://medicine.iupui.edu/clinpharm/ddis/clinicalTable.aspx

CYP1A2 has shown variability in its methylation levels [101,102]. *CYP2E1* methylation status fluctuates during development [103]. *CYP1B1* and *CYP2W1* have shown altered methylation levels in patients with cancer [101,104−106]. Furthermore, treatment of cancer cells with HDAC inhibitors 5-aza-2′-deoxycitidine and Trichostatin (TSA) has been associated with altered expression of *CYP3A4*, *CYP3A5*, and *CYP3A7* [107]. No biomarkers have been identified within epigenetically modified regions of *CYP2D6* and *CYP2C19*, the two most important genes regulating psychotropic drug metabolism. However, both genes have large CG dinucleotide-rich regions, suggesting a possible regulatory role for DNA methylation [99]. Therefore, potential epigenetic modifications of the cytochrome P450 enzymes could influence the metabolism of many prescribed drugs, including the psychotropic medications indicated in Table 6A.1.

EPIGENETIC MODIFICATIONS OF DRUG TRANSPORTERS

The multidrug resistance protein 1 gene, also known as the ATP-binding cassette subfamily B member 1 gene or *ABCB1*, encodes the transporter protein p-glycoprotein. P-glycoprotein acts to transport xenobiotic compounds out of cells, leading to multidrug resistance. Chemotherapeutic drugs induce histone modifications at the *ABCB1* locus, causing up-regulation of *ABCB1* [108]. Genetic polymorphisms in *ABCB1* have been shown to be associated with altered activity of p-glycoprotein, [109] and studies indicate that p-glycoprotein activity in the blood−brain barrier greatly limits the brain uptake of several atypical antipsychotic drugs [110,111]. Moreover, an inhibitory effect on p-glycoprotein-mediated activity occurs after administration of the antipsychotic drugs risperidone, olanzapine, clozapine, and flupentixol [112,113]. This interaction of p-glycoprotein with several antipsychotic agents and its susceptibility to epigenetic modification [114−116] make *ABCB1* a key target for future pharmacoepigenetic investigations. In fact, a polymorphism within the *ABCB1* gene is significantly associated with treatment response to olanzapine in female patients with schizophrenia [117]. Genetic variants in *ABCB1* have also been associated with differences in the clinical efficacy of antidepressants that are substrates for the *ABCB1*-transporter [118]. Although it is unknown whether these polymorphisms have been linked with epigenetic modification, it provides an example of *ABCB1* alterations playing a role in clinical treatment response and warrants further study of the gene's epigenetic status as it relates to drug transport.

EPIGENETICS AND ADVERSE DRUG REACTIONS

A new focus for research into adverse drug reactions (ADRs) has been the epigenetic modulation of enzymes involved in pharmacokinetic processes. Genetic and epigenetic modifications of enzymes and proteins involved in the absorption, distribution, metabolism, and excretion (ADME) of drugs play an important role in drug response [119]. Epigenetic modifications found in the *ABCB1* gene have been associated with increased tolerance to particular anticancer drugs. The *ABCB1* gene product, the p-glycoprotein transporter protein, has been shown to interact with several psychotropic drugs, indicating that changes involving methylation or histone modification of this gene may also have clinically relevant impacts on a patient's ability to process and transport these drugs. In addition, although the focus has been in the field of cancer research, epigenetic modifications within the cytochrome P450 family (CYP1A1, CYP1A2, CYP26C1, CYP26E1, etc.) are also likely to influence response to psychotropic drugs. Several studies have shown that patients with polymorphisms of *CYP2D6* showed an increased risk of adverse drug reactions upon administration of antidepressants [120]. An increased rate of adverse effects upon administration of antipsychotics such as risperidone, phenothiazine, and haloperidol has also been seen in patients with low *CYP2D6* metabolic activity [121,122]. A meta-analysis by Fleeman et al. also found evidence that *CYP2D6* genotype was associated with extrapyramidal side effects [97]. Therefore, the influence of epigenetic modifications on ADRs may provide predictive biomarkers of treatment compliance and efficacy.

LIMITATIONS OF EPIGENETIC BIOMARKERS IN PSYCHIATRY

Discovery of epigenetic markers for psychiatric diagnosis or the prediction of treatment response in psychiatric patients is in its early stages. We briefly summarize the challenges for successful biomarker development in this section.

Which Tissue Is Most Appropriate for Biomarker Development?

Brain tissue cannot be routinely sampled in a clinic from live psychiatric patients; therefore, the need arises to validate pharmacogenetic or pharmacoepigenetic markers in peripheral tissues, which can be removed by minimally invasive techniques. Biomarker discovery in psychiatric patients has therefore been limited to buccal epithelial cells, lymphocytes, platelets, fibroblasts, nasal epithelial cells, and cells from cerebrospinal fluid. One primary obstacle to the development of epigenetic biomarkers is that unlike most genetic variation, epigenetic variation differs by cell type [123]. Pharmacological or pharmacokinetic processes influencing treatment response are likely to occur in the central nervous system, gastrointestinal tract, liver, and kidneys. Therefore, extrapolating epigenetic variation measured in peripheral tissues to predict psychotropic treatment response is difficult to validate. Identifying a representative and accessible cell type for biomarker development is a challenge [124].

Studies have attempted to discover biomarkers in lymphocytes. Lymphocyte nuclear extracts from patients with schizophrenia were shown to have a reduced baseline level of acetylation of a particular histone, H3, and reduced response to HDAC inhibitor (TSA) treatment compared to control subjects [125]. Increases in baseline levels of dimethylation of the ninth lysine residue of histone H3 (H3K9me2) were also detected in schizophrenia. This combination of decreased acetylation and increased dimethylation of H3 would predict that lymphocyte chromatin is more restricted to transcriptional machinery in patients with schizophrenia compared with controls [126]. In another study of plasma lymphocyte nuclear protein extracts, administration of valproic acid for four weeks to both medicated and unmedicated psychotic patients produced a significant increase in acetylated histone H3 [84]. It should be noted that in each of these studies of lymphocytes, HDAC inhibitors (valproic acid and TSA) were less able to reduce the restrictive chromatin state in patients with schizophrenia or bipolar disorder than control subjects [84,125,126]. These observations illustrate that monitoring lymphocyte response to HDAC inhibitor treatment in patients may indirectly measure therapeutic efficacy of the drugs [127].

Variability in Pharmacoepigenetic Biomarkers Due to Environmental Factors

An obstacle facing the clinical use of pharmacoepigenetics is the reliability of epigenetic modifications as stable biomarkers because a variety of environmental factors have been shown to correlate with epigenetic states. Methylation may fluctuate [128] because several environmental and social factors play a key role in the activity of DNMTs and DNA CpG hypermethylation. For example, drugs such as nicotine [67], alcohol [129], and antipsychotics [19] in addition to diet [130,131], viral infections [132], and substance abuse [68,69,133] all contribute to DNA methylation. Marked differences in methylation have been observed between controls and heroin addicts [69]. In addition to the environmental influence of drugs and chemicals, maternal care has also been shown to affect methylation levels and behavioral responses to stress [33], and recent reports have shown that maternal diet may even alter the histones of offspring [134]. In addition, the environmental influence of drugs, food, stress, and subject variability due to age, gender, and ethnicity [69,135–138] have all been shown to correlate with epigenetic states such as methylation. Therefore, pharmacoepigenetic biomarkers are predicted to have similar limitations of ethnic variability to other biomarkers of drug response, such as metabolizer status due to CYP genotype (Table 6A.1). Furthermore, some studies indicate sex-specific differences in gene expression and metabolizer status [69,140]. Although a study of human chromosomes 6, 20, and 22 found no significant differences between age groups or sexes in methylation levels [69,139,140], a separate genome-wide study did find DNA methylation differences between the sexes [137]. The variability of epigenetic data could contribute to the disparate findings between several studies of pharmacogenetics and epigenetics in psychiatric disorders, which would be an obstacle to a reliable prediction of treatment response in the clinic.

CONCLUSION

The pharmacological basis of drug response for many psychiatric disorders is uncertain because psychotropic drugs have multiple binding sites and their efficacy is often not observed until several days or months after starting drug treatment [141–143]. Therefore, progress in identifying pharmacodynamic biomarkers of treatment response has been slow, even though some markers have been replicated, such as serotonin$_{2A}$ receptor polymorphisms predicting response to antidepressant treatment [144–146]. Although supporting data are scarce, it is likely that molecules regulating pharmacokinetic processes will provide the first useful targets for the development of pharmacoepigenetic biomarkers of psychotropic treatment response. Although the limitations should be considered, the quest for "perfect" pharmacogenetic or pharmacoepigenetic biomarkers should not preclude the discovery of those that are merely "good." Future progress towards individualized treatment in psychiatric disorders may lie within the complex, developing field of epigenetics.

QUESTIONS FOR DISCUSSION

1. What epigenetic processes are believed to contribute to the pathophysiology of psychiatric disorders?
2. Which drugs have been shown to modify epigenetic processes implicated in the pathophysiology of schizophrenia?
3. What are challenges with drug therapies that combat potential detrimental epigenetic processes in psychiatric disorders?
4. How might epigenetic variation of genes regulating pharmacokinetic processes influence response to treatment in patients with psychiatric disorders?
5. What limitations exist to applying pharmacoepigenetic biomarkers to predict response to treatment in psychiatric disorders?

References

[1] McGuffin P. Nature and nurture interplay: schizophrenia. Psychiatr Prax 2004;31(Suppl 2):S189—93.
[2] Brown AS. The environment and susceptibility to schizophrenia. Prog Neurobiol 2011;93:23—58.
[3] van Os J, Linscott RJ, Myin-Germeys I, Delespaul P, Krabbendam L. A systematic review and meta-analysis of the psychosis continuum: evidence for a psychosis proneness-persistence-impairment model of psychotic disorder. Psychol Med 2009;39:179—95.
[4] Caspi A, Sugden K, Moffitt TE, Taylor A, Craig IW, Harrington H, et al. Influence of life stress on depression: moderation by a polymorphism in the 5-HTT gene. Science 2003;301:386—9.
[5] Jackson V, Shires A, Chalkley R, Granner DK. Studies on highly metabolically active acetylation and phosphorylation of histones. J Biol Chem 1975;250:4856—63.
[6] Zee BM, Levin RS, Dimaggio PA, Garcia BA. Global turnover of histone post-translational modifications and variants in human cells. Epigenetics Chromatin 2010;3:22.
[7] Hazzalin CA, Mahadevan LC. Dynamic acetylation of all lysine 4-methylated histone H3 in the mouse nucleus: analysis at c-fos and c-jun. PLoS Biol 2005;3:e393.
[8] Sharma RP. Schizophrenia, epigenetics and ligand-activated nuclear receptors: a framework for chromatin therapeutics. Schizophr Res 2005;72:79—90.
[9] Weaver IC, Champagne FA, Brown SE, Dymov S, Sharma S, Meaney MJ, et al. Reversal of maternal programming of stress responses in adult offspring through methyl supplementation: altering epigenetic marking later in life. J Neurosci 2005;25:11045—54.
[10] Weaver IC, Meaney MJ, Szyf M. Maternal care effects on the hippocampal transcriptome and anxiety-mediated behaviors in the offspring that are reversible in adulthood. Proc Natl Acad Sci U S A 2006;103: 3480—5.
[11] McGowan PO, Suderman M, Sasaki A, Huang TC, Hallett M, Meaney MJ, et al. Broad epigenetic signature of maternal care in the brain of adult rats. PLoS One 2011;6:e14739.
[12] Korosi A, Shanabrough M, McClelland S, Liu ZW, Borok E, Gao XB, et al. Early-life experience reduces excitation to stress-responsive hypothalamic neurons and reprograms the expression of corticotropin-releasing hormone. J Neurosci 2010;30:703—13.
[13] Amir RE, Van den Veyver IB, Wan M, Tran CQ, Francke U, Zoghbi HY. Rett syndrome is caused by mutations in X-linked MECP2, encoding methyl-CpG-binding protein 2. Nat Genet 1999;23:185—8.
[14] Jones PL, Veenstra GJ, Wade PA, Vermaak D, Kass SU, Landsberger N, et al. Methylated DNA and MeCP2 recruit histone deacetylase to repress transcription. Nat Genet 1998;19:187—91.
[15] Nan X, Ng HH, Johnson CA, Laherty CD, Turner BM, Eisenman RN, et al. Transcriptional repression by the methyl-CpG-binding protein MeCP2 involves a histone deacetylase complex. Nature 1998;393:386—9.
[16] Weaving LS, Ellaway CJ, Gecz J, Christodoulou J. Rett syndrome: clinical review and genetic update. J Med Genet 2005;42:1—7.
[17] Akbarian S, Jiang Y, Laforet G. The molecular pathology of Rett syndrome: synopsis and update. Neuromolecular Med 2006;8:485—94.

[18] Chahrour M, Zoghbi HY. The story of Rett syndrome: from clinic to neurobiology. Neuron 2007;56:422–37.

[19] Mill J, Tang T, Kaminsky Z, Khare T, Yazdanpanah S, Bouchard L, et al. Epigenomic profiling reveals DNA-methylation changes associated with major psychosis. Am J Hum Genet 2008;82:696–711.

[20] Lubin FD, Roth TL, Sweatt JD. Epigenetic regulation of BDNF gene transcription in the consolidation of fear memory. J Neurosci 2008;28:10576–86.

[21] Levenson JM, Roth TL, Lubin FD, Miller CA, Huang IC, Desai P, et al. Evidence that DNA (cytosine-5) methyltransferase regulates synaptic plasticity in the hippocampus. J Biol Chem 2006;281:15763–73.

[22] Miller CA, Sweatt JD. Covalent modification of DNA regulates memory formation. Neuron 2007;53:857–69.

[23] Monsey MS, Ota KT, Akingbade IF, Hong ES, Schafe GE. Epigenetic alterations are critical for fear memory consolidation and synaptic plasticity in the lateral amygdala. PLoS One 2011;6:e19958.

[24] Spataro J, Mullen PE, Burgess PM, Wells DL, Moss SA. Impact of child sexual abuse on mental health: prospective study in males and females. Br J Psychiatry 2004;184:416–21.

[25] Rowan AB, Foy DW, Rodriguez N, Ryan S. Posttraumatic stress disorder in a clinical sample of adults sexually abused as children. Child Abuse Negl 1994;18:51–61.

[26] Read J, van Os J, Morrison AP, Ross CA. Childhood trauma, psychosis and schizophrenia: a literature review with theoretical and clinical implications. Acta Psychiatr Scand 2005;112:330–50.

[27] Rishel CW, Greeno CG, Marcus SC, Sales E, Shear MK, Swartz HA, et al. Impact of maternal mental health status on child mental health treatment outcome. Community Ment Health J 2006;42:1–12.

[28] Pilowsky DJ, Wickramaratne P, Talati A, Tang M, Hughes CW, Garber J, et al. Children of depressed mothers 1 year after the initiation of maternal treatment: findings from the STAR*D-Child Study. Am J Psychiatry 2008;165:1136–47.

[29] Verdeli H, Ferro T, Wickramaratne P, Greenwald S, Blanco C, Weissman MM. Treatment of depressed mothers of depressed children: pilot study of feasibility. Depress Anxiety 2004;19:51–8.

[30] Gunlicks ML, Weissman MM. Change in child psychopathology with improvement in parental depression: a systematic review. J Am Acad Child Adolesc Psychiatry 2008;47:379–89.

[31] Weaver IC, Cervoni N, Champagne FA, D'Alessio AC, Sharma S, Seckl JR, et al. Epigenetic programming by maternal behavior. Nat Neurosci 2004;7:847–54.

[32] McGowan PO, Sasaki A, D'Alessio AC, Dymov S, Labonte B, Szyf M, et al. Epigenetic regulation of the glucocorticoid receptor in human brain associates with childhood abuse. Nat Neurosci 2009;12:342–8.

[33] Meaney MJ, Szyf M. Environmental programming of stress responses through DNA methylation: life at the interface between a dynamic environment and a fixed genome. Dialogues Clin Neurosci 2005;7:103–23.

[34] Weaver IC. Epigenetic programming by maternal behavior and pharmacological intervention. Nature versus nurture: let's call the whole thing off. Epigenetics 2007;2:22–8.

[35] Roth TL, Lubin FD, Funk AJ, Sweatt JD. Lasting epigenetic influence of early-life adversity on the BDNF gene. Biol Psychiatry 2009;65:760–9.

[36] Cheng JC, Matsen CB, Gonzales FA, Ye W, Greer S, Marquez VE, et al. Inhibition of DNA methylation and reactivation of silenced genes by zebularine. J Natl Cancer Inst 2003;95:399–409.

[37] Cheng X, Blumenthal RM. Mammalian DNA methyltransferases: a structural perspective. Structure 2008;16:341–50.

[38] Poulter MO, Du L, Weaver IC, Palkovits M, Faludi G, Merali Z, et al. GABAA receptor promoter hypermethylation in suicide brain: implications for the involvement of epigenetic processes. Biol Psychiatry 2008;64:645–52.

[39] Ernst C, Deleva V, Deng X, Sequeira A, Pomarenski A, Klempan T, et al. Alternative splicing, methylation state, and expression profile of tropomyosin-related kinase B in the frontal cortex of suicide completers. Arch Gen Psychiatry 2009;66:22–32.

[40] Weickert CS, Hyde TM, Lipska BK, Herman MM, Weinberger DR, Kleinman JE. Reduced brain-derived neurotrophic factor in prefrontal cortex of patients with schizophrenia. Mol Psychiatry 2003;8:592–610.

[41] Yee BK, Keist R, von Boehmer L, Studer R, Benke D, Hagenbuch N, et al. A schizophrenia-related sensori-motor deficit links alpha 3-containing GABAA receptors to a dopamine hyperfunction. Proc Natl Acad Sci U S A 2005;102:17154–9.

[42] Volk DW, Austin MC, Pierri JN, Sampson AR, Lewis DA. Decreased glutamic acid decarboxylase67 messenger RNA expression in a subset of prefrontal cortical gamma-aminobutyric acid neurons in subjects with schizophrenia. Arch Gen Psychiatry 2000;57:237–45.

[43] Ongur D, Prescot AP, McCarthy J, Cohen BM, Renshaw PF. Elevated gamma-aminobutyric acid levels in chronic schizophrenia. Biol Psychiatry 2010;68:667–70.

[44] Woo TU, Walsh JP, Benes FM. Density of glutamic acid decarboxylase 67 messenger RNA-containing neurons that express the N-methyl-D-aspartate receptor subunit NR2A in the anterior cingulate cortex in schizophrenia and bipolar disorder. Arch Gen Psychiatry 2004;61:649—57.

[45] Veldic M, Caruncho HJ, Liu WS, Davis J, Satta R, Grayson DR, et al. DNA-methyltransferase 1 mRNA is selectively overexpressed in telencephalic GABAergic interneurons of schizophrenia brains. Proc Natl Acad Sci U S A 2004;101:348—53.

[46] Veldic M, Guidotti A, Maloku E, Davis JM, Costa E. In psychosis, cortical interneurons overexpress DNA-methyltransferase 1. Proc Natl Acad Sci U S A 2005;102:2152—7.

[47] Ruzicka WB, Zhubi A, Veldic M, Grayson DR, Costa E, Guidotti A. Selective epigenetic alteration of layer I GABAergic neurons isolated from prefrontal cortex of schizophrenia patients using laser-assisted microdissection. Mol Psychiatry 2007;12:385—97.

[48] Guidotti A, Auta J, Davis JM, Di-Giorgi-Gerevini V, Dwivedi Y, Grayson DR, et al. Decrease in reelin and glutamic acid decarboxylase67 (GAD67) expression in schizophrenia and bipolar disorder: a postmortem brain study. Arch Gen Psychiatry 2000;57:1061—9.

[49] Fatemi SH, Earle JA, McMenomy T. Reduction in reelin immunoreactivity in hippocampus of subjects with schizophrenia, bipolar disorder and major depression. Mol Psychiatry 2000;5:654—63. 571.

[50] Akbarian S, Huang HS. Molecular and cellular mechanisms of altered GAD1/GAD67 expression in schizophrenia and related disorders. Brain Res Rev 2006;52:293—304.

[51] Rodriguez MA, Caruncho HJ, Costa E, Pesold C, Liu WS, Guidotti A. Patas monkey, glutamic acid decarboxylase-67 and reelin mRNA coexpression varies in a manner dependent on layers and cortical areas. J Comp Neurol 2002;451:279—88.

[52] Costa E, Davis J, Grayson DR, Guidotti A, Pappas GD, Pesold C. Dendritic spine hypoplasticity and downregulation of reelin and GABAergic tone in schizophrenia vulnerability. Neurobiol Dis 2001;8:723—42.

[53] Tsankova N, Renthal W, Kumar A, Nestler EJ. Epigenetic regulation in psychiatric disorders. Nat Rev Neurosci 2007;8:355—67.

[54] Kalkman HO, Loetscher E. GAD(67): the link between the GABA-deficit hypothesis and the dopaminergic- and glutamatergic theories of psychosis. J Neural Transm 2003;110:803—12.

[55] Grayson DR, Jia X, Chen Y, Sharma RP, Mitchell CP, Guidotti A, et al. Reelin promoter hypermethylation in schizophrenia. Proc Natl Acad Sci U S A 2005;102:9341—6.

[56] Abdolmaleky HM, Cheng KH, Russo A, Smith CL, Faraone SV, Wilcox M, et al. Hypermethylation of the reelin (RELN) promoter in the brain of schizophrenic patients: a preliminary report. Am J Med Genet B Neuropsychiatr Genet 2005;134B:60—6.

[57] Grayson DR, Chen Y, Costa E, Dong E, Guidotti A, Kundakovic M, et al. The human reelin gene: transcription factors (+), repressors (-) and the methylation switch (+/-) in schizophrenia. Pharmacol Ther 2006;111: 272—86.

[58] Costa E, Chen Y, Dong E, Grayson DR, Kundakovic M, Maloku E, et al. GABAergic promoter hypermethylation as a model to study the neurochemistry of schizophrenia vulnerability. Expert Rev Neurother 2009;9:87—98.

[59] Tremolizzo L, Doueiri MS, Dong E, Grayson DR, Davis J, Pinna G, et al. Valproate corrects the schizophrenia-like epigenetic behavioral modifications induced by methionine in mice. Biol Psychiatry 2005;57:500—9.

[60] Leonard S, Gault J, Adams C, Breese CR, Rollins Y, Adler LE, et al. Nicotinic receptors, smoking and schizophrenia. Restor Neurol Neurosci 1998;12:195—201.

[61] Goff DC, Henderson DC, Amico E. Cigarette smoking in schizophrenia: relationship to psychopathology and medication side effects. Am J Psychiatry 1992;149:1189—94.

[62] Greeman M, McClellan TA. Negative effects of a smoking ban on an inpatient psychiatry service. Hosp Community Psychiatry 1991;42:408—12.

[63] Kirch DG. Where there's smoke ... nicotine and psychiatric disorders. Biol Psychiatry 1991;30:107—8.

[64] Menza MA, Grossman N, Van Horn M, Cody R, Forman N. Smoking and movement disorders in psychiatric patients. Biol Psychiatry 1991;30:109—15.

[65] Freedman R, Olincy A, Buchanan RW, Harris JG, Gold JM, Johnson L, et al. Initial phase 2 trial of a nicotinic agonist in schizophrenia. Am J Psychiatry 2008;165:1040—7.

[66] Mexal S, Berger R, Logel J, Ross RG, Freedman R, Leonard S. Differential regulation of alpha7 nicotinic receptor gene (CHRNA7) expression in schizophrenic smokers. J Mol Neurosci 2010;40:185—95.

[67] Satta R, Maloku E, Zhubi A, Pibiri F, Hajos M, Costa E, et al. Nicotine decreases DNA methyltransferase 1 expression and glutamic acid decarboxylase 67 promoter methylation in GABAergic interneurons. Proc Natl Acad Sci U S A 2008;105:16356−61.

[68] Shukla SD, Velazquez J, French SW, Lu SC, Ticku MK, Zakhari S. Emerging role of epigenetics in the actions of alcohol. Alcohol Clin Exp Res 2008;32:1525−34.

[69] Nielsen DA, Hamon S, Yuferov V, Jackson C, Ho A, Ott J, et al. Ethnic diversity of DNA methylation in the OPRM1 promoter region in lymphocytes of heroin addicts. Hum Genet 2010;127:639−49.

[70] Tsankova NM, Berton O, Renthal W, Kumar A, Neve RL, Nestler EJ. Sustained hippocampal chromatin regulation in a mouse model of depression and antidepressant action. Nat Neurosci 2006;9:519−25.

[71] Guidotti A, Dong E, Kundakovic M, Satta R, Grayson DR, Costa E. Characterization of the action of anti-psychotic subtypes on valproate-induced chromatin remodeling. Trends Pharmacol Sci 2009;30:55−60.

[72] Tremolizzo L, Carboni G, Ruzicka WB, Mitchell CP, Sugaya I, Tueting P, et al. An epigenetic mouse model for molecular and behavioral neuropathologies related to schizophrenia vulnerability. Proc Natl Acad Sci U S A 2002;99:17095−100.

[73] Kagan BL, Sultzer DL, Rosenlicht N, Gerner RH. Oral S-adenosylmethionine in depression: a randomized, double-blind, placebo-controlled trial. Am J Psychiatry 1990;147:591−5.

[74] Papakostas GI, Mischoulon D, Shyu I, Alpert JE, Fava M. S-adenosyl methionine (SAMe) augmentation of serotonin reuptake inhibitors for antidepressant nonresponders with major depressive disorder: a double-blind, randomized clinical trial. Am J Psychiatry 2010;167:942−8.

[75] Baldessarini RJ. Neuropharmacology of S-adenosyl-L-methionine. Am J Med 1987;83:95−103.

[76] Strous RD, Ritsner MS, Adler S, Ratner Y, Maayan R, Kotler M, et al. Improvement of aggressive behavior and quality of life impairment following S-adenosyl-methionine (SAM-e) augmentation in schizophrenia. Eur Neuropsychopharmacol 2009;19:14−22.

[77] Cohen SM, Nichols A, Wyatt R, Pollin W. The administration of methionine to chronic schizophrenic patients: a review of ten studies. Biol Psychiatry 1974;8:209−25.

[78] Goren JL, Stoll AL, Damico KE, Sarmiento IA, Cohen BM. Bioavailability and lack of toxicity of S-adenosyl-L-methionine (SAMe) in humans. Pharmacotherapy 2004;24:1501−7.

[79] Carney MW, Martin R, Bottiglieri T, Reynolds EH, Nissenbaum H, Toone BK, et al. Switch mechanism in affective illness and S-adenosylmethionine. Lancet 1983;1:820−1.

[80] Macritchie K, Geddes JR, Scott J, Haslam D, de Lima M, Goodwin G. Valproate for acute mood episodes in bipolar disorder. Cochrane Database Syst Rev 2003. CD004052.

[81] Johannessen CU. Mechanisms of action of valproate: a commentatory. Neurochem Int 2000;37:103−10.

[82] Tunnicliff G. Actions of sodium valproate on the central nervous system. J Physiol Pharmacol 1999;50:347−65.

[83] Szyf M. Epigenetics, DNA methylation, and chromatin modifying drugs. Annu Rev Pharmacol Toxicol 2009;49:243−63.

[84] Sharma RP, Rosen C, Kartan S, Guidotti A, Costa E, Grayson DR, et al. Valproic acid and chromatin remodeling in schizophrenia and bipolar disorder: preliminary results from a clinical population. Schizophr Res 2006;88:227−31.

[85] Kundakovic M, Chen Y, Guidotti A, Grayson DR. The reelin and GAD67 promoters are activated by epigenetic drugs that facilitate the disruption of local repressor complexes. Mol Pharmacol 2009;75:342−54.

[86] Dong E, Guidotti A, Grayson DR, Costa E. Histone hyperacetylation induces demethylation of reelin and 67-kDa glutamic acid decarboxylase promoters. Proc Natl Acad Sci U S A 2007;104:4676−81.

[87] Zarate Jr CA, Singh J, Manji HK. Cellular plasticity cascades: targets for the development of novel therapeutics for bipolar disorder. Biol Psychiatry 2006;59:1006−20.

[88] Bolden JE, Peart MJ, Johnstone RW. Anticancer activities of histone deacetylase inhibitors. Nat Rev Drug Discov 2006;5:769−84.

[89] Byrd JC, Marcucci G, Parthun MR, Xiao JJ, Klisovic RB, Moran M, et al. A phase 1 and pharmacodynamic study of depsipeptide (FK228) in chronic lymphocytic leukemia and acute myeloid leukemia. Blood 2005;105:959−67.

[90] Bug G, Ritter M, Wassmann B, Schoch C, Heinzel T, Schwarz K, et al. Clinical trial of valproic acid and all-trans retinoic acid in patients with poor-risk acute myeloid leukemia. Cancer 2005;104:2717−25.

[91] Citrome L, Shope CB, Nolan KA, Czobor P, Volavka J. Risperidone alone versus risperidone plus valproate in the treatment of patients with schizophrenia and hostility. Int Clin Psychopharmacol 2007;22:356−62.

[92] Citrome L. Schizophrenia and valproate. Psychopharmacol Bull 2003;37(Suppl. 2):74—88.

[93] Fournel M, Bonfils C, Hou Y, Yan PT, Trachy-Bourget MC, Kalita A, et al. MGCD0103, a novel isotype-selective histone deacetylase inhibitor, has broad spectrum antitumor activity in vitro and in vivo. Mol Cancer Ther 2008;7:759—68.

[94] Simonini MV, Camargo LM, Dong E, Maloku E, Veldic M, Costa E, et al. The benzamide MS-275 is a potent, long-lasting brain region-selective inhibitor of histone deacetylases. Proc Natl Acad Sci U S A 2006;103: 1587—92.

[95] Sharma RP, Gavin DP, Chase KA. Heterochromatin as an incubator for pathology and treatment non-response: implication for neuropsychiatric illness. Pharmacogenomics J 2012.

[96] Chang Y, Zhang X, Horton JR, Upadhyay AK, Spannhoff A, Liu J, et al. Structural basis for G9a-like protein lysine methyltransferase inhibition by BIX-01294. Nat Struct Mol Biol 2009;16:312—7.

[97] Fleeman N, Dundar Y, Dickson R, Jorgensen A, Pushpakom S, McLeod C, et al. Cytochrome P450 testing for prescribing antipsychotics in adults with schizophrenia: systematic review and meta-analyses. Pharmaco-genomics J 2011;11:1—14.

[98] Gomez A, Ingelman-Sundberg M. Pharmacoepigenetics: its role in interindividual differences in drug response. Clin Pharmacol Ther 2009;85:426—30.

[99] Ingelman-Sundberg M, Sim SC, Gomez A, Rodriguez-Antona C. Influence of cytochrome P450 poly-morphisms on drug therapies: pharmacogenetic, pharmacoepigenetic and clinical aspects. Pharmacol Ther 2007;116:496—526.

[100] Anttila S, Hakkola J, Tuominen P, Elovaara E, Husgafvel-Pursiainen K, Karjalainen A, et al. Methylation of cytochrome P4501A1 promoter in the lung is associated with tobacco smoking. Cancer Res 2003;63:8623—8.

[101] Nakajima M, Iwanari M, Yokoi T. Effects of histone deacetylation and DNA methylation on the constitutive and TCDD-inducible expressions of the human CYP1 family in MCF-7 and HeLa cells. Toxicol Lett 2003;144:247—56.

[102] Hammons GJ, Yan-Sanders Y, Jin B, Blann E, Kadlubar FF, Lyn-Cook BD. Specific site methylation in the 5'-flanking region of CYP1A2 interindividual differences in human livers. Life Sci 2001;69:839—45.

[103] Jones SM, Boobis AR, Moore GE, Stanier PM. Expression of CYP2E1 during human fetal development: methylation of the CYP2E1 gene in human fetal and adult liver samples. Biochem Pharmacol 1992;43:1876—9.

[104] Tokizane T, Shiina H, Igawa M, Enokida H, Urakami S, Kawakami T, et al. Cytochrome P450 1B1 is over-expressed and regulated by hypomethylation in prostate cancer. Clin Cancer Res 2005;11:5793—801.

[105] Gomez A, Karlgren M, Edler D, Bernal ML, Mkrtchian S, Ingelman-Sundberg M. Expression of CYP2W1 in colon tumors: regulation by gene methylation. Pharmacogenomics 2007;8:1315—25.

[106] Karlgren M, Gomez A, Stark K, Svard J, Rodriguez-Antona C, Oliw E, et al. Tumor-specific expression of the novel cytochrome P450 enzyme, CYP2W1. Biochem Biophys Res Commun 2006;341:451—8.

[107] Dannenberg LO, Edenberg HJ. Epigenetics of gene expression in human hepatoma cells: expression profiling the response to inhibition of DNA methylation and histone deacetylation. BMC Genomics 2006;7:181.

[108] Baker EK, Johnstone RW, Zalcberg JR, El-Osta A. Epigenetic changes to the MDR1 locus in response to chemotherapeutic drugs. Oncogene 2005;24:8061—75.

[109] Hoffmeyer S, Burk O, von Richter O, Arnold HP, Brockmoller J, Johne A, et al. Functional polymorphisms of the human multidrug-resistance gene: multiple sequence variations and correlation of one allele with P-glycoprotein expression and activity in vivo. Proc Natl Acad Sci U S A 2000;97:3473—8.

[110] Wang JS, Taylor R, Ruan Y, Donovan JL, Markowitz JS, Lindsay De Vane C. Olanzapine penetration into brain is greater in transgenic Abcb1a P-glycoprotein-deficient mice than FVB1 (wild-type) animals. Neuro-psychopharmacology 2004;29:551—7.

[111] Wang JS, Ruan Y, Taylor RM, Donovan JL, Markowitz JS, DeVane CL. The brain entry of risperidone and 9-hydroxyrisperidone is greatly limited by P-glycoprotein. Int J Neuropsychopharmacol 2004;7:415—9.

[112] El Ela AA, Hartter S, Schmitt U, Hiemke C, Spahn-Langguth H, Langguth P. Identification of P-glycoprotein substrates and inhibitors among psychoactive compounds—implications for pharmacokinetics of selected substrates. J Pharm Pharmacol 2004;56:967—75.

[113] Wang JS, Zhu HJ, Markowitz JS, Donovan JL, DeVane CL. Evaluation of antipsychotic drugs as inhibitors of multidrug resistance transporter P-glycoprotein. Psychopharmacology (Berl) 2006;187:415—23.

[114] Baker EK, El-Osta A. Epigenetic regulation of multidrug resistance 1 gene expression: profiling CpG meth-ylation status using bisulphite sequencing. Methods Mol Biol 2010;596:183—98.

[115] Sharma D, Vertino PM. Epigenetic regulation of MDR1 gene in breast cancer: CpG methylation status dominates the stable maintenance of a silent gene. Cancer Biol Ther 2004;3:549−50.

[116] Reed K, Hembruff SL, Sprowl JA, Parissenti AM. The temporal relationship between ABCB1 promoter hypomethylation, ABCB1 expression and acquisition of drug resistance. Pharmacogenomics J 2010;10: 489−504.

[117] Bozina N, Kuzman MR, Medved V, Jovanovic N, Sertic J, Hotujac L. Associations between MDR1 gene polymorphisms and schizophrenia and therapeutic response to olanzapine in female schizophrenic patients. J Psychiatr Res 2008;42:89−97.

[118] Uhr M, Tontsch A, Namendorf C, Ripke S, Lucae S, Ising M, et al. Polymorphisms in the drug transporter gene ABCB1 predict antidepressant treatment response in depression. Neuron 2008;57:203−9.

[119] Kacevska M, Ivanov M, Ingelman-Sundberg M. Perspectives on epigenetics and its relevance to adverse drug reactions. Clin Pharmacol Ther 2011;89:902−7.

[120] Kirchheiner J, Rodriguez-Antona C. Cytochrome P450 2D6 genotyping: potential role in improving treatment outcomes in psychiatric disorders. CNS Drugs 2009;23:181−91.

[121] Spina E, Ancione M, Di Rosa AE, Meduri M, Caputi AP. Polymorphic debrisoquine oxidation and acute neuroleptic-induced adverse effects. Eur J Clin Pharmacol 1992;42:347−8.

[122] de Leon J, Susce MT, Pan RM, Fairchild M, Koch WH, Wedlund PJ. The CYP2D6 poor metabolizer phenotype may be associated with risperidone adverse drug reactions and discontinuation. J Clin Psychiatry 2005;66:15−27.

[123] Lister R, Pelizzola M, Dowen RH, Hawkins RD, Hon G, Tonti-Filippini J, et al. Human DNA methylomes at base resolution show widespread epigenomic differences. Nature 2009;462:315−22.

[124] Carlquist JF, Anderson JL. Pharmacogenetic mechanisms underlying unanticipated drug responses. Discov Med 2011;11:469−78.

[125] Gavin DP, Kartan S, Chase K, Grayson DR, Sharma RP. Reduced baseline acetylated histone 3 levels, and a blunted response to HDAC inhibition in lymphocyte cultures from schizophrenia subjects. Schizophr Res 2008;103:330−2.

[126] Gavin DP, Rosen C, Chase K, Grayson DR, Tun N, Sharma RP. Dimethylated lysine 9 of histone 3 is elevated in schizophrenia and exhibits a divergent response to histone deacetylase inhibitors in lymphocyte cultures. J Psychiatry Neurosci 2009;34:232−7.

[127] Grayson DR, Kundakovic M, Sharma RP. Is there a future for histone deacetylase inhibitors in the pharma-cotherapy of psychiatric disorders? Mol Pharmacol 2010;77:126−35.

[128] Akbarian S. Epigenetics of schizophrenia. Curr Top Behav Neurosci 2010;4:611−28.

[129] Marutha Ravindran CR, Ticku MK. Changes in methylation pattern of NMDA receptor NR2B gene in cortical neurons after chronic ethanol treatment in mice. Brain Res Mol Brain Res 2004;121:19−27.

[130] Friso S, Choi SW, Girelli D, Mason JB, Dolnikowski GG, Bagley PJ, et al. A common mutation in the 5,10-methylenetetrahydrofolate reductase gene affects genomic DNA methylation through an interaction with folate status. Proc Natl Acad Sci U S A 2002;99:5606−11.

[131] Dashwood RH, Ho E. Dietary histone deacetylase inhibitors: from cells to mice to man. Semin Cancer Biol 2007;17:363−9.

[132] Matsukura S, Soejima H, Nakagawachi T, Yakushiji H, Ogawa A, Fukuhara M, et al. CpG methylation of MGMT and hMLH1 promoter in hepatocellular carcinoma associated with hepatitis viral infection. Br J Cancer 2003;88:521−9.

[133] Chang HW, Ling GS, Wei WI, Yuen AP. Smoking and drinking can induce p15 methylation in the upper aerodigestive tract of healthy individuals and patients with head and neck squamous cell carcinoma. Cancer 2004;101:125−32.

[134] Sandovici I, Smith NH, Nitert MD, Ackers-Johnson M, Uribe-Lewis S, Ito Y, et al. Maternal diet and aging alter the epigenetic control of a promoter-enhancer interaction at the Hnf4a gene in rat pancreatic islets. Proc Natl Acad Sci U S A 2011;108:5449−54.

[135] Mays-Hoopes L, Chao W, Butcher HC, Huang RC. Decreased methylation of the major mouse long inter-spersed repeated DNA during aging and in myeloma cells. Dev Genet 1986;7:65−73.

[136] Issa JP. CpG-island methylation in aging and cancer. Curr Top Microbiol Immunol 2000;249:101−18.

[137] Liu J, Morgan M, Hutchison K, Calhoun VD. A study of the influence of sex on genome wide methylation. PLoS One 2010;5:e10028.

[138] Liu J, Hutchison K, Perrone-Bizzozero N, Morgan M, Sui J, Calhoun V. Identification of genetic and epigenetic marks involved in population structure. PLoS One 2010;5:e13209.

[139] Eckhardt F, Lewin J, Cortese R, Rakyan VK, Attwood J, Burger M, et al. DNA methylation profiling of human chromosomes 6, 20 and 22. Nat Genet 2006;38:1378—85.

[140] Sarter B, Long TI, Tsong WH, Koh WP, Yu MC, Laird PW. Sex differential in methylation patterns of selected genes in Singapore Chinese. Hum Genet 2005;117:402—3.

[141] Tanti A, Belzung C. Open questions in current models of antidepressant action. Br J Pharmacol 2010;159: 1187—200.

[142] Katz MM, Bowden CL, Frazer A. Rethinking depression and the actions of antidepressants: uncovering the links between the neural and behavioral elements. J Affect Disord 2010;120:16—23.

[143] Zedkova I, Dudova I, Urbanek T, Hrdlicka M. Onset of action of atypical and typical antipsychotics in the treatment of adolescent schizophrenic psychoses. Neuro Endocrinol Lett 2011;32:667—70.

[144] McMahon FJ, Buervenich S, Charney D, Lipsky R, Rush AJ, Wilson AF, et al. Variation in the gene encoding the serotonin 2A receptor is associated with outcome of antidepressant treatment. Am J Hum Genet 2006;78:804—14.

[145] Wilkie MJ, Smith G, Day RK, Matthews K, Smith D, Blackwood D, et al. Polymorphisms in the SLC6A4 and HTR2A genes influence treatment outcome following antidepressant therapy. Pharmacogenomics J 2009;9:61—70.

[146] Kato M, Zanardi R, Rossini D, De Ronchi D, Okugawa G, Kinoshita T, et al. 5-HT2A gene variants influence specific and different aspects of antidepressant response in Japanese and Italian mood disorder patients. Psychiatry Res 2009;167:97—105.

[147] Levenson JM, Sweatt JD. Epigenetic mechanisms in memory formation. Nat Rev Neurosci 2005;6:108—18.

[148] Renthal W, Nestler EJ. Epigenetic mechanisms in drug addiction. Trends Mol Med 2008;14:341—50.

[149] Kumar A, Choi KH, Renthal W, Tsankova NM, Theobald DE, Truong HT, et al. Chromatin remodeling is a key mechanism underlying cocaine-induced plasticity in striatum. Neuron 2005;48:303—14.

[150] Levine AA, Guan Z, Barco A, Xu S, Kandel ER, Schwartz JH. CREB-binding protein controls response to cocaine by acetylating histones at the fosB promoter in the mouse striatum. Proc Natl Acad Sci U S A 2005;102:19186—91.

[151] Petronis A. Epigenetics and bipolar disorder: new opportunities and challenges. Am J Med Genet C Semin Med Genet 2003;123C:65—75.

[152] Costa E, Dong E, Grayson DR, Guidotti A, Ruzicka W, Veldic M. Reviewing the role of DNA (cytosine-5) methyltransferase overexpression in the cortical GABAergic dysfunction associated with psychosis vulnerability. Epigenetics 2007;2:29—36.

[153] Parsey RV, Hastings RS, Oquendo MA, Hu X, Goldman D, Huang YY, et al. Effect of a triallelic functional polymorphism of the serotonin-transporter-linked promoter region on expression of serotonin transporter in the human brain. Am J Psychiatry 2006;163:48—51.

[154] Jakovljevic M, Reiner Z, Milicic D, Crncevic Z. Comorbidity, multimorbidity and personalized psychosomatic medicine: epigenetics rolling on the horizon. Psychiatr Danub 2010;22:184—9.

[155] Nan X, Campoy FJ, Bird A. MeCP2 is a transcriptional repressor with abundant binding sites in genomic chromatin. Cell 1997;88:471—81.

[156] Yeivin A, Razin A. Gene methylation patterns and expression. EXS 1993;64:523—68.

[157] Autry AE, Monteggia LM. Epigenetics in suicide and depression. Biol Psychiatry 2009;66:812—3.

[158] Ehrlich M, Gama-Sosa MA, Huang LH, Midgett RM, Kuo KC, McCune RA, et al. Amount and distribution of 5-methylcytosine in human DNA from different types of tissues of cells. Nucleic Acids Res 1982;10:2709—21.

[159] Goto K, Numata M, Komura JI, Ono T, Bestor TH, Kondo H. Expression of DNA methyltransferase gene in mature and immature neurons as well as proliferating cells in mice. Differentiation 1994;56:39—44.

[160] Wilson CB, Rowell E, Sekimata M. Epigenetic control of T-helper-cell differentiation. Nat Rev Immunol 2009;9:91—105.

[161] Li E. Chromatin modification and epigenetic reprogramming in mammalian development. Nat Rev Genet 2002;3:662—73.

[162] Holliday R, Pugh JE. DNA modification mechanisms and gene activity during development. Science 1975;187:226—32.

[163] Riggs AD. X inactivation, differentiation, and DNA methylation. Cytogenet Cell Genet 1975;14:9—25.

[164] Razin A, Riggs AD. DNA methylation and gene function. Science 1980;210:604—10.

[165] Roth TL, Lubin FD, Sodhi M, Kleinman JE. Epigenetic mechanisms in schizophrenia. Biochim Biophys Acta 2009;1790:869—77.

[166] Starzyk RM, Koontz SW, Schimmel P. A covalent adduct between the uracil ring and the active site of an aminoacyl tRNA synthetase. Nature 1982;298:136—40.

[167] Gruenbaum Y, Cedar H, Razin A. Substrate and sequence specificity of a eukaryotic DNA methylase. Nature 1982;295:620—2.

[168] Calvanese V, Lara E, Kahn A, Fraga MF. The role of epigenetics in aging and age-related diseases. Ageing Res Rev 2009;8:268—76.

[169] Larsen F, Gundersen G, Lopez R, Prydz H. CpG islands as gene markers in the human genome. Genomics 1992;13:1095—107.

[170] Takai D, Jones PA. Comprehensive analysis of CpG islands in human chromosomes 21 and 22. Proc Natl Acad Sci U S A 2002;99:3740—5.

[171] Jones PA, Baylin SB. The fundamental role of epigenetic events in cancer. Nat Rev Genet 2002;3:415—28.

[172] Miranda TB, Jones PA. DNA methylation: the nuts and bolts of repression. J Cell Physiol 2007;213:384—90.

[173] Klose R, Bird A. Molecular biology. MeCP2 repression goes nonglobal. Science 2003;302:793—5.

[174] Qiu Z, Cheng J. The role of calcium-dependent gene expression in autism spectrum disorders: lessons from MeCP2, Ube3a and beyond. Neurosignals 2010;18:72—81.

[175] Chahrour M, Jung SY, Shaw C, Zhou X, Wong ST, Qin J, et al. MeCP2, a key contributor to neurological disease, activates and represses transcription. Science 2008;320:1224—9.

[176] Cohen S, Zhou Z, Greenberg ME. Medicine. Activating a repressor. Science 2008;320:1172—3.

[177] Wang H, Huang ZQ, Xia L, Feng Q, Erdjument-Bromage H, Strahl BD, et al. Methylation of histone H4 at arginine 3 facilitating transcriptional activation by nuclear hormone receptor. Science 2001;293:853—7.

[178] Yokochi T, Robertson KD. Doxorubicin inhibits DNMT1, resulting in conditional apoptosis. Mol Pharmacol 2004;66:1415—20.

[179] Christman JK. 5-Azacytidine and 5-aza-2'-deoxycytidine as inhibitors of DNA methylation: mechanistic studies and their implications for cancer therapy. Oncogene 2002;21:5483—95.

[180] Deng C, Lu Q, Zhang Z, Rao T, Attwood J, Yung R, et al. Hydralazine may induce autoimmunity by inhibiting extracellular signal-regulated kinase pathway signaling. Arthritis Rheum 2003;48:746—56.

[181] Lin X, Asgari K, Putzi MJ, Gage WR, Yu X, Cornblatt BS, et al. Reversal of GSTP1 CpG island hypermethylation and reactivation of pi-class glutathione S-transferase (GSTP1) expression in human prostate cancer cells by treatment with procainamide. Cancer Res 2001;61:8611—6.

[182] Scheinbart LS, Johnson MA, Gross LA, Edelstein SR, Richardson BC. Procainamide inhibits DNA methyltransferase in a human T cell line. J Rheumatol 1991;18:530—4.

[183] Moyers SB, Kumar NB. Green tea polyphenols and cancer chemoprevention: multiple mechanisms and endpoints for phase II trials. Nutr Rev 2004;62:204—11.

[184] Park OJ, Surh YJ. Chemopreventive potential of epigallocatechin gallate and genistein: evidence from epidemiological and laboratory studies. Toxicol Lett 2004;150:43—56.

7

The Role of Pharmacogenomics in Diabetes, HIV Infection, and Pain Management

Christina L. Aquilante*, Y.W. Francis Lam†

*Department of Pharmaceutical Sciences, University of Colorado
Skaggs School of Pharmacy and Pharmaceutical Sciences, Aurora, Colorado, USA
†Department of Pharmacology, University of Texas Health Science Center at San Antonio,
San Antonio, Texas, USA

OBJECTIVES

1. Identify key candidate genes and polymorphisms that influence the disposition, response, and adverse effects of the sulfonylureas, biguanides, and thiazolidinediones.

2. Discuss the utility of genome-wide association studies in identifying genes and polymorphisms associated with diabetes risk, pathophysiology, and response to antidiabetic medications.

3. Discuss the challenges and opportunities associated with the potential application of pharmacogenomic information to the clinical management of diabetes.

4. Discuss with examples how pharmacogenomics may have a role in treatment of HIV-1 infection and management of pain.

INTRODUCTION

Chapters 4 to 6 highlight some examples of how genotyping has been used to guide drug therapy decisions, especially in the areas of oncology and cardiology. This chapter will describe the accumulating evidence for the role of pharmacogenomics in other therapeutic areas. While the focus of this chapter is primarily on diabetes; it also includes an overview of HIV infection and pain management.

DIABETES OVERVIEW

Diabetes mellitus has emerged as one of the most alarming public health epidemics in the 21st century. By 2030, it is estimated that 552 million people worldwide will have diabetes, and 398 million people will be considered at high risk for developing the disease [1]. Type 2 diabetes is the most prevalent form of diabetes in adults, accounting for 90 to 95% of cases worldwide [1]. It is characterized by a relative deficiency in pancreatic β-cell insulin secretion and diminished tissue responsiveness to the normal action of insulin (i.e., insulin resistance) [2]. In contrast, type 1 diabetes is observed primarily in children and young adults and is characterized by an absolute deficiency in insulin secretion with minimal insulin resistance [2]. Both type 1 and type 2 diabetes are associated with periods of chronic hyperglycemia, which contributes to micro- and macrovascular complications such as retinopathy, nephropathy, neuropathy, cardiovascular disease, peripheral vascular disease, and stroke [2]. Given these deleterious clinical consequences, there is a critical need to identify optimal treatment strategies to achieve and maintain good glycemic control in this patient population.

Prior to 1995, sulfonylureas and insulin were the only pharmacologic agents available to treat diabetes. Since that time, the field has witnessed a dramatic increase in the number of antidiabetic medications (Table 7.1). Insulin remains the mainstay of therapy for type 1 diabetes. In contrast, numerous treatment modalities are available for type 2 diabetes, including: sulfonylureas, meglitinides, biguanides, thiazolidinediones, dipeptidyl peptidase-4 inhibitors, α-glucosidase inhibitors, bile acid sequestrants, dopamine receptor agonists, incretin mimetics, amylin mimetics, and insulin. This large treatment armamentarium has dramatically improved patient care. However, it is also associated with significant challenges—namely, the selection of the right drug for the right patient. Clinical diabetes guidelines have helped in this regard by providing a tiered approach to drug selection [3]. Yet even with clinical guidelines, it is difficult to predict which patients will derive the best efficacy or be predisposed to toxicity for a given antidiabetic medication [4]. As such, the potential of pharmacogenomics to aid in the selection of antidiabetic drug therapy for an individual patient has garnered considerable attention in the last decade.

Pharmacogenetic Lessons from Monogenic Forms of Diabetes

The potential use of an individual's genetic information to tailor antidiabetic drug therapy is not a new concept. In fact, this area has evolved from clinical experience with monogenic (i.e., single-gene) forms of diabetes such as neonatal diabetes and maturity-onset diabetes of the young (MODY) [5]. It is estimated that monogenic forms of diabetes account for 1 to 2% of all diabetes cases [6]. Neonatal diabetes, and two of the most common subtypes of MODY—MODY2 and MODY3—are discussed hereafter.

TABLE 7.1 Pharmacologic treatment options for diabetes

Class	Agents	Mechanism of action	Formulation	FDA-approved indication
Sulfonylureas[a]	Glyburide (glibenclamide) Glipizide Glimepiride	Stimulates insulin secretion from pancreatic β cells	Oral	Type 2 diabetes
Meglitinides	Nateglinide Repaglinide	Stimulates first-phase insulin secretion from pancreatic β cells	Oral	Type 2 diabetes
Biguanides	Metformin	Activates AMPK; suppresses hepatic glucose production, inhibits intestinal glucose absorption, and improves insulin sensitivity	Oral	Type 2 diabetes
Thiazolidinediones	Rosiglitazone[b] Pioglitazone	Agonists for peroxisome proliferator-activated receptor-γ; improves insulin sensitivity	Oral	Type 2 diabetes
Dipeptidyl peptidase-4 (DPP-4) inhibitors	Sitagliptin Saxagliptin Linagliptin	Competitively inhibits DPP-4, thereby slowing the inactivation of incretin hormones (e.g., GLP-1, GIP)	Oral	Type 2 diabetes
α-glucosidase inhibitors	Acarbose Miglitol	Competitively inhibits α-glucosidase enzymes, thereby delaying the breakdown of complex carbohydrates in the small intestine	Oral	Type 2 diabetes
Bile acid sequestrants	Colesevelam	The mechanism by which colesevelam improves glycemic control is not known.	Oral	Type 2 diabetes
Dopamine receptor agonists	Bromocriptine	The mechanism by which bromocriptine improves glycemic control is not known.	Oral	Type 2 diabetes
Incretin mimetics	Exenatide Liraglutide	GLP-1 receptor agonists; enhances glucose-dependent insulin secretion from pancreatic β cells; suppresses inappropriately elevated glucagon secretion; and delays gastric emptying	Subcutaneous	Type 2 diabetes

(Continued)

TABLE 7.1   Pharmacologic treatment options for diabetes—cont'd

Class	Agents	Mechanism of action	Formulation	FDA-approved indication
Amylin mimetics	Pramlintide	Synthetic analog of human amylin; modulates gastric emptying, prevents postprandial elevations in glucagon, and promotes satiety	Subcutaneous	Type 1 diabetes Type 2 diabetes
Insulin	Various	Regulates glucose metabolism by stimulating peripheral glucose uptake and inhibiting hepatic glucose production. Insulin also inhibits lipolysis and proteolysis.	Subcutaneous Intravenous	Type 1 diabetes Type 2 diabetes

[a]*Second- and third-generation agents, which are most commonly used in clinical practice.*
[b]*Rosiglitazone is available through a restricted-access program only.*
AMPK, AMP-activated protein kinase; DPP-4, dipeptidyl peptidase-4; GIP, glucose-dependent insulinotropic polypeptide; GLP-1, glucagon-like peptide-1.

Neonatal diabetes develops within the first six months of life and is associated with marked hyperglycemia. Neonatal diabetes is caused by activating mutations in the potassium inwardly rectifying channel, subfamily J, member 11 gene (*KCNJ11*) and the sulfonylurea receptor gene (*ABCC8*) [5]. *KCNJ11* and *ABCC8* encode the Kir6.2 and sulfonylurea receptor-1 (SUR1) subunits of pancreatic ATP-sensitive potassium (K_{ATP}) channels, respectively. Activating mutations in *KCNJ11* or *ABCC8* cause K_{ATP} channels to remain in the open state, thereby promoting hyperpolarization of the pancreatic β cell membrane and impairing insulin release [7,8]. Many patients with neonatal diabetes are initially diagnosed with type 1 diabetes and treated with insulin therapy. However, sulfonylureas have been shown to be particularly effective in patients with *KCNJ11* or *ABCC8* activating mutations. Sulfonylureas promote K_{ATP} channel closure, which results in pancreatic β cell membrane depolarization and insulin release [8–10]. As such, genetic testing in the neonatal period can reveal patients who can be successfully treated with sulfonylureas, rather than insulin, for diabetes management [8–10].

MODY is a general term used to describe a group of autosomal dominant forms of diabetes that are associated with pancreatic β cell dysfunction [11]. Patients with MODY present with clinical features similar to type 1 and/or type 2 diabetes, and as a result, often have delayed diagnosis or misdiagnosis of their disease [12]. Many subtypes of MODY exist (e.g., MODY 1 through 6), and they are differentiated by their primary gene defect(s) [2]. The most common subtype of MODY is due to mutations in the hepatocyte nuclear factor-1α gene (*HNF1A*), also called *HNF1A*-MODY or MODY3. *HNF1A* plays a key role in pancreatic β cell development and function. Patients with *HNF1A*-MODY have impaired first- and second-phase insulin secretion in response to glucose but are also very sensitive to the effects of sulfonylureas [11]. As such, it has been shown that patients with *HNF1A*-MODY can be successfully treated with low-dose sulfonylurea therapy [13,14]. Another common subtype of MODY is due to mutations in the glucokinase gene (GCK), also called GCK-MODY or MODY2. Glucokinase functions as the "glucose sensor" in pancreatic β cells and contributes to insulin secretion. Patients with GCK-MODY have impaired glucose sensing, which results in a higher glucose threshold needed to stimulate insulin secretion [11]. Many patients with GCK-MODY exhibit mild hyperglycemia and slow β cell deterioration over time. Often, these patients are effectively managed with lifestyle interventions rather than drug therapy [12].

These examples illustrate how genetic mechanisms underlying pancreatic β cell dysfunction in monogenic forms of diabetes can influence therapeutic management of these disorders. Thus, there has been great interest in using genetic information to tailor antidiabetic therapy in polygenic forms of diabetes, such as type 2 diabetes.

TYPE 2 DIABETES PHARMACOGENOMICS

Type 2 diabetes is a heterogeneous disorder characterized by defects in insulin secretion and/or insulin action. It is well appreciated that interindividual variability exists in the disposition (i.e., pharmacokinetics), response (i.e., pharmacodynamics), and adverse effects of medications used to treat type 2 diabetes. As such, pharmacogenomics is viewed as a promising tool to elucidate pharmacologic response variability among patients. However, compared to other chronic diseases, type 2 diabetes pharmacogenomics is in its infancy. To date, the candidate gene approach has been the primary means to assess genetic determinants of antidiabetic drug disposition and response. The candidate gene approach has focused on the following areas: (1) antidiabetic drug clinical pharmacology, that is, drug metabolizing enzymes, drug transporters, drug targets, and effector pathways; and (2) genomic markers underlying type 2 diabetes pathophysiology (i.e., disease risk alleles) [15]. This paradigm is shown in Figure 7.1. The candidate gene approach has posed some challenges, given that type 2 diabetes is a polygenic disease, and the biological pathways underlying its pathophysiology are numerous and complex. In order to overcome these challenges, genome-wide association studies (GWASes) have recently been applied to type 2 diabetes pharmacogenomics to discover novel genes and polymorphisms that underlie diabetes pathophysiology and drug response [15].

The following sections review major antidiabetic drug classes—sulfonylureas, biguanides, and thiazolidinediones—for which there exist a moderate amount of pharmacogenomic research. Within this framework, the most clinically relevant findings from candidate gene studies and/or GWASes are highlighted for each drug class (Table 7.2). The challenges and opportunities associated with the potential translation of pharmacogenomic information to the clinical management of diabetes are also discussed.

Sulfonylureas

Sulfonylureas have been a major component of Type 2 diabetes pharmacotherapy for more than 50 years. Mechanistically, sulfonylureas bind to SUR1 and stimulate insulin release from pancreatic β cells in a glucose-independent manner. Although the sulfonylureas are effective antihyperglycemic agents, interindividual variability exists in their pharmacokinetics,

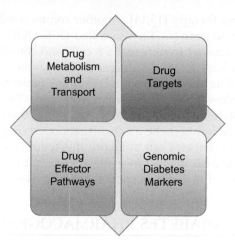

FIGURE 7.1   Candidate gene paradigm in type 2 diabetes pharmacogenomics.

pharmacodynamics, and adverse effects. It is estimated that 10 to 20% of patients have less than a 20 mg/dL decrease in fasting plasma glucose following initiation of sulfonylurea therapy, which is referred to as primary sulfonylurea failure [16]. In contrast, other patients have an adequate early response to sulfonylurea therapy but then later fail treatment. This represents secondary sulfonylurea failure and is estimated to occur at a rate of 5 to 7% per year [16]. Sulfonylureas also have a higher failure rate when given as monotherapy as compared to other antidiabetic agents such as metformin and thiazolidinediones [17]. Although some patients experience sulfonylurea failure, other patients appear to have increased sensitivity to the hypoglycemic effects of sulfonylureas. The United Kingdom Prospective Diabetes Study (UKPDS) found that mild hypoglycemia occurred in 31% of patients during the first year of glibenclamide (glyburide) therapy and the incidence of severe hypoglycemia was 1% per year [18]. Given the therapeutic challenges associated with sulfonylurea therapy, there has been great interest in determining the contribution of polymorphisms in drug metabolism, drug target, and diabetes risk genes to interindividual variability in sulfonylurea disposition, response, and adverse effects [19].

Drug Metabolism

Most sulfonylureas are primarily metabolized in the liver by the cytochrome P450 (CYP) 2C9 enzyme. Therefore, *CYP2C9* is a logical candidate gene to interrogate in relation to sulfonylurea clinical pharmacology. Studies have shown that the *CYP2C9*3* (Ile359Leu, I359L) and, to a lesser extent, *CYP2C9*2* (Arg144Cys, R144C) polymorphisms are associated with decreased oral clearance and increased plasma exposure of tolbutamide, glyburide, glipizide, and glimepiride [20]. For example, glyburide oral clearance in *CYP2C9*3* homozygotes was less than half that of wild-type *CYP2C9*1* homozygotes [21]. However, most early studies were conducted in healthy volunteers; therefore, the clinical consequences of *CYP2C9* polymorphisms in patients with type 2 diabetes were unknown until recently. Patient-focused studies have since begun to shed more light on this topic. A population-based study of

TABLE 7.2 Examples of major pharmacogenes of interest in type 2 diabetes

Drug class	Gene name	Protein	Role in clinical pharmacology
Sulfonylureas			
	CYP2C9	Cytochrome P450 2C9	Drug metabolism
	KCNJ11	Kir6.2 subunit of the K_{ATP} channel	Drug target; pancreatic β cell
	ABCC8	Sulfonylurea receptor-1 subunit of the K_{ATP} channel	Drug target; pancreatic β cell
	TCF7L2	WNT signaling pathway	Diabetes risk allele
Metformin			
	SLC22A1	Organic cation transporter-1	Drug transporter
	SLC22A2	Organic cation transporter-2	Drug transporter
	SLC47A1	Multidrug and toxin extrusion transporter-1	Drug transporter
	SLC27A2	Multidrug and toxin extrusion transporter-2	Drug transporter
	ATM	Ataxia-telangiectasia mutated gene	Drug target and response pathway
Thiazolidinediones			
	CYP2C8	Cytochrome P450 2C8	Drug metabolism
	PPARG	Peroxisome proliferator-activated receptor-γ	Drug target
	NFATC2, AQP2, SLC12A1	Putative genes involved in the pathophysiology of thiazolidinedione-induced edema	Adverse effect risk alleles

sulfonylurea users found that individuals with the CYP2C9*2/*2, *2/*3, or *3/*3 genotypes were 3.4 times more likely to achieve a hemoglobin 1c (HbA_{1c}) less than 7% compared with CYP2C9 wild-type homozygotes [22]. Other studies have shown lower sulfonylurea dose requirements in carriers of the CYP2C9*3 allele compared with wild-type homozygotes [23,24]. For example, in one study the daily dose of tolbutamide increased 279 mg from the first to 10th prescription in wild-type homozygotes but increased only 12 mg in CYP2C9*3 carriers [23]. Similar findings exist for glimepiride, where a trend was observed for a lower dose in CYP2C9*3 carriers (0.61 mg) versus wild-type homozygotes (1.01 mg) [24]. In terms of adverse effects, the CYP2C9*3 allele has been associated with an increased risk of hypoglycemia in patients treated with sulfonylureas [25,26]. For example, a small study found that individuals with CYP2C9*3/*3 or *2/*3 genotypes had 5.2 times the odds of a severe sulfonylurea-associated hypoglycemic event than those without these genotypes [25]. However, these results could not be replicated in a larger cohort of sulfonylurea-treated patients [27]. Possible reasons for the inconsistent reports in the literature include different definitions and assessments of hypoglycemia, clinical demographics of the sulfonylurea users (e.g., age, metabolic control), and possible variability in other genes and proteins involved in the complex and multifactorial process of hypoglycemia.

Currently, the clinical utility of *CYP2C9* genotyping for the prediction of sulfonylurea dose, response, or adverse effects is unclear. A multitude of clinical and genetic factors influence the glycemic response to sulfonylurea therapy. As such, the most practical application of *CYP2C9* genotyping may be to identify patients with a predisposition to sulfonylurea-induced hypoglycemia [28]. However, additional prospective studies are needed to more comprehensively define the role of *CYP2C9* genotyping in this area. Importantly, a consensus definition of hypoglycemia will need to be formulated and applied consistently in order to avoid potential discrepancies between studies.

Drug Targets

The complexity of glycemic response to antidiabetic drug therapy has prompted researchers to move beyond drug metabolism in the quest to identify genetic predictors of drug response. The two primary drug target candidate genes that have been studied in relation to sulfonylurea clinical pharmacology are *KCNJ11* and *ABCC8*. Interest in these as potential pharmacogenes came from clinical experience with monogenic forms of diabetes, such as neonatal diabetes. *KCNJ11* encodes the Kir6.2 subunit (pore) of the pancreatic K_{ATP} channel, and *ABCC8* encodes the SUR1 subunit of the K_{ATP} channel. It can be hypothesized that defects in the Kir6.2 and/or SUR1 subunits, as a result of genetic polymorphisms, may alter pancreatic β cell physiology, insulin secretion, and response to antidiabetic medications. The most widely studied polymorphisms in *KCNJ11* and *ABCC8* are Glu23Lys (E23K) and Ser1369Ala (S1369A), respectively. Of note, the *KCNJ11* Glu23Lys polymorphism has emerged as a type 2 diabetes risk allele in various cohorts [29–33]. The *KCNJ11* Glu23Lys and *ABCC8* Ser1369Ala polymorphisms are in strong linkage disequilibrium; therefore, most individuals who carry a *KCNJ11* Lys23 allele will also carry an *ABCC8* Ala1369 allele [33]. Recombinant human K_{ATP} channels with the Lys23/Ala1369 risk haplotype demonstrated increased sensitivity to gliclazide compared with wild-type K_{ATP} channels [34]. Subsequently, Ala1369 was determined to be the causative allele governing increased sulfonylurea sensitivity in the Lys23/Ala1369 haplotype [34].

Despite these informative *in vitro* findings, only a few studies have evaluated the impact of *KCNJ11* Glu23Lys and/or *ABCC8* Ser1369Ala polymorphisms on sulfonylurea efficacy in patients with type 2 diabetes. In the largest study to date, Chinese patients with the *ABCC8* Ala/Ala genotype had 2.2 times greater odds of responding to gliclazide treatment than patients with the Ser/Ser genotype over an eight-week period [35]. Although these results are interesting and in line with *in vitro* findings, *ABCC8* Ser1369Ala genotyping is not yet ready for translation to the clinic. Additional prospective studies are needed and should consider the following factors in their study designs: longer treatment durations (i.e., to assess genetic factors governing sulfonylurea failure after long-term treatment); clinically relevant glycemic endpoints (e.g., fasting plasma glucose, HbA_{1c}); and assessment of other race and ethnic groups. Recent *in vitro* work has also provided additional considerations for future research. Specifically, recombinant human K_{ATP} channels containing the Lys23/Ala1369 haplotype were sensitive to gliclazide inhibition, and K_{ATP} channels containing the Glu23/Ser1369 haplotype were sensitive to tolbutamide, chlorpropamide, and glimepiride inhibition [36]. These data suggest that sulfonylurea chemical structure—for example, a ring-fused pyrrole moiety on gliclazide—may influence the pharmacogenetic effects mediated by *KCNJ11* and *ABCC8* polymorphisms. Therefore, the type of sulfonylurea will undoubtedly be an important factor to consider in future pharmacogenetic studies.

Diabetes Risk Alleles

To date, more than 40 genetic loci have been associated with an increased risk of type 2 diabetes [37]. Although the magnitude of risk associated with each genetic marker tends to be modest, these loci provide key insights into the molecular mechanisms of the disease [38,39]. For example, most loci identified thus far have been genes involved in insulin secretion [38,39]. Thus, the future of type 2 diabetes pharmacogenomics will likely involve the use of risk alleles to aid in the molecular classification of type 2 diabetes and, subsequently, antidiabetic drug selection [15].

In terms of sulfonylurea pharmacogenomics, type 2 diabetes risk genes that influence processes such as insulin secretion, glucose homeostasis, or pancreatic β cell function, among others, could potentially contribute to variability in sulfonylurea response between patients. In this regard, the most intensively studied diabetes risk gene has been transcription factor 7-like 2 (*TCF7L2*). *TCF7L2* is a transcription factor involved in the WNT signaling pathway. The WNT signaling pathway is involved in glucose homeostasis, lipid metabolism, pancreatic β cell proliferation and function, and the production of glucagon-like peptide 1 [40]. Importantly, *TCF7L2* was associated with type 2 diabetes in the first GWAS evaluating novel risk loci for type 2 diabetes [41]. Since that time, the *TCF7L2* rs7903146 C>T polymorphism has been associated with impaired insulin secretion both *in vitro* and *in vivo* [42–44]. A few clinical studies have shown the *TCF7L2* rs7903146 variant T allele to be associated with an increased risk of sulfonylurea failure in patients with type 2 diabetes [45–47]. For example, individuals homozygous for the T allele had 1.73 times increased odds of sulfonylurea failure as compared with wild-type homozygotes [45]. In another study, the rs7903146 T allele occurred more frequently in patients who failed sulfonylurea treatment versus the control group (36% versus 26%; odds ratio, 1.57) [46]. It is important to note that the published studies have varied in design, sample size, treatment duration, and definition of sulfonylurea failure. Currently, the clinical utility of *TCF7L2* genotyping to identify sulfonylurea nonresponders is uncertain given that it appears that the observed effect is small and would not merit a standalone test. Nonetheless, additional studies are needed to determine whether *TCF7L2*, in combination with other pharmacogenes, could improve the antidiabetic drug selection process and clinical outcomes. Besides *TCF7L2*, other disease-related genes have been implicated in altered sulfonylurea response such as insulin receptor substrate-1 (*IRS1*), nitric oxide synthase 1 adaptor protein (*NOS1AP*), and *CDKAL1* [48–50]. However, in comparison to *TCF7L2*, most of these findings have not yet undergone replication in additional patient cohorts.

The previous examples have largely focused on single polymorphisms in known type 2 diabetes risk genes. However, findings from GWASes have allowed for a more comprehensive approach to assess the impact of diabetes risk alleles on drug response. This paradigm is best exemplified by a study that hypothesized that a panel of 20 type 2 diabetes risk alleles would influence sulfonylurea response [51]. Most of the risk alleles included in this panel were putative mediators of insulin secretion and had been repeatedly associated with type 2 diabetes risk in GWASes. The study found that patients who carried more than 17 diabetes risk alleles had a 1.7-fold decreased likelihood of achieving a stable sulfonylurea dose, suggesting a decreased response to sulfonylurea therapy [51]. One of the major limitations of this study was that it did not interrogate continuous endpoints of glycemic status (e.g., fasting plasma glucose, HbA_{1c}). However, this study does illustrate the proof of concept that the genetic underpinnings of type 2 diabetes may mediate differential response to antidiabetic drug therapy [51].

Biguanides: Metformin

Metformin is a biguanide antidiabetic agent that is recommended as first-line treatment for type 2 diabetes [3]. Metformin's mechanism of action and drug target(s) have largely remained a mystery since its inception [52]. However, it is now generally accepted that metformin activates adenosine monophosphate-activated protein kinase (AMPK) in the liver [52]. Metformin's major pharmacodynamic actions are to suppress hepatic glucose production, increase glucose uptake, improve insulin sensitivity in peripheral tissues, decrease fatty acid synthesis, and decrease intestinal glucose absorption [53]. Significant interindividual variability exists in metformin response. Clinical studies have demonstrated that approximately 36% of patients fail to achieve a fasting plasma glucose level of less than 140 mg/dL with metformin alone [17]. Furthermore, it is estimated that up to 50% of patients fail to achieve an HbA_{1c} of less than 7% after one year of metformin monotherapy [45]. There has been considerable research geared toward identifying the genetic predictors of metformin disposition and response. Although most research has primarily focused on drug transporter candidate genes, recent studies have taken a GWAS approach to broadly interrogate the pharmacogenomics of metformin response.

Drug Transporters

Metformin's pharmacokinetic disposition in the body is largely governed by drug transporters, particularly those involved in hepatic uptake and renal excretion. Organic cation transporter-1 (SLC22A1) mediates metformin uptake in the liver, and organic cation transporter-2 (SLC22A2) mediates active secretion of metformin in the kidney. Candidate gene studies in healthy volunteers and patients with type 2 diabetes have investigated the impact of polymorphisms within these drug transporter genes on the pharmacokinetics and pharmacodynamics of metformin [52,54]. Early studies reported that reduced-function SLC22A1 polymorphisms altered metformin pharmacokinetics and pharmacodynamics in healthy volunteers, most likely because of decreased hepatic uptake of metformin [55,56]. Subsequently, studies in patients with type 2 diabetes have shown inconsistent results with respect to the impact of reduced-function SLC22A1 alleles on metformin pharmacodynamics [57–59]. As such, it does not appear that SLC22A1 polymorphisms consistently explain a sufficient degree of metformin response variability to be used in clinical practice. Polymorphisms in SLC22A2, particularly c.808G>T, have been studied for their relationship with metformin renal clearance. This research is important, given that renal clearance of metformin is proposed to have a strong underlying genetic component [60]. Healthy volunteer studies have shown that individuals with the SLC22A2 c.808 T/T genotype have marked reductions in metformin renal clearance as compared with individuals who possess the

G/T or G/G genotypes [52]. Based on these data, one might expect an increased response to metformin with the T/T genotype. However, the effect of this polymorphism on metformin response in patients with type 2 diabetes remains to be determined.

The multidrug and toxin extrusion transporters, MATE1 (encoded by *SLC47A1*) and MATE2 (encoded by *SLC47A2*), are H^+/organic cation antiporters located on the brush-border of the renal epithelium and the canalicular membrane of hepatocytes [52]. They are important contributors to metformin disposition because they mediate the transport and excretion of metformin into the urine and bile. In a population-based cohort study, the *SLC47A1* rs2289669 G>A polymorphism was significantly associated with the reduction in HbA_{1c} following metformin therapy, implying that this polymorphic allele is associated with decreased transporter function [61]. Specifically, the decrease in HbA_{1c} was 0.3% greater per copy of the A allele, which may be a clinically important decrease in patients with type 2 diabetes [61]. Subsequently, this association was confirmed by the Diabetes Prevention Program (DPP), where the *SLC47A1* rs8065082 C>T polymorphism, which is in tight linkage disequilibrium with rs2289669, was associated with metformin response [62]. In terms of *SLC47A2*, a gain-of-function promoter polymorphism (rs12943590 G>A) was associated with a weaker glycemic response to metformin in patients with type 2 diabetes [63]. Specifically, Caucasian patients who were homozygous for the A allele had a smaller relative change in HbA_{1c} than G allele carriers (−0.03 versus −0.15) [63]. These results have yet to be replicated in other diabetes cohorts. Altogether, it appears that MATE transporters may be important determinants of metformin disposition and response. However, additional studies are needed to confirm previous associations and elucidate the pharmacologic alterations resulting from these genetic polymorphisms. In addition, recent data suggest that polymorphisms in genes encoding OCT and MATE may interact to influence metformin's pharmacodynamic effects [64]. As such, future studies will need to consider the interaction between different drug transporters, and genetic variation within these transporters, on metformin clinical pharmacology.

Drug Targets

Compared to drug transporters, less is known about the influence of drug target or effector pathway gene polymorphisms on metformin pharmacokinetics, pharmacodynamics, and adverse effects. This lack of knowledge is largely a result of metformin's complex mechanism of action. Therefore, large-scale candidate gene studies and/or GWASes are essential components of future metformin pharmacogenomic research. Metformin works, in part, by activating AMPK, which is a master regulator of cell and body energy homeostasis and glucose uptake in skeletal muscle [52]. A large-scale candidate gene study of the DPP trial showed nominally significant, but interesting, associations with metformin response for genes such as serine-threonine kinase 11 (*STK11*, which catalyzes the activation of AMPK) and catalytic subunits of AMPK (e.g., *PRKAA1*, *PRKAA2*, and *PRKAB2*) [62]. More recently, the GWAS approach has been applied to metformin pharmacogenomics [65]. Importantly, this was the first GWAS conducted for any antidiabetic medication. The GWAS identified a significant association between a polymorphism located in a locus containing the ataxia-telangiectasia mutated (*ATM*) and metformin response in patients with type 2 diabetes [65]. *ATM* is a DNA repair gene that has also been implicated in insulin signaling pathways, β-cell dysfunction, and AMPK activation [65]. Following *in vitro* experiments, the study further concluded that *ATM* acts upstream of AMPK and is required for metformin's action [65]. Although the *ATM* polymorphism only explained 2.5% of the variance in metformin response, this example illustrates the utility of GWASes for discovering previously unknown mediators of antidiabetic drug pharmacology.

Diabetes Risk Alleles

As discussed in the sulfonylurea section, an emerging area of pharmacogenomic research is the extent to which type 2 diabetes risk alleles influence response to antidiabetic drug therapy. Few studies have been conducted in this area with respect to metformin. However, researchers involved with the DPP trial devised a genetic risk score that was composed of 34 type 2 diabetes risk alleles [66]. The risk alleles were selected based on published reports of their individual association with type 2 diabetes at a genome-wide significance level ($p < 5 \times 10^{-8}$) [66]. No interaction between the genetic risk score and metformin treatment was observed in the study. Nonetheless, this "genetic risk score" approach will likely be studied more in the future as type 2 diabetes risk alleles continue to emerge through GWASes. In sum, no genetic markers have been identified that explain a sufficiently high percentage of variability in metformin response. However, compared with other antidiabetic agents, metformin pharmacogenomic research has made substantial progress in adopting a GWASes approach to identify novel genetic sources of drug disposition and response variability in patients with type 2 diabetes [15].

Thiazolidinediones

Thiazolidinediones are agonists for the peroxisome proliferator-activated receptor-γ (PPAR-γ) in the cell nucleus. As nuclear receptor agonists, thiazolidinediones regulate the transcription of numerous genes involved in fatty acid uptake and storage, glucose homeostasis, insulin sensitivity, and adipocyte differentiation [67]. Thiazolidinediones are commonly referred to as "insulin sensitizers" due to their ability to increase insulin sensitivity in muscle, liver, and fat. Although thiazolidinediones are useful in type 2 diabetes patients who exhibit a moderate to high degree of insulin resistance, treatment guidelines classify these agents as less well validated therapies [3]. In addition, there have been concerns about increased cardiovascular risk associated with rosiglitazone, and it has been placed in a restricted access program [68,69]. As such, the only thiazolidinedione that is widely available at this time is pioglitazone.

Interindividual variability in thiazolidinedione response and adverse effects have been demonstrated in clinical studies. Approximately 25% of patients with type 2 diabetes fail to achieve a greater than 10% decrease in fasting plasma glucose following pioglitazone therapy [70]. Along the same lines, 30% of patients at high risk for type 2 diabetes do not show an improvement in insulin sensitivity following thiazolidinedione treatment [71]. Edema and congestive heart failure are among the more troubling adverse effects associated with thiazolidinedione therapy [72,73]. In fact, edema occurs relatively frequently, with reported rates of 2 to 28% for pioglitazone [74]. Because of this, the use of thiazolidinediones in patients with heart failure is discouraged, especially for those with moderate to severe heart failure symptoms [75]. Clinical studies have sought to identify genetic determinants of thiazolidinedione pharmacokinetics, pharmacodynamics, and adverse effects [76]. However, compared to the pharmacogenetic progress of sulfonylureas and metformin, thiazolidinedione pharmacoge-netic studies have lagged behind in their approach, mostly focusing on only a limited number of candidate genes and polymorphisms.

Drug Metabolism

The thiazolidinediones are primarily metabolized by CYP2C8 and, to a lesser extent, by CYP2C9 (rosiglitazone) and CYP3A4 (pioglitazone) [77,78]. Although thiazolidinediones have a wide therapeutic index, alterations in plasma exposure may influence glycemic control, insulin sensitization, and the risk of concentration-dependent adverse effects (e.g., edema and weight gain). The polymorphism most often studied in relation to thiazo-lidinedione metabolism is *CYP2C8*3* (Arg139Lys, R139K; Lys399Arg, L399K). Healthy volunteer studies have shown that *CYP2C8*3* carriers have greater oral clearance and lower plasma exposure of rosiglitazone and pioglitazone than *CYP2C8*1* homozygotes [79–81]. Thus, *CYP2C8*3* appears to function as a high-activity allele for thiazolidinedione metabo-lism, resulting in a 25 to 35% decrease in plasma exposure. However, it remains to be deter-mined whether *CYP2C8*-mediated differences in thiazolidinedione plasma exposure translate into differences in glycemic control or insulin sensitization in patients with type 2 diabetes [76].

Drug Targets

Given that thiazolidinediones are PPAR-γ agonists, PPAR-γ (*PPARG*) is the most logical drug target candidate gene for this drug class. Pro12Ala (P12A) is the most frequently studied polymorphism in *PPARG*, and the Ala12 allele has been associated with an approx-imate 20% reduction in the risk of type 2 diabetes [82]. Most pharmacogenetic studies have shown no association between the *PPARG* Pro12Ala polymorphism and glycemic response or insulin sensitization following thiazolidinedione therapy [70,71,83]. Thus, *PPARG* Pro12Ala genotyping does not have a role in optimizing thiazolidinedione management. Beyond *PPARG*, many other drug target and effector pathway genes have been interrogated for their relationship with thiazolidinedione response including but not limited to adiponectin, lipo-protein lipase, lipin-1, perilipin, PPAR-γ coactivator-1, resistin, uncoupling protein 2, β3-adrenergic receptor, tumor necrosis factor-α, and voltage-gated potassium channel-1 [76,84]. Unfortunately, most of these studies had significant limitations such as lack of repli-cation cohorts; lack of study in race/ethnic groups besides Asians; small sample sizes; and failure to consider the complexity of thiazolidinedione response. In order to ultimately move thiazolidinedione pharmacogenomics to the clinic, it is imperative that a more compre-hensive discovery approach, such as GWASes, be undertaken in large thiazolidinedione-treated patient cohorts. This approach is especially important given the diverse genes and proteins known to mediate thiazolidinedione clinical pharmacology. In addition, little is known about the relationship between type 2 diabetes risk alleles and thiazolidinedione response. In GWAS, some insulin resistance genes have shown a signal for increased type 2 diabetes risk [85]. Thus, it can be hypothesized that patients with a type 2 diabetes subtype driven primarily by insulin resistance may derive a greater benefit from thiazolidinediones than other antidiabetic therapies (e.g., sulfonylureas).

Another potential application of pharmacogenomics to thiazolidinedione therapy may be in the prediction of adverse effects, especially edema. Some recent work in this area has yielded promising results. A genetic substudy of the Diabetes Reduction Assessment with Ramipril and Rosiglitazone Medication (DREAM) trial interrogated more than 30,000 polymorphisms among Europeans receiving rosiglitazone [86]. One polymorphism in the nuclear factor of activated T-cells cytoplasmic calcineurin-dependent 2 gene (*NFATC2*) was significantly associated with rosiglitazone-induced edema, yielding an odds ratio of 1.9 [86]. Another study in Chinese patients found significant associations between thiazolidinedione-induced edema and polymorphisms in the aquaporin 2 (*AQP2*) and the bumetanide-sensitive Na-K-2Cl cotransporter (*SLC12A1*) genes [74]. These researchers went on to develop a prediction model, which included age, sex, and *AQP2* and *SLC12A1* polymorphisms, to estimate the risk of thiazolidinedione-induced edema in type 2 diabetes patients [74]. Replication of these genetic findings and assessment of the clinical utility of this prediction model will need to be conducted in other populations. However, these findings demonstrate how genetic and nongenetic factors may be integrated into clinically applicable models to aid in the prediction of adverse effects. Along these lines, these types of algorithms may be useful in selecting pharmacologic strategies for the prevention of type 2 diabetes. Recently, pioglitazone, as compared with placebo, was associated with a dramatic 72% reduction in the risk of a patient converting from impaired glucose tolerance to type 2 diabetes [87]. However, pioglitazone was also associated with a significant increase in the incidence of edema and weight gain. Perhaps, in the future, algorithms containing clinical and genetic factors may be used to tailor pharmacologic prevention strategies in patients at high risk for type 2 diabetes in order to attenuate disease onset while minimizing adverse effects.

Other Antidiabetic Drugs

The pharmacogenomic literature for other antidiabetic drugs is scarce, with the exception of the meglitinides. The nonsulfonylurea meglitinides, repaglinide and nateglinide, stimulate insulin secretion in pancreatic β cells [3]. Polymorphisms in genes encoding drug metabolizing enzymes (e.g., *CYP2C8*, *CYP2C9*), drug transporters (e.g., *SLCO1B1*), and drug targets (e.g., *KCNJ11*, *ABCC8*) have been implicated in altered meglitinide disposition and/or response [84]. However, the meglitinides have limited use in clinical practice. Therefore, the clinical utility of pharmacogenomics is not likely to be pursued for these agents. Perhaps even more surprising is that virtually no pharmacogenomic information exists for incretin mimetics (i.e., exenatide and liraglutide) or DPP-4 inhibitors (i.e., sitagliptin, saxagliptin, and linagliptin). Exenatide and liraglutide have substantially changed the landscape of type 2 diabetes pharmacotherapy by placing more emphasis on the important role of gastrointestinal incretin hormones in type 2 diabetes pathophysiology [88]. In addition, the incretin-enhancing DPP-4 inhibitors have gained popularity due to their reasonable efficacy and tolerability profiles [89]. Interindividual variability exists in the pharmacodynamic effects of most of these newer agents. For example, a meta-analysis of the efficacy of DPP-4 inhibitor in patients with type 2 diabetes showed that approximately 60% of patients failed to achieve an HbA_{1c} of less than 7% with this therapy [90]. Along the same lines, another meta-analysis showed that 35% and 55% of patients failed to achieve an HbA_{1c} of less than 7% following liraglutide and exenatide therapy, respectively [91]. Taken together, there exists a substantial gap in knowledge regarding genetic and clinical predictors of response to these newer antidiabetic agents. Future research will need to consider how pharmacogenomics, along with the molecular mechanisms underlying type 2 diabetes phenotypes, can be used to optimally guide the use of these therapies.

CHALLENGES AND OPPORTUNITIES OF PHARMACOGENOMICS IN DIABETES

It is well appreciated that diabetes pharmacogenomics is in its early stages. With the exception of monogenic forms of diabetes [5], there exist no examples of genetically guided diabetes treatment algorithms that are currently being used in clinical practice. In order to move the field forward and foster the translation of genetic information to the clinical setting, several challenges and opportunities will need to be considered by the medical and scientific communities.

A major challenge facing the field is optimal study design. To date, most studies have been retrospective in nature, had small to moderate sample sizes, investigated only a limited number of genes and polymorphisms, and lacked statistical adjustment for multiple comparisons. These factors have likely contributed to the lack of replication of pharmacogenomic findings between study cohorts. Ideally, future studies should be conducted in large cohorts with adequate power to detect pre specified differences, include a comprehensive approach for gene interrogation, apply appropriate statistical adjustments for multiple comparisons, and consider both clinical and genetic factors. Accomplishing these goals, particularly in large diabetes populations, will not be an easy task. However, these obstacles may be overcome through collaborations among individuals in academia, federal and private grant agencies (e.g., National Institutes of Health), community settings, and the pharmaceutical industry [15]. Another challenge for the field is the variety of antidiabetic drug response definitions that have been used in clinical studies. The sulfonylureas are a good example of this situation, where several definitions exist for primary and secondary sulfonylurea failure. More uniform definitions of diabetes drug response and adverse effects will need to be applied to pharmacogenomic research in order to gain consistency in phenotype definitions between studies [92]. Along these lines, physiologically relevant endpoints (e.g., insulin sensitivity, hepatic glucose output) will need to be selected to define drug response, and attention paid to the methodology used to measure these endpoints [92,93]. Another area that has been lacking in diabetes pharmacogenomic research is the assessment of the interplay between environmental (e.g., lifestyle) and genetic factors in mediating diabetes risk and drug response phenotypes. These interactions are especially important when conducting studies in different race and ethnic groups, as metabolic pathophysiology (e.g., insulin resistance and obesity) can differ substantially between populations. For example, individuals of Asian ancestry may have metabolic risk factors despite only modest increases in waist circumference or body mass index [94]. Lastly, future studies will need to carefully quantify the predictive ability and cost-effectiveness of personalized medicine strategies [37,95]. For example, a cost/utility analysis of genetic testing for neonatal diabetes showed that genetic testing improved quality of life and produced a total cost savings of $12,528 at 10 years [95].

Despite these challenges, a substantial number of exciting opportunities exist for diabetes pharmacogenomics. The technological advances made possible through GWASes represent the most promising opportunity in the field. Although only one diabetes pharmacogenomic

GWAS has been published to date, it is certain that this approach will be used more often in the future. For example, genetic samples were collected as part of the Action to Control Cardiovascular Risk Factors in Diabetes (ACCORD) trial and will likely be used for genome-wide assessments of drug response. In a complex disease such as type 2 diabetes, GWASes have the potential to enhance our understanding of the disease and its treatment by: (1) providing a comprehensive approach to identify polymorphisms that govern drug response and adverse effects; (2) contributing novel insights into the mechanism of action of diabetes pharmacotherapy; (3) providing information on the molecular basis of diabetes subtypes; and (4) discovering novel metabolic targets for drug development [4,15,37]. To date, many GWASes have been conducted in European cohorts. However, in the future, it will be important to conduct GWASes in non-European cohorts in order to identify race/ethnicity-specific type 2 diabetes risk alleles that were not identified in previous European studies [96]. In order to be used in clinical practice, genomic markers will need to account for a sufficient amount of variability in drug response in order to be considered for future testing. Additionally, other technological strategies, such as next-generation sequencing, will need to be undertaken to identify less common variants that mediate differential drug response phenotypes between individuals.

Once promising markers are identified, a major opportunity for the field will be to develop prospective studies to determine how genetic information may be used to select or optimize type 2 diabetes pharmacotherapy and whether this strategy results in better outcomes than the non-genotype approach. Prospective studies will most certainly evaluate the impact of genetics on glycemic endpoints. However, a more important question will be whether genetically guided drug therapy and glycemic control significantly decrease the incidence of microvascular and macrovascular complications associated with the disease. Finally, another innovative opportunity for the field will be to use genetic information to predict diabetes risk, promote behavioral modifications, and devise individualized diabetes prevention strategies [97]. In the future, it can be envisioned that individuals with a high genetic risk score for type 2 diabetes may be subjected to earlier and more aggressive lifestyle or pharmacologic interventions to mitigate their risk of developing the disease [97,98].

PHARMACOGENOMICS AND HIV

Drug treatment of patients with the human immunodeficiency virus (HIV) serves as a good example of how genetic technology and pharmacogenomics can be applied for drug development and treatment of disease—in this case, HIV infection.

Chemokine Coreceptor (CCR) Antagonist

The HIV must enter the host cells, such as helper T cells and macrophages, where it eventually undergoes viral nuclei acid synthesis; production and assembly of new virions; and subsequent release from the cell to further infect other host cells. Entry into the host cell is mediated by the HIV interacting with cytokine receptors and chemokine coreceptors located on the host cell membrane. Different HIV strains utilize a specific cytokine receptor/chemokine coreceptor pair made of different proteins, each of which is encoded by a specific gene. For example, the more virulent strain, HIV type 1 (HIV-1), requires the CD4 receptor/CC-chemokine receptor-5 (CCR5) coreceptor pair. Essentially, this process involves interaction of the viral envelope protein gp120 with the CD4 molecule on the host cell, with subsequent conformational changes that allow interaction with the chemokine coreceptor CCR5. Mutation of the genes would cause changes in either expression or function of the protein and subsequently the function or structure of the cytokine receptor/chemokine coreceptor pair, potentially impairing the ability of the HIV strain to enter the host cell.

The observation that some patients have natural resistance to HIV infection despite repeated exposure to HIV-1 led to the discovery of mutations in the *CCR5* gene. Specifically, deletion of 32 base pair, or delta 32, from the coding region of the *CCR5* gene (*CCR5Δ32*) results in a frame shift and prevents the production of the full-length CCR5 protein. As with other genetic variants, ethnic differences in allele frequency exist. The *CCR5Δ32* allele occurs in up to 15% of Caucasians and about 2% of African Americans but is almost absent in East Asians and native Africans.

Although the lack of functional CCR5 receptors apparently results in no harmful effects per se in humans, homozygous carriers of the *CCR5Δ32* allele would present a nonfunctional CD4 receptor/CCR5 coreceptor pair that does not support membrane fusion, thereby preventing HIV-1 from entry into the host cell. As a result, these patients "acquire" natural protection from the infection. Heterozygous carriers of the *CCR5Δ32* allele would produce a reduced amount of functioning CD4 receptor/CCR5 coreceptor pair, resulting in them having relative protection from HIV-1 and possibly a slower progression to acquired immunodeficiency disease syndrome [99–101].

Since that discovery, efforts have focused on developing antiretroviral drugs that act as antagonists at the CCR5 receptor, including maraviroc. Maraviroc is a reversible antagonist that blocks binding of the HIV-1 envelope protein gp120 to the CCR5 receptor and its subsequent entry into the host's cell. Since the approval of maraviroc in 2007, several new CCR5 receptor antagonists, including vicriviroc, have been developed and tested in Phase II or III clinical trials. However, none of these compounds have yet been submitted to the Food and Drug Administration for approval.

Drug Selection Based on Molecular Genetic Test

The clinical use of maraviroc also illustrates how genetic technology can be used to optimize drug selection. Because the pharmacological action of maraviroc is dependent on HIV-1 recognizing the CCR5 coreceptor, an HIV tropism assay is required for confirmation of a susceptible HIV-1 strain, the so-called macrophage-tropic (M-tropic) strain that utilizes the CCR5 for entry. If the patient's HIV-1 is identified as the T-tropic strain that uses CXCR4, another chemokine receptor for entry, then maraviroc will not be effective and should not be prescribed.

Another example of application of pretreatment genetic testing is abacavir, a nucleoside reverse transcriptase inhibitor of HIV-1. The use of abacavir has been associated with severe, and sometimes fatal, hypersensitivity reactions in some patients. The abacavir hypersensitivity reaction (AHR) is strongly associated with the presence of the human leukocyte antigen (HLA) B*57 01 of the major histocompatibility complex [102,103]. Two recent studies suggest that the mechanism of AHR is a result of noncovalent binding of abacavir to the F-pocket of the peptide-binding groove of HLA B*57 01 and alteration of its binding specificities to different peptides, with subsequent abacavir-specific CD8(+) T-cell responses to antigens [104,105].

The prevalence of the *HLA B*57 01* variant varies among different ethnic groups and geographical regions. It is absent in Japan and rare in East Asian and African countries, but is common in western Europe and India with a prevalence of 5 to 7% and up to 10%, respectively. The incidence of AHR is higher in Caucasians and Hispanics than in patients of African ancestry. Several studies have provided strong evidence of reduction of AHR incidence with *HLA B*57 01* screening [106–109]. Currently, genetic testing for the *HLA B*57 01* variant is commercially available for screening abacavir candidate patients before prescribing the drug. Patients found to be at high risk for AHR based on genotype will be treated with other antiretroviral drug regimens. This screening recommendation has been incorporated into the abacavir package insert as well as treatment guidelines [110,111].

PHARMACOGENOMICS AND PAIN CONTROL

Opioid Analgesics

Over the years, several single-nucleotide polymorphisms (SNPs) have been discovered in genes that encode different analgesic drug targets, and association studies have been carried out in various pain phenotypes. Not surprisingly, nonreplication of findings is common. Therefore, the following brief overview will focus only on genes that have associations evaluated in multiple cohorts. The μ–opioid receptor (MOR) is the primary drug target for the opioid analgesics. With more than 100 variants of the μ–opioid receptor gene (*OPRM1*) identified [112], the most studied polymorphism is the c.118A>G (rs1799971) SNP in exon 1 of *OPRM1* that results in p.N40D, and lower mRNA expression and protein amount associated with the G allele [113]. MOR function was affected as a result of increased binding affinity for the endogenous opiate β-endorphin, thereby affecting opioid action at the receptor site, with decreased opioid potency by a factor of two to three [114].

A significant association has also been shown between the A118G SNP and decreased potency of morphine-6-glucuronide (M6G), the pharmacologically active metabolite of morphine. Using pharmacokinetic–pharmacodynamic modeling, the study showed that the effector site EC_{50} for M6G was 714 ± 197 nmol/L in six homozygous carriers of the wild-type, $1,475 \pm 424$ nmol/L in five heterozygotes, and 3,140 nmol/L in a homozygous carrier of the G allele [115]. Decreased clinical response to opioids had also been shown in carriers of the G allele. Klepstad et al. reported morphine dosage requirements in cancer patients differed among carriers of the wild-type versus the variant allele. Four homozygous carriers of the G allele required 225 ± 143 mg/day for effective pain control compared to 97 ± 89 mg/day in 78 wild-type homozygotes ($p = 0.006$). However, dosage requirement for heterozygotes was 66 ± 50 mg/day, so there was no evidence of a gene–dose effect [116]. Chou et al. also reported similar findings of different dosage requirements, at 24 and 48 hours after total knee arthroplasty, respectively, of 22.3 ± 10 mg and 40.4 ± 21 mg in homozygous carriers of the G allele versus 16 ± 8 mg and 25.3 ± 15.5 mg in wild-type homozygotes [117]. A "low" responder of morphine receiving 2 gm/day was identified as a carrier of the G allele [118].

Although MOP is essential for opiate action, other gene polymorphisms might also play a role in mediating opioid efficacy and toxicity. Among these polymorphisms, the one that has been quoted extensively in the literature is the *CYP2D6* gene polymorphism, resulting in either lack of efficacy for codeine and hydrocodone in the poor metabolizer phenotype or increased toxicity for codeine, hydrocodone, and tramadol in the ultra-rapid metabolizer phenotype (see chapter 1 for details). Despite strong evidence of a genotype effect on the pharmacokinetics of these opioids, the impact on dosage requirement is much less obvious. In this regard, the value of the *CYP2D6* genotype lies more with guiding the choice of the appropriate analgesic rather than genotype-based dosage recommendations [119].

The catechol-O-methyltransferase (COMT) enzyme is a key modulator of the adrenergic and dopaminergic systems. A functional SNP in the *COMT* gene, c.472G>A (rs4680), results in a p.V158M substitution with three- to fourfold decrease in enzyme activity. The reduced enzyme activity leads to decreased dopamine degradation and subsequent increases in norepinephrine and epinephrine levels that may be associated with exaggerated levels of pain. Down-regulation of endorphins with compensatory up-regulation of MOR has also been suggested to be a result of the SNP [120,121]. Cancer patients who are homozygous carriers of the M variant (high-pain-sensitivity patients) were reported to require more morphine (155 ± 160 mg/day) than heterozygotes (117 ± 10 mg/day) and homozygous carriers of the wild-type V allele (95 ± 99 mg/day) [122].

CONCLUSION

Significant strides have been made in diabetes pharmacogenomics over the last decade, and there is no doubt that the field will continue to expand at a rapid pace. It is reasonable to expect that in the years to come, GWASes will further uncover the molecular underpinnings of diabetes pathophysiology, drug response, and adverse effects. As researchers overcome challenges and take advantages of opportunities in the field, the translation of diabetes pharmacogenomic information to clinical practice will likely become a reality. Additional successes with pharmacogenomics have also been demonstrated in management of patients with HIV-1 infection and in shaping the response to opioid analgesic drugs.

QUESTIONS FOR DISCUSSION

1. Discuss the relevant polymorphisms that affect antidiabetic pharmacotherapy.
2. Describe how GWASes could offer additional insight into antidiabetic pharmacotherapy.
3. What unique feature of HIV-1 makes it a target for antiretroviral drug development?
4. Explain how morphine therapy can be influenced by pharmacogenomics.

References

[1] http://www.idf.org/diabetesatlas/5e/the-global-burden.
[2] Diagnosis and classification of diabetes mellitus. Diabetes Care 2012;35(Suppl. 1):S64–71.

[3] Nathan DM, Buse JB, Davidson MB, Ferrannini E, Holman RR, Sherwin R, et al. Medical management of hyperglycemia in type 2 diabetes: a consensus algorithm for the initiation and adjustment of therapy: a consensus statement of the American Diabetes Association and the European Association for the Study of Diabetes. Diabetes Care 2009;32:193—203.

[4] Smith RJ, Nathan DM, Arslanian SA, Groop L, Rizza RA, Rotter JI. Individualizing therapies in type 2 diabetes mellitus based on patient characteristics: what we know and what we need to know. J Clin Endocrinol Metab 2010;95:1566—74.

[5] Gloyn AL, Ellard S. Defining the genetic aetiology of monogenic diabetes can improve treatment. Expert Opin Pharmacother 2006;7:1759—67.

[6] Murphy R, Ellard S, Hattersley AT. Clinical implications of a molecular genetic classification of monogenic beta-cell diabetes. Nat Clin Pract Endocrinol Metab 2008;4:200—13.

[7] Gloyn AL, Pearson ER, Antcliff JF, Proks P, Bruining GJ, Slingerland AS, et al. Activating mutations in the gene encoding the ATP-sensitive potassium-channel subunit Kir6.2 and permanent neonatal diabetes. N Engl J Med 2004;350:1838—49.

[8] Babenko AP, Polak M, Cave H, Busiah K, Czernichow P, Scharfmann R, et al. Activating mutations in the ABCC8 gene in neonatal diabetes mellitus. N Engl J Med 2006;355:456—66.

[9] Sagen JV, Raeder H, Hathout E, Shehadeh N, Gudmundsson K, Baevre H, et al. Permanent neonatal diabetes due to mutations in KCNJ11 encoding Kir6.2: patient characteristics and initial response to sulfonylurea therapy. Diabetes 2004;53:2713—8.

[10] Pearson ER, Flechtner I, Njolstad PR, Malecki MT, Flanagan SE, Larkin B, et al. Switching from insulin to oral sulfonylureas in patients with diabetes due to Kir6.2 mutations. N Engl J Med 2006;355:467—77.

[11] Giuffrida FM, Reis AF. Genetic and clinical characteristics of maturity-onset diabetes of the young. Diabetes Obes Metab 2005;7:318—26.

[12] Thanabalasingham G, Owen KR. Diagnosis and management of maturity onset diabetes of the young (MODY). BMJ 2011;343:d6044.

[13] Pearson ER, Starkey BJ, Powell RJ, Gribble FM, Clark PM, Hattersley AT. Genetic cause of hyperglycaemia and response to treatment in diabetes. Lancet 2003;362:1275—81.

[14] Shepherd M, Shields B, Ellard S, Rubio-Cabezas O, Hattersley AT. A genetic diagnosis of HNF1A diabetes alters treatment and improves glycaemic control in the majority of insulin-treated patients. Diabet Med 2009;26:437—41.

[15] Huang C, Florez JC. Pharmacogenetics in type 2 diabetes: potential implications for clinical practice. Genome Med 2011;3:76.

[16] DeFronzo RA. Pharmacologic therapy for type 2 diabetes mellitus. Ann Intern Med 1999;131:281—303.

[17] Kahn SE, Haffner SM, Heise MA, Herman WH, Holman RR, Jones NP, et al. Glycemic durability of rosiglitazone, metformin, or glyburide monotherapy. N Engl J Med 2006;355:2427—43.

[18] Intensive blood-glucose control with sulphonylureas or insulin compared with conventional treatment and risk of complications in patients with type 2 diabetes (UKPDS 33). UK Prospective Diabetes Study (UKPDS) Group. Lancet 1998;352:837—53.

[19] Aquilante CL. Sulfonylurea pharmacogenomics in Type 2 diabetes: the influence of drug target and diabetes risk polymorphisms. Expert Rev Cardiovasc Ther 2010;8:359—72.

[20] Kirchheiner J, Roots I, Goldammer M, Rosenkranz B, Brockmoller J. Effect of genetic polymorphisms in cytochrome p450 (CYP) 2C9 and CYP2C8 on the pharmacokinetics of oral antidiabetic drugs: clinical relevance. Clin Pharmacokinet 2005;44:1209—25.

[21] Kirchheiner J, Brockmoller J, Meineke I, Bauer S, Rohde W, Meisel C, et al. Impact of CYP2C9 amino acid polymorphisms on glyburide kinetics and on the insulin and glucose response in healthy volunteers. Clin Pharmacol Ther 2002;71:286—96.

[22] Zhou K, Donnelly L, Burch L, Tavendale R, Doney AS, Leese G, et al. Loss-of-function CYP2C9 variants improve therapeutic response to sulfonylureas in type 2 diabetes: a Go-DARTS study. Clin Pharmacol Ther 2010;87:52—6.

[23] Becker ML, Visser LE, Trienekens PH, Hofman A, van Schaik RH, et al. Cytochrome P450 2C9 *2 and *3 polymorphisms and the dose and effect of sulfonylurea in type II diabetes mellitus. Clin Pharmacol Ther 2008;83:288—92.

[24] Swen JJ, Wessels JA, Krabben A, Assendelft WJ, Guchelaar HJ. Effect of CYP2C9 polymorphisms on prescribed dose and time-to-stable dose of sulfonylureas in primary care patients with Type 2 diabetes mellitus. Pharmacogenomics 2010;11:1517−23.

[25] Holstein A, Plaschke A, Ptak M, Egberts EH, El-Din J, Brockmoller J, et al. Association between CYP2C9 slow metabolizer genotypes and severe hypoglycaemia on medication with sulphonylurea hypoglycaemic agents. Br J Clin Pharmacol 2005;60:103−6.

[26] Ragia G, Petridis I, Tavridou A, Christakidis D, Manolopoulos VG. Presence of CYP2C9*3 allele increases risk for hypoglycemia in Type 2 diabetic patients treated with sulfonylureas. Pharmacogenomics 2009;10:1781−7.

[27] Holstein A, Hahn M, Patzer O, Seeringer A, Kovacs P, Stingl J. Impact of clinical factors and CYP2C9 variants for the risk of severe sulfonylurea-induced hypoglycemia. Eur J Clin Pharmacol 2011;67:471−6.

[28] Grant RW, Wexler DJ. Loss-of-function CYP2C9 variants: finding the correct clinical role for Type 2 diabetes pharmacogenetic testing. Expert Rev Cardiovasc Ther 2010;8:339−43.

[29] Gloyn AL, Hashim Y, Ashcroft SJ, Ashfield R, Wiltshire S, Turner RC. Association studies of variants in promoter and coding regions of beta-cell ATP-sensitive K-channel genes SUR1 and Kir6.2 with Type 2 diabetes mellitus (UKPDS 53). Diabet Med 2001;18:206−12.

[30] Gloyn AL, Weedon MN, Owen KR, Turner MJ, Knight BA, Hitman G, et al. Large-scale association studies of variants in genes encoding the pancreatic beta-cell KATP channel subunits Kir6.2 (KCNJ11) and SUR1 (ABCC8) confirm that the KCNJ11 E23K variant is associated with type 2 diabetes. Diabetes 2003;52:568−72.

[31] Nielsen EM, Hansen L, Carstensen B, Echwald SM, Drivsholm T, Glumer C, et al. The E23K variant of Kir6.2 associates with impaired post-OGTT serum insulin response and increased risk of type 2 diabetes. Diabetes 2003;52:573−7.

[32] Barroso I, Luan J, Middelberg RP, Harding AH, Franks PW, Jakes RW, et al. Candidate gene association study in type 2 diabetes indicates a role for genes involved in beta-cell function as well as insulin action. PLoS Biol 2003;1:E20.

[33] Florez JC, Burtt N, de Bakker PI, Almgren P, Tuomi T, et al. Haplotype structure and genotype-phenotype correlations of the sulfonylurea receptor and the islet ATP-sensitive potassium channel gene region. Diabetes 2004;53:1360−8.

[34] Hamming KS, Soliman D, Matemisz LC, Niazi O, Lang Y, Gloyn AL, et al. Coexpression of the type 2 diabetes susceptibility gene variants KCNJ11 E23K and ABCC8 S1369A alter the ATP and sulfonylurea sensitivities of the ATP-sensitive K(+) channel. Diabetes 2009;58:2419−24.

[35] Feng Y, Mao G, Ren X, Xing H, Tang G, Li Q, et al. Ser1369Ala variant in sulfonylurea receptor gene ABCC8 is associated with antidiabetic efficacy of gliclazide in Chinese type 2 diabetic patients. Diabetes Care 2008;31:1939−44.

[36] Lang VY, Fatehi M, Light PE. Pharmacogenomic analysis of ATP-sensitive potassium channels coexpressing the common type 2 diabetes risk variants E23K and S1369A. Pharmacogenet Genomics 2012;22(3):206−14.

[37] McCarthy MI. Genomics, type 2 diabetes, and obesity. N Engl J Med 2010;363:2339−50.

[38] Stolerman ES, Florez JC. Genomics of type 2 diabetes mellitus: implications for the clinician. Nat Rev Endocrinol 2009;5:429−36.

[39] Petrie JR, Pearson ER, Sutherland C. Implications of genome wide association studies for the understanding of type 2 diabetes pathophysiology. Biochem Pharmacol 2011;81:471−7.

[40] Jin T. The WNT signalling pathway and diabetes mellitus. Diabetologia 2008;51:1771−80.

[41] Sladek R, Rocheleau G, Rung J, Dina C, Shen L, Serre D, et al. A genome-wide association study identifies novel risk loci for type 2 diabetes. Nature 2007;445:881−5.

[42] Lyssenko V, Lupi R, Marchetti P, Del Guerra S, Orho-Melander M, Almgren P, et al. Mechanisms by which common variants in the TCF7L2 gene increase risk of type 2 diabetes. J Clin Invest 2007;117:2155−63.

[43] Saxena R, Gianniny L, Burtt NP, Lyssenko V, Giuducci C, Sjogren M, et al. Common single nucleotide polymorphisms in TCF7L2 are reproducibly associated with type 2 diabetes and reduce the insulin response to glucose in nondiabetic individuals. Diabetes 2006;55:2890−5.

[44] Florez JC, Jablonski KA, Bayley N, Pollin TI, de Bakker PI, et al. TCF7L2 polymorphisms and progression to diabetes in the Diabetes Prevention Program. N Engl J Med 2006;355:241−50.

[45] Pearson ER, Donnelly LA, Kimber C, Whitley A, Doney AS, McCarthy MI, et al. Variation in TCF7L2 influences therapeutic response to sulfonylureas: a GoDARTs study. Diabetes 2007;56:2178−82.

[46] Holstein A, Hahn M, Korner A, Stumvoll M, Kovacs P. TCF7L2 and therapeutic response to sulfonylureas in patients with type 2 diabetes. BMC Med Genet 2011;12:30.

[47] Schroner Z, Javorsky M, Tkacova R, Klimcakova L, Dobrikova M, Habalova V, et al. Effect of sulphonylurea treatment on glycaemic control is related to TCF7L2 genotype in patients with type 2 diabetes. Diabetes Obes Metab 2011;13:89—91.

[48] Becker ML, Aarnoudse AJ, Newton-Cheh C, Hofman A, Witteman JC, Uitterlinden AG, et al. Common variation in the NOS1AP gene is associated with reduced glucose-lowering effect and with increased mortality in users of sulfonylurea. Pharmacogenet Genomics 2008;18:591—7.

[49] Schroner Z, Javorsky M, Haluskova J, Klimcakova L, Babjakova E, Fabianova M, et al. Variation in CDKAL1 gene is associated with therapeutic response to sulphonylureas. Physiol Res 2012.

[50] Sesti G, Marini MA, Cardellini M, Sciacqua A, Frontoni S, Andreozzi F, et al. The Arg972 variant in insulin receptor substrate-1 is associated with an increased risk of secondary failure to sulfonylurea in patients with type 2 diabetes. Diabetes Care 2004;27:1394—8.

[51] Swen JJ, Guchelaar HJ, Baak-Pablo RF, Assendelft WJ, Wessels JA. Genetic risk factors for type 2 diabetes mellitus and response to sulfonylurea treatment. Pharmacogenet Genomics 2011;21:461—8.

[52] Zolk O. Disposition of metformin: variability due to polymorphisms of organic cation transporters. Ann Med 2012;44(2):119—29.

[53] Bailey CJ, Turner RC. Metformin. N Engl J Med 1996;334:574—9.

[54] Pacanowski MA, Hopley CW, Aquilante CL. Interindividual variability in oral antidiabetic drug disposition and response: the role of drug transporter polymorphisms. Expert Opin Drug Metab Toxicol 2008;4:529—44.

[55] Shu Y, Brown C, Castro RA, Shi RJ, Lin ET, Owen RP, et al. Effect of genetic variation in the organic cation transporter 1, OCT1, on metformin pharmacokinetics. Clin Pharmacol Ther 2008;83:273—80.

[56] Shu Y, Sheardown SA, Brown C, Owen RP, Zhang S, Castro RA, et al. Effect of genetic variation in the organic cation transporter 1 (OCT1) on metformin action. J Clin Invest 2007;117:1422—31.

[57] Zhou K, Donnelly LA, Kimber CH, Donnan PT, Doney AS, Leese G, et al. Reduced-function SLC22A1 polymorphisms encoding organic cation transporter 1 and glycemic response to metformin: a GoDARTS study. Diabetes 2009;58:1434—9.

[58] Shikata E, Yamamoto R, Takane H, Shigemasa C, Ikeda T, Otsubo K, et al. Human organic cation transporter (OCT1 and OCT2) gene polymorphisms and therapeutic effects of metformin. J Hum Genet 2007;52:117—22.

[59] Becker ML, Visser LE, van Schaik RH, Hofman A, Uitterlinden AG, Stricker BH. Genetic variation in the organic cation transporter 1 is associated with metformin response in patients with diabetes mellitus. Pharmacogenomics J 2009;9:242—7.

[60] Leabman MK, Giacomini KM. Estimating the contribution of genes and environment to variation in renal drug clearance. Pharmacogenetics 2003;13:581—4.

[61] Becker ML, Visser LE, van Schaik RH, Hofman A, Uitterlinden AG, et al. Genetic variation in the multidrug and toxin extrusion 1 transporter protein influences the glucose-lowering effect of metformin in patients with diabetes: a preliminary study. Diabetes 2009;58:745—9.

[62] Jablonski KA, McAteer JB, de Bakker PI, Franks PW, Pollin TI, et al. Common variants in 40 genes assessed for diabetes incidence and response to metformin and lifestyle intervention in the diabetes prevention program. Diabetes 2010;59:2672—81.

[63] Choi JH, Yee SW, Ramirez AH, Morrissey KM, Jang GH, Joski PJ, et al. A common 5'-UTR variant in MATE2-K is associated with poor response to metformin. Clin Pharmacol Ther 2011;90:674—84.

[64] Becker ML, Visser LE, van Schaik RH, Hofman A, Uitterlinden AG, et al. Interaction between polymorphisms in the OCT1 and MATE1 transporter and metformin response. Pharmacogenet Genomics 2010;20:38—44.

[65] Zhou K, Bellenguez C, Spencer CC, Bennett AJ, Coleman RL, Tavendale R, et al. Common variants near ATM are associated with glycemic response to metformin in type 2 diabetes. Nat Genet 2011;43:117—20.

[66] Hivert MF, Jablonski KA, Perreault L, Saxena R, McAteer JB, Franks PW, et al. Updated genetic score based on 34 confirmed type 2 diabetes Loci is associated with diabetes incidence and regression to normoglycemia in the diabetes prevention program. Diabetes 2011;60:1340—8.

[67] Yki-Jarvinen H. Thiazolidinediones. N Engl J Med 2004;351:1106—18.

[68] Nissen SE, Wolski K. Effect of rosiglitazone on the risk of myocardial infarction and death from cardiovascular causes. N Engl J Med 2007;356:2457—71.

[69] http://www.fda.gov/Drugs/DrugSafety/ucm255005.htm.

[70] Bluher M, Lubben G, Paschke R. Analysis of the relationship between the Pro12Ala variant in the PPAR-gamma2 gene and the response rate to therapy with pioglitazone in patients with type 2 diabetes. Diabetes Care 2003;26:825–31.

[71] Snitker S, Watanabe RM, Ani I, Xiang AH, Marroquin A, Ochoa C, et al. Changes in insulin sensitivity in response to troglitazone do not differ between subjects with and without the common, functional Pro12Ala peroxisome proliferator-activated receptor-gamma2 gene variant: results from the Troglitazone in Prevention of Diabetes (TRIPOD) study. Diabetes Care 2004;27:1365–8.

[72] Berlie HD, Kalus JS, Jaber LA. Thiazolidinediones and the risk of edema: a meta-analysis. Diabetes Res Clin Pract 2007;76:279–89.

[73] Lago RM, Singh PP, Nesto RW. Congestive heart failure and cardiovascular death in patients with prediabetes and type 2 diabetes given thiazolidinediones: a meta-analysis of randomised clinical trials. Lancet 2007;370: 1129–36.

[74] Chang TJ, Liu PH, Liang YC, Chang YC, Jiang YD, Li HY, et al. Genetic predisposition and nongenetic risk factors of thiazolidinedione-related edema in patients with type 2 diabetes. Pharmacogenet Genomics 2011;21:829–36.

[75] Nesto RW, Bell D, Bonow RO, Fonseca V, Grundy SM, Horton ES, et al. Thiazolidinedione use, fluid retention, and congestive heart failure: a consensus statement from the American Heart Association and American Diabetes Association. Circulation 2003;108:2941–8.

[76] Aquilante CL. Pharmacogenetics of thiazolidinedione therapy. Pharmacogenomics 2007;8:917–31.

[77] Baldwin SJ, Clarke SE, Chenery RJ. Characterization of the cytochrome P450 enzymes involved in the in vitro metabolism of rosiglitazone. Br J Clin Pharmacol 1999;48:424–32.

[78] Jaakkola T, Laitila J, Neuvonen PJ, Backman JT. Pioglitazone is metabolised by CYP2C8 and CYP3A4 in vitro: potential for interactions with CYP2C8 inhibitors. Basic Clin Pharmacol Toxicol 2006;99:44–51.

[79] Kirchheiner J, Thomas S, Bauer S, Tomalik-Scharte D, Hering U, Doroshyenko O, et al. Pharmacokinetics and pharmacodynamics of rosiglitazone in relation to CYP2C8 genotype. Clin Pharmacol Ther 2006;80: 657–67.

[80] Aquilante CL, Bushman LR, Knutsen SD, Burt LE, Rome LC, Kosmiski LA. Influence of SLCO1B1 and CYP2C8 gene polymorphisms on rosiglitazone pharmacokinetics in healthy volunteers. Hum Genomics 2008; 3:7–16.

[81] Tornio A, Niemi M, Neuvonen PJ, Backman JT. Trimethoprim and the CYP2C8*3 allele have opposite effects on the pharmacokinetics of pioglitazone. Drug Metab Dispos 2008;36:73–80.

[82] Ludovico O, Pellegrini F, Di Paola R, Minenna A, Mastroianno S, Cardellini M, et al. Heterogeneous effect of peroxisome proliferator-activated receptor gamma2 Ala12 variant on type 2 diabetes risk. Obesity (Silver Spring) 2007;15:1076–81.

[83] Florez JC, Jablonski KA, Sun MW, Bayley N, Kahn SE, Shamoon H, et al. Effects of the type 2 diabetes-associated PPARG P12A polymorphism on progression to diabetes and response to troglitazone. J Clin Endocrinol Metab 2007;92:1502–9.

[84] Manolopoulos VG, Ragia G, Tavridou A. Pharmacogenomics of oral antidiabetic medications: current data and pharmacoepigenomic perspective. Pharmacogenomics 2011;12:1161–91.

[85] Billings LK, Florez JC. The genetics of type 2 diabetes: what have we learned from GWAS? Ann N Y Acad Sci 2010;1212:59–77.

[86] Bailey SD, Xie C, Do R, Montpetit A, Diaz R, Mohan V, et al. Variation at the NFATC2 locus increases the risk of thiazolidinedione-induced edema in the Diabetes REduction Assessment with ramipril and rosiglitazone Medication (DREAM) study. Diabetes Care 2010;33:2250–3.

[87] DeFronzo RA, Tripathy D, Schwenke DC, Banerji M, Bray GA, Buchanan TA, et al. Pioglitazone for diabetes prevention in impaired glucose tolerance. N Engl J Med 2011;364:1104–15.

[88] Tahrani AA, Bailey CJ, Del Prato S, Barnett AH. Management of type 2 diabetes: new and future developments in treatment. Lancet 2011;378:182–97.

[89] Scheen AJ. DPP-4 inhibitors in the management of type 2 diabetes: a critical review of head-to-head trials. Diabetes Metab 2012;38(2):89–101.

[90] Esposito K, Cozzolino D, Bellastella G, Maiorino MI, Chiodini P, Ceriello A, et al. Dipeptidyl peptidase-4 inhibitors and HbA1c target of <7% in type 2 diabetes: meta-analysis of randomized controlled trials. Diabetes Obes Metab 2011;13:594–603.

[91] Zinman B, Schmidt WE, Moses A, Lund N, Gough S. Achieving a clinically relevant composite outcome of an HbA1c of <7% without weight gain or hypoglycaemia in type 2 diabetes: a meta-analysis of the liraglutide clinical trial programme. Diabetes Obes Metab 2012;14:77–82.

[92] Watanabe RM. Drugs, diabetes and pharmacogenomics: the road to personalized therapy. Pharmacogenomics 2011;12:699–701.

[93] Vella A. Pharmacogenetics for type 2 diabetes: practical considerations for study design. J Diabetes Sci Technol 2009;3:705–9.

[94] Alberti KG, Eckel RH, Grundy SM, Zimmet PZ, Cleeman JI, Donato KA, et al. Harmonizing the metabolic syndrome: a joint interim statement of the International Diabetes Federation Task Force on Epidemiology and Prevention; National Heart, Lung, and Blood Institute; American Heart Association; World Heart Federation; International Atherosclerosis Society; and International Association for the Study of Obesity. Circulation 2009;120:1640–5.

[95] Greeley SA, John PM, Winn AN, Ornelas J, Lipton RB, Philipson LH, et al. The cost-effectiveness of personalized genetic medicine: the case of genetic testing in neonatal diabetes. Diabetes Care 2011;34:622–7.

[96] Imamura M, Maeda S. Genetics of type 2 diabetes: the GWAS era and future perspectives [Review]. Endocr J 2011;58:723–39.

[97] Markowitz SM, Park ER, Delahanty LM, O'Brien KE, Grant RW. Perceived impact of diabetes genetic risk testing among patients at high phenotypic risk for type 2 diabetes. Diabetes Care 2011;34:568–73.

[98] de Miguel-Yanes JM, Shrader P, Pencina MJ, Fox CS, Manning AK, et al. Genetic risk reclassification for type 2 diabetes by age below or above 50 years using 40 type 2 diabetes risk single nucleotide polymorphisms. Diabetes Care 2011;34:121–5.

[99] Carrington M, Dean M, Martin MP, O'Brien SJ. Genetics of HIV-1 infection: chemokine receptor CCR5 polymorphism and its consequences. Hum Mol Genet 1999;8:1939–45.

[100] Liu R, Paxton WA, Choe S, Ceradini D, Martin SR, Horuk R, et al. Homozygous defect in HIV-1 coreceptor accounts for resistance of some multiply-exposed individuals to HIV-1 infection. Cell 1996;86:367–77.

[101] Samson M, Libert F, Doranz BJ, Rucker J, Liesnard C, Farber CM, et al. Resistance to HIV-1 infection in Caucasian individuals bearing mutant alleles of the CCR-5 chemokine receptor gene. Nature 1996;382: 722–5.

[102] Hetherington S, Hughes AR, Mosteller M, Shortino D, Baker KL, Spreen W, et al. Genetic variations in HLA-B region and hypersensitivity reactions to abacavir. Lancet 2002;359:1121–2.

[103] Mallal S, Nolan D, Witt C, Masel G, Martin AM, Moore C, et al. Association between presence of HLA-B*5701, HLA-DR7, and HLA-DQ3 and hypersensitivity to HIV-1 reverse-transcriptase inhibitor abacavir. Lancet 2002;359:727–32.

[104] Illing PT, Vivian JP, Dudek NL, Kostenko L, Chen Z, Bharadwaj M, et al. Immune self-reactivity triggered by drug-modified HLA-peptide repertoire. Nature 2012;486:554–8.

[105] Ostrov DA, Grant BJ, Pompeu YA, Sidney J, Harndahl M, Southwood S, et al. Drug hypersensitivity caused by alteration of the MHC-presented self-peptide repertoire. Proc Natl Acad Sci U S A 2012;109:9959–64.

[106] Mallal S, Phillips E, Carosi G, Molina JM, Workman C, Tomazic J, et al. HLA-B*5701 screening for hypersensitivity to abacavir. N Engl J Med 2008;358:568–79.

[107] Rauch A, Nolan D, Martin A, McKinnon E, Almeida C, Mallal S. Prospective genetic screening decreases the incidence of abacavir hypersensitivity reactions in the Western Australian HIV cohort study. Clin Infect Dis 2006;43:99–102.

[108] Zucman D, Truchis P, Majerholc C, Stegman S, Caillat-Zucman S. Prospective screening for human leukocyte antigen-B*5701 avoids abacavir hypersensitivity reaction in the ethnically mixed French HIV population. J Acquir Immune Defic Syndr 2007;45:1–3.

[109] Young B, Squires K, Patel P, Dejesus E, Bellos N, Berger D, et al. First large, multicenter, open-label study utilizing HLA-B*5701 screening for abacavir hypersensitivity in North America. Aids 2008;22:1673–5.

[110] Martin MA, Klein TE, Dong BJ, Pirmohamed M, Haas DW, Kroetz DL. Clinical pharmacogenetics implementation consortium guidelines for HLA-B genotype and abacavir dosing. Clin Pharmacol Ther 2012;91:734–8.

[111] Thompson MA, Aberg JA, Cahn P, Montaner JS, Rizzardini G, Telenti A, et al. Antiretroviral treatment of adult HIV infection: 2010 recommendations of the International AIDS Society-USA panel. J Am Med Assoc 2010;304:321–33.

[112] Ikeda K, Ide S, Han W, Hayashida M, Uhl GR, Sora I. How individual sensitivity to opiates can be predicted by gene analyses. Trends Pharmacol Sci 2005;26:311–7.

[113] Zhang Y, Wang D, Johnson AD, Papp AC, Sadee W. Allelic expression imbalance of human mu opioid receptor (OPRM1) caused by variant A118G. J Biol Chem 2005;280:32618–24.

[114] Bond C, LaForge KS, Tian M, Melia D, Zhang S, Borg L, et al. Single-nucleotide polymorphism in the human mu opioid receptor gene alters beta-endorphin binding and activity: possible implications for opiate addiction. Proc Natl Acad Sci U S A 1998;95:9608–13.

[115] Lotsch J, Skarke C, Grosch S, Darimont J, Schmidt H, Geisslinger G. The polymorphism A118G of the human mu-opioid receptor gene decreases the pupil constrictory effect of morphine-6-glucuronide but not that of morphine. Pharmacogenetics 2002;12:3–9.

[116] Klepstad P, Rakvag TT, Kaasa S, Holthe M, Dale O, Borchgrevink PC, et al. The 118 A > G polymorphism in the human mu-opioid receptor gene may increase morphine requirements in patients with pain caused by malignant disease. Acta Anaesthesiol Scand 2004;48:1232–9.

[117] Chou WY, Yang LC, Lu HF, Ko JY, Wang CH, Lin SH, et al. Association of mu-opioid receptor gene polymorphism (A118G) with variations in morphine consumption for analgesia after total knee arthroplasty. Acta Anaesthesiol Scand 2006;50:787–92.

[118] Hirota T, Ieiri I, Takane H, Sano H, Kawamoto K, Aono H, et al. Sequence variability and candidate gene analysis in two cancer patients with complex clinical outcomes during morphine therapy. Drug Metab Dispos 2003;31:677–80.

[119] Crews KR, Gaedigk A, Dunnenberger HM, Klein TE, Shen DD, Callaghan JT, et al. Clinical Pharmacogenetics Implementation Consortium (CPIC) guidelines for codeine therapy in the context of cytochrome P450 2D6 (CYP2D6) genotype. Clin Pharmacol Ther 2012;91:321–6.

[120] Berthele A, Platzer S, Jochim B, Boecker H, Buettner A, Conrad B, et al. COMT Val108/158Met genotype affects the mu-opioid receptor system in the human brain: evidence from ligand-binding, G-protein activation and preproenkephalin mRNA expression. Neuroimage 2005;28:185–93.

[121] Zubieta JK, Heitzeg MM, Smith YR, Bueller JA, Xu K, Xu Y, et al. COMT val158met genotype affects mu-opioid neurotransmitter responses to a pain stressor. Science 2003;299:1240–3.

[122] Rakvag TT, Klepstad P, Baar C, Kvam TM, Dale O, Kaasa S, et al. The Val158Met polymorphism of the human catechol-O-methyltransferase (COMT) gene may influence morphine requirements in cancer pain patients. Pain 2005;116:73–8.

Molecular Approaches, Models, and Techniques in Pharmacogenomic Research and Development

Wenbo Mu*, Wei Zhang[†, **, ‡]

*Department of Bioengineering, University of Illinois at Chicago, Chicago, Illinois, USA
†Department of Pediatrics
**Institute of Human Genetics
‡Cancer Center, University of Illinois at Chicago, Chicago, Illinois, USA

OBJECTIVES

1. Demonstrate an understanding of the fundamental knowledge of human genetic variation and the general strategies in pharmacogenomic research and development.

2. Become familiar with the statistical approaches and bioinformatic tools commonly used in pharmacogenomic research.

INTRODUCTION

The clinical response to prescribed drugs is often inconsistent across individual patients and may range from beneficial effects to lack of efficacy to even fatal adverse drug reactions (ADRs). Severe ADRs were found to account for more than 2.2 million hospitalizations and over 100,000 annual deaths in a meta-analysis of prospective studies, making these reactions between the fourth and sixth leading causes of hospitalization and death in the United States [1]. Hence, identifying patients at risk for severe ADRs to prescribed medicine prior to treatment would greatly improve the current clinical practice and patient care. Of particular concern are those drugs with a narrow therapeutic index. For example, warfarin, the most commonly prescribed anticoagulant, exhibits large interpatient variability in dose requirement, which if not properly monitored may increase either thrombosis risk (i.e., dose too low) or serious bleeding risk (i.e., dose too high). In oncology, many anticancer drugs also have a narrow therapeutic index (i.e., a narrow range of drug amount between anticancer effect and toxicity (e.g., nephrotoxicity)). A variety of genetic and nongenetic (e.g., environment, diet, lifestyle) factors can influence an individual's response to medicines. Therefore, there is often no simple way to determine whether a patient will respond well, badly, or not at all to a medication. For example, though patient-specific factors (e.g., age, body size, race, concurrent diseases) may explain some of the variability in the interindividual dosing requirement for warfarin, genetic factors (e.g., genetic variants in *CYP2C9* [encoding cytochrome P450, family 2, subfamily C, polypeptide 9] and *VKORC1* [encoding vitamin K epoxide reductase complex, subunit 1]) have been demonstrated to contribute a significant proportion of the variability in warfarin dose [2]. In fact, several studies have derived algorithms that integrate both genetic and nongenetic factors to predict warfarin dose [3,4]. Prospectively, personalized medicine has the ambitious goal of both maximizing effective therapy and avoiding adverse effects by tailoring medical care based on a patient's genetic makeup, together with consideration of other relevant clinical information.

With the completion of the first human genome sequences [5,6] and the launching of several large-scale population-based genetic variation mapping efforts such as the International HapMap Project [7—9] and the 1000 Genomes Project [10], the general patterns of human genetic variation and the genetic architectures of a variety of complex traits and phenotypes (e.g., gene expression, cytotoxicities to anticancer drugs) have begun to be elucidated using these models and data. For example, significant variation in the quantitative gene expression and gene regulation through genetic variants in the forms of single-nucleotide polymorphisms (SNPs) and copy number variants (CNVs) have been investigated both within and across human populations [11—18]. More recently, epigenetic markers including microRNAs [19] and DNA methylation [20,21] have emerged to be novel regulators of the quantitative gene expression phenotype, as well as crucial players in determining gene expression variation between populations [19,21,22].

Previous studies have demonstrated that drug response is likely to be a heritable phenotype, which may be partially attributed to genetic diversity. For example, it was estimated that genetic factors could determine approximately 47% of the susceptibility to cisplatin-induced cytotoxicity, indicating that sensitivity to the cytotoxic effects may be under substantial genetic influence [23]. Furthermore, there is evidence that drug response can be due to multiple genomic loci or regions, each of which may only contribute a small proportion of the variability of the drug response phenotype. For example, linkage analysis revealed 11 genomic regions on 6 chromosomes to be significantly associated with cisplatin-induced cytotoxicity in a panel of lymphoblastoid cell lines (LCLs) [24]. Therefore, similar to the quantitative gene expression phenotype, drug response is likely to be a complex trait, in the sense that both genetic/epigenetic variation in multiple genes and various nongenetic factors could contribute to the clinical variability of drug response. For example, genetic contribution to drug response may be through regulating the expression of relevant genes as eQTLs (expression quantitative trait loci), which are genomic loci (e.g., SNPs), acting either locally (i.e., *cis*-acting) or distantly (i.e., *trans*-acting), that are identified to be associated with gene expression.

Concurrent with the increasing knowledge of human genetic variation (e.g., linkage disequilibrium [LD] patterns between individual SNPs) and the advances in technologies (e.g., microarrays, genotyping, sequencing), the development of pharmacogenetic (i.e., focusing on individual genes) and pharmacogenomic (i.e., utilizing genome-wide approaches) research has paved the road to personalized medicine by investigating the relationships between human genetic variation and drug response. Given the complex nature of drug response and the scale of genetic variation data, pharmacogenetic and pharmacogenomic research rely on various bioinformatic tools and statistical approaches. Though many techniques and research approaches used in pharmacogenetic and pharmacogenomic research are similar to those used in other studies for disease susceptibility, unique situations in pharmacogenomic studies (e.g., difficulty in identifying normal controls in a case-control association study for toxic anticancer drugs) present certain challenges in study design and data analysis.

With the aim of providing a general overview of the research techniques and approaches, we start from an introduction to human genetic variation and a couple of important concepts that are the basis for pharmacogenetic and pharmacogenomic studies. The general strategies in pharmacogenomic research and development (i.e., candidate gene and whole-genome approaches) are then described with an emphasis on the data analysis pipelines and statistical approaches. In particular, for the whole-genome approach, a cell-based model using samples from the International HapMap Project [7−9] is introduced as an example, as significant progress in pharmacogenomic discovery has been made using this model [25,26] by taking advantage of the tremendous resources of the HapMap samples. In addition, because various bioinformatic resources have been developed to facilitate the use of human genetic variation data and the integration of data from different studies, we provide an introduction to several important pharmacogenomics-related bioinformatic resources. Finally, we briefly discuss the challenges raised by the burgeoning applications of the next-generation sequencing technologies [27,28] to pharmacogenomic research and development.

HUMAN GENETIC VARIATION

Human genetic variation refers to the observed differences in DNA sequences (i.e., polymorphisms or genetic variants) both within and across populations. It has been estimated that 99.9% genetic identity exists between two randomly chosen individuals. The 0.1% difference can be translated into approximately 3 million base-pair differences between any two randomly picked individuals. Of the approximately 0.1% of DNA that varies among individuals, approximately 85 to 90% of genetic variation is found within three major continental groups (Asians, Europeans, and Africans) and only an additional 5 to 15% of variation is found between any two populations [29]. Typically, individuals of African ancestry have higher levels of genetic diversity than non-African populations [7,8]. Though the extent of genetic contribution to drug response is still controversial, it is believed that at least some of the variation in drug response between individuals may be attributed to the 0.1% genetic difference. Therefore, the human genetic variation is the basis for pharmacogenetic and pharmacogenomic studies. Research techniques and approaches have been developed to utilize the features of human genetic variation (e.g., linkage between a genotyped genetic variant and the untyped causal variant) to identify polymorphisms associated with drug response in pharmacogenetic and pharmacogenomic studies. An overview of human genetic variations and a few important concepts will serve as the foundation before introducing the general strategies used in pharmacogenomic research and development.

Forms of Human Genetic Variation and Complex Traits

The most common form of genetic variation is an SNP, which refers to genetic variation in a DNA sequence that occurs when a single nucleotide (A: adenine, C: cytosine, T: thymine, or G: guanine) in a genome is altered. SNPs can be bialleic (i.e., two allelic variants segregating in the population) or multi-allelic (e.g., triallelic). The distribution of SNPs across the human genome is not homogeneous, with more SNPs located in noncoding regions than coding regions, reflecting the combined effects of natural selection, recombination, and mutation rates [30,31]. The current dbSNP database [32,33] (build 135)—that is, the National Center for Biotechnology Information (NCBI) database of short genetic variations—contains more than 40 million validated SNPs, including more than 35 million SNPs with frequency information (e.g., derived from the 1000 Genomes Project [10]) and more than 20 million SNPs in genic regions. For Mendelian diseases or traits, a single SNP in one gene could be the causal variant. In contrast, for common, complex diseases and traits including drug response, multiple SNPs—together with other nongenetic factors—may need to work in coordination to manifest a particular phenotype [34]. For example, gene expression, an intermediate phenotype between DNA sequence and cellular, whole-body phenotypes, has been demonstrated to be a complex and heritable trait that is partially contributed by genetic polymorphisms including SNPs (i.e., eQTLs) [18,35]. To date, association studies have implicated SNPs in the susceptibilities to a variety of common, complex diseases (e.g., cancer [36], type 2 diabetes [36], coronary artery disease [37], and Alzheimer's disease [38]), as well as traits such as adult height [39], body mass index [39], and hair color [40]. Genetic variants in the form of SNPs have also been identified for various therapeutic phenotypes (e.g., cytotoxicities to certain anticancer agents of different mechanisms [25,26]). In addition to SNPs, structural variation in the form of CNVs (i.e., genomic deletions, insertions, inversions, duplications) have also been demonstrated to account for a substantial fraction of natural variation in gene expression [41], as well as various complex disease traits [41] (e.g., systemic lupus erythematosus [43]) and drug response [42]. To identify the genomic contributors to pharmacogenomic phenotypes, it would be necessary to systematically explore the roles of both SNPs and other structural genetic variations (e.g., CNVs). Notably, most common CNVs typed by existing platforms would be indirectly explored through SNP studies, as they are well tagged by known SNPs [43]. We will, therefore, focus on SNPs in this chapter.

Alleles, Genotypes, and Frequencies

An allele refers to the form of nucleotide (A, C, T, or G) at a particular locus in the human genome. Because the majority of SNPs currently studied and genotyped are biallelic [47], we will discuss alleles and genotypes assuming two alleles at each polymorphic locus, segregating in the population, though the discussion can be generalized for SNPs with more than two alleles. For example, if we have two alleles $A = C$ and $a = G$ (i.e., a C/G SNP) at a genomic locus, there will be three genotypes (i.e., the specific allele makeup in an individual) in the population: that is, two homozygous genotypes, AA (CC) and aa (GG), and one heterozygous genotype, Aa (CG). The relative frequencies of these three genotypes, $P_{i,j}$ must add to one, that is,

$$P_{A,A} + P_{A,a} + P_{a,a} = 1 \tag{8.1}$$

where $P_{A,A}$ is the frequency of genotype AA, $P_{A,a}$ is the frequency of genotype Aa, and $P_{a,a}$ is the frequency of genotype aa. Besides genotype frequencies, allele frequencies play a particularly important role in human genetics, as many trait-associated variants have been shown to have a dosage effect in terms of the copy number of risk or causal alleles. For a biallelic SNP such as our C/G SNP (A allele = C, a allele = G), the frequency of the A allele in the population can be calculated from the corresponding genotype frequencies,

$$p = P_{A,A} + \frac{1}{2} \times P_{A,a} \tag{8.2}$$

while the frequency of the a allele is

$$q = 1 - p = P_{a,a} + \frac{1}{2} \times P_{A,a} \tag{8.3}$$

Between the two alleles of a particular SNP, the frequency of the less common allele is referred as minor allele frequency (MAF), which can be used to define common SNPs. For example, MAFs greater than 5% or in some cases greater than 1% are often used as cutoffs for defining common genetic variants. A popular hypothesis about the allelic architecture of common, complex diseases and traits proposes that most of the genetic contribution to these traits is likely due to common genetic variants, though the current knowledge of common diseases and traits also suggests alternative models (e.g., rare variants with large effects) [44]. Because a minor allele may not always be the same in all populations or phenotype groups (e.g., cases and controls), a reference allele may be arbitrarily chosen from the two alleles in the analysis. The frequency of the reference allele then can be evaluated for association with a particular trait or phenotype between populations or phenotype groups.

Hardy–Weinberg Equilibrium

Assuming random mating (i.e., no mating based on genotype preference) and an infinitely large population, the Hardy–Weinberg Equilibrium (HWE) law describes the state in which allele and genotype frequencies in a population remain constant, if without outside evolutionary forces (e.g., natural selection). In the case of a biallelic SNP, if the frequency of allele A is p and the frequency of allele a is q, the HWE law indicates that the new individuals in the following generation will have frequency of p^2 for the AA homozygotes, q^2 for the aa homozygotes, and $2pq$ for the Aa heterozygotes in the population. In other words, under the HWE law, the genotype and allele frequencies in the population will not change over time, thus remaining in equilibrium.

To identify the relationships between genetic variation and therapeutic phenotypes of interest, pharmacogenetic and pharmacogenomic studies rely on accurate genotypic data from genotyping assays using an SNP genotyping array or whole-genome profiling platform. Errors contained in genotypic data may greatly reduce the power of a genetic study and lead to spurious associations or false conclusions. Assuming no obvious violations of the

Hardy—Weinberg assumptions (e.g., no inbreeding, selection, or random genetic drift), deviation from the HWE in random samples may be used to indicate problematic genotyping calls [45]. Therefore, departure from the HWE is often tested in association studies for controlling genotyping quality, which is critical for inferring valid conclusions in pharmacogenomic research. For example, the deviation of the genotype distribution from the HWE law raised concerns from some investigators of the interpretability and validity of the results presented in clinical trials on the relationships between *CYP2D6* (cytochrome P450, family 2, subfamily D, polypeptide 6) polymorphisms and tamoxifen response in postmenopausal women with endocrine-responsive breast cancer [46]. Statistically, testing deviation from the HWE can be performed using Pearson's chi-squared test (or Fisher's exact test with small genotype counts), based on the observed genotype frequencies from the real data and the expected genotype frequencies derived based on the HWE law. A statistically significant deviation may be used to identify SNPs with possible genotyping error in a dataset, which can then be cleaned by removing these problematic variants.

Linkage Disequilibrium

Linkage disequilibrium (LD) is the nonrandom association of alleles at linked loci. LD plays a crucial role in the current methods for mapping complex disease or trait-associated genes [47]. In particular, association mapping takes advantage of the fact that LD may exist between a known marker locus and an unknown trait locus not directly genotyped. For a pair of biallelic variants (e.g., SNP1: alleles A and a; SNP2: alleles B and b) at two genomic loci, assuming the probability of generating a recombinant gamete after a round of meiosis (i.e., recombination rate) is r, and the genotype frequency of, for example, AB is $P_{A,B}$, then the frequency of AB after a round of random mating, $P'_{A,B}$, can be denoted by

$$P'_{A,B} = (1 - r) \times P_{A,B} + r\, p_1\, p_2 \tag{8.4}$$

based on the sum of probabilities of two independent events (i.e., with or without recombination), where p_1 is the frequency of allele A, and p_2 is the allele frequency of B (q_1 and q_2 for the corresponding a and b alleles, respectively). The change in the genotype frequency of AB after a round of random mating is then

$$\Delta P_{A,B} = P'_{A,B} - P_{A,B} = -r(P_{A,B} - p_1\, p_2) = -rD \tag{8.5}$$

where $P_{A,B} - p_1\, p_2$ is defined for linkage disequilibrium, D. Therefore, LD can be described as a measure of the difference between the frequency of a genotype (i.e., $P_{A,B}$) and the expected frequency derived from its allele frequencies assuming random association between the two alleles (i.e., $p_1\, p_2$). Similarly, the same D can be derived from the frequencies of other genotypes (i.e., Aa, Ab, and BB) in the population. Given the previous definition of LD, the value of D after a round of random mating, D', can be written as

$$D' = (1 - r)D \tag{8.6}$$

by taking advantage of the HWE law. Therefore, the change in D becomes

$$\Delta D = D' - D = (1 - r)D - D = -rD \tag{8.7}$$

D can be further generalized in the form of

$$D_t = (1 - r)^t D_0 \tag{8.8}$$

where D_0 is the initial measure of LD, and t is the generation time. Equation 8.8 suggests that the ultimate state of the population will be 0 for D, if given a long enough generation time. Because D can be very sensitive to the allele frequencies under consideration, another commonly used measure of association between alleles, r^2, may be used to represent LD. r^2 can be derived from:

$$r^2 = \frac{D^2}{p_1 q_1 p_2 q_2} \tag{8.9}$$

Large values of r^2 indicate stronger association between alleles, and lower values of r^2 indicate weaker association between alleles. A threshold based on r^2 (e.g., $r^2 \geq 0.8$) can be used to determine whether two SNPs are independent. In association studies, tag-SNPs may be selected based on r^2 (e.g., by software such as Tagger [48]) to represent genetic variation in a region of the genome with high LD or nonrandom association of alleles.

STRATEGIES IN PHARMACOGENOMIC RESEARCH AND DEVELOPMENT

Various bioinformatic tools and statistical approaches have been used to investigate the relationships between human genetic variation and variation in drug response. There are generally two categories of research approaches in pharmacogenomic research and development: the candidate gene approach and the whole-genome approach (Figure 8.1). Briefly, studies applying the candidate gene approach focus on a limited number of known candidate genes with *a priori* knowledge (e.g., genes involved in drug-metabolizing pathways or identified to be relevant from previous studies) for a therapeutic phenotype of interest. Considering that drug response is likely a complex trait affected by various genetic and nongenetic factors, studies applying the whole-genome approach aim to identify genes or genetic variants associated with a phenotype of interest using an unbiased, genome-wide scan (e.g., across all of the expressed genes or common genetic variants). It should be noticed that a combination of these two approaches is often necessary for the successful dissection of genetic contribution to the pharmacogenomic traits under investigation. Due to the different scales of these two general approaches, careful consideration must be taken in choosing appropriate bioinformatic tools and statistical approaches, especially the selection of cutoffs for statistical significance and control of false discovery rate (FDR). The general strategies and pipelines of these two approaches are described in the following sections.

The Candidate Gene Approach

Traditionally, pharmacogenetic studies have focused on well-characterized pharmacology-related genes such as those genes that are known to affect pharmacokinetics and pharmacodynamics (e.g., drug metabolizing enzyme, transporter, target genes). Successful examples applying the candidate gene approach include, for example, (1) the identification of genetic polymorphisms in *TPMT* (encoding thiopurine S-methyltransferase), which led to decreased

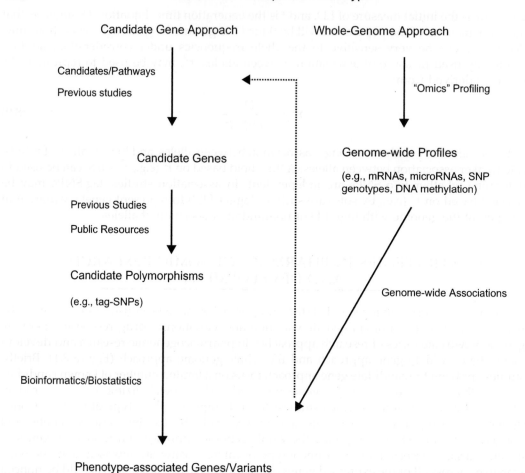

FIGURE 8.1   **Comparison of the candidate gene and whole-genome approaches.** The traditional candidate gene approach targets known genes and/or polymorphisms with *a priori* knowledge. In contrast, the whole-genome approach utilizes an unbiased, genome-wide strategy that is facilitated by the availability of whole-genome profiling technologies. The whole-genome approach may also provide novel targets for further studies using the candidate gene approach.

TPMT enzyme activity and subsequently increase 6-mercaptopurine toxicity [49]; (2) the identification of polymorphisms in *CYP2C9* and *VKORC1* associated with interindividual variability in the dose response of warfarin [50]; and (3) the finding of decreased activity of *UGT1A1*28* (encoding UDP glucuronosyltransferase 1 family, polypeptide A1) polymorphism associated with an increased risk of neutropenia during irinotecan treatment [51]. As a hypothesis-driven approach, the advantages of targeting polymorphisms in these candidate genes include cost-efficiency (due to the limited number of genetic variants needed to be genotyped) and potentially direct relevance to the phenotype of interest. Given the much smaller number of genetic variants and genes under investigation relative to the whole-genome approach, data analysis is often more straightforward for the candidate gene approach. For example, because a limited number of genetic variants are often interrogated in a particular study, less stringent statistical cutoffs for controlling FDR may be used to identify significant loci associated with a therapeutic phenotype. The general strategy used in the candidate gene approach is outlined in Figure 8.1.

Selection of Candidate Genes

A successful candidate gene study relies on the selection of reasonable candidate genes that are relevant to the therapeutic phenotype under investigation (e.g., pharmacokinetic parameters such as the area under plasma concentration-time curve [AUC], cytotoxicity to anticancer drugs). In general, candidate genes can be selected based on the relevance of a known pathway or evidence from previous studies. For example, to study the potential contribution of genetics to the pharmacokinetics of a particular drug, researchers may target those genes that have been implicated in drug absorption, distribution, biotransformations, as well as the effects and routes of excretion of the metabolites of the drug [52], for example drug metabolizing enzymes such as cytochrome P450 (CYP) and UDP-glucuronosyltransferase (UGT) enzymes. In comparison, to study the potential contribution of genetics to the pharmacodynamics of a particular drug, researchers may target those genes that are likely to have interactions with the administered drug. For example, to study the pharmacodynamics of *EGFR* (encoding epidermal growth factor receptor) tyrosine kinase inhibitors (TKIs), a class of targeted anticancer agents [53], genes involving the EGFR signaling pathway or other related pathways, such as *EGFR* [54] or *KRAS* (encoding v-Ki-ras2 Kirsten rat sarcoma viral oncogene homolog) [55], may be used as candidates for investigating the relationships between their genetic variation and EGFR TKI response.

Selection of Candidate Polymorphisms

Once a reasonable candidate gene (or a set of genes) is selected, to study the contribution of genetic variation to the therapeutic phenotype of interest, researchers must identify the genetic polymorphisms (e.g., common SNPs) that are to be investigated. Generally, there are several criteria that need to be considered in selecting candidate genetic polymorphisms (Table 8.1). The polymorphisms can be chosen based on previous studies. For example, candidate SNPs can be selected from previous significant findings for the purpose of validation in an independent cohort. Candidate SNPs may also be selected based on a general reference panel for human genetic variation, if there are no known, associated SNPs from previous studies. For example, to select candidate SNPs in a particular gene, public resources of human genetic variation, such as the International HapMap Project [7—9] and the 1000 Genomes Project [10], can be used to identify a set of genetic variants in candidate genes of interest. When selecting candidate polymorphisms using a public resource (e.g., the 1000 Genomes Project [10] data), two general strategies may be used, depending on the specific aims of a particular study. Either the tag-SNPs that aim to capture the general genetic variation in the candidate genes (e.g., by software such as Tagger [48]) or the potentially functional SNPs (e.g., SNPs located in promoters, untranslated regions [UTRs], nonsynonymous substitutions in exons, and splice sites) may be selected for further investigation in a candidate gene

TABLE 8.1   Common criteria used to prioritize candidate polymorphisms

Strategy	Type of Variant	Location	Functional Relevance
Previous studies of relevant phenotypes or diseases	SNPs	Candidate genes	SNPs in candidate genes from previous studies
SNP function	Nonsynonymous	Coding sequence	Changes amino acid sequence
	Promoter/regulatory regions	Promoter, 5′UTR, 3′UTR	May affect the level, timing, or location of gene expression
	Splice site	Exon/intron junction	Changes different isoforms
	Frameshift	Coding sequence	Changes the frame of the protein-coding region
	Nonsense	Coding sequence	Terminates amino acid sequence prematurely
Studies of general human genetics	eQTLs	cis- or trans-	SNPs associated with gene expression
	mSNPs	cis-	SNPs associated with DNA methylation
	miRSNPs	microRNAs	Changes microRNA expression or binding efficiency

Notes: UTR, untranslated region; eQTL, expression quantitative trait locus; mSNP, SNP associated with quantitative DNA methylation; miRSNP, SNP in microRNAs.

study. Polymorphisms can also be prioritized based on other criteria, such as whether a particular SNP has been identified as an eQTL. Because eQTLs were found to be enriched in complex trait-associated SNPs [56], these SNPs may be more likely to be relevant to other therapeutic phenotypes traits as well, such as chemotherapeutic drug susceptibility [57]. Other novel findings in human genetics, such as miRSNPs (i.e., polymorphisms in microRNAs) [58] and mSNPs (i.e., SNPs associated with DNA methylation) [20,21], may provide additional information for selecting candidate polymorphisms, as well as functional interpretation for the associated genetic variants with therapeutic phenotypes. Furthermore, because there is substantial genetic variation between human populations [9,10], when selecting candidate polymorphisms, special consideration should be taken for different populations. For example, for individuals of African ancestry, more variants might be necessary to capture sufficient genetic variation than in individuals of European ancestry. Another important consideration in selecting polymorphisms for a candidate gene study is whether there is significant LD in the candidate gene in the study population. LD estimation from a reference panel may be used to optimize SNP selection and can provide information about possible haplotypes for the association analysis.

Genotyping, SNP Discovery, and Imputation

Once candidate genes and polymorphism are selected, genotypes in the study cohort can be obtained using a genotyping platform. In contrast, if there is not enough knowledge of genetic variation in a candidate gene, sequencing may be used to identify novel SNPs in the study cohort. Besides these experimental approaches, *in silico* genotyping or imputation may be used to infer unobserved genotypes based on either the genotypes from related family members or, in case of unrelated samples, the haplotypes from a reference panel such as the 1000 Genomes Project [10]. Several bioinformatic tools, such as MACH [59] (Markov Chain–based haplotyper) and IMPUTE2 [60] have been developed to impute untyped genotypes for association studies by combining study samples and reference panels.

Statistical Tests Used in the Candidate Gene Approach

The traditional candidate gene approach involves testing association between individual genetic markers (e.g., SNPs) and a phenotype of interest, which can be a quantitative trait or a categorical phenotype. The choice of statistical approaches depends on the type of phenotype (i.e., quantitative or categorical). When a quantitative trait (e.g., IC_{50}, the concentration at which 50% of cell growth is inhibited) is being investigated, a typical test will be to evaluate the differences in the means of the phenotype of interest across different genotypes at a given locus. In contrast, when a categorical phenotype (e.g., respondents vs. nonrespondents) is being investigated, a typical test will be to test the allele or genotype frequency differences between the phenotype groups. Notably, these two general analysis approaches are also used in the whole-genome approach. However, because the candidate gene approach usually targets a limited number of SNPs and genes, as well as the availability of *a priori* knowledge of the functional relevance of the candidate genes, a much less stringent cutoff may be used to achieve the same FDR as in an unbiased, whole-genome scan. Several traditional statistical approaches for pharmacogenetic data analysis are outlined as follows. The implementations of these approaches are freely available from packages such as the R Statistical Package [61] and PLINK [62].

CONVENTIONAL ASSOCIATION TESTS

When the phenotype is a categorical variable, the conventional association tests may be used to test if an allele is associated with the phenotype group of interest. For example, if we are interested in the association between an SNP and the status of drug response (e.g., respondents vs. nonrespondents), the chi-squared test may be used to test whether the allele or genotype frequencies are different between the respondents and nonrespondents. A statistically significant result (e.g., p value < 0.01) may indicate association between the allele or genotype and drug response status. In contrast, Fisher's exact test may be applied to datasets with small sample sizes.

ANALYSIS OF VARIANCE

Analysis of Variance (ANOVA), which generalizes the t-test for more than two groups, can be used to test for statistical differences in the means of a quantitative pharmacological trait (e.g., IC_{50}) across individuals with different genotypes (e.g., AA, AB, BB) of a particular candidate polymorphism. The assumptions of ANOVA include independence of samples, normality of the data distribution, and homoscedasticity (i.e., equality of variance). If these assumptions (e.g., data normality) cannot be met, instead, the nonparametric Kruskal—Wallis test may be used to compare differences in the mean values of a phenotype across different genotypes.

LINEAR REGRESSION

An alternative to ANOVA is the linear regression approach, which can fit the quantitative trait under investigation based on different genetic models (i.e., additive, dominant, and recessive). The linear regression model can be expanded to adjust for covariates (e.g., gender, race, and other clinical covariates). Within the linear regression framework, the therapeutic phenotype of interest is the dependent variable, and the genotypes (properly coded based on a given genetic model, such as the additive model: $AA=0$, $AB=1$, $BB=2$; dominant model: $AA=0$, $AB/BB=1$; or recessive model: $AA/AB=0$, $BB=1$) and any covariates are independent variables. Furthermore, the linear regression approach can be used to explore the interactions between variables (e.g., epistasis, the interaction between two genetic variants), as well as models based on multiple genes or genetic variants.

The Whole-Genome Approach

Results from the previous candidate gene or pharmacogenetic studies suggest that drug response is likely to be a multigenic or genetically complex trait that is affected by a variety of genetic and nongenetic factors, as well as the interactions between these factors. Understandably, each gene or genetic variant may contribute with varying strength to a given therapeutic phenotype of interest. Though the candidate gene approach focuses on potential genes with functional relevance (e.g., drug-metabolizing enzymes for a pharmacokinetic phenotype), there still can be unknown genes that may help determine or modify the phenotypes, such as through interactions with those candidate genes. Furthermore, in the context of missing heritability, in which drug response phenotypes are often not explained sufficiently with known genes, novel genes and pathways unidentified in previous studies may also contribute significantly to therapeutic phenotypes. Therefore, to comprehensively elucidate the underlying mechanisms of the genetic contribution to drug response, as well as to identify novel targets for further candidate gene or functional studies, the whole-genome approach could be used to expand and complement the candidate gene approach to interrogate the entire human genome, transcriptome, or epigenome for the relationships between genes/genomic loci and therapeutic phenotypes. Technologically, with the advances in genome-wide profiling technologies (e.g., microarray-based platforms), it is now possible to perform cost-efficient, whole-genome profiling of various molecular targets (e.g., mRNAs, SNP genotypes, microRNAs, DNA methylation), thus facilitating pharmacogenomic research and development on the whole-genome scale. More recently, the next-generation, whole-genome sequencing technologies [27,28] opened the possibility to identify all genetic variants and mRNA transcripts (RNA sequencing), thus potentially allowing direct mapping of the causal genetic variants [63] (or transcripts) contributing to therapeutic phenotypes without relying on LD between genotyped variants and causal variants or known sequences of transcripts. Compared to the array-based approaches, the sequencing-based whole-genome approach and its data analysis is still at a relatively early stage. Therefore, we will focus on the array-based whole-genome approach, particularly the general data analysis procedure and pipeline, in this section.

Genome-Wide Profiling Platforms and Data Processing

A variety of microarray-based genome-wide profiling platforms are available for simultaneously genotyping common genetic variants or measuring gene expression (or microRNAs and DNA methylation) in cells under a particular condition (e.g., baseline or with drug treatment). Commonly used platforms include the Affymetrix GeneChip and the Illumina BeadChip platforms. For example, the Affymerix Human Exon 1.0ST Array (measuring >22,000 genes and >1.4 million exons) (exon array) was used to profile whole-genome gene expression in a collection of HapMap samples as a pharmacogenomic discovery and general human genetic research model [12,13]. Given the large amount of data (e.g., >22,000 genes in the exon array data [12,13]) generated from these profiling platforms, bioinformatic tools are needed for data preprocessing and association (or identifying differential genes) under different conditions. Typically, these array-based genome-wide data need to be properly preprocessed: for instance, using quantile normalization [64] to normalize gene expression data or calling genotypes from genotyping platforms before downstream data analysis. For example, the *affy* and *oligo* tools in the Bioconductor [65] suite, an open source software platform for bioinformatics based on the R Statistical Package [61], may be used to normalize and summarize gene expression data measured with the 3'-based Affymetrix GeneChip platforms and the exon array, respectively. For SNP genotyping platforms (e.g., the Affymetrix 500K, SNP5.0, and SNP6.0 platforms), the Birdseed tool in Birdsuite [66] or the Affymetrix Power Tools can be used to call genotypes from the raw output of the genotyping arrays. The details and instructions for these packages are freely available from the developers.

Statistical Tests Used in the Whole-Genome Approach

Unlike the candidate gene approach, the design of the whole-genome approach is to scan the complete genome (or transcriptome and epigenome) for associated loci (or gene expression phenotypes and epigenetic markers) with a therapeutic phenotype of interest. In terms of statistical approaches, however, the same analysis techniques are often used in the whole-genome approach—for example, ANOVA and its nonparametric alternative, linear regression, and conventional association tests (e.g., chi-squared-test-based) (see the earlier section "The Candidate Gene Approach"). For example, a linear regression model may be used to associate genetic variants (or gene expression phenotypes) with a quantitative phenotype (i.e., IC_{50}). Software packages such as the R Statistical Package [61] and PLINK [62] may be used to identify the relationships between these whole-genome data and therapeutic phenotypes using an appropriate statistical approach (e.g., linear regression). Bioinformatic tools such as WGAViewer [67] may be used to visualize genome-wide association results, as well as provide annotation information. In the following section, we will use a cell-based model to present the general procedure in pharmacogenomic studies using the whole-genome approach. Specifically, a genome-wide association study (GWAS) on the identification of genetic variants associated with the cytotoxicity to etoposide [68] (a topoisomerase II inhibitor) is presented as an example.

A Cell-Based Model for Pharmacogenomic Discovery

The International HapMap Project [7–9] provides genotypic data on more than 3.1 million SNPs (Phase 1/2) in a panel of LCLs derived from major continental populations (See the later section "Major Resources of the HapMap Samples"). Because the HapMap LCL samples are commercially available from the Coriell Institute for Medical Research (Camden, New Jersey), comprehensive phenotypic data, including whole-genome gene expression [11,12,15,16], microRNA expression [19], and DNA methylation [20–22] have been accumulated on these samples (Table 8.2). In particular, using this cell-based

TABLE 8.2   Major Resources of the HapMap Samples

Data Type	Resource/GEO	HapMap Samples	Platform	Reference
Gene expression	GSE7792 (gene-level)	CEU; YRI	Affymetrix Human Exon Array	[68]
	GSE9703 (exon-level)	CEU; YRI	Affymetrix Human Exon Array	[89]
	GSE6536	CEU; CHB; JPT; YRI	Illumina	[16]
	GSE5859	CEU; CHB; JPT	Affymetrix Human Focus Array	[11]
	GSE19480	YRI	RNA-sequencing	[86]
SNP genotypes	The HapMap Project	Phase 1/2: CEU; CHB; JPT; YRI Phase 3: ASW; CEU; CHB; CHD; GIH; JPT; LWK; MXL; MKK; TSI; YRI	Genome-wide SNP genotyping and targeted sequencing	[7–9]
	The 1000 Genomes Project*	ASW; CEU; CHB; JPT; LWK; MXL; TSI; YRI;	Next-generation sequencing	[10]
CNVs	The Sanger Institute CNV Project	CEU; YRI; CHB; JPT	Whole Genome TilePath Array	[81]
DNA methylation	GSE26133	YRI	Illumina HumanMethylation 27K Array	[20]
	GSE27146	CEU, YRI	Illumina HumanMethylation 27K Array	[21]
	GSE39672	CEU, YRI	Illumina HumanMethylation 450 Array	[22]
DNase I sensitivity	GSE31388	YRI	DNase-seq	[82]

Notes: only populations from the HapMap Project in the integrated phase 1 data are listed; ASW, African ancestry in Southwest USA; CEU, Utah residents with Northern and Western European ancestry from the CEPH collection; CHB, Han Chinese in Beijing, China; CHD, Chinese in Metropolitan Denver, Colorado; GIH, Gujarati Indians in Houston, Texas; JPT, Japanese in Tokyo, Japan; LWK, Luhya in Webuye, Kenya; MXL, Mexican ancestry in Los Angeles, California; MKK, Maasai in Kinyawa, Kenya; TSI, Tuscans in Italy; YRI, Yoruba in Ibadan, Nigeria; GEO, The NCBI Gene Expression Omnibus SNP, single nucleotide polymorphism; CNV, copy number variation.

model, cytotoxicity response phenotypes have been profiled for chemotherapeutic agents such as cisplatin [69], carboplatin [70], etoposide [68], daunorubicin [71], and cytarabine [72]. Taking advantage of this tremendous whole-genome resource, cytotoxicity-associated genomic loci and gene expression phenotypes have been identified for several anticancer agents of different mechanisms [25,26]. In addition, LCLs from unrelated individuals were shown to exhibit variation in growth inhibition and consistent profiles for a wide range of drugs (e.g., antidepressants, steroid drugs, and antipsychotics) sharing a similar pathway [73], thus suggesting that the LCL model could potentially be used to identify pharmacogenomic loci for drugs other than the anticancer agents as well.

Example: A Genome-Wide Approach to Identify Genetic Variants Contributing to Etoposide-Induced Cytotoxicity

The genome-wide identification of genetic variants associated with etoposide-induced cytotoxicity [68] was among the earliest pharmacogenomic studies using the HapMap cell-based model. Specifically, the cytotoxicities (IC_{50}) were phenotyped in the HapMap CEU (Caucasians from Utah, USA) and YRI (Yoruba people from Ibadan, Nigeria) samples (30 parents—child trios in each population) [68]. Utilizing the comprehensive SNP genotypic data from the International HapMap Project [7—9], a GWAS was performed to identify genetic variants contributing to etoposide-induced cytotoxicity. Data analysis of this project included the following major steps, showcasing the general pipeline and data analysis techniques using the whole-genome approach.

Specifically, genotype calls for approximately 2 million SNPs were downloaded from the International HapMap [7—9] database (release 21). To perform a high-quality whole-genome association study, the downloaded genotypic data were cleaned by several criteria (e.g., removing SNPs with Mendelian allele transmission errors and removing rare SNPs with very low allele frequencies). Notably, because this study used parents—child trios, the Mendelian allele transmission errors were checked to control the quality of genotypic data. In contrast, if unrelated samples were used, the departure from HWE could be used to identify variants with genotyping errors.

To identify genotype—cytotoxicity association in the parents-child trios, a quantitative transmission disequilibrium test (QTDT) was performed to identify genotype—cytotoxicity association relationships using the QTDT software [74,75]. Because of the possible heterogeneity between and within each population, QTDT was performed in the CEU and YRI separately by using gender as a covariate, and in the combined samples using gender and population as covariates. In contrast, if unrelated samples were used, a linear regression model (i.e., dependent variable: IC_{50}, and independent variable: SNP genotype under a given genetic model) could be used to identify SNPs associated with the drug response phenotype.

When using the whole-genome approach, a particularly challenging topic is how to address the problem of multiple comparisons, due to the large number of comparisons being tested simultaneously, in addition to other disadvantages such as high cost and technical limitations from the current whole-genome platforms (e.g., the selection of SNPs not population optimized). Several procedures were proposed to correct multiple comparisons in genome-wide studies, such as the Bonferroni correction, the QVLAUE [76], the

Benjamin—Hochberg procedure [77], and the permutation-based correction. A detailed description of these algorithms is out of the scope of this chapter, though it should be noted that different algorithms have different assumptions (e.g., correlation between genes), and the Bonferroni correction is often deemed the most conservative procedure for adjusting multiple comparisons. Bioinformatic tools to calculate these adjusted p-values are available or can be implemented conveniently using statistical software such as the R Statistical Package [61] and Bioconductor [65] (e.g., the *multtest* tool).

BIOINFORMATIC RESOURCES FOR PHARMACOGENOMIC RESEARCH AND DEVELOPMENT

Pharmacogenomic research and development relies on the progress in the general field of human genetics, as well as the advances of technologies for phenotyping and genotyping. In particular, the availability of detailed human genetic variation data provides the foundation of today's pharmacogenetic and pharmacogenomic research. In addition, given the complex association relationships identified in pharmacogenomic studies, web-based resources have been developed to serve the primary results from these studies. For example, the Pharmacogenomics and Cell database [78] PACdb) is a database that makes available relationships between SNPs, gene expression phenotypes, and cytotoxicities to a variety of anticancer agents (e.g., cisplatin [69], carboplatin [70]) in the HapMap cell-based model. It is expected that such "result databases" will offer opportunities for integrating different aspects (e.g., SNPs, CNVs, microRNAs, DNA methylation) of association relationships for specific therapeutic phenotypes of interest. However, with the aim of focusing on the general tools that may be useful for a wide range of studies, several important databases and websites related to human genetic variation and pharmacogenomic research will be briefly introduced to provide an overview of the current bioinformatic resources related to pharmacogenomic research and development.

Major Resources of the HapMap Samples

The HapMap LCLs have been used in pharmacogenomic discovery as a cell-based model [24,25], because tremendous resources have been accumulated for these samples [79]. The International HapMap Project provides genotypic data on more than 3.1 million SNPs in 270 LCLs derived from three major continental populations [9] including Asians (CHB: Han Chinese from Beijing, China; JPT: Japanese from Tokyo, Japan), Africans (YRI), and Europeans (CEU) (Phase 1/2 data). In addition, the Phase 3 HapMap Project provides genotypes of approximately 1.6 million SNPs in approximately 1,200 LCL samples derived from 11 global populations [79]. The HapMap website [80] provides bioinformatics tools for interrogating detailed information of these genotypic data (e.g., genotype and allele frequencies), selecting tag-SNP for use in association studies, viewing haplotypes graphically, and examining marker-to-marker LD patterns, as well as downloading of the raw genotypic data. Besides genotypes, other related resources have been accumulated for the HapMap samples by the research community, including gene expression [11,42,81], microRNAs [18], DNA methylation [19,20], CNVs [81], and DNase I sensitivity [82] (Table 8.2). The gene expression and DNA methylation data on these samples may be downloaded from the NCBI GEO (Gene Expression Omnibus) database.

The 1000 Genomes Project

The 1000 Genomes Project [10] which was launched in 2008, aims to provide the most detailed map of human genetic variation by sequencing about 2,500 genomes from about 25 global populations. The genetic variation data provided by this international collaboration will support genome-wide association studies of complex traits and phenotypes including pharmacogenomic phenotypes [83]. The 1000 Genomes Project [10] currently supports queries for the SNPs and deletions/insertions from the integrated phase 1 release (March 2012) that covers more than 1,000 samples in 14 global populations, including several HapMap populations (e.g., CEU, YRI), though not all of the HapMap samples were covered (Table 8.2). The variant calls can be downloaded from the 1000 Genomes Project [10] website, as well as from the NCBI and the European Bioinformatics Institute (EBI) servers. Given the large amount of data available from the 1000 Genomes Project [10], however, it can be a challenge for the pharmacogenetic research community to take advantage of these data. Specifically, a database of some pharmacogenes [84] of particular interest to pharmacogenetic researchers has been developed to provide a convenient portal for convenient utilization of the 1000 Genomes Project [10] data in pharmacogenetic studies.

dbSNP

The NCBI dbSNP [32,33] database is a public-domain archive for a broad collection of simple genetic polymorphisms, including SNPs, small-scale multibase deletions or insertions, and short tandem repeats. The current version (build 135) covers more than 40 million validated SNPs, including more than 35 million SNPs with frequency information. The genetic variation data from the HapMap Project [7–9] and the 1000 Genomes Project [10] are integrated in the dbSNP [32,33], which acts as a central portal for genetic variation. The dbSNP [32,33] supports both individual queries and bulk downloading of the genetic variation data.

SCANdb

The SNP and Copy number Annotation Database [85] (SCANdb) aims to facilitate the interpretation of the association results (e.g., from a GWAS) and the designing of follow-up experimental validations by providing publicly available physical and functional annotations of SNPs and CNVs, as well as multilocus LD annotations and, most important, the annotations of eQTLs, which were mapped in the full set of the HapMap CEU and YRI samples using the Affymetrix Human Exon 1.0 ST Array gene expression data [12,13] and the HapMap genotypic data [7–9]. Because complex trait-associated variants [56], including chemotherapeutic susceptibility-associated loci [57], have been found to be enriched in eQTLs, the SCANdb [85] may be utilized to prioritize results from associations studies by targeting potential eQTLs. Because gene expression profiled in these samples reflects a snapshot of the cellular environment at a given moment, other eQTL resources (e.g., eQTLs using

RNA-sequencing data on the YRI samples [86]) may also be valuable for cross-validation, but the SCANdb [85] provides links to much more related information (e.g., LD) rather than just making available the eQTL mapping results.

PharmGKB

The Pharmacogenomics Knowledgebase [87] (PharmGKB) is a comprehensive resource and portal that curates knowledge about the impact of genetic variation on drug response. The PharmGKB [87] also serves as a central database for several international pharmacogenomics consortia such as the International Warfarin Pharmacogenetics Consortium [88], the International Tamoxifen Pharmacogenomics Consortium, and the International Clopidogrel Pharmacogenomics Consortium, as well as large-scale pharmacogenomics projects such as the 1200 Patients Project at the University of Chicago. As a resource for pharmacogenetic and pharmacogenomic research, the PharmGKB [87] provides updated annotations for genetic variants and gene—drug—disease relationships via literature reviews and the summaries for important pharmacogenomic genes, associations between genetic variants and drugs, and drug pathways. The PharmGKB [87] also maintains a list of very important pharmacogenes (VIPs) such as transporters and drug-metabolizing enzymes.

CONCLUSION AND OUTLOOK

The launching of the Human Genome Project and several parallel efforts (e.g., the 1000 Genomes Project [10]) for elucidating human genetic variation provided the foundation for pharmacogenetic and pharmacogenomic research, which aims to understand the relationships between genetic variants and pharmacological phenotypes. Traditionally, the question of identifying genetic association with therapeutic phenotypes has relied on the candidate gene approach, which targets the polymorphisms in candidate genes with *a priori* knowledge of functional relevance. With the advances of genome-wide profiling technologies, the whole-genome approach may provide more comprehensive knowledge of the mechanisms underlying the variation in therapeutic phenotypes. In particular, the next-generation sequencing technologies opened the possibility of directly mapping causal genetic variants associated with complex traits including therapeutic phenotypes. It is expected that more detailed genetic variation data from these technologies will facilitate the next wave of research efforts, thus providing more insights into the current views of the genetic architecture of therapeutic phenotypes. Unlike the array-based genome-wide studies, the data analysis involving sequencing-based studies is still at a relatively early stage. Significant challenges for better analyzing the much larger amount of sequencing data must be met to fully take advantage of these technologies. Platforms such as the Galaxy Project represent the current efforts to make the analysis of next-generation sequencing data more accessible to biomedical investigators, including the pharmacogenomic research community. We expect that more novel bioinformatics tools and statistical approaches will become available in the near future to facilitate the breakthroughs in pharmacogenetic and pharmacogenomic research. Finally, with the availability of more pharmacogenetic and pharmacogenomic studies, future integration of various data and studies (e.g., through meta-analysis) will help pave the road to the ultimate goal of personalized medicine.

Acknowledgement

This work was supported, in part, by a grant from the NIH (R21HG006367 to WZ). The authors declare no conflicts of financial interests.

QUESTIONS FOR DISCUSSION

1. What is the minor allele frequency for an SNP with the following genotype frequencies: *AA* 0.64, *Aa* 0.32, and *aa* 0.04?
2. Why is it important to determine the Hardy–Weinberg Equilibrium?
3. What are the advantages and disadvantages of candidate gene and genome-wide approaches for pharmacogenetic studies?
4. What bioinformatic sources are available to assist with pharmacogenetic research?

References

[1] Lazarou J, Pomeranz BH, Corey PN. Incidence of adverse drug reactions in hospitalized patients: a meta-analysis of prospective studies. J Am Med Assoc 1998;279:1200–5.
[2] Kamali F, Wynne H. Pharmacogenetics of warfarin. Annu Rev Med 2010;61:63–75.
[3] Klein TE, Altman RB, Eriksson N, Gage BF, Kimmel SE, Lee MT, et al. Estimation of the warfarin dose with clinical and pharmacogenetic data. N Engl J Med 2009;360:753–64.
[4] Gage BF, Eby C, Johnson JA, Deych E, Rieder MJ, Ridker PM, et al. Use of pharmacogenetic and clinical factors to predict the therapeutic dose of warfarin. Clin Pharmacol Ther 2008;84:326–31.
[5] Lander ES, Linton LM, Birren B, Nusbaum C, Zody MC, Baldwin J, et al. Initial sequencing and analysis of the human genome. Nature 2001;409:860–921.
[6] Venter JC, Adams MD, Myers EW, Li PW, Mural RJ, Sutton GG, et al. The sequence of the human genome. Science 2001;291:1304–51.
[7] International HapMap Consortium. The International HapMap Project. Nature 2003;426:789–96.
[8] International HapMap Consortium. A haplotype map of the human genome. Nature 2005;437:1299–320.
[9] Frazer KA, Ballinger DG, Cox DR, Hinds DA, Stuve LL, Gibbs RA, et al. A second generation human haplotype map of over 3.1 million SNPs. Nature 2007;449:851–61.
[10] 1000 Genomes Project Consortium. A map of human genome variation from population-scale sequencing. Nature 2010;467:1061–73.
[11] Spielman RS, Bastone LA, Burdick JT, Morley M, Ewens WJ, Cheung VG. Common genetic variants account for differences in gene expression among ethnic groups. Nat Genet 2007;39:226–31.
[12] Zhang W, Duan S, Kistner EO, Bleibel WK, Huang RS, Clark TA, et al. Evaluation of genetic variation contributing to differences in gene expression between populations. Am J Hum Genet 2008;82:631–40.
[13] Duan S, Huang RS, Zhang W, Bleibel WK, Roe CA, Clark TA, et al. Genetic architecture of transcript-level variation in humans. Am J Hum Genet 2008;82:1101–13.
[14] Storey JD, Madeoy J, Strout JL, Wurfel M, Ronald J, Akey JM. Gene-expression variation within and among human populations. Am J Hum Genet 2007;80:502–9.
[15] Stranger BE, Nica AC, Forrest MS, Dimas A, Bird CP, Beazley C, et al. Population genomics of human gene expression. Nat Genet 2007;39:1217–24.
[16] Stranger BE, Forrest MS, Dunning M, Ingle CE, Beazley C, Thorne N, et al. Relative impact of nucleotide and copy number variation on gene expression phenotypes. Science 2007;315:848–53.

[17] Cheung VG, Conlin LK, Weber TM, Arcaro M, Jen KY, Morley M, et al. Natural variation in human gene expression assessed in lymphoblastoid cells. Nat Genet 2003;33:422—5.

[18] Morley M, Molony CM, Weber TM, Devlin JL, Ewens KG, Spielman RS, et al. Genetic analysis of genome-wide variation in human gene expression. Nature 2004;430:743—7.

[19] Huang RS, Gamazon ER, Ziliak D, Wen Y, Im HK, Zhang W, et al. Population differences in microRNA expression and biological implications. RNA Biol 2011;8:692—701.

[20] Bell JT, Pai AA, Pickrell JK, Gaffney DJ, Pique-Regi R, Degner JF, et al. DNA methylation patterns associate with genetic and gene expression variation in HapMap cell lines. Genome Biol 2011;12:R10.

[21] Fraser HB, Lam LL, Neumann SM, Kobor MS. Population-specificity of human DNA methylation. Genome Biol 2012;13:R8.

[22] Moen E, Mu W, Delaney S, Wing C, McQuade J, Godley L, et al. Differences in DNA methylation between the African and European HapMap populations. Proc Am Assoc Cancer Res 2012;72:5010.

[23] Dolan ME, Newbold KG, Nagasubramanian R, Wu X, Ratain MJ, Cook Jr EH, et al. Heritability and linkage analysis of sensitivity to cisplatin-induced cytotoxicity. Cancer Res 2004;64:4353—6.

[24] Shukla SJ, Duan S, Badner JA, Wu X, Dolan ME. Susceptibility loci involved in cisplatin-induced cytotoxicity and apoptosis. Pharmacogenet Genomics 2008;18:253—62.

[25] Welsh M, Mangravite L, Medina MW, Tantisira K, Zhang W, Huang RS, et al. Pharmacogenomic discovery using cell-based models. Pharmacol Rev 2009;61:413—29.

[26] Zhang W, Dolan ME. Use of cell lines in the investigation of pharmacogenetic loci. Curr Pharm Des 2009;15:3782—95.

[27] Mardis ER. The impact of next-generation sequencing technology on genetics. Trends Genet 2008;24:133—41.

[28] Mardis ER. Next-generation DNA sequencing methods. Annu Rev Genomics Hum Genet 2008;9:387—402.

[29] Watkins WS, Rogers AR, Ostler CT, Wooding S, Bamshad MJ, Brassington AM, et al. Genetic variation among world populations: inferences from 100 Alu insertion polymorphisms. Genome Res 2003;13:1607—18.

[30] Barreiro LB, Laval G, Quach H, Patin E, Quintana-Murci L. Natural selection has driven population differentiation in modern humans. Nat Genet 2008;40:340—5.

[31] Varela MA, Amos W. Heterogeneous distribution of SNPs in the human genome: microsatellites as predictors of nucleotide diversity and divergence. Genomics 2010;95:151—9.

[32] Sherry ST, Ward MH, Kholodov M, Baker J, Phan L, Smigielski EM, et al. dbSNP: the NCBI database of genetic variation. Nucleic Acids Res 2001;29:308—11.

[33] Smigielski EM, Sirotkin K, Ward M, Sherry ST. dbSNP: a database of single nucleotide polymorphisms. Nucleic Acids Res 2000;28:352—5.

[34] Frazer KA, Murray SS, Schork NJ, Topol EJ. Human genetic variation and its contribution to complex traits. Nat Rev Genet 2009;10:241—51.

[35] Cheung VG, Spielman RS, Ewens KG, Weber TM, Morley M, Burdick JT. Mapping determinants of human gene expression by regional and genome-wide association. Nature 2005;437:1365—9.

[36] Chung CC, Chanock SJ. Current status of genome-wide association studies in cancer. Hum Genet 2011;130:59—78.

[37] Wang Y, Zhang W, Li S, Song W, Chen J, Hui R. Genetic variants of the monocyte chemoattractant protein-1 gene and its receptor CCR2 and risk of coronary artery disease: a meta-analysis. Atherosclerosis 2011;219:224—30.

[38] Kamboh MI. Molecular genetics of late-onset Alzheimer's disease. Ann Hum Genet 2004;68:381—404.

[39] Frayling TM, Timpson NJ, Weedon MN, Zeggini E, Freathy RM, Lindgren CM, et al. A common variant in the FTO gene is associated with body mass index and predisposes to childhood and adult obesity. Science 2007;316:889—94.

[40] Han J, Kraft P, Nan H, Guo Q, Chen C, Qureshi A, et al. A genome-wide association study identifies novel alleles associated with hair color and skin pigmentation. PLoS Genet 2008;4:e1000074.

[41] Girirajan S, Campbell CD, Eichler EE. Human copy number variation and complex genetic disease. Annu Rev Genet 2011;45:203—26.

[42] Gamazon ER, Huang RS, Dolan ME, Cox NJ. Copy number polymorphisms and anticancer pharmacogenomics. Genome Biol 2011;12:R46.

[43] Craddock N, Hurles ME, Cardin N, Pearson RD, Plagnol V, Robson S, et al. Genome-wide association study of CNVs in 16,000 cases of eight common diseases and 3,000 shared controls. Nature 2010;464:713—20.

[44] Gibson G. Rare and common variants: twenty arguments. Nat Rev Genet 2012;13:135—45.

[45] Leal SM. Detection of genotyping errors and pseudo-SNPs via deviations from Hardy-Weinberg equilibrium. Genet Epidemiol 2005;29:204—14.

[46] Nakamura Y, Ratain MJ, Cox NJ, McLeod HL, Kroetz DL, Flockhart DA. Re: CYP2D6 genotype and tamoxifen response in postmenopausal women with endocrine-responsive breast cancer: The Breast International Group 1-98 trial. J Natl Cancer Inst 2012;104:1264.

[47] Jorde LB. Linkage disequilibrium and the search for complex disease genes. Genome Res 2000;10:1435—44.

[48] de Bakker PI, Yelensky R, Pe'er I, Gabriel SB, Daly MJ, et al. Efficiency and power in genetic association studies. Nat Genet 2005;37:1217—23.

[49] Zhou S. Clinical pharmacogenomics of thiopurine S-methyltransferase. Curr Clin Pharmacol 2006;1:119—28.

[50] Manolopoulos VG, Ragia G, Tavridou A. Pharmacogenetics of coumarinic oral anticoagulants. Pharmacogenomics 2010;11:493—6.

[51] Biason P, Masier S, Toffoli G. UGT1A1*28 and other UGT1A polymorphisms as determinants of irinotecan toxicity. J Chemother 2008;20:158—65.

[52] Lin JH, Lu AY. Role of pharmacokinetics and metabolism in drug discovery and development. Pharmacol Rev 1997;49:403—49.

[53] Yoshida T, Zhang G, Haura EB. Targeting epidermal growth factor receptor: central signaling kinase in lung cancer. Biochem Pharmacol 2010;80:613—23.

[54] Liu W, Wu X, Zhang W, Montenegro RC, Fackenthal DL, Spitz JA, et al. Relationship of EGFR mutations, expression, amplification, and polymorphisms to epidermal growth factor receptor inhibitors in the NCI60 cell lines. Clin Cancer Res 2007;13:6788—95.

[55] Garcia J, Riely GJ, Nafa K, Ladanyi M. KRAS mutational testing in the selection of patients for EGFR-targeted therapies. Semin Diagn Pathol 2008;25:288—94.

[56] Nicolae DL, Gamazon E, Zhang W, Duan S, Dolan ME, Cox NJ. Trait-associated SNPs are more likely to be eQTLs: annotation to enhance discovery from GWAS. PLoS Genet 2010;6:e1000888.

[57] Gamazon ER, Huang RS, Cox NJ, Dolan ME. Chemotherapeutic drug susceptibility associated SNPs are enriched in expression quantitative trait loci. Proc Natl Acad Sci U S A 2010;107:9287—92.

[58] Mu W, Zhang W. Bioinformatic resources of microRNA sequences, gene targets, and genetic variation. Front Genet 2012;3:31.

[59] Willer CJ, Sanna S, Jackson AU, Scuteri A, Bonnycastle LL, Clarke R, et al. Newly identified loci that influence lipid concentrations and risk of coronary artery disease. Nat Genet 2008;40:161—9.

[60] Howie B, Marchini J, Stephens M. Genotype imputation with thousands of genomes. G3 (Bethesda) 2011;1:457—70.

[61] R_Development_Core_Team. R: a language and environment for statistical computing, R Foundation for Statistical Computing. Vienna: Austria; 2011.

[62] Purcell S, Neale B, Todd-Brown K, Thomas L, Ferreira MA, Bender D, et al. PLINK: a tool set for whole-genome association and population-based linkage analyses. Am J Hum Genet 2007;81:559—75.

[63] Need AC, Goldstein DB. Next generation disparities in human genomics: concerns and remedies. Trends Genet 2009;25:489—94.

[64] Bolstad BM, Irizarry RA, Astrand M, Speed TP. A comparison of normalization methods for high density oligonucleotide array data based on variance and bias. Bioinformatics 2003;19:185—93.

[65] Gentleman RC, Carey VJ, Bates DM, Bolstad B, Dettling M, Dudoit S, et al. Bioconductor: open software development for computational biology and bioinformatics. Genome Biol 2004;5:R80.

[66] Korn JM, Kuruvilla FG, McCarroll SA, Wysoker A, Nemesh J, Cawley S, et al. Integrated genotype calling and association analysis of SNPs, common copy number polymorphisms and rare CNVs. Nat Genet 2008;40:1253—60.

[67] Ge D, Zhang K, Need AC, Martin O, Fellay J, Urban TJ, et al. WGAViewer: software for genomic annotation of whole genome association studies. Genome Res 2008;18:640—3.

[68] Huang RS, Duan S, Bleibel WK, Kistner EO, Zhang W, Clark TA, et al. A genome-wide approach to identify genetic variants that contribute to etoposide-induced cytotoxicity. Proc Natl Acad Sci U S A 2007;104:9758—63.

[69] Huang RS, Duan S, Shukla SJ, Kistner EO, Clark TA, Chen TX, et al. Identification of genetic variants contributing to cisplatin-induced cytotoxicity by use of a genomewide approach. Am J Hum Genet 2007;81:427—37.

[70] Huang RS, Duan S, Kistner EO, Hartford CM, Dolan ME. Genetic variants associated with carboplatin-induced cytotoxicity in cell lines derived from Africans. Mol Cancer Ther 2008;7:3038–46.

[71] Huang RS, Duan S, Kistner EO, Bleibel WK, Delaney SM, Fackenthal DL, et al. Genetic variants contributing to daunorubicin-induced cytotoxicity. Cancer Res 2008;68:3161–8.

[72] Hartford CM, Duan S, Delaney SM, Mi S, Kistner EO, Lamba JK, et al. Population-specific genetic variants important in susceptibility to cytarabine arabinoside cytotoxicity. Blood 2009;113:2145–53.

[73] Morag A, Kirchheiner J, Rehavi M, Gurwitz D. Human lymphoblastoid cell line panels: novel tools for assessing shared drug pathways. Pharmacogenomics 2010;11:327–40.

[74] Abecasis GR, Cardon LR, Cookson WO. A general test of association for quantitative traits in nuclear families. Am J Hum Genet 2000;66:279–92.

[75] Abecasis GR, Cookson WO, Cardon LR. Pedigree tests of transmission disequilibrium. Eur J Hum Genet 2000;8:545–51.

[76] Storey JD, Tibshirani R. Statistical significance for genomewide studies. Proc Natl Acad Sci U S A 2003;100:9440–5.

[77] Benjamini Y, Hochberg Y. Controlling the false discovery rate: a practical and powerful approach to multiple testing. J R Statist Soc B 1995;57:289–300.

[78] Gamazon ER, Duan S, Zhang W, Huang RS, Kistner EO, Dolan ME, et al. PACdb: a database for cell-based pharmacogenomics. Pharmacogenet Genomics 2010;20:269–73.

[79] Zhang W, Ratain MJ, Dolan ME. The HapMap resource is providing new insights into ourselves and its application to pharmacogenomics. Bioinform Biol Insights 2008;2:15–23.

[80] Thorisson GA, Smith AV, Krishnan L, Stein LD. The International HapMap Project web site. Genome Res 2005;15:1592–3.

[81] Redon R, Ishikawa S, Fitch KR, Feuk L, Perry GH, Andrews TD, et al. Global variation in copy number in the human genome. Nature 2006;444:444–54.

[82] Degner JF, Pai AA, Pique-Regi R, Veyrieras JB, Gaffney DJ, Pickrell JK, et al. DNase I sensitivity QTLs are a major determinant of human expression variation. Nature 2012;482:390–4.

[83] Zhang W, Dolan ME. Impact of the 1000 Genomes Project on the next wave of pharmacogenomic discovery. Pharmacogenomics 2010;11:249–56.

[84] Gamazon ER, Zhang W, Huang RS, Dolan ME, Cox NJ. A pharmacogene database enhanced by the 1000 Genomes Project. Pharmacogenet Genomics 2009;19:829–32.

[85] Gamazon ER, Zhang W, Konkashbaev A, Duan S, Kistner EO, Nicolae DL, et al. SCAN: SNP and copy number annotation. Bioinformatics 2010;26:259–62.

[86] Pickrell JK, Marioni JC, Pai AA, Degner JF, Engelhardt BE, Nkadori E, et al. Understanding mechanisms underlying human gene expression variation with RNA sequencing. Nature 2010;464:768–72.

[87] McDonagh EM, Whirl-Carrillo M, Garten Y, Altman RB, Klein TE. From pharmacogenomic knowledge acquisition to clinical applications: the PharmGKB as a clinical pharmacogenomic biomarker resource. Biomark Med 2011;5:795–806.

[88] Owen RP, Altman RB, Klein TE. PharmGKB and the International Warfarin Pharmacogenetics Consortium: the changing role for pharmacogenomic databases and single-drug pharmacogenetics. Hum Mutat 2008;29:456–60.

[89] Zhang W, Duan S, Bleibel WK, Wisel SA, Huang RS, Wu X, et al. Identification of common genetic variants that account for transcript isoform variation between human populations. Hum Genet 2009;125:81–93.

8A

Modeling the Pharmacogenetic Architecture of Drug Response

Yafei Lu*, Xin Li*, Sisi Feng*, Yongci Li†, Xiaofeng Zeng**, Mengtao Li**, Xinjuan Liu**, Rongling Wu‡

*Center for Computational Biology, Beijing Forestry University, Beijing, China
†College of Science, Beijing Forestry University, Beijing, China
**Department of Rheumatology, Peking Union Medical College Hospital, Chinese Academy of Medical Science & Peking Union Medical College, Beijing, China
‡Center for Statistical Genetics, Pennsylvania State University, Hershey, Pennsylvania, USA

OBJECTIVES

1. Review a general procedure for genetic mapping used to identify genes for drug response.

2. Assess the advantage of functional mapping in increasing statistical power and biological relevance of gene detection.

3. Pinpoint an extension of functional mapping to genome-wide association studies to comprehend the genetic architecture of drug response.

INTRODUCTION

The response of patients to a drug is complex in terms of its underlying genetic basis [1–5]. Drug response, either drug efficacy or adverse drug reactions, is usually controlled by a network of genes that operate independently or interactively [6]. Moreover, the effects triggered by the genes rely on the genetic background of a patient and his/her lifestyle as well as the change in the body's metabolic environment arising from drug administration [7]. An increasing body of molecular studies has been conducted to identify the genetic

Pharmacogenomics
http://dx.doi.org/10.1016/B978-0-12-391918-2.00017-2

variants that influence drug response, aiming at pretreatment selection and development of drugs that are effective and safe for individual patients according to their genetic blueprints [3,4].

Detailed genetic studies of drug response have largely benefited from the discovery of molecular markers, distinctive segments of DNA that serve as landmarks for a target gene. Because both markers and target genes are distributed through the genome, the latter can be inferred from the former through linkage analysis. The motive force that identifies individual target genes for complex traits, such as drug response, comes from the publication of Lander and Botstein's [8] seminal paper, in which the authors proposed a mixture model to define and map latent genes that control a quantitative trait, called quantitative trait loci (QTLs), based on the co-segregation of markers with QTLs. This so-called genetic mapping approach has been widely used to map and identify QTLs for drug response [9–11]. These studies have been instrumental for identifying the genetic contribution to interpersonal variability in pharmacological parameters and refreshing our knowledge about the pharmacogenetic architecture of drug response.

Although conventional QTL mapping is limited by its inability to elucidate a detailed picture of the genetic control of drug response, GWASes using high-throughput single-nucleotide polymorphisms (SNPs) have emerged as a powerful tool to study pharmacogenomics by identifying an entire set of genes that control a particular pharmacological trait [3,4]. The successful applications of GWASes to study the pharmacogenomics of drug response have been described in a recent review article [3]. For example, through GWASes, significant genetic loci have been detected for interferon-α [12–14] and clopidogrel response [15], and for anticoagulant dose requirement [16–18]. For adverse drug reactions, GWASes have also identified significant associations; for example, those for statin-induced myopathy [19] and flucloxacillin-induced liver injury [20]. Most of these studies have made important contributions to the field, with some potentially benefiting clinical practice [4].

This chapter reviews the current approaches for genetic mapping and GWASes used to map pharmacogenes. As it is much more challenging to obtain an adequate number of cases for pharmacogenomics studies [4], as compared with most studies on complex disease genetics, a typical problem of pharmacogenomics studies is small sample sizes that have insufficient power to detect small or moderately sized effects. We pinpoint that this problem can be overcome by incorporating dynamic models into genetic analysis based on repeated measures of pharmacological parameters at multiple time points or doses [21,22].

GENETIC MAPPING

The underlying QTLs for variation in drug response can be mapped to specific regions of the genome. The original model for genetic mapping capitalizes on the linkage between the markers and phenotypic traits to identify the QTLs that control the traits in a controlled cross between two parents that are segregating at the genetic loci [8]. This model has now been used as a routine tool to study the genetics of complex traits [23]. It can also be used to map QTLs for pharmacological traits by using the mouse as a model system that allows the cross population to be generated [9]. Genetic mapping has now been extended to natural populations based on the linkage disequilibrium approaches [24].

Linkage Mapping

Consider two contrasting mouse inbred lines in drug response from which a segregating population of size n, such as the backcross, F_2, or recombinant inbred lines, is produced. All n progeny is genotyped for a panel of molecular markers from which a genetic linkage map is constructed. Meanwhile, the progeny is also phenotyped for a pharmacological variable y after the mice are administered a particular drug. Linkage mapping allows the QTLs underlying drug response to be localized on the genetic linkage map.

Consider a QTL, with three genotypes QQ, Qq, and qq, located in the interval between two interval markers A and B in an F_2 population. The mean values of the three genotypes in drug response y are expressed as

$$
\begin{aligned}
\mu_1 &= \mu + a \quad \text{for } QQ \text{ (1)}\\
\mu_2 &= \mu + d \quad \text{for } Qq \text{ (2)}\\
\mu_3 &= \mu - a \quad \text{for } qq \text{ (3)}
\end{aligned}
\tag{8A.1}
$$

where μ is the overall mean, a is the additive effect of the QTL, and d is the dominant effect of the QTL.

In practice, the QTL cannot be observed directly, but it can be inferred from the markers that link with it. Let two interval markers A and B, jointly with nine genotypes $AABB$, $AABb$, $AAbb$, $AaBB$, $AaBb$, $Aabb$, $aaBB$, $aaBb$, and $aabb$, bracket the QTL. Denote r_1, r_2, and r to be the recombination fractions between the marker A and QTL, the QTL and marker B, and the two markers. The conditional probability of a particular QTL genotype j, conditional upon the marker genotype of progeny i (M_i), is expressed as a function of r_1, r_2, and r, that is,

$$
\pi_{j|i} = \text{Prob}(j|M_i) = g(r_1, r_2, r), \quad j = 1, 2, 3
\tag{8A.2}
$$

The specific form of $\pi_{j|i}$ can be derived for various types of mapping populations [23].

For a particular progeny i, we do not know its QTL genotype. To estimate the QTL genotype, we formulate a mixture model-based likelihood expressed as

$$
L(y) = \prod_{i=1}^{n} \left[\pi_{1|i} f_1(y_i; \mu_1, \sigma^2) + \pi_{2|i} f_2(y_i; \mu_2, \sigma^2) + \pi_{3|i} f_3(y_i; \mu_3, \sigma^2) \right]
\tag{8A.3}
$$

where $\pi_{j|i}$ ($j = 1, 2, 3$) is defined as above, and $f_j(y_i; \mu_j, \sigma^2)$ is the normal distribution density function with QTL genotype-specific mean μ_j and variance σ^2. Lander and Botstein [8] implemented the EM algorithm to obtain the maximum likelihood estimates (MLEs) of means μ_j (and therefore additive effect, a, and dominant effect, d) and variance σ^2 through an iterative procedure. The genomic position of the QTL, reflected by r_1, r_2, and r, can be estimated by a grid approach. This approach assumes the QTL at a particular position and scans it every 2 cm within a marker interval. By plotting the likelihood value calculated from (8A.3) against the linkage map, we can determine the position of the QTL at which the likelihood is maximum.

To test the significance of the QTL, two alternative hypotheses are formulated as

$$
H_0: \mu_1 = \mu_2 = \mu_3 = \mu
$$

H_1: At least one equality in the H_0 does not hold. $\tag{8A.4}$

After the likelihoods under the H_0 and H_1 are calculated, respectively, we calculate the log-likelihood ratio expressed as

$$LR = -2\left[L_0(\tilde{\mu}, \tilde{\sigma}^2) - L_1(\hat{\mu}_1, \hat{\mu}_2, \hat{\mu}_3, \hat{\sigma}^2)\right] \tag{8A.5}$$

where the tildes denote the MLEs of the parameters under the H_0 and the hats denote the MLEs of the parameters under the H_1. Because of the violation of regularity, the LR may not follow a chi-square distribution with a known degree of freedom. The permutation tests by shuffling the phenotypic values can be used to determine the critical threshold [25]. The LR values beyond the threshold are thought to harbor significant QTLs at the corresponding genomic locations.

After a significant QTL is detected, we need to estimate and test the additive and dominant effect of the QTL. The null hypothesis for testing these two effects is expressed, respectively, as

$$H_0 : a = 0$$
$$H_0 : d = 0$$

These tests allow the inheritance mode of the QTL to be studied and identified. The critical threshold of the tests of a and d can be determined from simulation studies.

Mapping Gene—Gene Interactions

Genetic interactions between different QTLs, epistasis, play an important role in contributing to interpersonal variation in drug response. Consider two QTLs **P** and **Q**, each located in a different marker interval, which form nine genotypes with values expressed as

	QQ (1)	Qq (2)	qq (3)
PP (1)	$\mu_{11} = \mu + a_1 + a_2 + i_{aa}$	$\mu_{12} = \mu + a_1 + d_2 + i_{ad}$	$\mu_{13} = \mu + a_1 - a_2 - i_{aa}$
Pp (2)	$\mu_{21} = \mu + d_1 + a_2 + i_{da}$	$\mu_{22} = \mu + d_1 + d_2 + i_{dd}$	$\mu_{23} = \mu + d_1 - a_2 - i_{da}$
pp (3)	$\mu_{31} = \mu - a_1 + a_2 - i_{aa}$	$\mu_{32} = \mu - a_1 + d_2 - i_{ad}$	$\mu_{33} = \mu - a_1 - a_2 + i_{aa}$

where $\mu = 1/4[(\mu_{11} + \mu_{31}) + (\mu_{13} + \mu_{33})]$ is the overall mean,

$$a_1 = 1/4[(\mu_{11} + \mu_{13}) - (\mu_{31} + \mu_{33})] \tag{8A.6}$$

and

$$a_2 = 1/4[(\mu_{11} + \mu_{31}) - (\mu_{13} + \mu_{33})] \tag{8A.7}$$

are the additive effects of QTLs **P** and **Q**, respectively,

$$d_1 = 1/2(\mu_{21} + \mu_{23}) - 1/4[(\mu_{11} + \mu_{31}) + (\mu_{13} + \mu_{33})] \tag{8A.8}$$

and

$$d_2 = 1/2(\mu_{12} + \mu_{32}) - 1/4[(\mu_{11} + \mu_{31}) + (\mu_{13} + \mu_{33})] \tag{8A.9}$$

are the dominant effects of QTLs **P** and **Q**, respectively, and

$$i_{aa} = 1/4[(\mu_{11} + \mu_{33}) - (\mu_{13} + \mu_{31})] \tag{8A.10}$$

$$i_{ad} = 1/2(\mu_{12} - \mu_{32}) - 1/4[(\mu_{11} + \mu_{13}) - (\mu_{31} + \mu_{33})] \tag{8A.11}$$

$$i_{da} = 1/2(\mu_{21} - \mu_{23}) - 1/4[(\mu_{11} + \mu_{31}) - (\mu_{13} + \mu_{33})] \tag{8A.12}$$

and

$$i_{dd} = \mu_{22} + 1/4(\mu_{11} + \mu_{31} + \mu_{13} + \mu_{33}) - 1/2(\mu_{12} + \mu_{21} + \mu_{23} + \mu_{32}) \tag{8A.13}$$

are the additive × additive, additive × dominant, dominant × additive, and dominant × dominant effects between QTLs **P** and **Q**, respectively.

The likelihood for mapping epistasis is expressed as

$$L(y) = \prod_{i=1}^{n} \sum_{j_1=1}^{3} \sum_{j_2=1}^{3} \left[\pi_{j_1 j_2 | i} f_{j_1 j_2} \left(y_i; \mu_{j_1 j_2}, \sigma^2 \right) \right] \tag{8A.14}$$

where the conditional probability of a particular QTL genotype $j_1 j_2$ ($j_1, j_2 = 1, 2, 3$), conditional upon the marker genotype of progeny i (M_i), is expressed as

$$\pi_{j_1 j_2 | i} = \text{Prob}(j_1 | M_{1i}) \otimes \text{Prob}(j_2 | M_{2i}), \quad j_1, j_2 = 1, 2, 3$$

and $f_{j_1 j_2}(y_i; \mu_{j_1 j_2}, \sigma^2)$ is the normal distribution with genotype-specific mean values $\mu_{j_1 j_2}$ and variance σ^2. The EM algorithm can be used to estimate these parameters. From the MLEs of $\mu_{j_1 j_2}$, we can estimate and test individual genetic effects including the additive, dominant, and epistatic effects.

Mapping Gene—Environment Interactions

The same QTL may be expressed differently, depending on the environment of patients. We divide the mapping population into two groups of size n_1 and n_2, each treated under a different environment. Let y_{1i_1} and y_{2i_2} denote the phenotypic value of a pharmacological trait for progeny i_1 ($i_1 = 1, \ldots, n_1$) under environment 1 and progeny i_2 ($i_2 = 1, \ldots, n_2$) under environment 2, respectively. A unifying likelihood for the phenotypic values determined by a putative QTL in the two environments is written as

$$L(y_1, y_2) = \prod_{i_1=1}^{n_1} \left[\pi_{1|i_1} f_{11} \left(y_{i_1}; \mu_{11}, \sigma_1^2 \right) + \pi_{2|i_1} f_{12} \left(y_{i_1}; \mu_{12}, \sigma_1^2 \right) + \pi_{3|i_1} f_{13} \left(y_{i_1}; \mu_{13}, \sigma_1^2 \right) \right]$$

$$\times \prod_{i_2=1}^{n_2} \left[\pi_{1|i_2} f_{21} \left(y_{i_2}; \mu_{21}, \sigma_2^2 \right) + \pi_{2|i_2} f_{22} \left(y_{i_2}; \mu_{22}, \sigma_2^2 \right) + \pi_{3|i_2} f_{23} \left(y_{i_2}; \mu_{23}, \sigma_2^2 \right) \right] \tag{8A.15}$$

where $\pi_{j|i_1}$ and $\pi_{j|i_2}$ are the conditional probability of the same genotype j given the marker genotype of progeny i_1 and i_2 under environments 1 and 2, respectively; and $f_{12}(y_{i_1}; \mu_{1j}, \sigma_1^2)$ and $f_{12}(y_{i_2}; \mu_{2j}, \sigma_2^2)$ are the normal distributions under environments 1 and 2, respectively, with QTL genotype-specific means μ_{1j} and μ_{2j} and variances σ_1^2 and σ_2^2.

It is straightforward to implement the EM algorithm to obtain the MLEs of the QTL genotype-specific means μ_{1j} and μ_{2j} under the two environments, from which we can test the effect of gene–environment interactions on drug response. By formulating the hypotheses:

$$H_0 : \mu_{11} - \mu_{21} = \mu_{12} - \mu_{22} = \mu_{31} - \mu_{32}$$
$$H_1 : \text{At least one equality in the } H_0 \text{ does not hold}$$

we can determine whether a significant gene–environment interaction occurs. The rejection of the H_0 suggests a significant gene–environment interaction. The critical threshold for the above test can be obtained from simulation studies.

Linkage Disequilibrium Mapping

Assume that we sample n subjects randomly from a natural human population at Hardy–Weinberg Equilibrium (HWE). Consider a marker, **A** (with alleles A and a), and QTL (with alleles B and b), which produce four haplotypes AB, Ab, aB, and ab, with haplotype frequencies expressed, respectively, as

$$p_{11} = pq + D$$
$$p_{10} = p(1-q) - D$$
$$p_{01} = (1-p)q - D$$
$$p_{00} = (1-p)(1-q) + D$$

where p and $1-p$ are the allele frequencies of A and a, q and $1-q$ are the allele frequencies of B and b, respectively, and D is the linkage disequilibrium between the marker and QTL.

In the population, there are a total of 10 diplotypes, $AB|AB$, $AB|Ab$, $AB|aB$, $AB|ab$, $Ab|Ab$, $Ab|aB$, $Ab|ab$, $aB|aB$, $aB|ab$, and $ab|ab$ arising from the random union of maternally and paternally derived haplotypes. Because $AB|ab$ and $Ab|aB$ are genotypically the same, the 10 diplotypes emerge as nine genotypes.

Because the QTL is unknown, we formulate a likelihood based on a mixture model to estimate the QTL based on the marker information. The likelihood has a similar form to (8A.3), but at this time the conditional probability of a particular QTL genotype j, conditional upon the marker genotype of progeny i (M_i), is expressed as a function of p_{11}, p_{10}, p_{01}, and p_{00}, that is,

$$\pi_{j|i} = \text{Prob}(j|M_i) = g(p_{11}, p_{10}, p_{01}, p_{00}), \quad j = 1, 2, 3$$

In Wang and Wu [24], specific forms of $\pi_{j|i}$ are given. They derived a closed form for the EM algorithm to obtain the MLEs of $(p_{11}, p_{10}, p_{01}, p_{00})$ and $(\mu_1, \mu_2, \mu_3, \sigma^2)$ and provided a procedure for testing the significance of QTL and the association between the QTL and marker.

Functional Mapping

Many biological and biomedical traits, such as drug response, are expected to arise as curves. To determine the dynamic changes of genetic effects, we need to measure trait values at a series of discrete time points or states and model how these values for different QTL genotypes change over time or state. In a traditional way, one can extend interval mapping to accommodate the multivariate nature of time-dependent traits. However, this extension is limited due to the following reasons:

1. This extension estimates expected means of QTL genotypes at all the points and all variances and covariances, resulting in a substantial computational burden, especially when the data dimension is high.
2. The result from this extension is biologically less meaningful because the underlying biological principle for trait dynamics is not incorporated.
3. This extension is not flexible enough to ask and answer novel biologically or biomedically interesting questions in pharmacological studies.

At present, a collection of statistical methods implemented with growth model theories have been proposed to map QTL that govern growth trajectories using molecular linkage maps [21,25,27]. This method, called *functional mapping*, expresses the genotypic means of a QTL at different time points in terms of a continuous growth function with respect to time. Under this principle, the parameters describing the shape of growth curves, rather than the genotypic means at individual time points, as expected in traditional mapping strategies, are estimated by a maximum likelihood approach. Also, functional mapping estimates the parameters that model the covariance structure among multiple time points, which largely reduces the number of parameters being estimated for variances and covariances, especially when the number of time points is large.

Let $\mathbf{y}_i = (y_i(t_1), \dots, y_i(t_T))$ denote a vector of phenotypic values for progeny i measured at T time points. We assume that there is a QTL that controls a pharmacological trait. A mixture model-based likelihood that incorporates the QTL's genetic effect on the longitudinally measured trait is expressed as

$$L(\mathbf{y}) = \prod_{i=1}^{n} \left[\pi_{1|i} f_1\left(\mathbf{y}_i; \boldsymbol{\mu}_1, \boldsymbol{\Sigma}\right) + \pi_{2|i} f_2\left(\mathbf{y}_i; \boldsymbol{\mu}_2, \boldsymbol{\Sigma}\right) + \pi_{3|i} f_3\left(\mathbf{y}_i; \boldsymbol{\mu}_3, \boldsymbol{\Sigma}\right) \right] \tag{8A.16}$$

where $\pi_{j|i}$ is defined as above, and $f_j(\mathbf{y}_i; \boldsymbol{\mu}_j, \boldsymbol{\Sigma})$ is the multivariate normal distribution density function with QTL genotype-specific mean vector $\boldsymbol{\mu}_j$ and covariance matrix $\boldsymbol{\Sigma}$. Functional mapping models time-dependent means by a growth equation, that is,

$$\boldsymbol{\mu}_j = \left(\mu_j(t_1), \dots, \mu_j(t_T)\right) \tag{8A.17}$$

$$= \left(\frac{a_j}{1 + b_j e^{-r_j t_1}}, \dots, \frac{a_j}{1 + b_j e^{-r_j t_T}}\right) \tag{8A.18}$$

where three growth parameters (a_j, b_j, r_j) describe the form of growth curves. Thus, for a set of QTL genotypes, if their growth parameters (a_j, b_j, r_j) are different, this suggests that the QTL controls the growth process. Ma et al. [26] implemented the EM algorithm with the likelihood (8A.16) to estimate genotype-specific growth parameters and test their differences among genotypes.

As an application of longitudinal data analysis, functional mapping models the longitudinal covariance structure,

$$\Sigma = \begin{pmatrix} \sigma_1^2 & \cdots & \sigma_{1T} \\ \vdots & \ddots & \vdots \\ \sigma_{T1} & \cdots & \sigma_T^2 \end{pmatrix} \tag{8A.19}$$

where $\sigma_1^2, \ldots, \sigma_T^2$ are the variance at different time points; and $\sigma_{tt'} = \sigma_{t't}$ are the covariance between different time points t and t'. Many statistical approaches, such as autoregressive and structured antedependence approaches, nonparametric fitting, and semiparametric fitting, have been used to model the covariance structure in the functional mapping [28].

Since its invention in 2002, functional mapping has been extensively expanded to a wide spectrum of biomedical and evolutionary fields. By combining the principle of functional mapping and mathematical models for HIV dynamics, Wang and Wu [24] and Zhu et al. [29] can quantify the genetic effects of host QTLs that govern dynamic changes of HIV viral loads in a human body. Lin et al. [30,31] implemented well-developed pharmacodynamic and pharmacokinetic models to study the genetic control of drug response. As a ubiquitous phenomenon that occurs in organ development, naturally occurring or programmed cell death (PCD) arises from a specialized cellular machinery controlled by genes. Functional mapping provides a new avenue for identifying QTLs that control the PCD in any organism [32].

Example

In this section, we review an example in which functional mapping was used to map genetic variants for drug response [31]. The example was derived from a pharmacogenetic study aimed at testing the medical response of a drug. Dobutamine can improve the heart function of patients when they are not able to exercise. The drug was injected into patients at an increasing dose: 0 (baseline), 5, 10, 20, 30, and 40 mcg/min. Heart rate was measured repeatedly at these doses. Candidate genes, $\beta 1AR$ and $\beta 2AR$, are shown to affect heart function, and two common polymorphisms for $\beta 1AR$ and $\beta 2AR$ were genotyped for all patients. These two polymorphisms are at codons 49 with two alleles A and G and 389 with two alleles C and G for the $\beta 1AR$ gene, and at codons 16 with two alleles A and G and 27 with two alleles C and G for the $\beta 2AR$ gene, respectively.

Liu et al. [33] developed a model for detecting risk haplotypes, composed of alleles from different loci on the same chromosome, that contribute to quantitative variation. Lin et al. [31] combined the haplotype model and functional mapping to characterize haplotype effects on the change of heart rate in response to different doses of dobutamine. The functional mapping used incorporates the Emax equation that specifies drug effect E at a particular dose C, expressed as

$$E(C) = E_0 + \frac{E_{max}C^H}{EC_{50}^H + C^H} \tag{8A.20}$$

where E_0 is the baseline heart rate, E_{max} is the asymptotic (limiting) effect, EC_{50} is the drug concentration that results in 50% of the maximal heart rate effect, and H is the slope parameter that determines the slope of the dobutamine concentration–heart rate response curve. The larger the value of H, the steeper will be the linear phase of the log-concentration effect curve. The phenotypic longitudinal data were normalized to remove the baseline so that drug effect at different levels of dosage is defined by three parameters. By estimating these parameters for different genotypes, we could draw the curves of drug response for each genotype and test how different genotypes vary in the form of curve.

Through a combinatory analysis, it was found that the data could be best fit by a haplotype model with risk haplotype GC of $\beta1AR$ interacting with risk haplotype GG of $\beta2AR$. The combination of risk haplotypes at $\beta1AR$ and $\beta2AR$ produces nine composite diplotypes [31]. The significance test indicates that these nine composite diplotypes display dramatic differences in the trajectory of heart rate in response to dobutamine (Figure 8A.1), drawn from the estimated parameters (E_{maxj}, EC_{50j}, H_j) for a particular diplotype j ($j = 1, 2, 3$). This result suggests that epistasis between these two candidate genes expressed at the haplotype level plays an important role in affecting heart rate change.

A further test was performed to investigate how different components of genetic effects affect drug response. Although the additive and dominant effects at each gene are significant, the additive × additive, additive × dominant, dominant × additive, and dominant × dominant epistasis were also found to affect the response of heart rate to dobutamine. Based on the trajectory of heart rate change (Figure 8A.1), we can further determine the dose at which a particular composite diplotype has a maximum rate of drug response. Through genotypic comparisons, the genetic mechanisms underlying this dose-dependent change can be characterized.

In a simulation study by Wu et al. [22], functional mapping was found to be more powerful for association detection than traditional approaches based on a single dose. The more doses used, the more power detected. This advantage of functional mapping results from the integrative use of information expressed at multiple dose levels.

FIGURE 8A.1 Changes of heart rate in response to different dosages of dobutamine associated with haplotype variation from the $\beta1AR$ and $\beta2AR$ genes. Gray lines are heart rate changes for raw data and dark lines present nine different composite diplotypes. *Adapted from Lin et al. [31].*

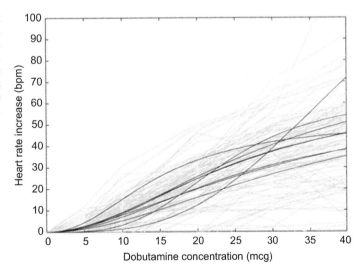

FUNCTIONAL GENOME-WIDE ASSOCIATION STUDY

Like complex disease studies, GWASes in pharmacogenomics have been increasingly growing during the past several years [3]. Some of these studies on drug response have already produced interesting findings, which can be potentially useful for developing an optimal drug for treatment or an optimal dose to use. With subsequent, larger studies, GWASes may play a more important role in deepening our knowledge about the genetic control of drug response and susceptibility to adverse drug reactions. However, the use of GWASes to study pharmacogenomics has also been criticized because of its limitation in obtaining adequate samples. We will show how this limitation can be overcome through a dynamic model of GWASes.

There are two types of design for GWASes: population-based design, in which a large cohort with genome profile is drawn from a natural population, measured for quantitative traits for each individual; and case-control design, in which patients (cases) with an outcome are selected, matched by independent controls without that outcome in terms of epidemiological factors. As an example, the association between the *SLCO1B1* gene and statin-induced myopathy was identified through a GWAS using a case-control design, as described in detail in chapters 1 and 5. In pharmacogenomics GWASes, the population-based design is often used.

Because the subjects of GWASes are from a natural population, it is likely that these samples are very heterogeneous in terms of genetic background, environmental factors, gender, race, and demographic factors. All these factors may confound the estimation of the genetic effects exerted by SNPs and should therefore be removed from data analysis. Traditional GWAS analysis is based on the association between SNP genotypes and phenotypic values measured at a single point, failing to connect the dynamic relationship of genotypes with trait changes. Functional GWAS, or *f*GWAS, integrates functional mapping and GWASes to estimate and test the dynamic pattern of genetic control [22,34].

In pharmacogenomics, *f*GWAS incorporates pharmacodynamic mechanisms to study the genetic control of drug response. For a particular pharmacogenomics GWAS datum, we construct a regression model as

$$y_i(C_{i\tau}) = \sum_{k=1}^{m} Z_{ik} g_k(C_{i\tau}) + \sum_{r=1}^{R} \alpha_r u_{ir} + \sum_{s=1}^{S} \sum_{l=1}^{L_s} x_{isl} v_{sl} + e_i(C_{i\tau}) \qquad (8A.21)$$

where $y_i(C_{i\tau})$ is the phenotypic value of a pharmacological trait for patient i at dose $C_{i\tau}$; $g_k(C_{i\tau})$ is the genotypic value of patient i who carries SNP genotype k at dose $C_{i\tau}$; m is the number of genotypes ($m = 3$ for the full model, 2 for the additive or dominant/recessive model); z_{ik} is an indicator variable of patient i defined as 1 if this patient carries a genotype considered and 0 otherwise; u_{ir} ($r = 1, \dots, R$) is the value of the rth continuous covariate, such as age and BMI, for subject i; α_r is the effect of the rth continuous covariate; v_{sl} ($l = 1, \dots, L_s$, $s = 1, \dots, S$) is the effect of the lth level for the sth discrete covariate, such as race, gender, and treatment, with $\sum_{l=1}^{L_s} v_{sl} = 0$ where L_s is the number of levels for the sth discrete covariate; x_{isl} is an indicator variable of subject i who receives the lth level of the sth discrete covariate; and $e_i(C_{i\tau})$ is a random error.

We used the Emax Equation 8A.20 to model dose-dependent genotypic values for each SNP genotype k by pharmacological parameters (E_{maxk}, EC_{50k}, H_k). A maximum likelihood is formulated to estimate these parameters and covariate effects shown in Equation 8A.21. Considering the autocorrelation feature of the random error, the autoregressive regression model is used to estimate the across-dose covariance structure. Other approaches for modeling covariance structure are available in the literature [28], allowing a choice of the best fit model for a practical data set. In the situation where the same SNP is detected to be significant by genotypic, additive, and dominant/recessive models, we use likelihood and/or BIC values to choose an optimal model.

fGWAS is found to display much greater power for association detection, as compared with traditional analysis. In a simulation study [22], a set of GWAS subjects was produced with SNP genotypes and pharmacological measures at multiple doses (say, 4) for each subject. The traditional approach associates SNP genotypes with phenotypic values separately for individual doses, from which the power to detect significant associations is calculated. The same data are analyzed by an fGWAS that uses phenotypic data from all doses. The result shows that an fGWAS produces greater power by 20 to 30% than traditional GWAS analysis when the sample size is about 200 and heritability is 0.1. This conclusion has a potential implication for the design of GWASes in pharmacogenomics in which sample size is particularly low. Wu et al. [22] also provided a procedure for testing the pattern of genetic control of drug response through fGWAS.

GWASes often confront a major challenge in analyzing thousands of thousands of SNPs (p) on much fewer samples (n). This challenge is not an issue when SNPs are analyzed individually using a simple regression model. However, such individual SNP analysis cannot capture the correlation information among SNPs, reducing the precision and power of association detection and also failing to comprehend the picture of the genetic control of drug response. The best way such a picture is illustrated is to analyze all SNPs simultaneously. In the current statistical literature, several models of variable selection have been available to handle such a high-dimension data characterized by big p and small n, of which the least absolute shrinkage and selection operator (lasso) has proved to be the most powerful and computationally feasible [35].

Li et al. [36] proposed a two-stage procedure for multi-SNP modeling and analysis in GWASes by first producing a "preconditioned" response variable using a supervised principle component analysis and then formulating Bayesian lasso to select a subset of significant sites. The Bayesian lasso is implemented with a hierarchical model, in which scale mixtures of normal are used as prior distributions for the genetic and epigenetic effects and exponential priors are considered for their variances, and then solved by using the Markov chain Monte Carlo (MCMC) algorithm. This approach obviates the choice of the lasso parameter by imposing a diffuse hyperprior on it and estimating it along with other parameters and is particularly powerful for selecting the most relevant SNPs where the number of predictors exceeds the number of observations.

DISCUSSION

Variation in drug response is complex in terms of its multifactorial and interactive nature [1,3–5,37]. A good deal of evidence has indicated that genes and their interactions expressed at various levels contribute to a significant extent of variation. Thus, the identification of the genetic contribution to drug response has become one of the most active and promising areas in modern genomics. Gene identification has been typically made from genetic mapping, but this approach is limited in revealing a comprehensive picture of the genetic architecture of drug response.

Recent progress in dissecting genetic contribution in drug response through the use of functional mapping [26,38,39] has been made, facilitating our understanding of the mechanistic basis of drug response. Functional mapping integrates pharmacodynamic and pharmacokinetic changes of gene expression into a mixture model framework using mathematical aspects of biological or clinical processes [31]. Functional mapping estimates and tests the effect of genetic variants for dynamic phenotypes based on a small set of parameters that define the shape and pattern of drug responses. This treatment dramatically enhances statistical precision and biological relevance in the genetic mapping of drug response.

Functional mapping has now been integrated with powerful GWASes, leading to the birth of a novel GWAS model called functional GWAS or *f*GWAS [34]. *f*GWAS combines the strength of functional mapping in detecting clinically meaningful results and GWAS in characterizing an entire set of genetic actions and interactions through the genome. Statistical models are now available to tackle high-dimensional SNPs on much fewer samples by variable selection approaches such as lasso [36]. The implications of *f*GWAS for pharmacogenomics studies could shed light on the genetic architecture of interpatient variation in complex drug responses.

In clinical practice, a ubiquitous problem is missing or incomplete response; that is, there are always some patients who have to drop out early due to physiological side effects or limited duration [40]. By including the dropout time as a covariate for the longitudinal development, we can extend *f*GWAS to study the genetic control of longitudinal curve more precisely. This extension should be useful for providing scientific guidance about the design of efficient and effective clinical trials for biomedical and health studies, greatly helping to translate genetic information into clinical trials and promote the practical use of "clinical genomics."

QUESTIONS FOR DISCUSSION

1. How can quantitative trait loci (QTL) be utilized to identify genetic contributions to drug response?
2. What are the advantages of genome-wide association studies over conventional QTL mapping?
3. How can functional mapping be used to map genetic variants of drug response?

Acknowledgements

This work is supported by NIH/UL1RR0330184.

References

[1] Weinshilboum R. Inheritance and drug response. N Engl J Med 2003;348:529–37.
[2] Weinshilboum RM, Wang L. Pharmacogenetics and pharmacogenomics: development, science, and translation. Annu Rev Genomics Hum Genet 2006;7:223–45.

[3] Daly AK. Genome-wide association studies in pharmacogenomics. Nat Genet Rev 2010;11:241–6.

[4] Bailey KR, Cheng C. Genome-wide association studies in pharmacogenetics research debate. Pharmacogenomics 2010;11(3):305–8.

[5] Wang LW, McLeod HL, Weinshilboum RM. Genomics and drug response. N Engl J Med 2011;364:1144–53.

[6] Liou YJ, Bai YM, Lin E, et al. Gene-gene interactions of the INSIG1 and INSIG2 in metabolic syndrome in schizophrenic patients treated with atypical antipsychotics. Pharmacogenomics J 2012;12:54–61.

[7] Baye TM, Wilke RA. Mapping genes that predict treatment outcome in admixed populations. Pharmacogenomics J 2010;10:465–77.

[8] Lander ES, Botstein D. Mapping Mendelian factors underlying quantitative traits using RFLP linkage maps. Genetics 1989;121:185–99.

[9] Grisel JE, Belknap JK, O'Toole LA, et al. Quantitative trait loci affecting methamphetamine responses in BXD recombinant inbred mouse strains. J Neurosci 1997;17(2):745–54.

[10] Watters JW, Kraja A, Meucci MA, et al. Genome-wide discovery of loci influencing chemotherapy cytotoxicity. Proc Natl Acad Sci U S A 2004;101:11809–14.

[11] Choy E, Yelensky R, Bonakdar S, et al. Genetic analysis of human traits in vitro: drug response and gene expression in lymphoblastoid cell lines. PLoS Genet 2008;4(11):e1000287.

[12] Ge D, Fellay J, Thompson AJ, et al. Genetic variation in IL28B predicts hepatitis C treatment-induced viral clearance. Nature 2009;461:399–401.

[13] Suppiah V, Moldovan M, Ahlenstiel G, et al. IL28B is associated with response to chronic hepatitis C interferon-α and ribavirin therapy. Nat Genet 2009;41:1100–4.

[14] Tanaka Y, Nishida N, Sugiyama M, et al. Genome-wide association of IL28B with response to pegylated interferon-α and ribavirin therapy for chronic hepatitis C. Nat Genet 2009;41:1105–9.

[15] Shuldiner AR, O'Connell JR, Bliden KP, et al. Association of cytochrome P450 2C19 genotype with the anti-platelet effect and clinical efficacy of clopidogrel therapy. J Am Med Assoc 2009;302:849–57.

[16] Cooper GM, Johnson JA, Langaee TY, et al. A genome-wide scan for common genetic variants with a large influence on warfarin maintenance dose. Blood 2008;112:1022–7.

[17] Takeuchi F, McGinnis R, Bourgeois S, et al. A genome-wide association study confirms VKORC1, CYP2C9, and CYP4F2 as principal genetic determinants of warfarin dose. PLoS Genet 2009;5. e1000433.

[18] Teichert M, Eijgelsheim M, Rivadeneira F, et al. A genome-wide association study of acenocoumarol maintenance dosage. Hum Mol Genet 2009;18:3758–68.

[19] Link E, Parish S, Armitage J, et al. SLCO1B1 variants and statin-induced myopathy—a genomewide study. N Engl J Med 2008;359:789–99.

[20] Daly AK, Donaldson PT, Bhatnagar P, et al. HLA-B*5701 genotype is a major determinant of drug-induced liver injury due to flucloxacillin. Nat Genet 2009;41:816–9.

[21] Wu RL, Lin M. Statistical and Computational Pharmacogenomics. London: Chapman & Hall/CRC; 2008.

[22] Wu RL, Tong CF, Wang Z, et al. A conceptual framework for integrating pharmacodynamic principles into genome-wide association studies for pharmacogenomics. Drug Discov Today 2011;16:884–90.

[23] Wu RL, Ma CX, Casella G. Statistical Genetics of Quantitative Traits: Linkage, Maps, and QTL. New York: Springer-Verlag; 2007.

[24] Wang ZH, Wu RL. A statistical model for high-resolution mapping of quantitative trait loci determining human HIV-1 dynamics. Stat Med 2004;23:3033–51.

[25] Churchill GA, Doerge RW. Empirical threshold values for quantitative trait mapping. Genetics 1994;138:963–71.

[26] Ma CX, Casella G, Wu RL. Functional mapping of quantitative trait loci underlying the character process: a theoretical framework. Genetics 2002;161:1751–62.

[27] Li Y, Wu RL. Functional mapping of growth and development. Bio Rev 2010;85:207–16.

[28] Yap J, Fan J, Wu RL. Nonparametric modeling of covariance structure in functional mapping of quantitative trait loci. Biometrics 2009;65:1068–77.

[29] Zhu Y, Hou W, Wu RL. A haplotype block model for fine mapping of quantitative trait loci regulating HIV-1 pathogenesis. J Theor Med 2004;5:227–34.

[30] Lin M, Aqvilonte C, Johnson JA, et al. Sequencing drug response with HapMap. Pharmacogenomics J 2005;5:149–56.

[31] Lin M, Hou W, Li HY, et al. Modeling sequence-sequence interactions for drug response. Bioinformatics 2007;23:1251–7.

[32] Cui YH, Zhu J, Wu RL. Functional mapping for genetic control of programmed cell death. Physiol Genomics 2006;25:458—69.

[33] Liu T, Johnson JA, Casella G, et al. Sequencing complex diseases with HapMap. Genetics 2004;168:503—11.

[34] Das K, Li JH, Wang Z, et al. A dynamic model for genome-wide association studies. Hum Genet 2011;129:629—39.

[35] Li JH, Das K, Fu GF, et al. The Bayesian lasso for genome-wide association studies. Bioinformatics 2011;27:516—23.

[36] Tibshirani R. Regression shrinkage and selection via the lasso. J R Stat Soc Ser B 1996;58:267—88.

[37] Roden DM, George Jr AL. The genetic basis of variability in drug responses. Nat Rev Drug Discov 2002;1:37—44.

[38] Wu RL, Lin M. Functional mapping—how to map and study the genetic architecture of dynamic complex traits. Nat Rev Genet 2006;7:229—37.

[39] Ahn K, Luo JT, Berg A, et al. Functional mapping of drug response with pharmacodynamic-pharmacokinetic principles. Trends Pharmacol Sci 2010;31:306—11.

[40] Li HY, Wu RL. A pattern-mixture model for functional mapping of quantitative trait nucleotides with non-ignorable dropout data. Stat Sinica 2012;22:337—57.

Study Designs in Clinical Pharmacogenetic and Pharmacogenomic Research

Julia Stingl (formerly Kirchheiner)*, Jürgen Brockmöller[†]

*Institute of Pharmacology of Natural Products and Clinical Pharmacology, University Ulm, Ulm, Germany

[†]Department of Clinical Pharmacology, University Göttingen, Göttingen, Germany

OBJECTIVES

- Explaining the grounds of study designs and developing a fascination for the beauty of a good study design.

- Understanding how poor study design can result in wrong conclusions and showing the ways to more effective pharmacogenetic research in the future.

- Understanding the role of evidence-based medicine for human therapies but also understanding that individual therapy

  decisions often cannot be based on clinical trial evidence.

- Understanding the arguments of why, when, and how to perform pharmacogenetic and pharmacogenomic research in drug development.

- Stimulating development of new or improved study designs based upon solid knowledge of existing study design concepts.

INTRODUCTION

The Difficulties of Introducing Individualized Medicine into Clinical Practice

Individualized medicine is a rational concept, with potentials for major improvements in the efficacy and safety of drug therapies in specific subgroups of patients. Polymorphisms in gene coding for drug metabolism enzymes and drug membrane transporters may affect drug bioavailability, drug plasma concentrations, and drug availability at target sites. Genetic polymorphisms in signaling pathways may have an even greater impact on the response to drug therapies. In addition, genetic polymorphisms can explain why some patients suffer from idiosyncratic adverse drug reactions mediated by immune reactions or other mechanisms.

Most aspects discussed in this section will be presented with examples of candidate-gene-based approaches (pharmacogenetics) using one or a few polymorphisms. However, almost all study design concepts presented here can be applied similarly in broad genome-wide approaches considering a wide range of genes and proteins (pharmacogenomics).

Although there are numerous examples of how pharmacogenetics may make drug therapies safer, individualized drug selection and dosing based on pharmacogenetic biomarkers is at present mostly an individual decision and not uniformly applied [1,2]. To understand the gap between the availability of extensive pharmacogenetics knowledge and its scarce adoption in clinical practice, one must first acknowledge that genomic variation is only one of the numerous factors that determine drug response (Figure 9.1). It is unrealistic to assume that

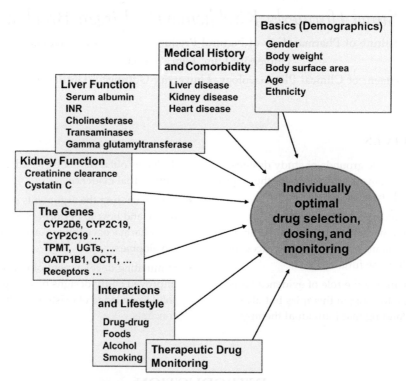

FIGURE 9.1   Multiple factors to be considered in individually optimal drug selection, drug dosing, and therapeutic monitoring. Clearly, one should never forget in medical therapy that genotype is only one factor among several influencing drug response. Consequently, pharmacogenetic studies have to be rather large and the data analysis has to be rather complex. In addition, many genes often statistically interact with multiple other factors. Such interactions may mean, for instance, that a genetic polymorphism is medically relevant only in the presence of some other environmental factors depicted here but not in the absence (gene−environment interactions) of such factors.

genetic variation dominates all other factors. Obviously, unrealistic expectations have to be dampened and clinical studies have to consider multifactorial determinants of drug response (Figure 9.1). Concerning study designs, there are a number of consequences from such multifactorial influences: sample size usually will have to be relatively large, all relevant other factors will have to be adequately documented, and gene–environment interactions (and gene–gene interactions) have to be considered in the data analysis.

Before discussing study designs for evidence gain in pharmacogenetic and pharmacogenomic research, it may be important to clarify how pharmacogenetic evidence in principle may influence clinical practice or clinical decisions. Table 9.1 lists typical scenarios and mechanisms in which pharmacogenetic variability may influence therapy decisions—if enough evidence would be available [2–4]. As illustrated in Table 9.1, genetic variability can influence drug therapy in a specific manner depending on the relevant polymorphisms and the drugs of interest. This complexity with different scenarios concerning application of pharmacogenetic diagnostics for the respective drug may be one factor slowing down the widespread adoption of pharmacogenetic diagnostic by clinicians.

During development of a drug known to have response affected by genetic variation, a decision might be made to stop further development of that drug and promote development of a drug not affected by genetic variation. However, an alternative drug may not exist in many instances, and codevelopment of the original drug with a pharmacogenetic test is an appropriate choice. Alternatively, the result of additional research may reveal that the relevant genotype does not need to be considered. The impact of genetic polymorphisms may greatly differ depending on other factors such as drug interactions, very low or very high age (e.g., those below 5 or above 75 years of age), or impaired organ function. According to a recent European guideline [5], drug companies are advised to elaborate the worst cases that might happen from such interactions between genotype and other factors. The exaggerated bioactivation of codeine to morphine by carriers of the CYP2D6 gene duplication may serve as an example for that type of case. Although it is less likely that adult ultra-rapid metabolizers will suffer from respiratory arrest, fatal outcomes have been reported in newborns [6].

It is a major challenge to prove the value of individualized approaches like those summarized in Table 9.1 with the commonly accepted principles of evidence-based medicine (EBM) [7]. The commonly accepted study designs in EBM are based on comparisons between two or more groups with sufficiently large sample sizes, but many functionally relevant genetic polymorphisms are too rare to allow for sufficient statistical power, especially with limited resources available for such studies (Figure 9.2). Even for frequent variants with strong effects, such as the CYP2D6 polymorphism that consistently and strongly affects the pharmacokinetics of tricyclic antidepressants [2], a major benefit from pharmacogenetic diagnostics can only be expected in the outliers (i.e., poor metabolizers or ultra-rapid metabolizers among 10% of Caucasians). Also, among these 10%, it is unlikely that the genotype-based dose adjustments will improve therapy in every subject. Optimistically, we may achieve clinical benefit in 50% of the outliers, and thus only 50 of 1,000 subjects from the target population can contribute to the statistical power in a study on CYP2D6 genotyping for therapy with tricyclic antidepressants. This simple estimation shows that even for that example, a sample size of about 1,000 would be required to prove with reasonable statistical power that genotype-based individualized therapy is really better than conventional therapy.

TABLE 9.1   Translation of pharmacogenetics to individualized therapy

Category	Problem	Concrete actions and citations
Pharmacokinetics		
	Pharmacologically active drugs metabolized to inactive metabolites by genetically polymorphic enzymes	Increase or reduce dose proportionally to increased or reduced drug clearance known to result from the specific genotypes.
	Pharmacologically active drugs metabolized to active metabolites by genetically polymorphic enzymes	No general recommendations possible; drug-specific actions according to studies with valid surrogate markers or clinical endpoints.
	Pharmacologically inactive prodrugs bioactivated to the active moiety by genetically polymorphic enzymes	Adjust doses in the other genotypes accordingly to achieve comparable concentrations of the active moiety; alternative drugs in completely deficient bioactivation.
	Drugs affected by genetically polymorphic drug transporters	At present, no specific general recommendations possible. Do close clinical monitoring and effect site therapeutic drug monitoring.
Pharmacodynamics		
	Drugs acting on a genetically polymorphic target site (receptor, enzyme)	Adjust dose according to studies with valid surrogate markers or clinical endpoints to achieve comparable drug effects in all genotypes (e.g., warfarin dose adjustment according to *VKORC1* genotypes).
Idiosyncratic and other types of adverse drug reactions		
	Immune polymorphisms predicting high risk for allergic adverse drug reaction	If the odds ratio is reproducibly above 5, do genotyping and generally do not administer the drug to risk carriers. Consider similar approaches even with lower relative risks. Examples include carbamazepine and *HLA-B*1502* [72], abacavir and *HLA-B*5701* [55], flucloxacillin and *HLA-B*5701* [43], and allopurinol and *HLA-B*5801* [73,74].
	Polymorphism predisposing to vulnerability to a certain side effect	Avoid drug with the respective side effect in risk-gene carriers (example may be estrogen-containing hormonal contraceptives in factor V Leiden carriers).

This problem of insufficient statistical power and failure of traditional study design approaches will further increase in the near future with the routine use of massive parallel sequencing (next-generation sequencing) techniques. These techniques will reveal millions of rare but potentially medically relevant inherited variants with frequencies below 1% in the human genome. And we already know that the rare variants explain a lot of the

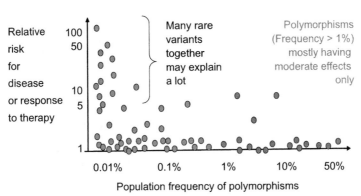

Population frequency of polymorphisms

FIGURE 9.2 Relationship between frequency of genomic variants and functional or medical impact. Traditionally in pharmacogenetics, polymorphisms are those variants in the human population with frequencies above 1%. In genome-wide association studies (GWASes), the focus is on polymorphisms with frequencies above 5% for reasons of statistical power. But now, the almost routine use of genomic resequencing will reveal millions of rare genetic variants that together may have significant impact on the outcomes of drug treatment, and it is a major task to find reliable ways to consider such information in medical care. There are hints that indeed rare variants or, concerning populations, the sum of rare variants do play a bigger role for a number of common diseases than frequent genetic polymorphisms. For instance, we know from twin and family studies that schizophrenia has a relatively high heritability, but thus far, candidate gene studies or GWAS approaches have elucidated only a minor part of genetics in schizophrenia (the so-called mystery of missing heritability).

variability beyond that explained by frequent polymorphisms as shown, for instance, in hypertension [8,9] and dilated cardiomyopathy (Figure 9.2). In a recent study, investigators found 67 mostly unique nonsense, splicing, frameshift, or copy number variations in 312 patients with dilated cardiomyopathy, but only three such mutations in hypertrophic cardiomyopathy [10]. Thus, it is no longer just one specific variant but the sum of rare and even very rare variants that explains clinically relevant frequent phenotypes. Thus, next-generation sequencing could reveal a genetic basis even in those instances in which research on frequent genetic polymorphisms in the past decades has failed to elucidate the genetic background. However, concerning our main question about when and how to use pharmacogenetic or pharmacogenomic diagnostics in individualized medicine, there are many unanswered questions as to how such rarely or even singularly occurring mutations may be used in therapeutic decisions.

The low frequency of most relevant polymorphisms is not the only point to consider in the design of clinical pharmacogenomic studies. Other points of consideration include the moderate effect size of most polymorphisms and the combined synergistic or antagonistic effects (interactions) with other polymorphisms (gene—gene interactions) and with environmental factors (gene—environment interactions). Both gene—gene interactions and gene—environment interactions may differ between populations and particularly between different ethnic groups because of significant variability in genetic background and lifestyles. Consequently, an effect of a polymorphism shown in one ethnic group may not necessarily exist in other ethnic groups. With all of these factors, it is not too surprising that many published associations between genotypes and drug therapy outcomes are either unconfirmed or not tested further in subsequent studies.

A poor knowledge of clinical pharmacogenetic research methodology may also contribute to a lack of replicable data in pharmacogenetics [11] and, consequently, to a lack of clinical translation. Therefore, this chapter should help investigators to find the optimal study design for a given question. It should also help to distinguish between preliminary findings that clearly need further confirmation versus credible results that may be used in individualized therapies.

When and Why to Perform Pharmacogenetic Research in Drug Development

Sampling of biomaterial for pharmacogenetic analysis is almost standard nowadays in clinical drug development. Well-designed and executed preclinical and clinical pharmacogenetic research may provide well-substantiated and carefully reasoned decisions, ranging from stopping further development because of a serious pharmacogenetic problem, codeveloping the drug with pharmacogenetic testing, or ignoring pharmacogenetics because of negligible effects. In addition, newly discovered pharmacogenetic issues may help to improve the benefit–risk ratio of marketed drugs.

However, in drug development, except for a limited number of valid pharmacogenetic biomarkers, there is no consensus regarding when, how, and what genes should be analyzed. The distinction between valid biomarkers and all other genetic polymorphisms is important in order to understand controversies in the pharmacogenetic literature and to differentiate between what must be included in drug studies and what may be included. Therefore, we list some valid biomarkers in Table 9.2. In general, valid biomarkers determine a significant fraction of the great interindividual variation in the phenotype of interest [12,13]. If valid biomarkers like those listed in Table 9.2 may be relevant in a given clinical trial, their consideration is mandatory [4]. On the other hand, the functional and medical impact of the vast majority of polymorphisms appears much less clear. Analysis of such polymorphisms may optionally be included as integral parts or as add-on studies in clinical trials.

Often, there is no consensus about whether to require biomaterial sampling for pharmacogenetic analysis from every study participant versus optional voluntary sampling of biomaterial. However, with voluntary sampling typically resulting in pharmacogenetic data in much fewer than 80% of study participants, there is a substantial risk with this approach that the whole pharmacogenetic analysis becomes meaningless or even misleading [14]. Thus, there are good reasons to propose that in such cases biomarker analysis should become an integral part of the main study. On the other hand, voluntary add-on research and corresponding voluntary broad consent may have an important place in possible future applications of biomaterials outside the specific context of the present study.

Preclinical Phase of Pharmacogenetic and Pharmacogenomic Research

The impact of inherited pharmacogenomic variation for a drug may be predicted from *in vitro* and *in vivo* studies of relevant genetically polymorphic drug metabolizing enzymes, transporters, drug targets, signaling pathways, and immune system proteins prior to Phase I human studies.

Defined inherited genetic variations in individuals can be studied in model cell lines using recombinant gene technology, whereas cell lines derived from large numbers of human

TABLE 9.2   Examples of valid pharmacogenetic biomarkers

Gene	Specific variants and comments	Examples of drugs where it might be a concern
CYP2C9	Amino acid variants with significantly reduced activity	Warfarin, phenytoin, most NSAIDs, many sulfonylurea antidiabetics
CYP2C19	Null, active, and ultra-rapid genotypes	Omeprazole (most proton pump inhibitors), voriconazol, clopidogrel (bioactivation)
CYP2D6	Null, active, and ultra-rapid genotypes	Numerous psychotropic and cardiovascular drugs, codeine (bioactivation), tramadol (bioactivation), and tamoxifen (bioactivation)
CYP3A5	Null and active genotypes	Tacrolimus
Factor V	High thrombosis risk genotype	Estrogen-containing oral contraceptives, anticoagulants
HLA-B*5701	High allergy risk genotype	Abacavir, flucloxacillin
NAT2	Very low, intermediate, and high active genotypes	Isoniazide, hydralazine
OATP1B1	Low and highly active genotypes	Pravastatin, simvastatin (many statins)
OCT1	Null, intermediate, and active genotypes	Metformin, tramadol, tropisetron
TPMT	Null, intermediate, and active genotypes	Azathioprin, 6-mercaptopurine
UGT1A1	Null, intermediate, and active genotypes	Irinotecan
VKORC1	Low and high expression genotypes	Acenocoumarol, phenprocoumon, warfarin

Valid biomarkers may be defined as having a highly reproducible functional impact that is consistent between *in vitro* preclinical data and clinical data. Typically, there are no or almost no contradictory data in the literature concerning the functional impact. Most valid pharmacogenetic biomarkers have null-activity or almost null-activity genotypes/phenotypes versus active genotypes/phenotypes in the population. In drug research analysis of such biomarkers, if involved in the drug of interest, is almost mandatory. Consequently, most valid biomarkers listed here are included in FDA drug labeling.

beings can be used to study the entire scope of human genetic variation (Table 9.3, Figure 9.3). Primary cell lines, such as peripheral blood mononuclear cells or subtypes of these, fibroblasts, hair follicle cells, or epithelial cells from buccal swabs may be easily obtained, but the small amount of such primary cells may limit the number of possible experiments. Peripheral blood B lymphocytes may be immortalized by Epstein–Barr virus transfection. Such cell lines have already been extensively used, although they reflect only one type of tissue, and viral transfection may confer significant artifacts unrelated to human genetic variation [15]. Significant progress in functional pharmacogenetic and pharmacogenomic research is expected from the generation of organ-specific cell types from induced

TABLE 9.3 Human cell models in pharmacogenetic and pharmacogenomic research

Cell model	Description	Advantages and disadvantages	Suggested applications
Peripheral blood mononuclear cells	Lymphocytes and monocytes isolated from centrifugation of human blood	Easy to achieve, relatively robust but heterogeneous cell types, and limited replication	Functional analysis of polymorphisms in the TGF beta signaling pathway [75]
Subtypes of peripheral blood mononuclear cells	Specific cell subtypes isolated from human blood using affinity media or FACS	More homogeneous model, more laborious procedures, and smaller cell numbers	ABCB1 polymorphisms functionally relevant specifically in CD52 blood cells [76]
EBV immortalized lymphocytes	Peripheral blood mononuclear cells transfected with Epstein–Barr virus, resulting in unlimited growth	Infinite number of cells but artifacts resulting from viral transformation	Cytostatic drug pharmacogenetics (e.g., etoposide) [77]
Fibroblasts	Connective tissue cells growing from (small) human skin biopsies	Robust cell model for numerous applications but requires a skin biopsy	Used in radiation biology [78] and DNA repair research [79] but thus far only infrequently in pharmacogenetics
Keratinocytes	Primary cells from hair follicles	Special cell model obtained from a few plucked hairs but limited replication of the primary keratinocytes	May be an optimal model to study mechanisms of EGFR inhibitor-related skin toxicity
Inducible pluripotent stem cells	Primary cell lines are transformed by overexpression of stem-cell associated proteins (Oct3/4, Sox2, Klf4, and c-Myc) into inducible pluripotent stem cells from which almost all cell types can be generated	Allows principally also the study of those cell types that can otherwise be obtained from humans only under exceptional circumstances but difficult to generate and numerous artifacts possible	Use to study inherited forms of the long QT syndrome [80]

This table lists cell models of human genomic variation as variation occurs in the respective donors of the cell lines. Not listed here are cell models modified by recombinant gene technology to express genetically polymorphic genes, which play a major role in pharmacogenetic research. EBV, Epstein–Barr virus; TGF, transforming growth factor; FACS, fluorescence-activated cell sorting; ABCB1, ATP-binding cassette, subfamily B, member 1; EGFR, epidermal growth factor receptor; Oct, octamer-binding transcription factor; Klf, Krueppel-like factor.

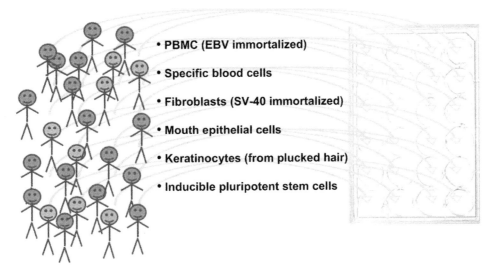

- PBMC (EBV immortalized)
- Specific blood cells
- Fibroblasts (SV-40 immortalized)
- Mouth epithelial cells
- Keratinocytes (from plucked hair)
- Inducible pluripotent stem cells

FIGURE 9.3 Human genetic polymorphisms (depicted by different colors) can be studied by deriving appropriate cell models from the studied human beings. The *ex vivo* approach illustrated here allows studying inherited genomic variants in the complex real human genomic background. Peripheral blood mononuclear cells (PBMC) may be used with or without Epstein–Barr virus immortalization to get sufficient numbers of cells, whereas with fibroblasts, mostly immortalization will not be necessary. Inducible pluripotent stem (IPS) cells have the potential to investigate also poorly accessible cell types, but a lot of research will still be required before we can robustly use this IPS approach and understand when and how the cells derived from such IPS cells reflect the corresponding individual human tissues.

pluripotent cells (IPS) obtained from individual blood samples [16]. A detailed discussion of cell biology approaches to pharmacogenetics is beyond the scope of the chapter, but it is evident that the enormous task of studying genomic variation, as illustrated in Figure 9.1, in future clinical therapies can only be solved by intelligent combinations of *in vitro* and *ex vivo* research combined with appropriate clinical studies.

Why Worry about Pharmacogenetic Study Design—Isn't It Trivial?

In reviewing the literature, we often find insufficient information regarding the rationale of choosing a particular study design, and too frequently the actual study design is described without correct terminology and reference to the appropriate methodical references [17]. In addition, the same pharmacogenetics question is often analyzed in several studies of different design. This replication would make sense if there were a rational order of studies, beginning with small studies with an exploratory design to the bigger confirmatory studies. Yet such an order is often not found. Compounding the problem is the fact that up to 90% of pharmacogenetic associations cannot be replicated [18] and even fewer pharmacogenetic or pharmacogenomic findings make it to clinical application. Primary problems of any pharmacogenetic and pharmacogenomic research include major risks for false-positive results particularly due to multiple testing on one hand and lack of statistical power often due to the low frequency of genotypes of interest and the moderate penetrance of typical pharmacogenetic

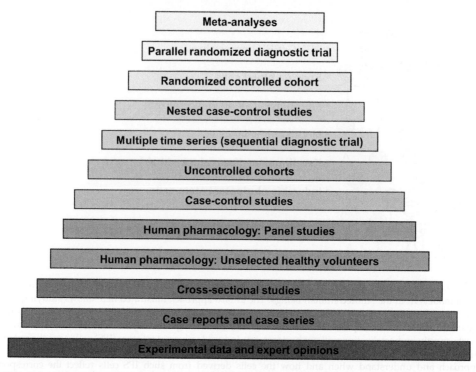

FIGURE 9.4 The hierarchy of study designs. Definitions of the different study designs are given in the text sections of this chapter and in Table 9.4. On top of this hierarchy are the study designs with the highest validity concerning assessment of clinical utility. Of course, the value of any single study also depends on details of the specific study design and how the study was performed. Notably, the order of study designs concerning the potential for new discoveries, for innovation, and for explanation may be the reverse [19], and thus the hierarchy is not referring to the value of the study designs for medical research in general but only to the value concerning introduction of a certain procedure, drug, or pharmacogenetic diagnostic principle into medical practice.

polymorphisms on the other. Such problems have to be well known to everyone conducting pharmacogenetic research, and our chapter should help select and organize appropriate designs to master these primary problems of pharmacogenetic research.

In general, the hierarchy of pharmacogenetic study designs (Figure 9.4) should be known. All study designs (levels in Figure 9.4) can contribute to new knowledge generation [19], but for constituting the basis for clinical application, the highest feasible level according to Figure 9.4 should be chosen. Depending on specific study issues and the effect sizes found in the studies, modification of study designs may be necessary [20].

CLINICAL STUDY DESIGNS TO DISCOVER AND CONFIRM THE IMPACT OF HUMAN PHARMACOGENETIC VARIATION

In the following sections, a comprehensive list of study designs useful for clarifying the medical impact of pharmacogenetics is presented. We begin this presentation with human pharmacological studies performed often with healthy volunteers and focusing on surrogate endpoints. We then continue with pharmacogenomic studies typically performed in clinical Phases II to IV of drug development. This list is followed by discussion of observational study designs typically conducted after drug approval. Finally, we provide an introduction of study designs specifically aimed to prove the added value of pharmacogenetic diagnostics. Concerning the last part, it is important to realize that proof of clinical differences between genotype-defined subgroups does not necessarily mean that considering the pharmacogenetics biomarker in therapy decisions will improve therapeutic outcome.

Human Pharmacology Studies

In this category, we summarize all studies in which drugs are administered to healthy humans in a nontherapeutic context, and pharmacokinetics, drug effects, and side effects are measured. All measurements of pharmacogenetic parameters in clinical Phase I studies in drug development also fall into this category.

Pharmacogenetic Studies in Samples of Healthy Volunteers

In Phase I studies, new drug entities (NDEs) are administered the first time to healthy volunteers for evaluation of pharmacokinetics, drug effects, and safety. Usually there are data from preclinical studies suggesting what type of pharmacogenetic variation in drug metabolizing enzymes, transporters, and/or targets could potentially influence the pharmacokinetics and pharmacodynamics of the NDE. Pharmacogenetic analysis in unselected healthy volunteers could serve as a first step in evaluating the extent of interindividual variation and the potential impact of a single genetic polymorphism. For a single polymorphism with presumed strong effect size, the typical sample size would be a minimum of about 20 to 100 subjects. However, several hundred or even thousands of subjects would be necessary if one were aiming to explore the population for more rare variants or variants with less strong effect size or if the investigator wanted to study the interaction between multiple polymorphisms. Although the homogeneous population and the controlled study design provide a good estimate of how big the effects of the studied polymorphisms on drug pharmacokinetics or pharmacodynamics may be, only surrogate parameters of efficacy or side effects can be measured. This approach of studying unselected healthy volunteers in Phase I—like clinical trials is not limited to studies performed during clinical drug development prior to approval. Even decades later such studies may be used, and historically major insights in pharmacogenetics have been obtained by such research in unselected healthy volunteers (e.g., concerning the *TPMT*, *NAT2*, *CYP2C19*, or *CYP2D6* polymorphisms with their bimodal or trimodal frequency distributions in the studied populations) [21–23].

Panel Studies as a Typical Study Design in Pharmacogenetics

A further refinement of the Phase I pharmacogenetic sampling and analysis design would be to evaluate the pharmacokinetics and pharmacodynamics in preselected panels of healthy volunteers with specific genotypes from the larger population of unselected samples. The focused quasi-experimental design allows for a reduction of sample size requirements of

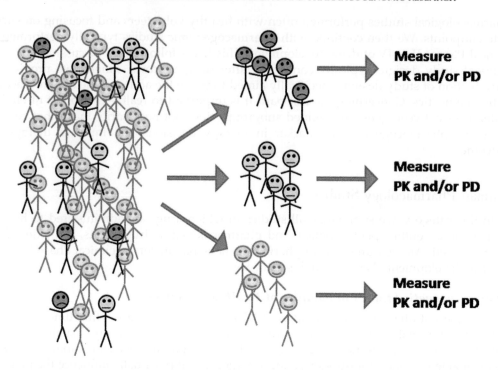

Measure PK and/or PD

Measure PK and/or PD

Measure PK and/or PD

FIGURE 9.5 Schematic figure of the design of a pharmacogenetic panel study: subjects with different physiognomy may represent carriers of different genotypes. It would not add much statistical power (value in clarification of the impact of that genotype) if all carriers of the genotype denoted by the smiling face would be included. Therefore, prior to the study all subjects will be genotyped, and from each genotype only a randomized subset will be included. With rare genotypes, often all carriers of those rare genotypes will be included.

the drug study by about twofold (as illustrated in Figure 9.5) or as much as 100-fold when screening larger populations for rare variation. As a case example, for studying the impact of the CYP2D6 ultra-rapid metabolizer genotype, which has a frequency of about 2% in unselected Caucasian populations, approximately 500 subjects would need to be included to provide at least 10 ultra-rapid metabolizers for comparison with extensive metabolizers, intermediate metabolizers, or poor metabolizers. However, by preselection, studies with only about 20 participants are possible and essentially achieve the same statistical power [24–27], illustrating a 25-fold reduction of sample size in this example. Of course, the same approach may be used to study genetic polymorphisms affecting pharmacodynamics, such as polymorphisms in signaling pathways.

In general, results of pharmacogenetic panel studies show a high between-study agreement, as illustrated with *CYP2C19* or *CYP2D6* [2]. During drug development, similar numbers of subjects with normal, moderately impaired, or severely impaired organ function are included in Phase I studies to generate data for appropriate labeling of dosage requirements in patients with various degrees of impaired liver or kidney function. Therefore, a Phase I panel pharmacogenetic study, in a similar way, might provide relevant pharmacokinetic parameters for dosing recommendations to minimize toxicity and maximize efficacy if valid surrogate markers are included in the study. In addition, enriching a Phase I study with subjects preselected by genotype [4,28] would alleviate the concern of desirable sample size and the mostly limited financial funding available for pharmacogenetic clinical research. With appropriate screening, one may even consider applying this approach to studies with clinical outcomes, but we do not yet know case examples for preselecting typical germline pharmacogenetic genotypes before including patients in an outcome study. In this approach, investigators must be aware that screening large patient populations for genotypes is often already part of a clinical trial and may thus incur significant costs. In addition, there may be ethical concerns if we know from the prescreening about risk genotypes but nevertheless do not consider this knowledge in all participants for scientific reasons. Still, these problems are not unknown in medical research, and the enrichment approach may be a most promising and significantly money-saving tool not only in Phase I–like research but also in studies on clinical outcomes.

Family and Twin Studies in Pharmacogenetics

In many areas of drug research, pharmacogenetic and pharmacogenomic studies are performed in spite of lack of firm evidence that genetic variation indeed explains variability in response or adverse effects. It was suggested that high interindividual and low intra-individual variation are indicators of heritability [29], but higher inter- than intra-individual variation does not at all prove a genetic contribution. However, twin and family studies can be used to quantify the contribution of heritability to interindividual variability in pharmacokinetics and pharmacodynamics. Numerous studies performed by Vesell and colleagues have shown how much interindividual variation in pharmacokinetics and pharmacodynamics of substances such as ethanol, phenylbutazone, antipyrine, or warfarin is heritable [30]. Several studies have shown that pharmacokinetics of and response to lithium in bipolar (manic depressive) disorder is heritable [31]. Thus far, the value of such pharmacogenetic twin and family studies has not been not reflected in a clear identification of the specific underlying genes and polymorphisms but in the proof that genetics does play a big role in phenotypes and outcomes related to a specific drug. For identification of the underlying polymorphisms, other study designs (described in Table 9.4) in large, unrelated populations appear to be more powerful.

Concerning the question of whether to prefer family or twin studies, many drug effects are age dependent, and therefore (given the age dependency of most drug effects), studies of monozygotic versus dizygotic twins are likely more simple to interpret and more statistically powerful than family studies. On the other hand, twins rarely suffer from the same disease at the same time. Therefore, twin studies have the same limitation of surrogate marker analyses as preselected panel studies. Looking to the future, twin studies with massive parallel sequencing and pharmacogenomic biomarker measurements may spark a renaissance in twin pharmacogenetics, and several studies have already significantly contributed to our understanding of epigenetics [32,33], which may be relevant for interindividual variation in the response to drugs and adverse drug effects.

Pharmacogenetic Studies within Clinical Drug Development

In this section, we summarize pharmacogenetic and pharmacogenomic analyses in studies that are, at their core, mostly designed to assess the efficacy and the benefit–risk ratio of new

TABLE 9.4 Key features of important pharmacogenetic study designs

Category or study design	Characteristics	Some notable advantages and disadvantages	Examples
Human pharmacology studies			
Healthy volunteer	Specific phenotype(s) (e.g., pharmacokinetics, drug effects) measured in unselected healthy volunteers.	Good to study genetic variants in relation to biological phenotypes. Impossible to study clinical outcomes.	[81]
Panel studies (enrichment design)	Specific phenotypes (e.g., pharmacokinetics, drug effects) measured in participants who are preselected (enriched with rare genotypes) to achieve high power in the analysis of genotypes of interest.	As above, but major increase in statistical power possible. Only the predefined genes can be studied with that enhanced power.	[82]
Family and twin studies	Specific phenotypes (e.g., pharmacokinetics, drug effects) or outcomes of drug treatment (response, adverse effects) are analyzed in families in order to achieve firm estimates of heritability.	Clear estimation of heritability, usually without very high statistical power to identify the specific causative polymorphisms (family and twin studies also possible as observational studies using designs described below).	[83]
Within clinical drug development Phases II, III, and IV			
Uncontrolled cohort	One group identified and defined before administration of one defined drug treatment scheme is then analyzed concerning the subsequent outcomes* of drug treatment in relation to molecular biomarkers.	Differentiation of prognostic and predictive biomarkers is not possible. Practical feasible design, but limited validity.	[34]
Randomized controlled cohort (biomarker stratified designs)	Two groups (one with new drug and one with placebo or control drug) are comparatively analyzed concerning the subsequent outcomes of drug treatment in relation to the molecular biomarkers.	Concerning study of drug effects, currently undisputedly highest quality design. Differentiation of prognostic and predictive biomarkers is possible. The design often requires strict exclusion criteria resulting in limited generalizability (external validity).	[36]
Nested case control	Within a cohort study typically all cases (e.g., nonresponders or patients affected by adverse effects) are compared with controls from the same cohort not having those outcomes.	Saving money and efforts, because only a limited selection of the possibly very large cohort has to be analyzed without losing relevant statistical power.	[37]

Observational (noninterventional)

Case reports and case series	Single cases (or series of cases) reported with unusual phenotypes or outcomes related to a specific drug and specific gene(s).	Cannot prove anything but may stimulate further thinking and research.	[38]
Cross-sectional	Collecting data about phenotypes and genotypes at one specific point in time (descriptive, like a census).	Impossible to draw any conclusions concerning causality.	[39]
Case-control	Retrospective design comparing frequencies of genotypes in cases (e.g., nonresponders or sufferers from an adverse event) with controls (responders or those not having the adverse event). Different frequencies of the genotype may indicate causal involvement.	Numerous types of systematic error (bias) possible. Good design for rare outcomes.	[43]
Naturalistic uncontrolled cohort	One sample is defined prior to treatment (any typical treatment) and is then analyzed concerning the subsequent outcomes of drug treatment in relation to molecular biomarkers.	Compared with the uncontrolled cohort above, even more feasible and better external validity (generalizability), but limited internal validity due to heterogeneous treatments.	[44]

Pharmacogenetic diagnostic trials

Time series (sequential pre–post designs)	Outcomes are sequentially measured before and after implementation of a pharmacogenetic diagnostic test.	Good external validity but many types of systematic error are difficult to control.	[47]
Parallel randomized diagnostic trials	Outcomes are measured in parallel between one group treated according to a pharmacogenetic diagnostic test and another group treated as usual.	Highest quality design according to current common understanding, but very expensive design and quite artificial setting. Thus a lot of patients from the true target population may have to be excluded (limited external validity).	[54]
Community intervention	Outcomes are compared between communities (or other larger units) that have implemented pharmacogenetic diagnostics versus those that have not implemented pharmacogenetic diagnostics.	Naturalistic setting but poorly controlled for systematic error due to other factors, which may differ irrespective of the pharmacogenetics diagnostics.	[56]

* "Outcomes" may be all relevant consequences of treatment including clinically relevant surrogate markers (such as blood pressure or HbA1C reduction) or clinical endpoints (such as mortality) or adverse drug effects.

therapies. Pharmacogenetic or pharmacogenomic parameters are analyzed as predictors of drug efficacy or toxicity. Typically such studies are performed as Phase II to Phase IV studies in drug development.

Uncontrolled Cohort Pharmacogenetics Studies

With the single-arm uncontrolled cohort pharmacogenetics study, the cohort of study subjects is defined before drug treatment or intervention (Figure 9.6). The term "uncontrolled" merely refers to the lack of a control or comparison group regarding the drug treatment of interest. Otherwise, the treatment or intervention received by the entire cohort is well controlled by the study protocol. The definition of the cohort must have been made at the beginning of the study and then, in the following weeks, months, or years, all outcomes are recorded. Pharmacogenetic or pharmacogenomic parameters within the existing cohorts may be analyzed at any time even if decisions about such types of analyses were not made at initiation of the cohort study. Disadvantages of the cohort design are the costs and time required. However, if pharmacogenetic analyses are performed as ancillary (add-on) studies to existing clinical drug trials, the additional costs may be relatively low.

Selection bias, defined as systematic differences in the group that participated in the pharmacogenetic study compared with the group that was not included, may be one problem compromising validity of conclusions from pharmacogenetic add-on studies. This is the case particularly if less than about 80 percent of the cohort [14] elects to participate or if DNA from specific subgroups (e.g., those who dropped out or died early) is not available. In the past, pharmacogenetic studies in drug development were usually done as voluntary add-on studies with participation rate often much below 80%, and such data may be of little scientific or medical value [14].

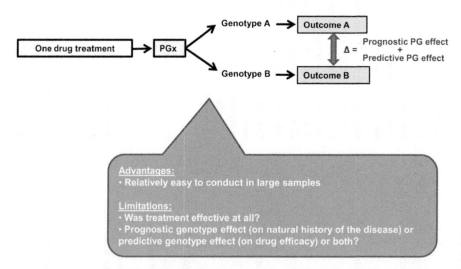

FIGURE 9.6   The concept of a prospective pharmacogenetic cohort study. In this design, one group of subjects is defined prior to drug treatment and then followed up for the outcomes of drug treatment (response to therapy and adverse drug effects). Effects of one or several genotypes on the outcomes of drug treatment are analyzed.

As shown in Figure 9.6, analysis of pharmacogenetics or pharmacogenomics in a single-arm cohort study in which all patients received a treatment will usually not allow for distinguishing between parameters modulating prognosis per se (i.e., independently from a specific therapy (so-called prognostic parameters)) and those parameters modulating the response to a specific therapy (so-called predictive parameters). As an example, antidepressant drug therapy outcomes were correlated with genotypes in the Sequenced Treatment Alternatives to Relieve Depression (STAR*D) study [34], yet it is impossible to differentiate between predictive markers that explain drug response and prognostic markers that remain relevant regardless of the specific drug administered for the study.

The two types of pharmacogenetic biomarkers, the prognostic markers and the predictive markers, are both medically relevant, but concerning the therapeutic consequences a clear differentiation is most important. A prognostic biomarker indicates a good prognosis of the subgroup of patients carrying that marker, but it may not necessarily mean that carriers of that genotype have a good response specifically to that therapy. Instead, it may define a subtype of disease with better prognosis or other inherited factors contributing to better prognosis. Thus, prognostic factors do not help in differential selection of specific drugs (but possibly in selecting between more or less aggressive therapies). In contrast, predictive biomarkers may result in the selection of specific therapies. For instance, Her2 or estrogen receptor expression in breast cancer results in the selection of trastuzumab or estrogen antagonistic therapies.

Randomized Controlled Cohort Pharmacogenetic Studies (Biomarker-Stratified Design)

Differentiation between prognostic and predictive genotypes can be achieved with integration of a randomized controlled cohort pharmacogenetics design within a randomized control clinical study (Figure 9.7), in which patients are randomly assigned to a treatment or control group. This design is also known as biomarker-stratified design [35]. These studies are the typical Phase II and Phase III studies in drug development, and pharmacogenetic genotypes are analyzed in all participants (in both the treatment group and the control group), allowing differentiation between prognostic effects (independent from the specific drug therapy) and predictive effects (specific for a drug or drug class). As illustrated in Figure 9.7, randomization may be performed regardless of biomarker status, which often may be the more feasible option, but additional randomization performed separately in the groups with or without the biomarker can increase statistical power concerning biomarker analysis. Using the term cohort studies, we emphasize that—as in all pharmacogenetic research—unlike drug treatment, the genotype is a feature that can never be randomized. This fact should be kept in mind when evaluating the causalities behind predictive biomarkers. A frequently cited example for that discrimination between prognostic and predictive biomarkers is the IPASS study, which showed that the epidermal growth factor receptor (EGFR) tyrosine kinase inhibitor gefitinib significantly improved survival in EGFR mutation-positive patients but not in EGFR mutation-negative patients [36]. This finding alone might be interpreted either as a prognostic or as a predictive finding, but the same study also had a control group without tyrosine kinase inhibitor treatment, and indeed, in this control group, the survival did not differ between EGFR mutation-positive and -negative patients [36]. Thus, here EGFR mutations turned out to be truly predictive biomarkers.

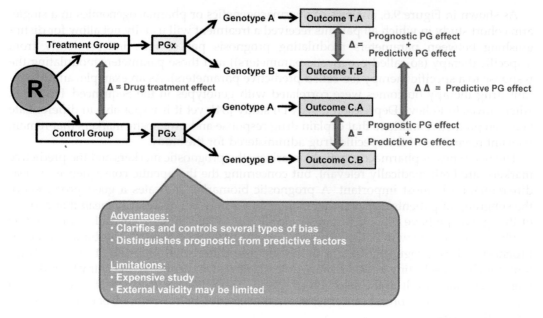

FIGURE 9.7   The concept of a randomized controlled cohort pharmacogenetic study being integrated within a randomized controlled clinical drug trial (RCT). The letter "R" refers to randomization of treatment, PGx refers to pharmacogenetic diagnostics, the Δ (delta) refers to differences depending on one factor, either drug treatment or genotype, and the ΔΔ (double delta) refers to the difference in outcome depending on genotype for that specific treatment (predictive effect). As a modification of this design, stratified inclusion into the treatment group or into the control group may be performed by genotyping prior to randomization. This design has also been termed biomarker-stratified design [35].

An important point to consider is that extensive multiple testing in any pharmacogenetic research may cancel out even the best study design. That means that our optimism about the randomized controlled cohort design providing the conclusive evidence of the impact of genetic polymorphisms on drug therapy is usually true only if the biomarker was prespecified in the study protocol. In addition to a type I error, significant bias may emerge if the biomarker is not analyzed in the entire study population, as discussed earlier for the uncontrolled cohort studies.

Nested Case-Control Studies

An important modification of the pharmacogenetic cohort study design is the nested case-control study, or case-cohort study. Although the name might imply a retrospective study design, the study is based on cohorts that are prospectively followed and therefore of similar validity as the randomized controlled cohort pharmacogenetic study. With the case-cohort study, all study participants who had the outcome of interest (cases) and a randomly selected subgroup of participants who did not have the outcome of interest (controls) are evaluated. Although one control per case is typically chosen, statistical power can be increased by inclusion of up to four controls per case. Because analysis does not occur for the entire cohort, study cost may be significantly lower. As importantly, many serious types of bias associated with case-control studies do not exist in nested case-control studies. The recent report of statin-induced myopathy [37] is an example of the nested case-control study design.

Observational Study Designs

In contrast to experimental study designs discussed previously, in which details of the drug treatments are defined by the investigators or via randomization, in observational studies the decisions about drug treatment or other types of intervention are made as they occur (or are observed) in medical practice. In this context, the recording of the data or biomaterial sampling and genotyping would not be considered as an intervention. Examples of observational studies include a wide variation of designs such as case reports, cross-sectional studies, case-control studies, and naturalistic cohort studies. Most of these are performed with approved and marketed drugs.

Case Reports or Case Series

Publishing observations in a single patient (case reports) or in groups of patients with similar outcomes (case series) is a traditional mode of medical knowledge propagation. Within the subdisciplines of pharmacogenetics and pharmacogenomics, case reports such as those on fatal outcomes in carriers of inherited TPMT deficiency [38] stimulate further research through hypothesis generation. However, beyond that, the uncontrolled environment associated with case reports does not lend itself to confirmation of any research hypotheses. Case series may be medically important signals—for instance, if case series for adverse drug events are reported—but beyond that no firm conclusions can be based on case series. In contrast to cohort designs or other higher-ranking designs, there is no control for selective reporting in case series.

Cross-Sectional Pharmacogenetic Studies

In a cross-sectional pharmacogenetic study, the medical problem and relevant biomarkers are recorded and analyzed at one specific time point. This step may provide descriptive information about frequencies of specific genotypes and medical problems in a sample of the population, and this type of study is relatively easy and quick to perform. This type of study has provided us with information regarding the important role of CYP2D6 polymorphism in psychiatric and geriatric patients [39] and may thus help to prioritize areas in which pharmacogenetics should be analyzed because of potentially great medical impact. But beyond providing such descriptive information about disease frequencies, status, or medical care in a population, or frequencies of genetic variation, usually causalities cannot be unambiguously proven in cross-sectional studies because the temporal relationships between presumed causes and presumed effects cannot be analyzed here. As an additional problem, subjects with severe side effects or therapeutic failure may be underrepresented in such studies due to their higher likelihood of having stopped the medication or having died from the disease of interest before having had a chance to be enrolled into the study. Thus, within the context of clinical pharmacogenetics study designs, cross-sectional studies usually can serve only as preliminary exploratory data for hypothesis generation.

Case-Control Pharmacogenetic Studies

To further elucidate the role of genetic polymorphism in the predisposition to a particular therapeutic outcome, such as an adverse drug reaction, case-control pharmacogenetics studies can be conducted. Genotype frequencies in the case group (subjects with specific therapeutic outcome or disease state) are compared to those in a concurrently sampled, usually age- and gender-matched control group. Members of the control group may not have the outcome of interest (Figure 9.8). Although in this type of retrospective study design the control group does not need to have had a comparable (drug) exposure, validity and reliability of the study result may be improved if the control group did have had a comparable (drug) exposure. In addition, since drug exposure is usually well documented in pharmacogenetic case-control studies, they may in general be more valid than case-control studies in the genetic epidemiology discipline [40], and only moderate resources are needed to achieve high statistical power.

Case-control studies are particularly prone to systematic error, also called bias [41,42]. Consequently, as a rule of thumb, only if the risk estimates (odds ratios) for a certain research question are reproducibly above 5, may results from case control studies be considered as clinically applicable evidence. Otherwise, data obtained in case-control studies would require replication by studies with a different design. Potential types of bias should be identified and controlled as much as possible for the study and subsequent replication efforts. In general, results from case-control studies should be supported by other studies with more definitive designs, that is, those that occupy a higher position in the study hierarchy diagram (Figure 9.4). As illustrated with the rare and idiosyncratic flucloxacillin-induced liver injury [43],

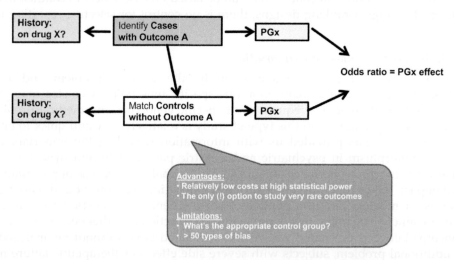

FIGURE 9.8 The pharmacogenetic case-control study. Patient sampling starts with identification of patients with a particular outcome; usually, for case-control studies this should be a rare outcome (e.g., a rare severe adverse drug effect). This case is then included into the analysis if prior exposure to drug X was found. As a next step, controls are identified, which are usually humans from a similar age group and with a similar proportion of male and female as in the case group. The control group must not have the outcome of interest, but it is advisable that the control group members have been exposed to the same drug.

case-control studies may present the sole means of analyzing the pharmacogenetics of rare and severe adverse events. This example optimally illustrates the rare-disease prerequisite justifying application of the case-control study design.

Naturalistic Uncontrolled Cohort Studies

Study design for the naturalistic uncontrolled cohort study is essentially the same as the uncontrolled cohort pharmacogenetics study explained in the previous section. Both are considered "uncontrolled" because there is no control (comparison) group. However, in the naturalistic uncontrolled cohort design, the drug treatment is not specified by the study protocol, and this design is termed "naturalistic" because it better reflects the real medical world. Thus, there may be a wide variety of concurrently administered medications and doses in the naturalistic cohort. As discussed earlier, the effects of genotypes on the study outcome are analyzed. For instance, in a naturalistic cohort of patients treated as usual with a variety of antidepressant medications, we found that carriers of one specific genotype in the dopamine transporter had a sixfold increased risk (odds ratio of 6) for poor response [44].

This study design thus is advantageous in that it reflects true medical reality, is relatively easy to perform, and may produce less-biased results than the case-control design for frequently occurring outcomes for a number of reasons. One important reason is that there is a lower risk of selection bias. Selection bias, a significant problem that affects both case and control groups for case-control research, is much less relevant in naturalistic cohort studies (because by definition all subjects fulfilling the inclusion criteria should be included in the cohort study). In addition, when compared with the more standardized uncontrolled cohort design discussed above, these naturalistic uncontrolled cohort studies may much better reflect medical reality because broad inclusion criteria may be possible (e.g., all patients treated for a specific disease). Compared with the uncontrolled cohort design, the naturalistic uncontrolled design has lower internal validity (correctness of causal inferences but possibly relevant only for the specific selection of study subjects) but higher external validity (generalizability or validity concerning relevance for typical target populations). However, the heterogeneous nature (drug treatments, disease stages, and severity) of the study sample of naturalistic uncontrolled cohorts results in a risk of misinterpretation and also has low statistical power, and thus large sample sizes will be required. Similar to the uncontrolled cohort pharmacogenetics study, this design cannot provide differentiation between predictive and prognostic genotypes and thus cannot serve as the only basis of pharmacogenetics-guided therapy recommendations.

Pharmacogenetic Diagnostic Trials

When properly carried out, the study designs described thus far should reveal the value of pharmacogenetic diagnostics. However, in reality, such possible value of pharmacogenetic diagnostics to improve the outcomes of drug therapies may not be apparent in medical practice for numerous reasons. Such reasons may include limited time availability or lack of knowledge about correct interpretation of pharmacogenetic diagnostic tests. Therefore, studies specifically conducted to evaluate the value of pharmacogenetic diagnostics may be necessary. This study type is termed pharmacogenetic diagnostic trial. With prospective pharmacogenetic diagnostic trial designs, the pharmacogenetic diagnostics are treated as interventions and performed during the study. Such designs may be particularly relevant to measure clinical utility of the pharmacogenetic diagnostic test. Clinical utility is not easily defined [45] but is nevertheless considered most important for deciding whether to perform and how much to pay for pharmacogenetic diagnostics. In general, pharmacogenetic diagnostic trials may be classified as either experimental studies or as quasi-experimental studies. An experimental study is characterized by a concurrent control group and by random assignment of study participants to the new intervention (i.e., new diagnostics or new treatment) group or to the control group. This design controls for various types of systematic error.

However, such types of experimental clinical diagnostic studies are often impractical and may sometimes be even unethical. In such instances, quasi-experimental studies may help. Typically, in quasi-experimental studies there is no random assignment of the study participants. Thus, quasi-experimental studies have a place between purely observational studies and randomized controlled clinical trials. As a typical quasi-experimental trial, we present below the sequential pre−post crossover design.

Time Series or Sequential Pre−Post Designs

In the sequential pre−post (before−after) one-sequence individual crossover design, one group of patients receives drug therapy initially without genotyping, and subsequently a comparable group receives drug therapy according to genotyping results (Figure 9.9). Intra-individual sequential comparisons will often not be possible because after the first attempt of drug treatment, physicians and patients will have learned how an individual patient responds to a drug. As a very weak mode of evidence gain in medicine, historical data were or are compared with the outcomes of a current new therapy, but the pre−post design performs this comparison in a more structured way, usually within a defined time schedule. This pre−post design is relatively easy to perform and—compared to parallel

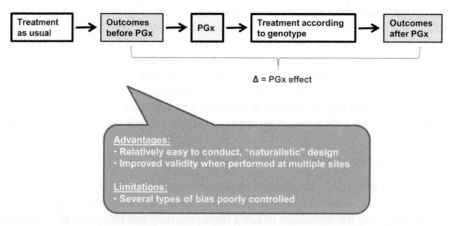

FIGURE 9.9 The pharmacogenetic time series or pretest−posttest study design. Approaches like that presented in this figure are frequently also termed quasi-experimental designs because a lot of factors are controlled, but there is no randomized assignment of study subjects to the diagnostics group versus a control group. One possible type of systematic error in the pre−post design is that over time, many medical procedures improve, so that improvement of outcomes after PGx compared with the outcomes before PGx may not unambiguously related to the PGx intervention. That problem may be solved by the multiple time series modification with starting points at different study sites at different times.

diagnostic trial designs introduced below—requires fewer patients. Such a design may reveal that a new mode of treatment is indeed better than the usual therapy [46]. An example from the pharmacogenetics field is the before–after comparison concerning the value of *CYP2C9* and *VKORC1* (see Table 9.2) genotyping in warfarin therapy on hospitalization frequency, where the authors indeed could show that hospitalization rates were significantly lower after genotyping compared with the year before [47].

The before–after study design is mostly not considered as a very definitive design. Due to positive anticipations (here as placebo effects of genotyping), experimental new therapies may often appear better than other therapeutic modalities even if they are truly not better [48]. Often, there are residual pharmacological effects, side effects, and treatment knowledge from the first (pregenotyping) period, which may affect physicians' actions and patients' responses in the second (postgenotyping) period. In addition, patients may be sicker at study initiation than other times later during the study, thus introducing a bias known as regression to the mean. Therefore, the before–after design may provide credible evidence regarding the value of pharmacogenetic diagnostics only if pharmacogenetic diagnostics confers a large improvement in the measured outcome. Still, before–after comparisons are the basis of medical practice in many important therapeutic areas, [49] and such studies may greatly help to familiarize physicians with the practice of individualized medicine and pharmacogenetic diagnostics, which is important for their successful implementation.

Some of the weaknesses of the before–after design may be resolved by introducing the intervention at different times at multiple centers [50], a design also termed multiple time series [51]. There are numerous other variations of such quasi-experimental designs using preplanned additions or removals of the interventions to be analyzed. However, a brief glance at the medical literature will show that such quasi-experimental designs are not very frequently used in high-ranking medical journals at present. Nevertheless, with knowledgeable use, these quasi-experimental designs may be useful tools in the development and introduction of pharmacogenetic diagnostics particularly when low genotype frequencies, moderate effect sizes related to pharmacogenetic biomarkers, and limited financial recourses make application of other designs unfeasible [52].

Parallel Randomized Diagnostics Trials

Some of the limitations associated with the sequential pre–post designs, or with quasi-experimental designs in general, can be abolished with inclusion of a concurrently studied (parallel) comparison group (Figure 9.10). In a parallel randomized pharmacogenetic diagnostics trial, subjects are assigned to either the pharmacogenetic diagnostic arm or the control arm receiving standard therapy. This design has also been termed biomarker-strategy design, and Freidlin et al. [35] explained why this design may even result in lower statistical power than the controlled cohort pharmacogenetics study—because only a subgroup (e.g., 50%) of the sample is genotyped, and thus there is no information about potential benefits of pharmacogenetic diagnostics in the other 50%. However, statistical power is only one point to consider when introducing new types of diagnostics and interventions into medicine. In addition to providing data on how pharmacogenetics-guided therapy can improve outcome, randomized parallel diagnostic trials could also further quantify the information with the number needed to genotype (the number of patients who have to be genotyped to achieve a significant benefit for one patient).

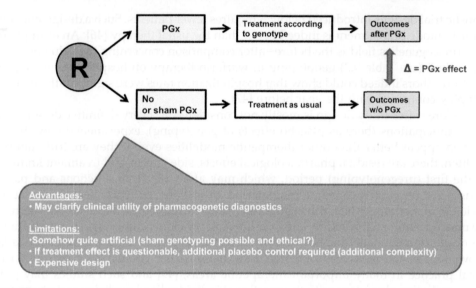

FIGURE 9.10   The parallel randomized pharmacogenetic trial. An essential feature of this design, in contrast to the pre—post design discussed earlier, is the randomized allocation to either the PGx arm or the "therapy as usual" arm. This design may prove the clinical utility of PGx, but because this randomized allocation may not be possible in many people for safety reasons or because of their willingness to participate in randomization, the external validity of this experimental approach may be lower than that of observational or quasi-experimental approaches.

The disadvantages of parallel randomized diagnostic trials are primarily related to their complexity and cost. The sample sizes required might need to be higher than in pivotal Phase III drug trials because the drug effect, pharmacogenetic effect, and drug—diagnostics interaction all need to be analyzed. Further, even a placebo/nocebo effect of genotyping may have to be considered in designing the specific details of this type of study [53]. One solution to these problems may be randomization by study center or even by the community intervention design (see below) rather than individual randomization of each patient. With pharmacogenetic biomarkers presumed to play a very strong role in therapy, it may be possible to realize the value of such studies. As a typical example of this parallel randomized diagnostic trial, we would like to cite the study of Furuta et al. [54], in which the authors randomized 300 patients to receive either individualized dosing for *H. pylori* eradication according to biomarkers including *CYP2C19* genotype or standard therapy with the same dose for all. Another example is the parallel randomized diagnostic trial on *HLA-B*5701* screening for hypersensitivity to abacavir [55]. In this example, there was little doubt about the genetic association with drug hypersensitivity, so it was critically asked whether a randomized diagnostic trial was necessary. More generally and irrespective of the discussions about the specific study of Mallal et al. [55]: if there are very strong predictors of drug therapy response or adverse drug effects, ethical and legal concerns might prohibit this type of scientifically stringent study.

In summary, despite the importance of parallel randomized pharmacogenetic diagnostic trials in proving the clinical utility of pharmacogenetics, it is unlikely that this complex and expensive design will be used often to help implement individualized medicine into medical practice.

Community Intervention Trials

With this study design, pharmacogenetic evaluation and analysis for a specific drug are performed at randomly chosen hospitals or geographic regions and the results are compared with control institutions that do not perform pharmacogenetics. We do not know whether this approach has been used in pharmacogenetic diagnostic research, but the study of Soumerai et al. [56] on the consequences of restrictions in medication reimbursement provides a good example of this design.

A Rational Sequence of Study Designs

A common framework of important steps in evaluating clinical application of pharmacogenetics is the ACCE, with A standing for analytical validity, the first C standing for clinical validity (i.e., the polymorphism has some medical or functional impact), the second C standing for clinical utility (i.e., testing improves outcomes), and E standing for ethical, legal, and societal issues [57,58]. However, given that analyses are performed according to good laboratory practice, analytical validity is not a relevant problem in pharmacogenetic diagnostics. Therefore, any clinically oriented pharmacogenetic research may be differentiated into three phases: first, hypothesis generation with any study design discussed here; second, replication and confirmation in additional studies with greater power or more valid designs; and third, implementation with studies providing a risk–benefit assessment for pharmacogenetic diagnostics under real medical conditions [59].

The exploratory phase with hypothesis generation is the most innovative step in experimental and clinical pharmacogenetic and pharmacogenomic research. However, from billions of possible and reasonable pharmacogenetic hypotheses, only a few will turn out to be medically relevant and reproducible. Thus, replication and confirmation is most essential. Based on reproducibility, the credibility of peer-reviewed clinical pharmacogenetic publications, even for those showing strong genetic association, is estimated to be between 10 and 60 percent [18]. A number of criteria for judging the credibility of pharmacogenetic study results may be adopted from the molecular epidemiology of disease genetics [40].

By definition, confirmatory studies must be adequately powered in order to replicate previous studies and must also be performed in a study sample different from that of the initial exploratory study. If evidence from exploratory studies is confirmed, a decision must be made regarding whether genotyping should be introduced into clinical practice at this stage or whether additional confirmatory studies, such as randomized interventional diagnostic trials, should be performed. Such interventional trials can yield the ultimate proof that genotype-based therapy is better than standard therapy. Figure 9.11 summarizes the study designs discussed in this chapter. The studies were classified according to their ability to control for systematic errors in the hierarchy of study designs (Figure 9.4). It is important to acknowledge that the purpose of the hierarchy is more for orientation than for being a closed and undisputable system, and mostly studies with observational designs (lower levels in the hierarchy; Figure 9.4) came to the same conclusions as randomized controlled designs [60]. Particularly if effect size is strong (odds ratios above 5 or even above 10), randomized controlled trials may be superfluous.

Study design	Costs	Exploration, Hypothesis generation	Confirmation	Medical decision driving
Meta-analysis	> 30,000 €			+
Parallel randomized diagnostic trial	> 1,000,000 €		+	+
Randomized controlled cohort PGx	> 1,000,000 €	+	+	+
Nested case-control PGx	> 1,000,000 €	+	+	+
Multiple time series	> 300,000 €		+	+
Uncontrolled cohort PGx	> 300,000 €	+	+	
Case-control studies	> 100,000 €	+	+	+
Human pharmacology: Panel studies	> 100,000 €		+	
Human pharmacology: Unselected	> 100,000 €	+	+	
Cross-sectional Studies	> 50,000 €	+		
Case reports or case series	> 1 €	+		
Scientific reasoning	> 1 €	+		

FIGURE 9.11   Rational sequence of study designs. Initial evidence will mostly come from approaches marked in the exploration column. For confirmation, another study in another sample should be performed, typically with a design marked in the confirmation column. In some instances, it may be advisable to perform specific diagnostic trials, such as the multiple time series (sequential) diagnostic design or the parallel randomized diagnostic trial. Costs given in the costs column cannot be a basis for any real cost calculation—amounts given there may be the minimum—but in all lines estimations refer to the entire study (thus not only the costs for a pharmacogenetic add-on study).

The Need for Innovation in the Design and Interpretation of Pharmacogenetic Studies

The potentials of individualized medicine using pharmacogenetic biomarkers are insufficiently investigated and used in medical practice. Few data provide good evidence about how much pharmacogenetic variation quantitatively accounts for nonresponse or adverse events to drug therapy, although the quantitative impact may be relatively large [61]. Examples of other underexplored areas include the impact of genetic variation on drug efficacy and side effects in the elderly, the impact of genetic variation in drug–drug interactions, and the combined effects of several gene variants (gene–gene interaction, also know as epistasis).

At the molecular and cellular levels, the impact of inherited pharmacogenetic variation on signal transduction, regulation of cell differentiation and cell growth, and the interaction between cells are still mostly unknown.

Thus, if pharmacogenetics and pharmacogenomics as part of individualized medicine are to be integrated into clinical practice, new approaches to study designs are apparently needed. Some thoughts concerning study design modifications and other concepts to support clinical application of pharmacogenetic diagnostics are summarized below.

Enrichment Approaches

One approach is to perform studies in subject subgroups that may particularly benefit from pharmacogenetic testing. Examples for this approach include the conduct of pharmacogenetic studies specifically in the poorest responders to drug therapies or in those with extreme metabolic genotypes such as the *CYP2D6* ultra-rapid metabolizer (UM) genotype and in patients with severe, unexplained adverse drug reactions. Thus, future interventional trials in pharmacogenetic diagnostics may focus particularly on difficult-to-treat patients for all major illnesses. Because the UM phenotype is very frequent in some Arabian and Northern African populations, it may be a good idea to initiate studies with relevant CYP2D6 substrates in these populations. The same approach would apply to patients in whom adverse drug reactions cannot be explained by inadequate dosing or other clinical data, and more studies with molecular pharmacogenetic analysis should be performed as part of postmarketing pharmacovigilance.

By definition, pharmacogenetic diagnostics is of particular value for subgroups, as many medically relevant pharmacogenetic genotypes occur at a low frequency. Demonstrating improvement in the outcomes of enriched and/or genotype-preselected subgroups would require significantly smaller studies than those that demonstrate improvement in the overall outcome of the entire study population [62,63], most of whom would not benefit from the intervention. Enrichment by genotype in panel studies [4] may make studies of rare genotypes more feasible. However, examples of how this approach can be successfully used are currently limited to Phase I—like pharmacogenetic trials, and modification of such a design as discussed earlier is needed for Phase II and III trials.

Adaptive Clinical Trial Designs

Adaptive clinical trial designs may significantly reduce sample size in studies on pharmacogenetic biomarkers, although there are thus far only few examples demonstrating how this design has contributed to the field. Even with a sample size reduction of 20%, significant study cost reduction could be achieved without ethical concerns. A typical question in Phase III trials may be whether a subgroup with a pharmacogenetic biomarker responds better than the total study sample. In such a case, the overall significance level may be split into the overall drug effect for the entire studied sample and the drug effect in the biomarker-positive subgroup [64]. Consultants of the European medical agencies found the concept interesting but there are currently no examples utilizing the adaptive design in pharmacogenetics or pharmacogenomics [65].

Systematic Meta-Analysis, Informatics, and Education

The complexity and amount of pharmacogenetics data dictate that the information needs to be explained and disseminated to medical practitioners and other health care providers in an organized manner, such a throughs meta-analyses. First, the huge amount of data has to be gathered and sorted in electronic databases. A notable example of a centralized resource for pharmacogenomic information is the Pharmacogenomics Knowledge Base [66] discussed in chapter 2. Second, data must be systematically analyzed and evaluated according to pre-defined quality criteria similar to the assessments made available by the Cochrane Collaboration. The Cochrane database currently does not include conclusions concerning therapy optimization by pharmacogenetics, but the first steps in this direction are being applied [1,67]. Third, the research data must be translated into concrete therapy recommendations [4,68] and must be made available in an easy-to-understand manner. Currently, several initiatives are developing databases and information portals for that purpose [1,67,69]. One of these is the Clinical Pharmacogenetic Implementation Consortium (CPIC), which scores evidence linking genetic information to drug therapy decisions and provides consensus-based guidelines on how to apply genetic test results to clinical practice. Fourth, additional educational efforts may be necessary to familiarize physicians with the potentials of individualized medicine [70]. Studies like those cited here are of course very important to convince health care providers about the value of the individualized therapy approach.

Evaluated Introduction

Although strict reasoning according to EBM may dictate that the value of new therapeutic approaches be proven before patients can be exposed to them, the medical reality, particularly concerning diagnostics, was and still is different. The value of many currently useful diagnostic tests was never determined by a randomized diagnostic trial, but rather the tests were just introduced into practice where their value was proven. Alternatively, the test disappeared from the scene if it proved not useful. Also, in drug development there is low external validity (generalizability) of many Phase III clinical drug trial findings, and the value of a new drug—or the limitations of a drug—become most apparent only after approval and broad use in medicine. Thus, one should also be open to something like an evaluated introduction of pharmacogenetic diagnostic tests into medical practice. This means that physicians should be open to using pharmacogenetic diagnostics, with the consequences documented and analyzed in a noninterventional observational manner.

Individualized Therapies beyond Evidence-Based Medicine (EBM)?

The study designs reviewed may help to understand the impact and prove the value of human genomic variation in individualized drug therapy. However, we think that it is a misconception that the medical value of every individualized therapy should be proven according to current concepts of EBM. Indeed, for every currently available drug, there are numerous subpopulation-specific recommendations that are not based on evidence but rather on biomarker data and pathophysiologic and mechanistic considerations. Such parameters or subgroups include patient age, body weight, renal function, liver function, and even the status of pregnancy and lactation. It is important to realize that EBM never meant and can never mean that every therapeutic decision is based on a randomized controlled clinical or diagnostic trial. Nobody has performed—or indeed, even asked for—a randomized controlled trial (RCT) to investigate whether doses administered to small children should be lower than those administered to adults. Similarly, the value of dose modifications in cancer therapy to minimize significant adverse effects has never been proven in

RCTs. As such, one can reason that if there are strong data indicating that pharmacogenetic diagnostics can save lives, it may be inadvisable to ask for randomized trials in every instance [71].

Thus, pharmacogenetics-based drug therapy in many instances should not be considered from the perspective of proof of better efficacy in the respective entire patient population but rather as a means for ensuring subgroups within the population would respond to treatment and/or do not experience severe side effects. In addition to ethical constraints, the low frequencies of many genotypes may almost preclude controlled prospective studies of drug effects in subgroups with these rare genotypes. Thus, the RCT cannot always be required in order to recommend pharmacogenetic genotyping and appropriate therapy adjustments.

Quality Criteria

If our perspective is that most RCTs cannot be performed to prove the value of pharmacogenetic diagnostics, then proper planning and effective evaluation of nonrandomized pharmacogenetic research would be critical. Although development of a full catalogue of quality criteria and a full discussion of how to grade the validity of clinical pharmacogenomic research is beyond the scope of this review, a few key assessment questions are summarized in Table 9.5.

Such questions should be asked (a) when initiating a pharmacogenetic study, (b) by peer reviewers and editors when publishing pharmacogenetic data, and (c) by the patients and their physicians if pharmacogenetic data are to be used for treatment decisions.

TABLE 9.5 Brief list of key questions concerning credibility of publications on pharmacogenomic biomarkers

	Credibility low if	Credibility high if
Study design	Not clearly defined and discussed in the publication	Defined and discussed in the publication
Mechanistic evidence	Contrary to or not supported by experimental data	Fits well with biochemical or cellular experimental data
Clinically reproduced in an independent sample	No published independent replication	Several replications published
Clinically contrary results	Reported contrary results in another study sample	No contrary data published
Sample size of the study	Not defined in the study protocol	Defined prior to starting the study
Study hypothesis and main outcome parameter	Not defined in the study protocol	Defined prior to starting the study
Multiple comparisons	Not mentioned but apparently (extensively) practiced	Clear discussion and control for multiple comparisons
Effect size (expressed as odds ratio or relative risks)	< 2	> 5

CONCLUSION AND PERSPECTIVES

This review may serve as a guideline for when and why to apply one of the numerous possible study designs when considering pharmacogenetics in drug development. We wanted to stimulate further discussions and research on the question of how to individualize drug therapy based on parameters that, because of their low frequencies and moderate effect sizes, cannot be proven by randomized controlled trials according to conventional understanding of evidence-based medicine. We anticipate that 10 or 20 years from now, the number of daily used medical therapies based on individual pharmacogenetic data will significantly increase. Part of this new type of medicine will then be proven by studies ranking on top of the pyramid (Figure 9.4), but in many other instances we will have to rely on observational and quasi-experimental designs. In addition, many decisions in individualized therapies will necessarily have to be based on rational approaches covering the many factors summarized in Figure 9.1.

QUESTIONS FOR DISCUSSION

1. What are the advantages and disadvantages of different study designs for pharmacogenetic/pharmacogenomic research?
2. What are the factors that need to be considered when interpreting results from a pharmacogenetic or pharmacogenomic study?
3. What is the limitation of applying the concept of evidence-based medicine to address pharmacogenetic/pharmacogenomic study design for personalized medicine?

References

[1] Swen JJ, Huizinga TW, Gelderblom H, de Vries EG, Assendelft WJ, Kirchheiner J, et al. Translating pharmacogenomics: challenges on the road to the clinic. PLoS Med 2007;4(8):e209.
[2] Kirchheiner J, Nickchen K, Bauer M, Wong ML, Licinio J, Roots I, et al. Pharmacogenetics of antidepressants and antipsychotics: the contribution of allelic variations to the phenotype of drug response. Mol Psychiatry 2004;9(5):442–73.
[3] Kirchheiner J, Brockmöller J. Clinical consequences of cytochrome P450 2C9 polymorphisms. Clin Pharmacol Ther 2005;77(1):1–16.
[4] Kirchheiner J, Fuhr U, Brockmöller J. Pharmacogenetics-based therapeutic recommendations—ready for clinical practice? Nat Rev Drug Discov 2005;4(8):639–47.
[5] EMA. Guideline on the use of pharmacogenetic methodologies in the pharmacokinetic evaluation of medicinal products. Committee for Medicinal Products for Human Use (CHMP); 2010. p. 13.
[6] Madadi P, Ross CJ, Hayden MR, Carleton BC, Gaedigk A, Leeder JS, et al. Pharmacogenetics of neonatal opioid toxicity following maternal use of codeine during breastfeeding: a case-control study. Clin Pharmacol Ther 2009;85(1):31–5.
[7] Stingl Kirchheiner JC, Brockmöller J. Why, when, and how should pharmacogenetics be applied in clinical studies? Current and future approaches to study designs. Clin Pharmacol Ther 2011;89(2):198–209.
[8] Ji W, Foo JN, O'Roak BJ, Zhao H, Larson MG, Simon DB, et al. Rare independent mutations in renal salt handling genes contribute to blood pressure variation. Nat Genet 2008;40(5):592–9.
[9] Acuna, R., L. Martinez-de-la-Maza, J. Ponce-Coria, N. Vazquez, P. Ortal-Vite, D. Pacheco-Alvarez, et al. Rare mutations in SLC12A1 and SLC12A3 protect against hypertension by reducing the activity of renal salt cotransporters. J Hypertens 29(3):475–83.
[10] Herman DS, et al. Truncations of titin causing dilated cardiomyopathy. N Engl J Med 2012;366(7):619–28.

[11] Jorgensen AL, Williamson PR. Methodological quality of pharmacogenetic studies: issues of concern. Stat Med 2008;27(30):6547—69.

[12] Nebert DW, Zhang G, Vesell ES. From human genetics and genomics to pharmacogenetics and pharmaco-genomics: past lessons, future directions. Drug Metab Rev 2008;40(2):187—224.

[13] Rodriguez-Antona C, Ingelman-Sundberg M. Cytochrome P450 pharmacogenetics and cancer. Oncogene 2006;25(11):1679—91.

[14] Wang SJ, O'Neill RT, Hung HJ. Statistical considerations in evaluating pharmacogenomics-based clinical effect for confirmatory trials. Clin Trials 2010;7(5):525—36.

[15] Choy E, Yelensky R, Bonakdar S, Plenge RM, Saxena R, De Jager PL, et al. Genetic analysis of human traits in vitro: drug response and gene expression in lymphoblastoid cell lines. PLoS Genet 2008;4(11):e1000287.

[16] Loh YH, Agarwal S, Park IH, Urbach A, Huo H, Heffner GC, et al. Generation of induced pluripotent stem cells from human blood. Blood 2009;113(22):5476—9.

[17] Cobos A, Sanchez P, Aguado J, Carrasco JL. Methodological quality in pharmacogenetic studies with binary assessment of treatment response: a review. Pharmacogenet Genomics 21(5):243—50.

[18] Ioannidis JP. Commentary: grading the credibility of molecular evidence for complex diseases. Int J Epidemiol 2006;35(3):572—8. Discussion 593—6.

[19] Vandenbroucke JP. Observational research, randomised trials, and two views of medical science. PLoS Med 2008;5(3):e67.

[20] Woodcock J, Lesko LJ. Pharmacogenetics—tailoring treatment for the outliers. N Engl J Med 2009;360(8):811—3.

[21] Weinshilboum RM, Sladek SL. Mercaptopurine pharmacogenetics: monogenic inheritance of erythrocyte thi-opurine methyltransferase activity. Am J Hum Genet 1980;32(5):651—62.

[22] Cascorbi I, Brockmöller J, Mrozikiewicz PM, Muller A, Roots I. Arylamine N-acetyltransferase activity in man. Drug Metab Rev 1999;31(2):489—502.

[23] Brockmöller J, Kirchheiner J, Meisel C, Roots I. Pharmacogenetic diagnostics of cytochrome P450 poly-morphisms in clinical drug development and in drug treatment. Pharmacogenomics 2000;1(2):125—51.

[24] Kirchheiner J, Heesch C, Bauer S, Meisel C, Seringer A, Goldammer M, et al. Impact of the ultrarapid metabolizer genotype of cytochrome P450 2D6 on metoprolol pharmacokinetics and pharmacodynamics. Clin Pharmacol Ther 2004;76(4):302—12.

[25] Kirchheiner J, Henckel HB, Franke L, Meineke I, Tzvetkov M, Uebelhack R, et al. Impact of the CYP2D6 ultra-rapid metabolizer genotype on doxepin pharmacokinetics and serotonin in platelets. Pharmacogenet Genomics 2005;15(8):579—87.

[26] Kirchheiner J, Keulen JT, Bauer S, Roots I, Brockmöller J. Effects of the CYP2D6 gene duplication on the pharmacokinetics and pharmacodynamics of tramadol. J Clin Psychopharmacol 2008;28(1):78—83.

[27] Kirchheiner J, Schmidt H, Tzvetkov M, Keulen JT, Lotsch J, Roots I, et al. Pharmacokinetics of codeine and its metabolite morphine in ultra-rapid metabolizers due to CYP2D6 duplication. Pharmacogenomics J 2007;7(4):257—65.

[28] Temple RJ. Enrichment designs: efficiency in development of cancer treatments. J Clin Oncol 2005;23(22):4838—9.

[29] Kalow W, Tang BK, Endrenyi L. Hypothesis: comparisons of inter- and intra-individual variations can substitute for twin studies in drug research. Pharmacogenetics 1998;8(4):283—9.

[30] Vesell ES, Page JG. Genetic control of dicumarol levels in man. J Clin Invest 1968;47(12):2657—63.

[31] Cruceanu C, Alda M, Turecki G. Lithium: a key to the genetics of bipolar disorder. Genome Med 2009;1(8):79.

[32] Bell JT, Spector TD. A twin approach to unraveling epigenetics. Trends Genet 2011;27(3):116—25.

[33] Coolen MW, Statham AL, Qu W, Campbell MJ, Henders AK, Montgomery GW, et al. Impact of the genome on the epigenome is manifested in DNA methylation patterns of imprinted regions in monozygotic and dizygotic twins. PLoS One 2011;6(10):e25590.

[34] Lin E, Chen PS. Pharmacogenomics with antidepressants in the STAR*D study. Pharmacogenomics 2008;9(7):935—46.

[35] Freidlin B, McShane LM, Korn EL. Randomized clinical trials with biomarkers: design issues. J Natl Cancer Inst 2010;102(3):152—60.

[36] Mok TS, Wu YL, Thongprasert S, Yang CH, Chu DT, Saijo N, et al. Gefitinib or carboplatin-paclitaxel in pulmonary adenocarcinoma. N Engl J Med 2009;361(10):947—57.

[37] Link E, Parish S, Armitage J, Bowman L, Heath S, Matsuda F, et al. SLCO1B1 variants and statin-induced myopathy—a genomewide study. N Engl J Med 2008;359(8):789—99.

[38] Schütz E, Gummert J, Mohr F, Oellerich M. Azathioprine-induced myelosuppression in thiopurine methyltransferase deficient heart transplant recipient. Lancet 1993;341(8842):436.

[39] Mulder H, Heerdink ER, van Iersel EE, Wilmink FW, Egberts AC. Prevalence of patients using drugs metabolized by cytochrome P450 2D6 in different populations: a cross-sectional study. Ann Pharmacother 2007;41(3):408—13.

[40] Ioannidis JP, Boffetta P, Little J, O'Brien TR, Uitterlinden AG, Vineis P, et al. Assessment of cumulative evidence on genetic associations: interim guidelines. Int J Epidemiol 2008;37(1):120—32.

[41] Sackett DL. Bias in analytic research. J Chronic Dis 1979;32(1—2):51—63.

[42] Weiss ST, Silverman EK, Palmer LJ. Case-control association studies in pharmacogenetics. Pharmacogenomics J 2001;1(3):157—8.

[43] Daly AK, Donaldson PT, Bhatnagar P, Shen Y, Pe'er I, Floratos A, et al. HLA-B*5701 genotype is a major determinant of drug-induced liver injury due to flucloxacillin. Nat Genet 2009;41(7):816—9.

[44] Kirchheiner J, Nickchen K, Sasse J, Bauer M, Roots I, Brockmöller J. A 40-basepair VNTR polymorphism in the dopamine transporter (DAT1) gene and the rapid response to antidepressant treatment. Pharmacogenomics J 2007;7(1):48—55.

[45] Lesko LJ, Zineh I, Huang SM. What is clinical utility and why should we care? Clin Pharmacol Ther 2010;88(6): 729—33.

[46] Lasagna L. Sounding Boards. Historical controls: the practitioner's clinical trials. N Engl J Med 1982;307(21): 1339—40.

[47] Epstein RS, Moyer TP, Aubert RE, O Kane DJ, Xia F, Verbrugge RR, et al. Warfarin genotyping reduces hospitalization rates results from the MM-WES (Medco-Mayo Warfarin Effectiveness study). J Am Coll Cardiol 2010;55(25):2804—12.

[48] Sacks HS, Chalmers TC, Smith Jr H. Sensitivity and specificity of clinical trials: randomized v historical controls. Arch Intern Med 1983;143(4):753—5.

[49] Norris SL, Atkins D. Challenges in using nonrandomized studies in systematic reviews of treatment interventions. Ann Intern Med 2005;142(12 Pt 2):1112—9.

[50] Cervical cancer screening programs. I. Epidemiology and natural history of carcinoma of the cervix. Can Med Assoc J 1976;114(11):1003—12.

[51] Fletcher RH, Fletcher SW. Clinical Epidemiology—The Essentials. Philadelphia: Lippincott, Williams & Wilkins; 2005. p. 252.

[52] Harris AD, Lautenbach E, Perencevich E. A systematic review of quasi-experimental study designs in the fields of infection control and antibiotic resistance. Clin Infect Dis 2005;41(1):77—82.

[53] Haga SB, Warner LR, O'Daniel J. The potential of a placebo/nocebo effect in pharmacogenetics. Public Health Genomics 2009;12(3):158—62.

[54] Furuta T, Shirai N, Kodaira M, Sugimoto M, Nogaki A, Kuriyama S, et al. Pharmacogenomics-based tailored versus standard therapeutic regimen for eradication of H. pylori. Clin Pharmacol Ther 2007;81(4):521—8.

[55] Mallal S, Phillips E, Carosi G, Molina JM, Workman C, Tomazic J, et al. HLA-B*5701 screening for hypersensitivity to abacavir. N Engl J Med 2008;358(6):568—79.

[56] Soumerai SB, Ross-Degnan D, Avorn J, McLaughlin T, Choodnovskiy I. Effects of Medicaid drug-payment limits on admission to hospitals and nursing homes. N Engl J Med 1991;325(15):1072—7.

[57] Sanderson S, Zimmern R, Kroese M, Higgins J, Patch C, Emery J. How can the evaluation of genetic tests be enhanced? Lessons learned from the ACCE framework and evaluating genetic tests in the United Kingdom. Genet Med 2005;7(7):495—500.

[58] Ozdemir V, Joly Y, Knoppers BM. ACCE, pharmacogenomics, and stopping clinical trials: time to extend the CONSORT statement? Am J Bioeth 2011;11(3):11—3.

[59] Tatsioni A, Zarin DA, Aronson N, Samson DJ, Flamm CR, Schmid C, et al. Challenges in systematic reviews of diagnostic technologies. Ann Intern Med 2005;142(12 Pt 2):1048—55.

[60] Benson K, Hartz AJ. A comparison of observational studies and randomized, controlled trials. N Engl J Med 2000;342(25):1878—86.

[61] Pirmohamed M, James S, Meakin S, Green C, Scott AK, Walley TJ, et al. Adverse drug reactions as cause of admission to hospital: prospective analysis of 18,820 patients. Br Med J 2004;329(7456):15—9.

[62] Elston RC, Idury RM, Cardon LR, Lichter JB. The study of candidate genes in drug trials: sample size considerations. Stat Med 1999;18(6):741—51.

[63] Kelly PJ, Stallard N, Whittaker JC. Statistical design and analysis of pharmacogenetic trials. Stat Med 2005;24(10):1495–508.

[64] Freidlin B, Korn EL. Biomarker-adaptive clinical trial designs. Pharmacogenomics 2010;11(12):1679–82.

[65] (CHMP) C.f.M.P.f.H.U. Reflection paper on methodical issues associated with pharmacogenomic biomarkers in relation to clinical development and patient selection; 2011; London. p. 1–21.

[66] Altman RB. PharmGKB: a logical home for knowledge relating genotype to drug response phenotype. Nat Genet 2007;39(4):426.

[67] Swen JJ, Nijenhuis M, de Boer A, Grandia L, Maitland-van der Zee AH, Mulder H, et al. Pharmacogenetics: from bench to byte—an update of guidelines. Clin Pharmacol Ther 2011;89(5):662–73.

[68] Kirchheiner J, Brosen K, Dahl ML, Gram LF, Kasper S, Roots I, et al. CYP2D6 and CYP2C19 genotype-based dose recommendations for antidepressants: a first step towards subpopulation-specific dosages. Acta Psychiatr Scand 2001;104(3):173–92.

[69] Relling MV, Klein TE. CPIC: Clinical Pharmacogenetics Implementation Consortium of the Pharmacogenomics Research Network. Clin Pharmacol Ther 2011;89(3):464–7.

[70] Gurwitz D, Lunshof JE, Dedoussis G, Flordellis CS, Fuhr U, Kirchheiner J, et al. Pharmacogenomics education: International Society of Pharmacogenomics recommendations for medical, pharmaceutical, and health schools deans of education. Pharmacogenomics J 2005;5(4):221–5.

[71] Smith GC, Pell JP. Parachute use to prevent death and major trauma related to gravitational challenge: systematic review of randomised controlled trials. Br Med J 2003;327(7429):1459–61.

[72] Chung WH, Hung SI, Hong HS, Hsih MS, Yang LC, Ho HC, et al. Medical genetics: a marker for Stevens-Johnson syndrome. Nature 2004;428(6982):486.

[73] Tassaneeyakul W, Jantararoungtong T, Chen P, Lin PY, Tiamkao S, Khunarkornsiri U, et al. Strong association between HLA-B*5801 and allopurinol-induced Stevens-Johnson syndrome and toxic epidermal necrolysis in a Thai population. Pharmacogenet Genomics 2009;19(9):704–9.

[74] Lonjou C, Borot N, Sekula P, Ledger N, Thomas L, Halevy S, et al. A European study of HLA-B in Stevens-Johnson syndrome and toxic epidermal necrolysis related to five high-risk drugs. Pharmacogenet Genomics 2008;18(2):99–107.

[75] Schirmer MA, Hoffmann AO, Campean R, Janke JH, Zidek LM, Hoffmann M, et al. Bioinformatic and functional analysis of TGFBR1 polymorphisms. Pharmacogenet Genomics 2009;19(4):249–59.

[76] Hitzl M, Drescher S, van der Kuip H, Schaffeler E, Fischer J, Schwab M, et al. The C3435T mutation in the human MDR1 gene is associated with altered efflux of the P-glycoprotein substrate rhodamine 123 from CD56+ natural killer cells. Pharmacogenetics 2001;11(4):293–8.

[77] Welsh M, Mangravite L, Medina MW, Tantisira K, Zhang W, Huang RS, et al. Pharmacogenomic discovery using cell-based models. Pharmacol Rev 2009;61(4):413–29.

[78] Alsbeih G, El-Sebaie M, Al-Harbi N, Al-Buhairi M, Al-Hadyan K, Al-Rajhi N. Radiosensitivity of human fibroblasts is associated with amino acid substitution variants in susceptible genes and correlates with the number of risk alleles. Int J Radiat Oncol Biol Phys 2007;68(1):229–35.

[79] Emmert S, Slor H, Busch DB, Batko S, Albert RB, Coleman D, et al. Relationship of neurologic degeneration to genotype in three xeroderma pigmentosum group G patients. J Invest Dermatol 2002;118(6):972–82.

[80] Malan, D., S. Friedrichs, B.K. Fleischmann, and P. Sasse. Cardiomyocytes obtained from induced pluripotent stem cells with long-QT syndrome 3 recapitulate typical disease-specific features in vitro. Circ Res 109(8): 841–7.

[81] Sachse C, Brockmöller J, Bauer S, Roots I. Cytochrome P450 2D6 variants in a Caucasian population: allele frequencies and phenotypic consequences. Am J Hum Genet 1997;60(2):284–95.

[82] Sehrt D, Meineke I, Tzvetkov M, Gultepe S, Brockmoller J. Carvedilol pharmacokinetics and pharmacodynamics in relation to CYP2D6 and ADRB pharmacogenetics. Pharmacogenomics 2011;12(6):783–95.

[83] Birkenfeld AL, Jordan J, Hofmann U, Busjahn A, Franke G, Kruger N, et al. Genetic influences on the pharmacokinetics of orally and intravenously administered digoxin as exhibited by monozygotic twins. Clin Pharmacol Ther 2009;86(6):605–8.

9A

Incorporating Pharmacogenomics in Drug Development: A Perspective from Industry

Ophelia Yin

Oncology Clinical Pharmacology, Novartis Pharmaceutical Corporation,
East Hanover, New Jersey, USA

OBJECTIVES

1. Discuss the role of pharmacogenomics in every key stage of the drug development process.

2. Identify major issues and challenges of applying pharmacogenomics in drug development.

3. Propose potential strategies towards increased success of pharmacogenomic research in the pharmaceutical industry.

INTRODUCTION

A recent analysis of the development procedures for new molecular and biological entities shows that the average time required for a product from the start of clinical testing to acquiring regulatory approval is 7.2 years. However, the likelihood that a compound that enters clinical testing will eventually reach the market continues to be low, averaging 16% across different therapeutic areas. The long development time and low success rate result in the high costs of developing drugs. In 2008, overall research and development (R&D) costs for the pharmaceutical sector exceeded $50 billion in the United States [1]. The onerous times, costs, and risks associated with new drug development pose a big challenge to the pharmaceutical industry. As has been well recognized, innovative approaches to bring new drug products to market in a more efficient and cost-effective manner are urgently needed, and pharmacogenomic approaches might provide a valuable contribution to this.

Pharmacogenomics
http://dx.doi.org/10.1016/B978-0-12-391918-2.00018-4

The primary processes in the development of new therapeutic modalities are evaluation and optimization of the pharmacokinetics, pharmacodynamics, efficacy, and safety (PK-PD-E-S) of drugs. In the past, the conventional approach to determine PK-PD-E-S was to use the general subject cohorts that met designated inclusion/exclusion criteria. With the technological advancements in pharmacogenomics and molecular diagnostics, the conventional "one-size-fits-all" drug development model is no longer desirable, and a new paradigm of incorporating pharmacogenomics in drug development is emerging.

Generally speaking, pharmacogenomics can aid drug development through the identification and validation of markers in the following four areas: patient stratification, drug exposure, efficacy, and safety.

- Stratification markers can identify a subset of patients who may uniquely benefit from a given therapy.
- Exposure markers can help determine whether the desired target tissue of a subject has been exposed to a drug at physiological concentrations.
- Efficacy markers provide molecular evidence above and beyond traditional clinical endpoints.
- Toxicity markers help predict and circumvent adverse drug effects.

PHARMACOGENOMICS IN DIFFERENT PHASES OF DRUG DEVELOPMENT

There are an increasing number of examples demonstrating that pharmacogenomics can play an important role at every key stage of the drug development process, from target identification and validation and preclinical animal toxicology studies to Phase I, II, and III trials, as well as in postmarketing studies (Table 9A.1). By incorporating pharmacogenomics into

TABLE 9A.1 Application of pharmacogenomics in different stages of drug development

Drug development phase	Applications
Target screening/identification	• Identification/characterization of the gene encoding the drug target • Assessment of drug target variability
Preclinical/animal toxicity	• Identification of safety markers • Provision of potential early safety indicators or warning signals
Phase I	• Explanation of outliers and interpatient variability in pharmacokinetics • Patient selection—inclusion/exclusion criteria • Bridge to other ethnic populations
Phases II and III	• Patient selection—inclusion/exclusion criteria • Dose-range selection and dose modification • Interpretation of trial results based on pharmacogenomic tests
Phase IV and patient therapy optimization	• Analysis of reported adverse effects with pharmacogenomic tests • Identification of patients at high risk for adverse drug effects • Identification of drug responders and nonresponders

these various processes, one has the potential to improve the efficiency and productivity of drug development by:

- Validating more genomically diverse and higher-quality drug targets.
- Eliminating unsuitable drug candidates and therapeutic targets at the earlier and cheaper stages of development.
- Accelerating clinical development by facilitating the design of better trials that more clearly show improved efficacy and safety.
- Optimizing risk-benefit profiles of drugs in targeted patient populations.

DRUG TARGET SCREENING AND IDENTIFICATION

Prospective screening and targeting a well-defined patient population can reduce the risk of failure and increase the likelihood of success of new drug development. For example, many cancer patients are benefiting from targeted therapy, such as imatinib for chronic myeloid leukemia (CML) with the presence of the Philadelphia chromosome cytogenetic abnormality and trastuzumab for breast cancer with overexpression of the human epidermal growth factor receptor 2 (HER2) proteins (Table 9A.2).

TABLE 9A.2 Examples in Which pharmacogenomic tests have been used in the prospective identification of drug target and patient populations for drug development

Biomarker/PGx test	Drug	Indication
CCR5 Δ32	Maraviroc	Human immunodeficiency virus type 1
C-KIT expression	Imatinib	Inoperable or metastatic gastrointestinal stromal tumors
EGFR expression	Cetuximab	Colorectal cancer
EGFR expression	Panitumumab	Metastatic colorectal carcinoma
EGFR expression	Erlotinib	Non−small cell lung cancer/metastatic pancreatic cancer
HER2 overexpression	Lapatinib	Advanced/metastatic breast cancer
HER2 overexpression	Trastuzumab	Metastatic breast cancer
LDL-R	Atorvastatin	Familial hypercholesterolemia
Philadelphia chromosome translocation	Imatinib	Chronic myeloid leukemia
Philadelphia chromosome translocation	Dasatinib	Acute lymphocytic leukemia and chronic myeloid leukemia
Philadelphia chromosome translocation	Nilotinib	Chronic myeloid leukemia

CCR5 Δ32: 32-bp deletion in C-C chemokine receptor 5; C-KIT: CD117 cytokine receptor; EGFR: epidermal growth factor receptor; HER2: human epidermal growth factor receptor 2; LDL-R: low density lipoprotein receptor

Imatinib

Chronic myeloid leukemia (CML) is characterized by a chromosomal translocation that leads to the chimeric Philadelphia chromosome (Ph) in a hematopoietic stem cell. This translocation fuses the V-abl Aberlson murine leukemia viral (*ABL*) gene from chromosome 9 onto the breakpoint cluster region (*BCR*) gene on chromosome 22, resulting in an oncogene that encodes the BCR-ABL protein having constitutively activated tyrosine kinase activity that causes CML [2, 3]. Thus the BCR-ABL tyrosine kinase domain is a rational target for therapeutic treatment in patients with CML.

Based on this understanding of the genomic basis of CML, imatinib was developed as an inhibitor of the tyrosine kinase activity of BCR-ABL and was among the first targeted anticancer agents to specifically inhibit an aberrant oncogenic signaling pathway in tumor cells, instead of nonspecifically targeting all rapidly dividing cells. As the first tyrosine kinase inhibitor (TKI) to target BCR-ABL, imatinib has been proven to be a remarkably successful treatment for CML. With a follow-up of eight years after the start of the original pivotal International Randomized Trial of Interferon and STI571 (IRIS), imatinib continues to show clinical benefit, with cumulative complete cytogenetic response (CCyR) rates achieved in 83% of patients and an overall survival rate of 85% [4]. This clinical success totally validated the concept of targeted therapy with TKIs in the treatment of CML, and consequently inspired the search for and development of second generation TKIs that are more potent than imatinib in treating newly diagnosed CML or may be effective in patients who fail imatinib.

Imatinib was subsequently discovered to be an inhibitor of the tyrosine kinase activity associated with the KIT receptor, and as such, imatinib is also indicated for the treatment of unresectable and/or metastatic malignant gastrointestinal stromal tumors caused by a gain-of-function mutation in KIT [5].

Trastuzumab

Trastuzumab is another excellent example that illustrates the development of a targeted anticancer drug based upon a genomic test for patient selection. Overexpression or amplification of HER2 occurs in 25 to 30% of breast cancer patients. The increased level of HER2 is associated with poor clinical prognosis, aggressive tumor, greater risk of cancer recurrence, and relative resistance to some types of chemotherapy, resulting in shorter disease-free and overall survival from breast cancer [6,7]. Thus, the discovery of HER2 provided a basis for searching for treatments targeting this receptor. Trastuzumab, which binds with high affinity and specificity to the extracellular domain of the HER2 receptor [8], was the first HER2 targeted agent for the treatment of HER2 positive metastatic breast cancer. Increasing clinical evidence supports the clinical benefits of trastuzumab treatment in cases of breast cancer with HER2 overexpression or amplification. As such, confirmation of HER2 test positive status becomes a prerequisite for the use of trastuzumab.

PRECLINICAL ANIMAL TOXICOLOGY STUDIES

In preclinical animal toxicology studies, the use of validated predictive safety biomarkers can enhance the understanding of toxicity mechanisms, aid the selection of drug candidates

TABLE 9A.3 Examples of consortia using pharmacogenomics for drug safety biomarker assessment

Consortium	Species and study design
CEBS program of the U.S. NIEHS proteomics	Rat/mouse; primary focus on liver toxicity
Japanese Toxicogenomics Project	Rat plus in vitro; primary focus on liver toxicity
FDA and BG Medicine Liver Toxicity Biomarker Study	Rat; focus on liver toxicity
C-Path Institute: Predictive Safety Testing Consortium	Rat and human markers being sought for liver and kidney toxicities, myopathy, vasculitis, and carcinogenicity
HESI Genomics Committee	Rat and in vitro markers being sought for kidney and heart toxicities and genotoxicity
InnoMed PredTox	Rat; focus on liver and kidney toxicities

CEBS: chemical effects in biological systems; NIEHS: National Institute of Environmental Health Sciences; FDA: Food and Drug Administration; HESI: Health and Environmental Sciences Institute

that are more likely to be tolerated in humans, potentially reduce the cost and time required for preclinical evaluation, and ultimately reduce late-phase failures. Thus, toxicogenomic profiling via the use of DNA microarray-based approaches has provided the most striking advances in understanding both disease mechanisms and the effects of drug treatment. Several consortia have emerged in recent years through partnerships among industry, academic, and other nonprofit groups to build toxicogenomic profiling platforms for drug safety assessment (Table 9A.3) [9]. For example, the Predictive Safety Testing Consortium (PSTC), including the membership of 15 pharmaceutical companies and led by Critical Path Institute (C-Path), was established with goals to [1] validate predictive animal model—based biomarkers aimed at reducing the cost and time involved in conducting nonclinical safety studies; [2] provide potential early indicators of clinical safety in drug development and postmarketing surveillance; and [3] provide new tools to assist in regulatory decision making. One of the main focuses of PSTC is qualification of candidate biomarkers of hepatotoxicity, nephrotoxicity, vasculitis, myopathy, as well as genotoxic and nongenotoxic carcinogens [10]. This group has recently developed a set of seven new nephrotoxicity biomarkers, which were qualified for nonclinical application in drug safety evaluation using the rat as a model. As part of the InnoMed PredTox project, a panel of novel biomarkers for improved detection of liver injury and renal toxicity in preclinical toxicity studies was also reported recently [11,12]. These biomarkers have become useful tools in accelerating decision making (i.e., the interpretation of isolated histopathological findings in some drug safety—evaluation animal models) and risk assessment choices [13].

PHASE I STUDIES

Both extrinsic (i.e., drug—food and drug—drug interactions) and intrinsic (i.e., demographics and organ function) factors may influence the pharmacokinetic and pharmacodynamic properties of drugs in humans. Among the important intrinsic factors is genetic makeup, such as the genetic polymorphisms in drug-metabolizing enzymes and transporters. If a drug is shown to be primarily metabolized by an enzyme that is subject to major genetic polymorphisms, considerable intersubject variability in the pharmacokinetics of the drug may be anticipated. Subjects with different genetic makeup may show variable exposure to the drug and experience different degrees of drug–drug interactions (i.e., poor metabolizers [PMs] may be insensitive to the drug–drug interactions predicted from *in vitro* studies, whereas ultra-rapid metabolizers may show a greater extent of interaction than extensive metabolizers [EMs]). Performing genotype and phenotype determinations to establish genotype–phenotype relationships in Phase I pharmacokinetic studies will enable an assessment of the effects of specific polymorphisms on the pharmacokinetics of drugs and provide explanations for outliers or intersubject variability in response. The information thus derived, together with other properties of drug (i.e., safety margin and pharmacodynamic curves for both safety and efficacy) and type of disease, will then allow an appropriate evaluation of the clinical consequences of the observed pharmacokinetic variability. These evaluations may ultimately assist in the design of clinical trials, for example through the inclusion or exclusion of specific subjects or the use of different dosing regimens in different subgroups of patients.

PHASE II AND III STUDIES

There are several different aspects of applying pharmacogenomics in Phase II and III studies. If Phase I pharmacokinetic studies have shown that the pharmacokinetics of a drug are variable due to genetic factors, pharmacogenomic data are often collected from patients during Phase II/III efficacy and safety trials, as illustrated in the development course of atomoxetine.

On the other hand, as an ongoing process during clinical development, the continuous assessment of relevant emerging data (including genetic association) provides new information about the drug. Sometimes the new information obtained may warrant the revision of the clinical trial design or patient stratification to match the right drug with the right patients, so that the risk of failure is reduced and the likelihood of success is increased. This integrated process of assessing emerging pharmacogenomic data during clinical trials can also confirm the relevance of the target for the disease (proof of concept) and enhance the confidence in continuing the clinical development program. Gefitinib and maraviroc are two typical examples in this regard.

Atomoxetine

Atomoxetine is a selective norepinephrine re uptake inhibitor for treatment of attention-deficit hyperactivity disorder. Atomoxetine is primarily metabolized in the liver via the cytochrome P4502D6 (CYP2D6) enzyme. Plasma clearance of atomoxetine was reported to be 0.35 L/hr/kg in CYP2D6 EMs and 0.03 L/hr/kg in CYP2D6 PMs, with the area under the concentration versus time curve (AUC) being approximately tenfold and steady-state peak plasma concentration (Cmax) being fivefold greater in PMs than in EMs. At the same dose level, atomoxetine AUC in PMs was similar to that observed in EMs with concomitant administration of strong CYP2D6 inhibitors [14]. Given the anticipated impact of *CYP2D6* genotype on atomoxetine pharmacokinetics, pharmacogenomic data were collected in efficacy and safety trials of atomoxetine. Such a database allowed a retrospective analysis, which revealed that some adverse drug effects occurred twice as frequently or statistically significantly more frequently in PM patients when compared with EM patients. Such adverse effects included decreased appetite (23% of PMs, 16% of EMs); insomnia (13% of PMs, 7% of EMs); sedation (4% of PMs, 2% of EMs); depression (6% of PMs, 2% of EMs); tremor (4% of PMs, 1% of EMs); early morning awakening (3% of PMs, 1% of EMs); pruritus (2% of PMs, 1% of EMs); and mydriasis (2% of PMs, 1% of EMs). Although pharmacogenetic testing is not mandated before prescribing atomoxetine, the updated product label suggests that dose adjustment of the drug may be necessary when administered to patients known to be CYP2D6 PMs or when coadministered with potent CYP2D6 inhibitors [15].

Gefitinib

Gefitinib inhibits the tyrosine kinase activity associated with the epidermal growth factor receptor (EGFR) and thus blocks intracellular signal transduction pathways emanating from this receptor implicated in the proliferation and survival of cancer cells. In mouse xenograft models, gefitinib shows a significant inhibition on tumor growth in a dose-dependent manner [16]. In Phase I trials in patients with non–small cell lung cancer (NSCLC), gefitinib was found to be well tolerated. Inhibition of EGFR and its related downstream signaling is achieved at a dosage of 250 /day, and the maximum tolerated dose of gefitinib is 700 mg/ day [17]. Accordingly, randomized Phase II trials (IDEAL-1 and IDEAL-2) were conducted to evaluate the activity of gefitinib at two dose levels, 250 and 500 mg/day, in pretreated patients with advanced NSCLC [18,19]. Similar results were obtained from both studies. Response rate of gefitinib was moderate and similar for the 250 mg versus 500 mg doses (18.4% versus 19% in IDEAL-1 and 12.0% versus 9.0% in IDEAL-2). However, from these trials important evidence emerged showing major efficacy of gefitinib in some specific subgroups of patients, such as females, those with adenocarcinoma histological subtype, and those of Asian ethnicity. In IDEAL-1, the odds of responding (i.e., having complete or partial responses) was approximately 3.5-fold higher for patients with adenocarcinoma than for patients with other tumor histology ($P = 0.021$), 2.5-fold higher for females than males ($P = 0.017$), and 1.6-fold higher for Japanese than non-Japanese patients ($P = 0.25$) [18]. In IDEAL-2, the response rate was greater in adenocarcinoma than in other histologies (13% versus 4%, $P = 0.046$), and greater in female than in male patients (19% versus 3%, $P = 0.001$) [19].

On the basis of these data, several Phase III trials were launched. The Iressa Survival Evaluation in Lung Cancer (ISEL) study enrolled 1,692 patients with advanced NSCLC who had previously received chemotherapy [20]. Median survival did not differ significantly between gefitinib versus placebo groups, either in the overall population (5.6 months versus 5.1 months, $P = 0.087$) or among the patients with adenocarcinoma (6.3 months versus 5.4 months, $P = 0.089$). However, preplanned subgroup analyses showed significantly longer survival in the gefitinib group than the placebo group for nonsmokers (8.9 months versus 6.1 months, $P = 0.012$) and patients of Asian origin (9.5 months versus 5.5 months, $P = 0.01$) [20]. In addition, analysis of ISEL tumor biopsy samples suggested that high

EGFR gene copy number was predictive of gefitinib-related effect on survival. Patients with *EGFR* mutations obtained higher response rates than those with wild-type *EGFR* genotype (37.5% versus 2.6%) [21].

To further understand the role of clinicopathologic features versus molecular selection, the Phase III Iressa Pan-Asia Study (IPASS) used several clinicopathologic criteria to identify a group of patients who may derive further benefit from gefitinib therapy [22]. The study included Asian, chemotherapy-naïve patients who never smoked and had adeno-carcinoma of the lung. In this carefully selected population, progression-free survival (PFS) was found to be superior for gefitinib as compared to carboplatin—paclitaxel. In the subgroup of patients who were positive for *EGFR* mutation, PFS was significantly longer among those who received gefitinib than among those who received carboplatin—paclitaxel, whereas in the subgroup of patients who were negative for *EGFR* mutation, PFS was significantly longer among those who received carboplatin—paclitaxel. Thus, this was the first study that definitively identified *EGFR* mutation status as an important predictive marker for gefitinib therapy. The use of first-line gefitinib in a selected patient population was further supported by other Phase III trials, in which only patients with chemotherapy-naïve advanced NSCLC harboring *EGFR* mutations were enrolled [23,24].

In reviewing additional Phase II and III trials [25—35] that were conducted with gefitinib as a first-line treatment of NSCLC (patients were either unselected or selected based on clinical characteristics or *EGFR* mutation) (Table 9A.4), it becomes obvious that gefitinib, when used in unselected patients, produces only a modest response rate of 10 to 20%. A greater benefit appears to be obtained in clinically selected subgroups of patients, such as nonsmokers, Asians, and patients with adenocarcinoma histology. Although clinical characteristics may identify potential candidates for gefitinib therapy, the most predictive marker is the presence of *EGFR* gene mutations, which are present in approximately 10 to 20% of NSCLC patients and more frequently found in nonsmokers, Asians, and patients with adeno-carcinoma. The exact mechanism for the enhanced sensitivity of *EGFR* mutation to gefitinib is unclear. One explanation is the so-called oncogene addiction. For mutation-positive tumors, the *EGFR* mutation is a key factor driving the malignant transformation. Thus elimination of such mutation consequently leads to inhibition of tumor growth. The other possible explanation is that gefitinib may be more potent in inhibiting mutated form than the wild-type enzyme. It has been shown that gefitinib binds 20 times more tightly to the L858R mutation than to the wild-type enzyme [36].

It is worthwhile to note that the important association between *EGFR* mutation and treatment outcome with gefitinib also offers an explanation for the failure of gefitinib in combination with chemotherapy (gemcitabine—cisplatin or paclitaxel—carboplatin) in previous trials. For example, two randomized Phase III trials, Iressa NSCLC Trial Assessing Combination Treatment-1 and -2 (INTACT-1 and INTACT-2), evaluated the drug in combination with chemotherapy as a first-line treatment. Both studies failed to demonstrate a survival advantage when gefitinib was administered with chemotherapy [37,38]. One of the most likely reasons could be that patients were not selected based on any of the criteria later found to be associated with a sensitivity to gefitinib. Thus, in both trials, the population that was most likely to receive a real benefit from the treatment (*EGFR* mutation) was not large enough to statistically change the overall results.

TABLE 9A.4 Phase II and III trials of gefitinib (250 mg/day) in first-line treatment of NSCLC

Trial (ref)	Patient selection criteria	Number of patients	RR (%)	PFS/TTP (months)	OS (months)	One-year survival (%)
PHASE II TRIALS						
Spigel et al. [25]	Unselected	70	4.0	3.7	6.3	24
Inoue et al. [26]	EGFR mutation	16	75	9.7	—	—
Lee et al. [27]	Clinically selected: non smokers with adenocarcinoma	72	55.6	6.8	19.7	—
Niho et al. [28]	Unselected	40	30			55
Reck et al. [29]	Unselected	58	5.2	1.8	7.3	—
Suzuki et al. [30]	Unselected	34	26.5			58.2
Sequist et al. [31]	EGFR mutation	31	55	9.2	—	—
Tamura et al. [32]	EGFR mutation	28	75	11.5		79
Yang et al. [33]	Unselected	106	50.9	5.5	22.4	—
Inoue et al. [34]	EGFR mutation	30	66	6.5	—	—
PHASE III TRIALS						
Mok et al. [22]	Clinically selected: non smokers or former light smokers with adenocarcinoma	Total ($n = 1217$) Gefitinib ($n = 609$) Carboplatin–paclitaxel ($n = 608$)	43*** 32.2	5.7*** 5.8	18.6 17.3	— —
	EGFR mutation subgroup	Total ($n = 261$) Gefitinib ($n = 132$) Carboplatin–paclitaxel ($n = 129$)	71.2*** 47.3	9.5*** 6.3	— —	— —

(Continued)

TABLE 9A.4 Phase II and III trials of gefitinib (250 mg/day) in first-line treatment of NSCLC—cont'd

Trial (ref)	Patient selection criteria	Number of patients	RR (%)	PFS/TTP (months)	OS (months)	One-year survival (%)
Lee et al. [35]	Clinically selected: non smokers with adenocarcinoma	Total ($n = 309$)	53.5	6.1	21.3	—
		Gefitinib ($n = 159$)	45.3	6.6	23.3	—
		Cisplatin–gemcitabine ($n = 150$)				
	EGFR mutation subgroup	Total ($n = 42$)	84.6**	8.5	30.6	—
		Gefitinib ($n = 26$)	37.5	6.7	26.5	—
		Carboplatin–paclitaxel ($n = 16$)				
Maemondo et al. [24]	EGFR mutation	Total ($n = 230$)	73.7***	10.8***	30.5	—
		Gefitinib ($n = 115$)	30.7	5.4	23.6	—
		Carboplatin–paclitaxel ($n = 115$)				
Mitsudomi et al. [23]	EGFR mutation	Total ($n = 177$)	62.1***	9.2***	—	—
		Gefitinib ($n = 88$)	32.2	6.3	—	—
		Cisplatin–docetaxel ($n = 89$)				

** p < 0.01 for gefitinib versus chemotherapy;
*** p < 0.001 for gefitinib versus chemotherapy;
NSCLC: non–small cell lung cancer; EGFR: epidermal growth factor receptor; RR: response rate; PFS: progression free survival; TTP: time to progression; OS: overall survival

Maraviroc

Maraviroc is the first of a new class of antiretroviral agents approved by the U.S. Food and Drug Administration (FDA) in August 2007 [39]. Following the recognition that individuals with a homozygous mutation in the human chemokine receptor 5 (*CCR5*) gene, designated as the *CCR5-Δ32* mutation, have a natural resistance to HIV infection [40,41], antagonism of CCR5 was established as a desirable target for drug therapy. Maraviroc was subsequently developed as a CCR5 receptor antagonist that selectively and reversibly blocks the binding of gp120 to the CCR5 receptor, which prevents the conformational changes necessary for entry of CCR5 tropic HIV-1 into CD4 cells [42]. During the clinical development of this compound, many publications reported a genetic association between *CCR5* and different diseases and phenotypes. In particular, association between *CCR5-Δ32* and the more severe hepatitis C virus (HCV) infection raised a potential concern about the safety of CCR5 antagonism in patients coinfected with HIV and HCV. Such a concern, if truly existing, could have affected the clinical development program and, specifically, the design of clinical trials of maraviroc. For example, patients can be placed at risk if true safety concerns are ignored. On the other hand, unwarranted emphasis on studies highlighting potential safety concerns may result in denial of new therapeutic options to patient subgroups (e.g., patients with HIV and HCV). A meta-analysis using published non-HIV genetic association studies was conducted to quantitatively assess the risk of susceptibility to HCV with relation to *CCR5-Δ32* [43]. The analysis demonstrated generally compatible findings among studies in general populations and showed a lack of evidence of an association with *CCR5-Δ32* and susceptibility to HCV infection. Although confirmation of the lack of such association requires real clinical data, this analysis provided some confidence in continuing the recruitment of HCV-coinfected HIV patients into the ongoing maraviroc trials [43]. Later on, based on the collected safety and efficacy data for coinfected patients in two major trials, Maraviroc versus Optimized Therapy in Viremic Antiretroviral Treatment-Experienced Patients-1 and -2 (MOTIVATE 1 and MOTIVATE 2), the incidence of all causality treatment−emergent adverse events was found to be similar in coinfected patients treated with maraviroc or placebo [44].

PHASE IV STUDIES

In the Phase IV stage of drug development, pharmacogenomics can be used to optimize therapy with currently approved drugs, such as predicting adverse drug effects with pharmacogenomic tests or identifying patients at high risk of adverse drug effects based on genotype. For example, the discovery of a strong association between the human leukocyte antigen B*5701 allele (*HLA-B*5701*) and abacavir-induced hypersensitivity reaction in HIV-infected patients led to the clinical prescreening of abacavir candidates for *HLA-B*5701*, with avoidance of abacavir in *HLA-B*5701*-positive patients consequently reducing the occurrence of abacavir-induced hypersensitivity reactions. The identification of the significant impact of uridine disphosphate-glucuronosyltransferases 1A1*28 (*UGT1A1*28*) polymorphism on irinotecan-associated toxicity provides guidance for selecting an appropriate irinotecan starting dose in individual patients.

Abacavir

Abacavir is a nucleoside analog inhibitor of HIV type-1 (HIV-1) reverse transcriptase approved for the treatment of HIV-1 infection in both treatment-naïve and treatment-experienced patients. When administered in combination with other antiretroviral drugs, usually as a fixed-dose combination of abacavir/lamivudine given together with a non-nucleoside reverse transcriptase inhibitor or a ritonavir-boosted protease inhibitor, abacavir has good efficacy against susceptible HIV-1 isolates [45,46].

Early in the clinical development of abacavir, it was observed that the drug caused a hypersensitivity reaction in some patients. This reaction typically appears within six weeks of initiation of therapy [47]. Symptoms resolve within a few days after discontinuation of abacavir but rechallenge results in the rapid onset of an overwhelming immediate-type hypersensitivity reaction that can lead to hypotension, respiratory failure, and death. The incidence of abacavir-induced hypersensitivity was shown to be higher in white patients than in black patients [48], indicating a potential genetic basis for susceptibility to this hypersensitivity reaction.

Subsequently, retrospective analyses identified a significant association between the presence of HLA-B*5701 allele and risk for abacavir-induced hypersensitivity [49,50]. Prospective screening for HLA-B*5701 in patients who were candidates for abacavir-containing antiretroviral therapy showed a reduction in the incidence of abacavir hypersensitivity reaction. In a large, prospective, double-blind, randomized trial conducted to validate the use of genetic testing to prevent abacavir hypersensitivity (PREDICT-1 study), 1,956 patients were randomly assigned to undergo prospective HLA-B*5701 screening, with exclusion of HLA-B*5701-positive patients from abacavir treatment (prospective-screening group) or to undergo a standard-of-care approach of abacavir use without prospective HLA-B*5701 screening (control group). A hypersensitivity reaction was clinically diagnosed in 93 patients, with a significantly lower incidence in the prospective-screening group than in the control group (3.4% versus 7.8%, $P < 0.001$). These results established the effectiveness of prospective HLA-B*5701 screening to prevent the hypersensitivity reaction to abacavir [51]. The abacavir package insert and antiretroviral treatment guidelines recommend screening for HLA B*5701 before prescribing abacavir [52].

Irinotecan

Irinotecan is a topoisomerase I inhibitor approved as a single agent for second-line treatment and in combination with 5-fluorouracil (5-FU) and leucovorin for first-line treatment in metastatic colorectal cancer. *In vivo*, irinotecan is hydrolyzed by carboxylesterase to its active metabolite, 7-ethyl-10-hydroxycamptothecin (SN-38), which is 1,000-fold more cytotoxic than the parent drug [53,54]. SN-38 is inactivated via glucuronidation catalyzed primarily by the UGT1A1 isoform. The formed conjugated product, SN-38G, is eliminated in the bile and can be deconjugated back to SN-38 via the action of intestinal β-glucuronidase enzyme [55]. The presence of a dinucleotide (TA) insertion in the TATA box of the UGT1A1 promoter results in a 70% reduction in enzyme expression and, in several studies, it was suggested that UGT1A1*28 polymorphism was significantly associated with decreased glucuronidation of SN-38 [56–59].

The use of irinotecan has been associated with severe grade 3 and 4 toxicities, primarily neutropenia and diarrhea, in a considerable number of patients [60]. It is generally considered that this toxicity is mediated by the active metabolite of irinotecan, SN-38. Based on a prospective study in 66 cancer patients who received irinotecan monotherapy, grade 4 neutropenia occurred in 50% and 12.5% of patients who were homozygous and heterozygous carriers of *UGT1A1*28*, respectively, whereas no grade 4 neutropenia occurred in patients who had the *UGT1A1* wild-type genotype [61]. Data from a large clinical trial (North Central Cancer Treatment Group N9741) in 520 patients with metastatic colorectal cancer showed similar results. The overall risk of grade 4 neutropenia was higher in homozygous *UGT1A1*28* patients than in those with the other genotypes: 36.2% in the homozygous *UGT1A1*28* group versus 18.2% in the heterozygous *UGT1A1*28* group and 14.8% in the homozygous wild-type group [62].

Based on these findings and other reports, the irinotecan label was modified in 2005 to indicate the role of *UGT1A1*28* polymorphism in the metabolism of irinotecan and the associated increased risk of severe neutropenia. The label modifications also include recommendations for lower starting doses of irinotecan in patients homozygous for the *UGT1A1*28* polymorphism [63]. Subsequent to the label update, a genetic test kit for *UGT1A1*28* was approved by the FDA.

Genetic testing to identify patients at increased risk for severe irinotecan-induced neutropenia and for susceptibility to abacavir hypersensitivity reaction are outstanding examples of how pharmacogenomics can improve patient care in the era of personalized medicine. These examples can be applied to other drugs, for which the development and operationalization of genetic tests may be useful to predict and circumvent adverse drug effects. On the other hand, pharmacogenomics can optimize therapy through identification of responders, suboptimal responders, or nonresponders. This process can be illustrated by the example of imatinib.

Imatinib

Although imatinib shows outstanding results in the treatment of CML, nonresponse or disease progression occurs in some patients. The emergence of new BCR-ABL tyrosine kinase domain mutations and clonal evolution are the known mechanisms for the acquired drug resistance [64,65]. However, increasing evidence also suggests that for a substantial number of patients, resistance may be apparent (pseudoresistance), and other factors such as drug transporters and imatinib plasma levels may play a contributing role to the therapeutic outcome in imatinib-treated CML patients. Thomas et al. were the first to show that inhibition of the organic cation transporter 1 (OCT1) in peripheral blood leukocytes from CML patients caused a decrease in intracellular imatinib uptake [66]. This finding was confirmed by others, showing that influx of imatinib into the CML cell is mediated by OCT1 [67]. Subsequent studies demonstrated that low OCT1 activity is a major determinant of suboptimal response to imatinib. More patients who had high OCT1 activity achieved major molecular response (major molecular response was defined as a BCR-ABL transcript level $\leq 0.1\%$, measured by real-time quantitative reverse transcriptase polymerase chain reaction and expressed on the International Scale) by five years compared with patients who had low OCT1 activity (89% versus 55%; $P = 0.007$). Moreover, a low OCT1 activity was significantly associated with lower event-free survival (48% versus 74%, $P = 0.03$) and overall survival (87% versus 96%, $P = 0.02$) following five years of treatment with imatinib (Figure 9A.1) [68]. A recent analysis further suggested that the combination of low OCT-1 activity and low-trough imatinib levels defines a group of patients who achieve the lowest rates of major molecular response by 24 months when compared to all other patients. These patients are also at the highest risk for imatinib failure when compared to all other patients (Table 9A.5) [69].

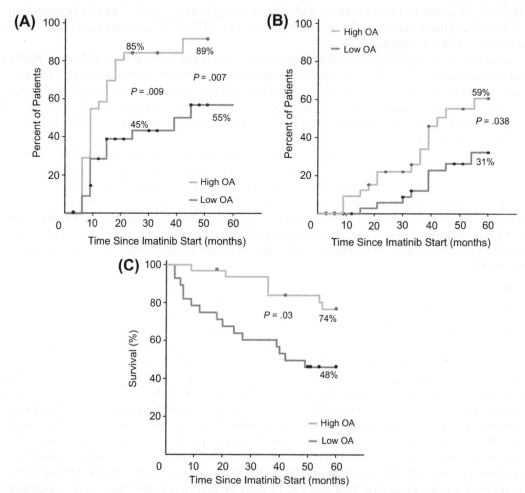

FIGURE 9A.1 The percentage of patients achieving (A) major molecular response (MMR), (B) complete molecular response (CMR), or (C) event-free survival on the basis of low and high organic cation transporter-1 (OCT-1) activity groups. Kaplan-Meier curves demonstrate that a significantly greater proportion of patients who had high OCT-1 activity achieve MMR and CMR by five years when compared with patients who had low OCT-1 activity. There is also significant event-free survival advantage for patients with high OCT-1 activity (*adopted from [68]*).

TABLE 9A.5 Impact of OCT1 activity and imatinib plasma levels on patient responses following 24 months of imatinib therapy [69]

Imatinib plasma level (ng/mlL)	% of patients achieving MMR at 24 months (n)			
	Total	Low OCT-1 activity	High OCT-1 activity	P value
Quartile 1 (<1200 ng/mL)	64 [25]	47 [15]	90 [10]	0.007
Quartile 2 (≥1200 and <1700 ng/mL)	80 [25]	71 [17]	100 [8]	0.064
Quartile 3 (≥1700 and <2800 ng/mL)	78 [25]	88 [19]	100 [8]	0.081
Quartile 4 (≥2800 ng/mL)	84 [25]	75 [16]	100 [9]	0.396

MMR: major molecular response; OCT1: organic cation transporter 1

Imatinib is primarily metabolized by CYP3A4/5, and drug transporters such as ABCB1 and ABCG2 are also reported to be involved in imatinib pharmacokinetics [70,71]. Because of the observed wide interpatient variability in imatinib plasma levels at a given dose [72], attempts were made to explore the impact of genetic polymorphisms of CYP3A4/5, ABCB1, and ABCG2 on the pharmacokinetics of imatinib (Table 9A.6). Studies by several groups consistently suggested a lack of significant effect of *CYP3A4/5* polymorphisms on imatinib clearance or concentrations [73–76]. Gurney et al. reported that patients with a TTT haplotype in *ABCB1* 1236C>T, 2677G>T/A, and 3435C>T loci had significantly higher estimated imatinib clearances (74). However, this finding conflicted with the study by Dulucq et al. that suggested higher imatinib trough plasma levels in patients with TTT haplotype [77]. In several other studies, no apparent association was identified between *ABCB1* polymorphisms and imatinib pharmacokinetics [75,76]. Results reported in the literature with regard to the impact of *ABCG2* polymorphism are also controversial. A lower imatinib clearance and a higher plasma concentrations were found in CML patients who were *ABCG2* 421 C>A mutant carriers [75,78], whereas in another study no difference in imatinib pharmacokinetic parameters was observed with regard to *ABCG2* 421 C>A mutation status [73].

Thus, in summary, a number of genetic studies have been conducted since the approval of imatinib. The *CYP3A4/5* polymorphism was not identified as a significant factor contributing to the interpatient variability in imatinib pharmacokinetics and systemic exposure. Results on *ABCB1* and *ABCG2* genotype associations with imatinib pharmacokinetics are inconclusive for translating into clinical management of patients treated with imatinib. However, the observed combined effect of OCT-1 activity and imatinib plasma levels on treatment outcomes in CML patients may have some clinical implications. For poor responses seen in patients with low OCT1 activity, increasing the initial dose of imatinib may provide a strategy to overcome low OCT1 activity. Alternatively, second-generation tyrosine kinase inhibitors, such as nilotinib and dasatinib, may be the drugs of choice in this situation, because influx of both nilotinib and dasatinib into the CML cell is independent of OCT1 [79,80].

TABLE 9A.6   Studies assessing polymorphisms of CYP3A5, ABCB1, and ABCG2 on imatinib

Studies (ref)	Population	Polymorphism	Clinical consequence
CYP3A5			
Gardner et al. [73]	Pts. with GIST	CYP3A4*1B CYP3A5*3	No change in imatinib CL/F No change in imatinib CL/F
Gurney et al. [74]	Pts. with CML or GIST	CYP3A5*3	No change in imatinib CL/F
Takahashi et al. [75]	Pts. with CML	CYP3A5*3	No change in dose-adjusted imatinib Cmin
Yamakawa et al. [76]	Pts. with CML	CYP3A5*3	No change in imatinib CL/F
ABCB1			
Gurney et al. [74]	Pts. with CML or GIST	1236C>T 2677G>T/A 3435C>T	Higher imatinib CL/F in TT genotype at each of 3 positions
Dulucq et al. [77]	Pts. with CML	1236 C>T	Higher imatinib Cmin in TT genotype
Takahashi et al. [75]	Pts. with CML	1236C>T 2677G>T/A 3435C>T	No change in dose-adjusted imatinib Cmin for all three variants
Yamakawa et al. [76]	Pts. with CML	1236C>T 2677G>T/A 3435C>T	No change in imatinib PK No change in imatinib PK Lower imatinib CL/F in CT and TT genotype
ABCG2			
Gardner et al. [73]	Pts. with GIST	421 C>A	No difference in imatinib PK parameters
Petain et al. [78]	Pts. with GIST and solid malignancies	421 C>A	Lower imatinib CL/F
Takahashi et al. [75]	Pts. with CML	421 C>A	Higher dose-adjusted imatinib Cmin

CML: chronic myeloid leukemia; CL/F: oral clearance; Cmin: trough plasma concentration; GIST: gastrointestinal stromal tumor; PK: pharmacokinetics

CHALLENGES IN APPLYING PHARMACOGENOMICS IN DRUG DEVELOPMENT

Numerous advances have been made in applying pharmacogenomics to all aspects of the drug development process. At an early stage of development, knowledge about the potential polymorphic metabolism of a drug is important in aiding the "go" or "no-go" decision and the design regarding a rational clinical development plan. As a common practice for the pharmaceutical industry, if a compound is identified to be metabolized extensively by an enzyme exhibiting genetic polymorphisms and consequently predicted to show extremely high interindividual variability, it may be abandoned for further development. If the compound is continued, identification of specific metabolizer groups and/or different dosing regimens would become crucial to allow optimal efficacy/safety results and thus maximize the probability of a successful late-phase trial.

Using genomic markers to prescreen patients for a preregistration clinical study is likely to restrict the indication and consequently the market potential of the approved drug. This issue may be a disincentive for companies to develop therapies for less severe diseases. However, excluding potential nonresponders will increase the risk-benefit ratio for the drug. As illustrated with gefitinib clinical development, earlier trials without carefully selected patient populations have resulted in disappointing outcomes, whereas restriction of study participation to the likely responders (i.e., patients with an *EGFR* mutation) has made the drug viable. For trastuzumab, it has been estimated that five times as many patients would have been needed to show clinical benefit for trastuzumab if the patients had not been screened for HER2 overexpression in clinical trials [81]. In these cases, incorporating pharmacogenomics into clinical trial design allowed a reduction in sample size and trial duration and assisted in increasing the likelihood of therapy success. In addition, pharmacogenomics can help manage postapproval risks, as is well demonstrated in the examples of abacavir and irinotecan, in which genetic testing is used to identify patients at high risk for serious adverse effects and to assist with treatment decisions in those patients.

Although the benefits of applying pharmacogenomics in drug development have been increasingly recognized in recent years and pharmacogenomic studies are becoming an integral part of drug development, there are a number of scientific and practical challenges: [1] lack of existing hypotheses; [2] inadequate sample size; and [3] feasibility and logistic issues.

Lack of existing hypotheses is one of the major obstacles in applying pharmacogenomics in drug development. In many cases, because of the complex nature of disease, the role of genetics may not be as specific to select or prioritize candidate genes, such as drug targets or disease- or pathway-related genes, for evaluation. The mechanism of action of a drug may be multifactorial, and many factors may contribute to the variation in drug response. The functional impact of genetic variants on candidate genes may not have been previously characterized. These issues add to the difficulties in choosing an appropriate hypothesis to test.

Ensuring adequate sample size is another challenge. In planning a pharmacogenomic study, both the size of the expected drug response and the size of the relevant genetic effect are important considerations [82]. When there is no clear existing genetic hypothesis, pharmacogenomic study evaluation is planned as a supplemental component or add-on to clinical trials. In this setting, the sample size of the trial is determined by the primary objective, such as the expected drug response. As a result, depending on the frequency of genotype/phenotype in the study population, the statistical power needed to detect a gene effect may not be adequate, leading to uninterpretable or inconclusive results. Even when the initial clinical studies with limited numbers of subjects generated positive correlation results, in many cases they are hypothesis generating and still required confirmation in a larger population.

Feasibility and logistic issues also present hurdles in implementing pharmacogenomics in clinical trials. Although most pharmaceutical companies now routinely include an optional pharmacogenomic sample collection in clinical trials, the overall collection rates remain low and variable. According to a recent survey conducted by the Industry Pharmacogenomic Working Group, the majority of companies reported that DNA samples were collected in only 10 to 49% of Phase I–III trials and in less than 10% of Phase IV trials [83]. Reasons for the low rate of sample collection include perceptions among institutional review boards (IRBs) and ethics committees (ECs) that collection of DNA samples is associated with greater privacy violation risks and issues with the wording of the informed consent form to use DNA sample results (i.e., informed consent form does not meet local requirement for a specific research plan). These could consequently lead to IRB/EC rejection of the pharmacogenomic substudy, resulting in the sponsor having to remove the pharmacogenomic component. Because of the concerns that inclusion of DNA sampling may cause IRB/EC rejection of the protocol and thus delay the start of trial, some sponsors may choose not to include a pharmacogenomic substudy in the clinical trial. In addition, investigators' willingness to present optional DNA sampling to patients varies, and their perceptions are influenced by the concerns regarding pressure that it may place on their patients or potential problems with health insurance agencies [83,84]. When a pharmacogenomic substudy is optional, investigators may not be sufficiently motivated to enroll patients and may choose not to participate.

From a logistics point of view, the lack of availability of a biomarker assay with adequately defined analytical performance characteristics also limits the ability to embed a prospective pharmacogenomic hypothesis within a clinical study. Qualification of a biomarker is a complex process that requires a detailed understanding of the biological pathways of the disease, large and well-characterized patient populations, cost-effective and high-throughput genotyping techniques, and sophisticated computational methodologies. Such a process involves considerable time and resource investment. It is not always feasible to obtain validated biomarkers in the early stage of drug development in order for them to be incorporated in late-phase trials for clinical validation and qualification. Last, when a biomarker is used for patient stratification/selection in clinical trials, companion diagnostic kit development is mandated by health authorities, and full premarket approval may be required for a codevelopment product. However, at present, the paths for drug and diagnostic codevelopment are not very clear, and many companies are proceeding with codevelopment programs and seeking advice from health authorities on a case-by-case basis [85].

FUTURE PERSPECTIVE

Pharmacogenomics hold the promise to aid drug development in a more efficient and effective manner. Although this promise seems still far away, there is clear evidence that the conventional "one-size-fits-all" model is eroding in favor of a new paradigm with biomarker-assisted drug development. Like many other new technologies, incorporating pharmacogenomics into drug development is associated with a number of challenges; some of them are being addressed, and others still need to be addressed. The following strategies may be considered to overcome some of the challenges.

When the mechanism of action of a drug is known, systematic pharmacogenomic investigation of drug targets and related pathways may be performed during the drug development program. When the mechanism of action is unknown, and a hypothesis is lacking, data-driven approaches such as genome-wide association studies and whole-genome/exome sequencing may be considered.

From the study design perspective, in situations in which it is not practical or ethical to conduct a new prospective clinical study, prospective–retrospective analysis of biomarkers may provide a viable alternative option for generating data to support regulatory submission and/or labeling [86]. The utility of this approach was demonstrated in the example of KRAS as a negative selection biomarker for the anti-EGFR monoclonal antibodies (mAbs), cetuximab, and panitumumab. KRAS mutation is present in 30 to 50% of colorectal cancers. Exploratory analyses of single-arm clinical studies suggested that tumors harboring mutant KRAS were predictive of lack of efficacy for the anti-EGFR mAbs, cetuximab, and panitumumab. However, at the time these data became available, pivotal Phase III studies of these drugs in first-line and refractory metastatic colorectal cancer were already underway or fully recruited. Subsequently, a systemic evaluation of the utility of KRAS as a predictive biomarker by using banked samples obtained from completed Phase III trials was performed, and the results supported the KRAS hypothesis and labeling claim [87,88].

With the idea that many new treatments are designed for specific targets and because personalized medicine is focused on patient subpopulations, novel trial designs—such as those with smaller sample sizes and faster enrollment—may facilitate availability of new treatment. Interestingly, a recent simulation study of a series of trial designs comparing the superiority of a novel treatment to standard of care suggested that the current risk-averse trial design strategy may slow long-term progress. When a disease is rare and the trial accrual is small or the trial accrual rate cannot be increased, a strategy of conducting more trials with smaller sample sizes and relaxed evidential criteria compared with those required under traditional trial design may contribute to important gains in cancer patient survival [88]. Although confirmative large-scale clinical trial with sufficient statistical power continues to remain the gold standard for drug approval, smaller and faster trials with targeted therapy maybe useful to provide earlier insight into whether a therapy is efficacious and whether a confirmative trial is warranted.

Improved investment in preclinical efforts can also facilitate understanding of the mechanism of action of a drug early in development and consequently aid the identification of relevant genetic markers for efficacy or safety. As the genetic data set becomes more and more complex, exploration of sophisticated analytical approaches (i.e., multilocus pathway analysis and multivariate analysis) in combination with bioinformatics and biostatistics tools will be important to facilitate the biomarker qualification.

Collaborations between industry, academia, and regulatory agencies are also essential for improving the success of applying pharmacogenomics in drug development. Several examples discussed in this chapter have highlighted the importance of precise understanding of disease biology and genetics in implementing a personalized approach in drug therapy, reorientation of clinical development with a focus on genetically stratified trials, availability of validated real-time genetic diagnostics, and the establishment of a network for molecular screening of patients. Achievement of these would be impossible without a close interaction between scientists from industry, academia, and regulatory agencies.

The U.S. FDA is encouraging pharmacogenomic work and has taken a wide range of initiatives to put a regulatory framework around the integration of pharmacogenomics into drug development. These are discussed in chapter 3. Briefly, these have included issuing white papers and guidances, establishing the voluntary genomic data submission process, organizing workshops, and developing online tools. The voluntary genomic data submission program has been shown to lead to mutually beneficial and effective interactions between sponsors and the FDA, and in some cases, it has positively affected the sponsors' subsequent regulatory applications [89]. The European Medicines Agency (EMA) has also been active in this area. Several EU regulatory guidances on EU experience in oncology, inclusion of pharmacogenomics in pharmacokinetic/pharmacodynamic studies, and codevelopment of drugs/companion diagnostics were released in recent years. Scientific advice and protocol assistance were made available to sponsors/developers through Scientific Advice Working Party (SAWP) or Pharmacogenomics Working Party (PGxWP). Recognizing the predictability and applicability issues of biomarkers in drug development, the Committee of Medicinal Products (CHMP), in collaboration with the SAWP and PGxWP, made efforts to provide guidance on the biomarker qualification process (91). In spite of these efforts, there is still a lack of clear guidance on the paths for codevelopment of drugs and diagnostics. Many companies are proceeding with a codevelopment program and seeking regulatory advice on a case-by-case basis. The standards required to validate genotype–phenotype associations in order to determine optimal dosing for updating product labels remain debatable. Additionally, harmonizing regulatory frameworks among different countries/regions is necessary to enable the drug development process to proceed efficiently worldwide.

DISCUSSION QUESTIONS

1. How would pharmacogenomic investigations benefit different phases of drug development?
2. What are the barriers for the pharmaceutical industry to fully incorporate pharmacogenomics?
3. Are there any instances in which pharmacogenomic investigations provide no value for the industry?
4. How could regulatory agencies further advance the incorporation of pharmacogenomic investigations into drug development?

References

[1] Kaitin KI. Deconstructing the drug development process: the new face of innovation. Clin Pharmacol Ther 2010;87:356–61.

[2] Deininger MW, Goldman JM, Melo JV. The molecular biology of chronic myeloid leukemia. Blood 2000;96:3343–56.

[3] Goldman JM, Melo JV. BCR-ABL in chronic myelogenous leukemia—how does it work? Acta Haematol 2008;119:212–7.

[4] Deininger M, O'Brien SG, Guihot F, Goldman JM, Hochhaus A, Hughes TP, et al. International randomized study of interferon vs. STI571 (IRIS) 8-year follow up: sustained survival and low risk for progression or events in newly diagnosed chronic myeloid leukemia in chronic phase treated with imatinib. Blood 2009;114:1126.

[5] Hirota S, Isozaki K, Moriyama Y, Hashimoto K, Nishida T, Ishiguro S, et al. Gain-of-function mutations of c-kit in human gastrointestinal stromal tumors. Science 1998;279:577–80.

[6] Slamon DJ, Clark GM, Wong SG, Levin WJ, Ullrich A, McGuire WL. Human breast cancer: correlation of relapse and survival with amplification of the HER-1/heu oncogene. Science 1987;235:177–82.

[7] Slamon DJ, Godolphin W, Jones LA, Holt JA, Wong SG, Keith DE, et al. Studies of the HER-2/neu proto-oncogene in human breast and ovarian cancer. Science 1989;244:707–12.

[8] Carter P, Presta L, Gorman CM, Ridgway JB, Henner D, Wong WL, et al. Humanization of an anti-p185[HER2] antibody for human cancer therapy. Proc Natl Acad Sci USA 1992:4285–9.

[9] Gallagher WM, Tweats D, Koenig J. Omic profiling for drug safety assessment: current trends and public-private partnerships. Drug Discov Today 2009;14:337–42.

[10] Woosley RL, Myers RT, Goodsaid F. The Critical Path Institute's approach to precompetitive sharing and advancing regulatory science. Clin Pharmacol Ther 2010;87:530–3.

[11] Hoffmann D, Adler M, Vaidya VS, Rached E, Mulrane L, Gallagher WM, et al. Performance of novel kidney biomarkers in preclinical toxicity studies. Toxicol Sci 2010;116:8–22.

[12] Adler M, Hoffmann D, Ellinger-Ziegelbauer H, Hewitt P, Matheis K, Mulrane L, et al. Assessment of candidate biomarkers of drug-induced hepatobiliary injury in preclinical toxicity studies. Toxicol Lett 2010;196:1–11.

[13] Goodsaid FM, Mendrick DL. Translational medicine and the value of biomarker qualification. Sci Transl Med 2010;2:47ps44.

[14] Sauer JM, Ponsler GD, Mattiuz EL, Long AJ, Witcher JW, Thomasson HR, et al. Disposition and metabolic fate of atomoxetine hydrochloride: the role of CYP2D6 in human disposition and metabolism. Drug Metab Dispos 2003;31:98–107.

[15] Strattera [package insert]. 2011. Eli Lilly and Company, Indianapolis, IN, USA.

[16] Wakeling AE, Guy SP, Woodburn JR, Ashton SE, Curry BJ, Barker AJ, et al. ZD1839 (Iressa): an orally active inhibitor of epidermal growth factor signaling with potential for cancer therapy. Cancer Res 2002;62:5749–54.

[17] Ranson M, Hammond LA, Ferry D, Kris M, Tullo A, Murray PI, et al. ZD1839, a selective oral epidermal growth factor receptor-tyrosine kinase inhibitor, is well tolerated and active in patients with solid, malignant tumors: results of a phase I trial. J Clin Oncol 2002;20:2240–50.

[18] Fukuoka M, Yano S, Giaccone G, Tamura T, Nakagawa K, Douillard JY, et al. Multi-institutional randomized phase II trial of gefitinib for previously treated patients with advanced non-small-cell lung cancer. J Clin Oncol 2003;21:2237–46.

[19] Kris MG, Natale RB, Herbst RS, Lynch Jr TJ, Prager D, Belani CP, et al. Efficacy of gefitinib, an inhibitor of the epidermal growth factor receptor tyrosine kinase, in symptomatic patients with non-small cell lung cancer: a randomized trial. J Am Med Assoc 2003;290:2149–58.

[20] Thatcher N, Chang A, Parikh P, Rodrigues PJ, Ciuleanu T, von Pawel J, et al. Gefitinib plus best supportive care in previously treated patients with refractory advanced non-small-cell lung cancer: results from a randomised, placebo-controlled, multicentre study (Iressa Survival Evaluation in Lung Cancer). Lancet 2005;366. 1527–1237.

[21] Hirsch FR, Varella-Garcia M, Bunn Jr PA, Franklin WA, Dziadziuszko R, Thatcher N, et al. Molecular predictors of outcome with gefitinib in a phase III placebo-controlled study in advanced non-small-cell lung cancer. J Clin Oncol 2006;24:5034–42.

[22] Mok TS, Wu YL, Thongprasert S, Yang CH, Chu DT, Saijo N, et al. Gefitinib or carboplatin-paclitaxel in pulmonary adenocarcinoma. N Engl J Med 2009;361:947–57.

[23] Mitsudomi T, Morita S, Yatabe Y, Negoro S, Okamoto I, Tsurutani J, et al. Gefitinib versus cisplatin plus docetaxel in patients with non-small-cell lung cancer harbouring mutations of the epidermal growth factor receptor (WJTOG3405): an open label, randomised phase 3 trial. Lancet Oncol 2010;11:121–8.

[24] Maemondo M, Inoue A, Kobayashi K, Sugawara S, Oizumi S, Isobe H, et al. Gefitinib or chemotherapy for non-small-cell lung cancer with mutated EGFR. N Engl J Med 2010;362:2380–8.

[25] Spigel DR, Hainsworth JD, Burkett ER, Burris HA, Yardley DA, Thomas M, et al. Single-agent gefitinib in patients with untreated advanced non-small-cell lung cancer and poor performance status: a Minnie Pearl Cancer Research Network Phase II Trial. Clin Lung Cancer 2005;7:127–32.

[26] Inoue A, Suzuki T, Fukuhara T, Maemondo M, Kimura Y, Morikawa N, et al. Prospective phase II study of gefitinib for chemotherapy naive patients with advanced non-small-cell lung cancer with epidermal growth factor receptor gene mutations. J Clin Oncol 2006;24:3340–6.

[27] Lee DH, Han JY, Yu SY, Kim HY, Nam BH, Hong EK, et al. The role of gefitinib treatment for Korean never-smokers with advanced or metastatic adenocarcinoma of the lung: a prospective study. J Thorac Oncol 2006;1:965–71.

[28] Niho S, Kubota K, Goto K, Yoh K, Ohmatsu H, Kakinuma R, et al. First-line single agent treatment with gefitinib in patients with advanced non-small-cell lung cancer: a phase II study. J Clin Oncol 2006;24. 64–9.

[29] Reck M, Buchholz E, Romer KS, Krutzfeldt K, Gatzemeier U, Manegold C. Gefitinib monotherapy in chemotherapy-naive patients with inoperable stage III/IV non-small-cell lung cancer. Clin Lung Cancer 2006;7:406—11.

[30] Suzuki R, Hasegawa Y, Baba K, Saka H, Saito H, Taniguchi H, et al. A phase II study of single-agent gefitinib as first-line therapy in patients with stage IV non-small-cell lung cancer. Br J Cancer 2006;94:1599—603.

[31] Sequist LV, Martins RG, Spigel D, Grunberg SM, Spira A, Jänne PA, et al. First-line gefitinib in patients with advanced non-small-cell lung cancer harboring somatic EGFR mutations. J Clin Oncol 2008;26:2442—9.

[32] Tamura K, Okamoto I, Kashii T, Negoro S, Hirashima T, Kudoh S, et al, West Japan Thoracic Oncology Group. Multicentre prospective phase II trial of gefitinib for advanced non-small cell lung cancer with epidermal growth factor receptor mutations: results of the West Japan Thoracic Oncology Group trial (WJTOG0403). Br J Cancer 2008;98:907—14.

[33] Yang CH, Yu CJ, Shih JY, Chang YC, Hu FC, Tsai MC, et al. Specific *EGFR* mutations predict treatment outcome of stage IIIB/IV patients with chemotherapy-naive non-small-cell lung cancer receiving first-line gefitinib monotherapy. J Clin Oncol 2008;26:2745—53.

[34] Inoue A, Kobayashi K, Usui K, Maemondo M, Okinaga S, Mikami I, et al, North East Japan Gefitinib Study Group. First-line gefitinib for patients with advanced non-small-cell lung cancer harbouring epidermal growth factor receptor mutations without indication for chemotherapy. J Clin Oncol 2009;27:1394—400.

[35] Lee JS, Park K, Kim SW, Lee DH, Kim HT, Han JY, et al. A randomized phase III study of gefitinib (IRESSATM) versus standard chemotherapy (gemcitabine plus cisplatin) as a first-line treatment for never-smokers with advanced or metastatic adenocarcinoma of the lung [abstract]. J Thor Oncol 2009;4(Suppl. 1):S283.

[36] Kumar A, Petri ET, Halmos B, Boggon TJ. Structural and clinical relevance of the epidermal growth factor receptor in human cancer. J Clin Oncol 2008;26:1742—51.

[37] Giaccone G, Herbst RS, Manegold C, Scagliotti G, Rosell R, Miller V, et al. Gefitinib in combination with gemcitabine and cisplatin in advanced non-small-cell lung cancer: a phase III trial—INTACT 1. J Clin Oncol 2004;22:777—84.

[38] Herbst RS, Giaccone G, Schiller JH, Natale RB, Miller V, Manegold C, et al. Gefitinib in combination with paclitaxel and carboplatin in advanced non-small-cell lung cancer: a phase III trial—INTACT 2. J Clin Oncol 2004;22:785—94.

[39] MacArthur RD, Novak RM. Reviews of anti-infective agents: maraviroc: the first of a new class of antiretroviral agents. Clin Infect Dis 2008;47:236—41.

[40] Liu R, Paxton WA, Choe S, Ceradini D, Martin SR, Horuk R, et al. Homozygous defect in HIV-1 coreceptor accounts for resistance of some multiply-exposed individuals to HIV-1 infection. Cell 1996;86:367—77.

[41] Samson M, Libert F, Doranz BJ, Rucker J, Liesnard C, Farber CM, et al. Resistance to HIV-1 infection in Caucasian individuals bearing mutant alleles of the CCR-5 chemokine receptor gene. Nature 1996;382:722—5.

[42] Dorr P, Westby M, Dobbs S, Griffin P, Irvine B, Macartney M, et al. Maraviroc (UK-427,857), a potent, orally bioavailable, and selective small-molecule inhibitor of chemokine receptor CCR5 with broad-spectrum anti-human immunodeficiency virus type 1 activity. Antimicrob Agents Chemother 2005;49:4721—32.

[43] Wheeler J, McHale M, Jackson V, Penny M. Assessing theoretical risk and benefit suggested by genetic association studies of CCR5: experience in a drug development programme for maraviroc. Antiviral Ther 2006;12:233—45.

[44] Fätkenheuer G, Nelson M, Lazzarin A, Konourina I, Hoepelman AI, Lampiris H, et al. Subgroup analyses of maraviroc in previously treated R5 HIV-1 infection. N Engl J Med 2008;359:1442—55.

[45] Moyle GJ, DeJesus E, Cahn P, Castillo SA, Zhao H, Gordon DN, et al. Abacavir once or twice daily combined with once-daily lamivudine and efavirenz for the treatment of antiretroviral-naive HIV-infected adults: results of the Ziagen Once Daily in Antiretroviral Combination Study. J Acquir Immune Defic Syndr 2005;38:417—25.

[46] Staszewski S, Keiser P, Montaner J, Raffi F, Gathe J, Brotas V, et al. Abacavir-lamivudine-zidovudine vs. indinavir-lamivudine-zidovudine in antiretroviral-naive HIV-infected adults: a randomized equivalence trial. J Am Med Assoc 2001;285:1155—63.

[47] Hewitt RG. Abacavir hypersensitivity reaction. Clin Infect Dis 2002;34:1137—42.

[48] Symonds W, Cutrell A, Edwards M, Steel H, Spreen B, Powell G, et al. Risk factor analysis of hypersensitivity reactions to abacavir. Clin Ther 2002;24:565—73.

[49] Mallal S, Nolan D, Witt C, Masel G, Martin AM, Moore C, et al. Association between presence of HLA-B*5701, HLA-DR7, and HLA-DQ3 and hypersensitivity to HIV-1 reverse-transcriptase inhibitor abacavir. Lancet 2002;359:727—32.

[50] Hetherington S, Hughes AR, Mosteller M, Shortino D, Baker KL, Spreen W, et al. Genetic variations in HLA-B region and hypersensitivity reactions to abacavir. Lancet 2002;359:1121—2.

[51] Mallal S, Phillips E, Carosi G, Molina JM, Workman C, Tomazic J, et al. HLA-B*5701 screening for hypersensitivity to abacavir. N Engl J Med 2008;358:568—79.

[52] Ziagen [package insert]. 2010. GlaxoSmithKline, Research Triangle Park, NC, USA.

[53] Slatter JG, Su P, Sams JP, Schaaf LJ, Wienkers LC. Bioactivation of the anticancer agent CPT-11 to SN-38 by human hepatic microsomal carboxylesterases and the in vitro assessment of potential drug interactions. Drug Metab Dispos 1997;25:1157—64.

[54] Humerickhouse R, Lohrbach K, Li L, Bosron WF, Dolan ME. Characterization of CPT-11 hydrolysis by human liver carboxylesterase isoforms hCE-1 and hCE-2. Cancer Res 2000;60:1189—92.

[55] Takasuna K, Hagiwara T, Hirohashi M, Kato M, Nomura M, Nagai E, et al. Inhibition of intestinal microflora beta-glucuronidase modifies the distribution of the active metabolite of the antitumor agent, irinotecan hydrochloride (CPT-11) in rats. Cancer Chemother Pharmacol 1998;42:280—6.

[56] Iyer L, Das S, Janisch L, Wen M, Ramírez J, Karrison T, et al. UGT1A1*28 polymorphism as a determinant of irinotecan disposition and toxicity. Pharmacogenomics J 2002;2:43—7.

[57] Mathijssen RH, Marsh S, Karlsson MO, Xie R, Baker SD, Verweij J, et al. Irinotecan pathway genotype analysis to predict pharmacokinetics. Clin Cancer Res 2003;9:3246—53.

[58] Paoluzzi L, Singh AS, Price DK, Danesi R, Mathijssen RH, Verweij J, et al. Influence of genetic variants in UGT1A1 and UGT1A9 on the in vivo glucuronidation of SN-38. J Clin Pharmacol 2004;44:854—60.

[59] Sai K, Saeki M, Saito Y, Ozawa S, Katori N, Jinno H, et al. UGT1A1 haplotypes associated with reduced glucuronidation and increased serum bilirubin in irinotecan-administered Japanese patients with cancer. Clin Pharmacol Ther 2004;75:501—15.

[60] Rothenberg ML. Efficacy and toxicity of irinotecan in patients with colorectal cancer. Semin Oncol 1998;25:39—46.

[61] Innocenti F, Undevia SD, Iyer L, Chen PX, Das S, Kocherginsky M, et al. Genetic variation in the UDP-glucuronosyltransferase 1A1 gene predict the risk of severe neutropenia of irinotecan. J Clin Oncol 2004;22:1382—8.

[62] McLeod HL, Parodi L, Sargent DJ, Marsh S, Green E, Abreu P, et al. UGT1A1*28, toxicity and outcome in advanced colorectal cancer: results from trial N9741. J Clin Oncol 2006;24(18S):3520.

[63] Camptosar [package insert]. 2008. Pfizer Inc., New York, NY, USA.

[64] Hochhaus A, Kreil S, Corbin AS, La Rosée P, Müller MC, Lahaye T, et al. Molecular and chromosomal mechanisms of resistance to imatinib (STI571) therapy. Leukemia 2002;16:2190—6.

[65] O'Hare T, Eide CA, Deininger MW. Bcr-Abl kinase domain mutations, drug resistance, and the road to a cure for chronic myeloid leukemia. Blood 2007;110:2242—9.

[66] Thomas J, Wang L, Clark RE, Pirmohamed M. Active transport of imatinib into and out of cells: implications for drug resistance. Blood 2004;104:3739—45.

[67] White DL, Saunders VA, Dang P, Engler J, Zannettino AC, Cambareri AC, et al. OCT-1-mediated influx is a key determinant of the intracellular uptake of imatinib but not nilotinib (AMN107): reduced OCT-1 activity is the cause of low in vitro sensitivity to imatinib. Blood 2006;108:697—704.

[68] White DL, Dang P, Engler J, Frede A, Zrim S, Osborn M, et al. Functional activity of the OCT-1 protein is predictive of long-term outcome in patients with chronic-phase chronic myeloid leukemia treated with imatinib. J Clin Oncol 2010;28:2761—7.

[69] White DL, Radich J, Soverini S, Saunders VA, Frede A, Dang P, et al. Chronic phase chronic myeloid leukemia patients with low OCT-1 activity randomised to high-dose imatinib achieve better responses, and lower failure rates, than those randomized to standard-dose. Haematologica 2011;97:907—14.

[70] Hamada A, Miyano H, Watanabe H, Saito H. Interaction of imatinib mesilate with human P-glycoprotein. J Pharmacol Exp Ther 2003;307:824—8.

[71] Burger H, van Tol H, Boersma AW, Brok M, Wiemer EA, Stoter G, et al. Imatinib mesylate (STI571) is a substrate for the breast cancer resistance protein (BCRP)/ABCG2 drug pump. Blood 2004;104:2940—2.

[72] Peng B, Hayes M, Resta D, Racine-Poon A, Druker BJ, Talpaz M, et al. Pharmacokinetics and pharmacodynamics of imatinib in a phase I trial with chronic myeloid leukemia patients. J Clin Oncol 2004;22:935—42.

[73] Gardner ER, Burger H, van Schaik RH, van Oosterom AT, de Bruijn EA, Guetens G, et al. Association of enzyme and transporter genotypes with the pharmacokinetics of imatinib. Clin Pharmacol Ther 2006;80:192—201.

[74] Gurney H, Wong M, Balleine RL, Rivory LP, McLachlan AJ, Hoskins JM, et al. Imatinib disposition and ABCB1 (MDR1, P-glycoprotein) genotype. Clin Pharmacol Ther 2007;82:33–40.

[75] Takahashi N, Miura M, Scott SA, Kagaya H, Kameoka Y, Tagawa H, et al. Influence of CYP3A5 and drug transporter polymorphisms on imatinib trough concentration and clinical response among patients with chronic phase chronic myeloid leukemia. J Hum Genet 2010;55:731–7.

[76] Yamakawa Y, Hamada A, Nakashima R, Yuki M, Hirayama C, Kawaguchi T, et al. Association of genetic polymorphisms in the influx transporter SLCO1B3 and the efflux transporter ABCB1 with imatinib pharmacokinetics in patients with chronic myeloid leukemia. Ther Drug Monit 2011;33:244–50.

[77] Dulucq S, Bouchet S, Turcq B, Lippert E, Etienne G, Reiffers J, et al. Multidrug resistance gene (MDR1) polymorphisms are associated with major molecular responses to standard-dose imatinib in chronic myeloid leukemia. Blood 2008;112:2024–7.

[78] Petain A, Kattygnarath D, Azard J, Chatelut E, Delbaldo C, Geoerger B, et al. Population pharmacokinetics and pharmacogenetics of imatinib in children and adults. Clin Cancer Res 2008;14:7102–9.

[79] Hiwase DK, Saunders V, Hewett D, Frede A, Zrim S, Dang P, et al. Dasatinib cellular uptake and efflux in chronic myeloid leukemia cells: therapeutic implications. Clin Cancer Res 2008;14:3881–8.

[80] Giannoudis A, Davies A, Lucas CM, Harris RJ, Pirmohamed M, Clark RE. Effective dasatinib uptake may occur without human organic cation transporter 1 (hOCT1): implications for the treatment of imatinib-resistant chronic myeloid leukemia. Blood 2008;112:3348–54.

[81] Press and Seelig, Target Medicine, NYC, Nov 2004. http://www.fda.gov/downloads/Drugs/ScienceResearch/ResearchAreas/Pharmacogenetics/ucm085827.pdf. [accessed on 27.06.12].

[82] Bromley CM, Close S, Cohen N, Favis R, Fijal B, Gheyas F, et al, Industry Pharmacogenomics Working Group. Designing pharmacogenetic projects in industry: practical design perspectives from the Industry Pharmacogenomics Working Group. Pharmacogenomics J 2009;9:14–22.

[83] Warner AW, Bhathena A, Gilardi S, Mohr D, Leong D, Bienfait KL, et al. Challenges in obtaining adequate genetic sample sets in clinical trials: the perspective of the industry pharmacogenomics working group. Clin Pharmacol Ther 2011;89:529–36.

[84] Rogausch A, Prause D, Schallenberg A, Brockmöller J, Himmel W. Patients' and physicians' perspectives on pharmacogenetic testing. Pharmacogenomics 2006;7:49–59.

[85] Hinman LM, Carl KM, Spear BB, Salerno RA, Becker RL, Abbott BM, et al. Development and regulatory strategies for drug and diagnostic co-development. Pharmacogenomics 2010;11:1669–75.

[86] Patterson SD, Cohen N, Karnoub M, Truter SL, Emison E, Khambata-Ford S, et al. Prospective-retrospective biomarker analysis for regulatory consideration: white paper from the industry pharmacogenomics working group. Pharmacogenomics 2011;12:939–51.

[87] Amado RG, Wolf M, Peeters M, Van Cutsem E, Siena S, Freeman DJ, et al. Wild-type KRAS is required for panitumumab efficacy in patients with metastatic colorectal cancer. J Clin Oncol 2008;26:1626–34.

[88] Karapetis CS, Khambata-Ford S, Jonker DJ, O'Callaghan CJ, Tu D, Tebbutt NC, et al. *K-ras* mutations and benefit from cetuximab in advanced colorectal cancer. N Engl J Med 2008;359:1757–65.

[89] Amur S, Frueh FW, Lesko LJ, Huang SM. Integration and use of biomarkers in drug development, regulation and clinical practice: a US regulatory perspective. Biomark Med 2008;2:305–11.

[90] Prasad K, Breckenridge A. Pharmacogenomics: a new clinical or regulatory paradigm? European experiences of pharmacogenomics in drug regulation and regulatory initiatives. Drug Discov Today 2011;16:867–72.

The Importance of Ethnicity Definitions and Pharmacogenomics in Ethnobridging

Elsa Haniffah Mejia Mohamed*,†, *Lou Huei-xin***,
Jalene Poh‡, *Dorothy Toh*¶, *Edmund Jon Deoon Lee*†

*Pharmacogenomics Laboratory, Department of Pharmacology, University Malaya, Malaysia
†Pharmacogenetics Laboratory, Department of Pharmacology, Yong Loo Lin School of Medicine,
National University of Singapore, Singapore
**Premarket Division, Health Products Regulation Group,
Health Sciences Authority, Singapore
‡Pharmaceuticals and Biologics Branch, Health Products Regulation Group,
Health Sciences Authority, Singapore
¶Vigilance Branch, Health Products Regulation Group, Health Sciences Authority, Singapore

OBJECTIVES

1. Emphasize the complexity of defining ethnicity and the need for a standardized definition of scientific research.

2. Outline the ethnic intrinsic and extrinsic factors in determining outcome of treatment.

3. Identify the factors considered in acceptability of foreign clinical data for drug approval.

4. Highlight the role of pharmacogenomics in global drug development.

Pharmacogenomics
http://dx.doi.org/10.1016/B978-0-12-391918-2.00010-X

INTRODUCTION

The ethnicity effect on variability of pharmacological outcome is well established and has been described extensively. This is one of the major considerations for any local drug regulatory authority when considering applications for new drug approval using foreign clinical data in their region. Bridging studies were proposed to provide supplemental data on a drug's pharmacokinetics/pharmacodynamics, safety, efficacy, dosage, and dosage regimen in a new country or region to determine whether the foreign clinical data can be applicable to the new region. Discussions on the (de)merits of doing bridging studies based on ethnicity are not new. Many arguments have been put forth for various reasons, and one of the main reasons for dispute is that the terms "ethnicity" and "race" are very poorly defined. The International Conference on Harmonization of Technical Requirements for the Registration of Pharmaceuticals for Human Use (ICH) E5 guideline [1] was drafted to facilitate registration of medicines in ICH-affiliated countries by providing a framework to evaluate ethnic factors on the outcome of medical treatment(s). It was intended to hasten the drug approval process while minimizing clinical trial duplication with swift delivery of new medicines to needy patients. Although fully adopted by some countries in East Asia, not all have followed suit, especially in developing nations like the Southeast Asian nations, due to inadequacies in experience and resources. Even in the countries where it has been adopted, there are appreciable differences in the way the bridging concept has been applied. An approach that may be used as a solution to this dilemma is the multiregional clinical trials (MRCT) or the parallel bridging method. Apart from having the advantage of reducing the lag time in drug approval significantly, this approach allows for prospective bridging of the data in various regions. The integration of pharmacogenomics and biomarkers in drug development is also seen as a positive attempt to better characterize ethnic factors. In the end, it is hoped that better risk-benefit consideration for drug treatment can be achieved through more relevant population data.

In this chapter, the relevance of ethnic-based bridging studies is examined in three different parts. The first part reviews the difficulties in the definition of ethnicity and the ambiguous ways in which the term has been applied. Next, ethnic factors as defined by the ICH E5 contributing to variability in drug response are examined, with emphasis on the effects of diet and use of herbal medicines in various population groups. Finally, approaches by different nations to acceptability of foreign clinical data for drug marketing approval are reviewed.

ETHNICITY

Ethnicity is a sociocultural construct with very vague scientific definitions. On many occasions, however, ethnicity becomes politically defined or nationality based rather than representing true ethnicity. The tendency to overgeneralize ethnic population groups also reduces the validity of many analyses based on ethnic stratification.

The relationship between self-identified race or ethnicity and disease risk has been depicted as a series of surrogate relationships between genetic and nongenetic factors [2]. The nongenetic component includes social, cultural, educational, and economic variables, all of which can influence disease risk. The distinction between the terms race and ethnicity is also very controversial and varied, as highlighted in an editorial: Census, Race and Science published in the year 2000 [3]. Several examples of dictionary definitions of race and ethnicity were exemplified as follows:

Race:

- A vague, unscientific term for a group of genetically related people who share certain physical characteristics
- A distinct ethnic group characterized by traits that are transmitted through their offspring
- Each of the major divisions of humankind having distinct physical characteristics
- A group of individuals who are more or less isolated geographically or culturally, who share a common gene pool, and whose allele frequencies at some loci differ from those of other populations

Ethnic group:

- A population of individuals organized around an assumption of common cultural origin
- Individuals with a common national or cultural tradition
- A social group or category of the population that, in a larger society, is set apart and bound together by commonalities of race, language, nationality, or culture

Although the definitions are ambiguous, it can be deduced that race and ethnicity give an insight to cultural, historical, and perhaps socioeconomic and political status, as well as ancestral geographic origins. Because race and ethnicity are different mixtures of biological with social constructs, they are highly dynamic in nature.

Defining race and ethnicity accurately is obviously difficult, given their dynamic nature and the fact that the genetic pool for a population would most likely be heterogeneous in nature. This issue makes any assumptions about purity and accuracy of the definitions fallacious.

Ethnicities Are Rarely Homogeneous in Any Nation

There is often a presumption that certain nations are ethnically homogeneous in nature and that extrapolation of clinical data can be made to certain nations this way. In most circumstances, such references to population groups are not actually ethnicity based but rather nation based. Even though this is understandable because regulatory agencies are nation-based entities, such practices are fundamentally erroneous, given the ethnically heterogeneous nature of almost all nations. Some examples are discussed in this section.

Japan

The word "Japanese" is used collectively for three ethnic groups known as the Yamato (Hondo-Japanese, or mainland Japanese), Ainu, and Ryukyuan (Okinawan). Even in the national census [4], actual ethnicity (minzoku) is not measured but rather Japanese nationality (kokuseki). However, the Ainu people in particular, who are regarded as the aborigines of Japan living in Hokkaido, differ from the rest of the broader Japanese group physically, linguistically, and culturally [5]. Genetic studies evaluating ancestral origins of the Japanese people have shown that they originate from two distinct groups: the Ainu and Ryukyuan populations are direct descendants of the Neolithic Jomon people, and the Hondo people are derived from the northeast of continental Asia [6]. Ancient mitochondrial analyses suggested that the gene flow was from South Eastern Siberia to the Jomon/Epi-Jomon people of Hokkaido, Sakhalin, and the Kuril archipelago, with the Okhotsk people being intermediaries [7]. Phylogenetic analyses comparing the three ethnic groups also suggested that the Ainu and Ryukyuan samples are clustered together, and the Hondo-Japanese and Koreans were clustered together in the neighbor-joining genetic tree [6].

When a Japanese person leaves Japan and migrates to some other country, the definition of Japanese assumes a slightly different context. This issue is important to recognize, especially if recruiting for bridging studies. The Japanese then will be categorized as [8]:

- First-generation Japanese (Issei): Subject born in Japan
- Second-generation (Nissei): Subject born elsewhere, both parents born in Japan
- Third-generation (Sansei): Subject and one or both parents born elsewhere, grandparents born in Japan

To be eligible for the studies, a Japanese person must have at least all four grandparents born in Japan with no mixed descent. Thus, a Japanese person of up to the third generation would be eligible to represent the Japan—Japanese populations [8].

China

Although Han Chinese is the largest ethnic group in China, there are 55 officially recognized ethnic minority groups [9] totaling about 105 million people in China. An equally diverse number of languages—up to 200—are being spoken in China, from seven linguistic families: Altaic, Austroasiatic, Austronesian, Daic, Hmong Mien, Sino-Tibetan, and Indo-European [9]. Furthermore, even within the Han Chinese ethnic group, there is significant genetic heterogeneity.

Han Chinese is possibly the largest ethnic group in the world, making up about 20% of the human population. It is also the most prevalent ethnic group in China, accounting for more than 90% of its total population [10]. Although the Han people are now spread all over the country, the formation of the Han people began with the ancient Huaxia tribe in northern China, which spread southward over 200 years ago [9]. Expansion of the Han ethnic group is the result of integration of multiple tribes and ethnic groups [11]. Studies on Han-Chinese mitochondrial DNA (mtDNA) from six provinces in China have suggested interesting variations within the Han population with obvious geographical differentiation, primarily on a north—south axis [11]. As a consequence, it may be necessary to cluster the Han Chinese population according to geographic origins. Today, there are several subgroups of Han Chinese, speaking different dialects or, arguably, different languages and in fact living in different parts of the world. It is uncertain whether a clinical study done with Han Chinese from the northern part of China can adequately represent the rest of the global Chinese ethnic group. Similarly, it is uncertain that clinical data derived from overseas Chinese who are primarily derived from a southern Chinese group can be taken to represent a generic Han Chinese population in China.

Apart from the heterogeneity in the Han Chinese, there is evidence of west Eurasian and northeast Asian (along the Silk Road) genetic admixture with the northwest Chinese populations [12]. It is further confirmed in later mtDNA studies from ethnic groups in Xinjiang that Central Asia is the main location of genetic admixture between east and west [13].

Malaysia

Malaysia is a multiethnic nation, consisting of Malays (42.1%), Chinese (24.6%), Indians (7.4%), and native Sabah and Sarawak (24.8%). The majority of the Sabah and Sarawak native ethnic groups are the Iban and Kadazan/Dusun ethnic groups [14].

Similar to the Han Chinese in China, heterogeneity can be observed within the Malays in the Peninsular Malaysia itself. The Malay race has been defined as members of indigenous people inhabiting the Malay archipelago and nearby islands, which consists of Malaysia, Singapore, Indonesia, and the Philippines. The migration history of the Malays suggests the Malays have several ancestral origins [15]. The Melayu Bugis (Bugis Malay) and Melayu Jawa (Javanese Malay), who are mainly in the southern peninsula, as well as the Melayu Minang (Minang Malay) in the western peninsula, are historically and culturally related to Indonesia (Sumatra and Java), and the Melayu Kelantan are related to Thailand (Siam). In fact, the states of Kelantan and Terengganu were part of the ancient Siam kingdom, and the rest of the Malay peninsula were part of the Majapahit and then Srivijaya empire, which were strongly influenced by Hinduism and Buddhism, in contrast to the Malacca empire, which was essentially an Islamic empire with Arab influences [16]. Genetic admixture with the Chinese and Indians also occurred when they were brought in large scale by the British colonizing Peninsular Malaysia. These are the main ethnic population groups that may play important parts in the genetic heterogeneity in the Malays [17] of Peninsular Malaysia. Genetic differences among the four Malay subethnic groups within Peninsular Malaysia were studied by Hatin et al., using 54,794 genome-wide single-nucleotide polymorphisms (SNPs) from the four Malay subethnic groups and compared to the genetic profile of 11 other populations' data obtained from the Pan Asian database [17].

The study showed that the Kelantanese Malays were genetically distinct from the other three groups of Malays, who showed high resemblance to the Indonesian Malays. Not surprisingly, the results showed that the Malays could be assigned to three different clusters, with the Melayu Minang and Melayu Bugis in Cluster I, Melayu Jawa in Cluster II, and Melayu Kelantan in Cluster III. The Melayu Kelantan formed an independent clade, suggesting a more divergent ancestry compared to the other two clusters.

There are also instances of political and religious bias in the definition of ethnicity. Take, for example, the definition of the Malay ethnicity in Malaysia. The Malaysian Constitution (article 160) [18] defines a Malay as:

- Malaysian citizen born to a Malaysian citizen
- Professes to be Muslim
- Habitually speaks the Malay language
- Adheres to Malay customs
- Resides in Malaysia or Singapore

It is obvious that the Malay ethnicity definition in Malaysia is a completely sociocultural and political construct and does not have any biological reference whatsoever. Apart from that, a Malay person in Malaysia is no longer considered Malay by law if the person converts out of Islam.

It should be noted, though, that only the difficulties in defining the Malays have been described here. Any attempts to carry out ethnobridging studies in Malaysia must also take into account the other ethnic populations, as described earlier. Although the situation in neighboring Singapore is not discussed here, it should be noted that the ethnic definition of Malay differs between the two countries and complicates attempts to generalize ethnic data from one country to another.

Indonesia

Indonesia is the fourth most populous country in the world. It is made up of 17,000 islands and is home to more than 240 million people with an immensely diverse admixture of more than 750 languages and 300 ethnic groups [19]. To promote nationalism and sense of unity in the postcolonization era, the *Pancasila* ideology was introduced in 1945, which underlines the national identity as culturally neutral, along with the use of *Bahasa Indonesia* (the Indonesian language) by all people in Indonesia. National census between 1961 and 1990 was devoid of ethnicity data because ethnicity and race were deemed too sensitive to be discussed or recorded [20]. The first ethnicity data were recorded only in the year 2000 census. According to this census, there are more than 100 self-identified ethnic groups in Indonesia. However, most of these ethnic groups are small in number, and only 15 groups have more than one million people [20]. The majority of the Indonesian people are of the Javanese ethnic group, which constitutes 41.7% of the total population. Table 10.1 lists the major ethnic groups in Indonesia, which makes up more than 81% of the total population percentage.

The majority of the nonnative Indonesians are of Chinese ethnicity. In the year 2000 census, it was reported that less than 1% of the population described themselves as Indonesian Chinese. It is believed that the figure was a gross underestimate and that the real figure was perhaps three to four times larger [21]. With the advent of assimilation of the Chinese with the native Indonesians, there is doubt that even the 1% self-reported Chinese were homogeneous, with no genetic admixture with the native Indonesians.

TABLE 10.1   List of major ethnic groups in indonesia [20]

Ethnic group	Percentage
Javanese	41.71
Sundanese	15.41
Malay	3.47
Madurese	3.37
Batak	3.02
Minangkabau	2.72
Betawi	2.51
Buginese	2.49
Bantenese	2.05
Banjarese	1.74
Balinese	1.51
Sasak	1.30

Thailand

Thailand became the official name for the kingdom once known as Siam in 1939. It is often perceived as unique and homogeneous in culture and ethnicity [22]. The 2000 census collected spoken language and religion data but not ethnicity data [23]. The official demographics report stated that 94% of the population are Thai-speaking Buddhists and 4% are Muslims [24]. In Thailand, there are over 30 distinct ethnic groups, including the Chinese, who migrated into Thailand in the nineteenth century to form significant urban communities [22]. Many have become significant political and economic figures and assimilated into the Thai society. Due to the lack of official figures, it is exceptionally difficult to get a reliable estimate of the actual numbers of ethnic groups and their composition in Thailand. The ethnic groups in Thailand can be categorized into five large groups based on the language groups spoken, as below [25]. It is not clear to what extent language groups reflect different genetic heritages, but this population diversity needs to be acknowledged.

- Tai-Kadai [26]
  - Yuan, Lue, Khuen, Yong, Thai-Korat, Thai-Khon Kaen, Thai-Chiang Mai, Phu-Thai, Lao-Song
- Austroasiatic [26]
  - Mon, Lawa, Paluang, Blang, H'tin, Khmer, Ching, Thai-Korat
- Sino-Tibetan [26]
  - Lisu, Mussur, Han-Yunnan, Han-Guangdong, Han-Wuhan, Han-Qingdao, Han-Liaoning, Han-Xinjiang, Tibetan-Qinghai
- Hmong-Mien [25]
  - Hmong, Mien
- Austronesian [25]
  - Malay, Cham

Ethnicity and Race as Defined by the U.S. FDA

There is a pressing need for standardization of terminologies for the collection of ethnicity and race information in biomedical research. Differences in response to medical products have also been observed in racially and ethnically distinct subgroups of the U.S. population. The U.S. Food and Drug Administration (FDA), mandated by the National Institutes of Health (NIH), prepared the *Guideline for Industry: Collection of Race and Ethnicity Data in Clinical Trials* [27], which is relevant in determining the safety and effectiveness of a drug or medical product, as well as addressing the issue of lack of inclusion of woman and minority groups in NIH-sponsored clinical research. However, much criticism has arisen over the racial and ethnic categories, as they are not anthropological or scientifically based designations but sociocultural categories as described by the Office of Management and Budget (OMB). In this guideline there are

- Five race categories: American Indian or Alaska Native, Asian, Black or African American, Native Hawaiian or other Pacific Islander, and White
- Two ethnicity categories: Hispanic or Latino and Not Hispanic or Latino

The OMB standards also mention explicitly that "the racial and ethnic categories set forth in the standard should not be interpreted as being primarily biological or genetic in reference" [28]. Thus, when these categories are used in a defined biological or genetic context, it creates confusion in the biological and sociocultural meanings of race and ethnicity [28]. The broad categorization of race—for example, Asian—is also arbitrary, as it combines very heterogeneous groups of people together (although the guideline does allow for more detailed information such as Japanese or Indian). Likewise, "White" is defined as a person having descent from any of the original peoples of Europe, North Africa, or the Middle East [8]. This definition includes Scottish, Greek, Welsh English, Moroccan, and Iranian, which completely abandons ethnic or cultural definitions altogether, instead grouping these populations together based on skin color.

The rapid rise in immigration and interracial marriages are on the increase, not only in the USA but also globally. This development creates populations with a very wide geographical and sociocultural background. If ethnicity and race were to serve, to a certain extent, as a surrogate for genetic variations, grouping people along racial and ethnic lines to study functional differences in their drug metabolizing capacity or to interpret results of research and clinical trials relating to various drugs has proven to be complex and challenging [29].

Conclusion

Ethnicity is a highly ambiguous and imprecise sociocultural and sociopolitical construct. Careful considerations should be given so as to not confuse ethnicity and nationality definitions because doing so could lead to fallaciously drawn conclusions, especially when considering risks and benefits of pharmacological treatment based on more biologically related processes. A useful way to describe ethnicity in scientific research would perhaps be to combine elements of geographic origins and the sociocultural context. For example, an ethnic group could be identified as Indonesian Chinese, Malaysian Chinese, American Chinese, and so on. This dual element in defining ethnicity would be better able to take into consideration the *extrinsic* components of ethnicity interacting with the *intrinsic* components.

ETHNIC FACTORS AFFECTING DRUG RESPONSE

The definition of ethnicity has both elements of biology and the environment. These have been described in the ICH E5 document as intrinsic and extrinsic factors [1]. Figure 10.1 summarizes the intrinsic and extrinsic factors, as outlined in the ICH E5 guidelines. It is often difficult to ascertain which of the intrinsic or extrinsic factors are causing differences in drug response. All intrinsic and extrinsic factors play important roles in influencing pharmacological outcome to treatment. In the following section, the genetics aspect of the internal factor and the food and traditional medication intake as well as medical practices aspect of extrinsic factors are discussed.

Intrinsic Factors: (Pharmaco)genetics

Interethnic differences have been demonstrated in the allele frequencies of various drug metabolizing enzymes, transporters, and pharmacologic targets. The clinical relevance for the known variants is not fully understood—some have very clear relevance, and some

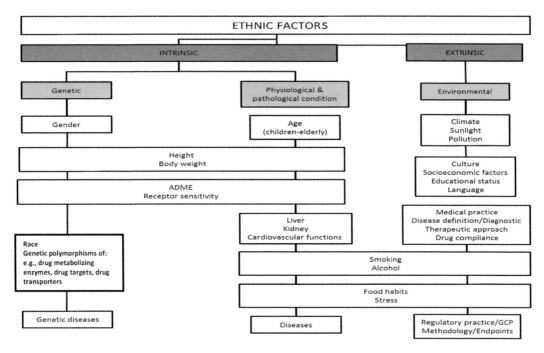

FIGURE 10.1 Classification of intrinsic and extrinsic factors [1].

are still unknown. Pharmacogenetics of drug target, drug metabolism, drug transport, disease susceptibility, and drug safety have been discussed extensively elsewhere [30]. Thus, this chapter will only briefly review some examples of pharmacogenetics of drug metabolizing enzymes (DMEs) and pharmacologic targets.

CYP2D6

CYP2D6 is involved in the metabolism of approximately 25% of all drugs [31]. The first CYP polymorphism was discovered for *CYP2D6*, which is perhaps one of the most studied and best characterized CYP genes. More than 50 alleles have been described for this gene [32], with approximately 20 affecting metabolism of CYP2D6 substrates. *CYP2D6* polymorphisms results in four phenotypes: poor metabolizers (PMs), intermediate metabolizers (IMs), extensive metabolizers (EMs), and ultra-rapid metabolizers (UMs) [33]. Allelic variants that have been associated with the phenotypes are listed in Table 10.2.

The bioavailability, systemic exposure, area under the curve (AUC), and half-life of relevant drugs for the PMs relative to the EMs have been reported to be between two- to sixfold, with metabolite clearance between 0.1- to 0.5-fold [34]. Meanwhile, UMs experience the extreme opposite, rapidly accumulating metabolites at the highest possible doses. Clinical effects of CYP polymorphisms have been reported for various drugs and are particularly serious with the use of tricyclic antidepressants, which are primarily metabolized by this enzyme. Tricyclic antidepressants are very toxic drugs, with potentially fatal adverse effects secondary to cardiac complications [35]. In vulnerable subpopulations like the CYP2D6 PMs,

TABLE 10.2   Effect of CYP2D6 variant allele phenotype on metabolism and potential clinical consequences [33]

Phenotype	Effect on metabolism and potential clinical consequences	Variant allele examples
Poor metabolizers (PM)	• Slowed drug metabolism • Greater potential for drug–drug interactions and adverse events • Slower conversion to active metabolites • Potentially lower efficacy	CYP2D6*3 CYP2D6*4 CYP2D6*5 CYP2D6*6 CYP2D6*10 CYP2D6*17
Extensive metabolizers (EM)	• "Normal" activity	CYP2D6*2 CYP2D6*1
Ultra-rapid metabolizers (UM)	• Accelerated drug metabolism • Greater rates of drug elimination • Potentially lower efficacy	CYP2D6*35 CYP2D6*41

TABLE 10.3   Clinical consequences for PM and UM phenotypes of CYP2D6 [34]

Poor metabolizers: increased risk for toxicity		Ultra-rapid metabolizers: increased risk for toxicity	
Drug	*Toxicity risk*	*Drug*	*Toxicity risk*
Debrisoquine	Postural hypotension and physical collapse	Encainide	Possibility of proarrythmias
Flecainide	Possibility of ventricular arrhythmias	Codeine	Morphine toxicity
Nortriptyline	Hypotension and confusion	*Ultra-rapid metabolizers: failure to respond*	
Thioridazine	Excessively prolonged QT interval	Nortriptyline	Need higher dose to be effective*
Tramadol	Hyper-anticoagulation from warfarin	Propafenone	Need higher dose to be effective*
Propafenone	CNS toxicity and bronchoconstriction	Tropisetron	Need higher dose to be effective*
		Ondansetron	Need higher dose to be effective*
Poor metabolizers: failure to respond		(* = Ineffective at regular doses)	
Codeine	Poor analgesic efficacy		
Tramadol	Poor analgesic efficacy		
Opiates	Protection from oral opiate dependence		

as well as the elderly and adolescents, very low initial doses are recommended [36]. Table 10.3 lists some of the clinical consequences of the use of CYP2D6 substrates for persons with the PM and UM phenotypes.

Interethnic differences in *CYP2D6* allelic frequencies and phenotypes have been shown in many studies. The PM phenotype occurs in about 7 to 10% of European populations compared to 1% of East Asians, with *3,*4, and *5 being most commonly implicated in this phenotype [37]. The *4 variant allele is the most common variant allele in Caucasians, with almost 21% frequency and, interestingly, the *4 variant is almost absent in Chinese. The most common variant allele in Chinese is the *10 (~50%), which is virtually absent in Caucasians. Other examples of differing CYP2D6 allele variant frequencies include the CYP2D6*3 allele (no enzyme activity phenotype), which is not found to be present in the Eastern to Southern Asian regions [38,39,40] but is present in Western Europeans with frequencies from 0.9 to 1.7% [41,42,43]. However, in some populations—for example, Japanese, Koreans, and Chinese—studies have found small differences in the allele frequencies for most of the CYP2D6 variants (<10% difference), except CYP2D6*10 between Japanese and Chinese, with 14.7% difference. For the same variant, the differences between Japanese and Koreans as well as between Koreans and Japanese are 7.6% and 7.1%, respectively [44].

CYP2C9 and VKORC1: Warfarin

Warfarin is one of the most widely used oral anticoagulants globally. It acts by interrupting the regeneration of dihydroxyquinone (KH2), the reduced, active form of vitamin K, by targeting vitamin K epoxide reductase complex 1 (VKORC1), leading to decreased carboxylation and activation of the vitamin K-dependent clotting factors. Warfarin use is hampered by the more than tenfold variability in dosing requirements [45,46] to achieve the target international normalized ratio (INR) in different patients. Overcoagulation causes bleeding episodes, with intracranial hemorrhage being one of the most catastrophic. The effects of genetic polymorphisms in its metabolizing enzyme, CYP2C9, as well as the VKORC1 gene's sensitivity to warfarin have been shown in many studies to significantly affect the dosing requirements [47,48,49,50,51], with the CYP2C9*1/*1 genotype associated with a higher maintenance dose compared to the genotype containing CYP2C9*2 or *3 alleles. There are about 28 known polymorphisms for the VKORC1 gene to date, with the 1639G>A polymorphism being most clinically significant. The −1639 AA genotype is associated with a significantly lower warfarin dose requirement [36]. Subsequent clinical studies incorporating the use of a pharmacogenomics algorithm containing CYP2C9 and VKORC1 genotypes demonstrated better overall predictions for the appropriate warfarin dosage needed to achieve the target INR versus standard management approaches [52,53,54]. Table 10.4 shows allele distribution for CYP2C9 and VKORC1 variants and stabilized warfarin dose according to ethnicity [51].

CYP2C19: Clopidogrel

Clopidogrel is an antiplatelet agent of the thienopyridine group used in the secondary prevention of myocardial infarction and ischemic stroke and also in existing peripheral arterial disease, among other indications. It is a prodrug, which is activated mainly by the hepatic enzyme CYP2C19, although other CYP enzymes such as the CYP1A2, CYP2B6, CYP2C9, and CYP3A4 are also involved. Apart from polymorphisms in CYP2C19, effectiveness of clopidogrel has also been associated with polymorphisms in the P-glycoprotein (P-gp) efflux pump, ABCB1—particularly the c.C3435T variant [56], which will not be discussed here. Variant CYP2C19 alleles have been associated with a range of differing activities, including UM, EM, IM, and PM. This finding has clinical implications whereby, for loss of function alleles, there is reduced activity of CYP2C19, rendering clopidogrel ineffective. In contrast, alleles with increased enzyme activity such as *17 may be associated with increased risk of bleeding. The CYP2C19*1 allele is the allele related to normal functionality, and the

TABLE 10.4   Stabilized warfarin dose according to ethnicity, allele distributions for CYP2C9, genotype distributions for VKORC1 [55]

Characteristic	African American	Caucasian	Hispanic American	Asian
Mean dose (mg/day)	5.2	4.3	4.0	2.7
CYP2C9*1 %(95% CI)	94 (89−99)	74 (66−82)	93 (85−100)	95 (89−100)
CYP2C9*2 %(95% CI)	1 (0−3)	19 (12−26)	0	0
CYP2C9*3 %(95% CI)	1 (0−3)	6 (2−10)	7 (0−15)	5 (0−10)
VKORC1 GG %(95% CI)	82 (74−90)	37 (28−46)	32 (18−46)	7 (0−13)
VKORC1 GA %(95% CI)	12 (6−18)	45 (36−54)	41 (25−56)	30 (18−42)
VKORC1 AA %(95% CI)	6 (1−11)	18 (11−25)	27 (14−49)	63 (51−75)

*2 (c.681G>A; rs4244285) is most commonly associated with loss of function and is observed in up to 30% of Europeans and Africans (3 to 4% homozygotes) and 70% of Asians (10 to 15% homozygotes) [57]. The cumulated evidence for CYP2C19 genotyping from clinical studies appears confusing, possibly due to differences in the role of CYP2C19 in patients with varying degree of disease risk, with greater importance in those at higher risk for poor outcomes, such as patients undergoing percutaneous coronary intervention (PCI) [57]. Studies that enrolled patients at lower cardiovascular risk (e.g., atrial fibrillation or acute coronary syndrome managed medically) mainly found no association between CYP2C19 genotype and treatment outcome with clopidogrel [58,59], and those who selected patients at higher risk found significant association [60,61]. The FDA approved a boxed warning on the product label regarding the risk of lack of clinical effect of clopidogrel in patients who are poor metabolizers in March 2010 [57]. The labeling does not, however, mandate genotyping for all patients who will be prescribed the medication. Several institutions are starting to embrace genotyping for loss-of-function alleles for all possible clopidogrel candidates [62].

Extrinsic Factors

Although intrinsic factors, especially genetic factors, are critical determinants of drug response, the impact of extrinsic factors may also be profound. Nutritional, dietary factors, intake of over-the-counter drugs, as well as use of traditional or alternative medicines all have the potential to alter treatment outcome with drugs. Not surprisingly, drugs with narrow therapeutic index and high potency are among those that have been documented to give rise to significant pharmacokinetics and pharmacodynamics alterations.

Food

Many different types of food and drugs are substrates of *CYPs* enzyme, particularly CYP3A4 as well as the membrane efflux transporter protein, P-gp. Both CYP3A4 and P-gp are constituently expressed in the enterocytes and, as such, affect the bioavailability of many drugs such as digoxin, cyclosporine, midazolam, and verapamil. As chemicals contained in foods are present in high concentrations in the gut, food types that affect CYP3A4 as well as P-gp would have a significant effect on the bioavailability of many drugs [63]. Expressions of P-gp and CYP enzymes at target tissues could also be affected by food—drug interactions in similar manner. Table 10.5 lists examples of food—drug interactions.

Cruciferous vegetables such as cabbage, broccoli, cauliflower, and Brussels sprouts are consumed by people worldwide. They are rich in glucosinolates, which can endogenously be converted to biologically active indoles, such as indole-3-carbinol (I3C) and sulforaphane (SFN) [64]. I3C, in nontoxic doses, has been shown to enhance chemoresistant K562 human leukemic cells in an experimental *in vitro* study [65]. The K562 cells were also cross-resistant to other chemotherapeutic drugs, such as doxorubicin and vincristine. The Western blot analysis in this study further showed that the P-gp expression was down-regulated when the cells were treated with I3C, suggesting that I3C could alleviate chemoresistance in patients taking these drugs for treatment. This development could potentially give rise to differences in chemoresistance profiles among populations consuming high amounts of I3C-containing foods versus those with lower consumption. The Koreans are among the largest cabbage consumers worldwide. The Korean population consumes a traditional fermented, spicy cabbage dish, *kimchi*, almost on a daily basis, totaling about 56.5 kg/person/year [66]—more than 10 times the average consumption of Americans, who consume an average of 4.2 kg/person/year [67]. Similarly, there are marked differences in some parts of the world with regard to consumption of various foods with the ability of modulating Pg-p function.

Grapefruit and grapefruit juice intake and drug interactions have been widely studied. Interactions with many drugs are mediated mainly through physical interactions with CYP P450 inhibition, specifically the intestinal CYP3A4, resulting in complete inactivation of this enzyme. This causes prolonged inhibition of the intestinal clearance of specific drug substrates of this enzyme, such as felodipine [68], as well as other drugs metabolized by this pathway. Furthermore, grapefruit has also been shown to inhibit P-gp mediated efflux, potentiating drugs used in HIV treatment and chemotherapy [69]—for example, vinblastine and saquinavir [70]. However, due to significant overlap in the substrates for P-gp and CYP3A4, studies to particularly isolate P-pg mediated interactions have been challenging. The consumption of grapefruit is highest in Eastern Asia, with Japan making up a significant portion, followed by the Americas and European Union nations. However, it should be noted that CYP3A4/P-pg is affected by a large number of phytochemicals. The potential differential effects in different regions should also be considered when potential drugs are being evaluated for market registration. The consumption of such phytochemicals, which are not all related to grapefruit, are highly ethnic specific as they relate to dietary exposures. In some parts of Asia, although grapefruit is not consumed regularly, the pomelo, a related citrus fruit, is consumed in great abundance. The CYP3A4 inhibitory effect of the pomelo has been reported to be as potent as that of the grapefruit [71].

TABLE 10.5   Examples of food—drug interactions and their mechanisms

Food type	Drug interaction	Mechanism of interaction	References
Piperine (black pepper constituent)	*In vitro*: Digoxin Cyclosporine A Verapamil	Pg-p efflux inhibitor CYP3A4 inhibitor	[126]
Capsaicin (red chili constituent)	*In vitro*: Digoxin	Pg-p efflux inhibitor	[127]
Curcumin (turmeric constituent)	*In vitro*: Digoxin	P-gp efflux inhibitor	[128,129]
Green tea	*In vitro*: Doxorubin, Vinblastine	P-gp efflux inhibitor	[130—133]
Grapefruit juice	*In vitro*: Vinblastine, Vincristine	CYP 1A2, 3A4 inhibition; P-gp modulation	[134,135]
Orange juice	*In vitro*: Vincristine	P-gp efflux inhibitor	[135]
Pomelo juice	*In vitro*: Tacrolimus *In vivo* case report	CYP3A4 inhibitor P-gp modulation	[136,137]
Russian green sweet pepper (Anastasia Green)	Verapamil	P-gp inhibitor	[138]
Seville orange	*In vitro*: vinblastine, fexofenadine, and glibenclamide In humans: atenolol, ciprofloxacine, cyclosporine, celiprolol, levofloxacin, and pravastatin	Inhibits: CYP3A4, P-glycoprotein, OATP-A, OATP-B	[139—147]
Tangerine	*In vitro*: Nifedipine, digoxin	Stimulates CYP3A4 activity and inhibits P-gp	[148—150]
Grapes	In humans: Cyclosporine	Inhibits: CYP3A4 and CYP2E1	[151,152]
Cranberry	*In vitro*: Diclofenac In humans: warfarin	Inhibits: CYP3A and CYP2C9	[153—156]
Pomegranate	In animals: carbamazepine	Inhibits: CYP3A and phenolsulfotransferase activity	[157,158]
Mango	*In vitro*: midazolam, diclofenac, chlorzoxazone, verapamil	Inhibits: CYP1A1, CYP1A2, CYP 3A1, CYP2C6, CYP2E1, P-gp (ABCB1)	[159—161]
Guava	No data available; possible P-gp mediated drug uptake inhibition	Inhibits: P-gp (ABCB1)	[162]
Black raspberry	*In vitro*: midazolam	Inhibits: CYP3A	[163]
Black mulberry	*In vitro*: midazolam; glibenclamide	Inhibits: CYP3A and OATP-B	

TABLE 10.5 Examples of food—drug interactions and their mechanisms—cont'd

Food type	Drug interaction	Mechanism of interaction	References
Apple	*In vitro*: fexofenadine	Inhibits: CYP1A1, OATP family	[164]
Broccoli, cauliflower	No data available; possible ABC transporters and *CYP 1A1, CYP2B1/2, CYP3A4, CYP2E1* substrates modulation	Inhibits: CYP1A1, CYP2B1/2, CYP3A 4, CYP2E1, hGSTA1/2, MRP-1, MRP-2, BCRP, UDP, glucorosytransferases, sulfotransferases, quinone reductases, phenolsulfotransferases Induces: UDP-glucuronosyltransferases (UGTs), sulfotransferases (SULTs), and quinone reductases (QRs)	[165—167]
Watercress	In humans: chlorzoxazone	Inhibits: CYP2E1, P-gp, MRP1, MRP2, and BCRP	[167,168]
Spinach	*In vitro*: heterocyclic aromatic amines	Possible inhibition of CYP1A2	[169]
Tomato	*In vitro*: diethylnitrosamine, N-methyl-N-nitrosourea, and 1,2-dimethylhydrazine	Inhibits: CYP1A1, CYP1B1, UGP (Wang and Leung 2010) Increases: UGT and CYP2E1	[170]

Use of Alternative, Complementary, and Traditional Medicines

According to the World Health Organization (WHO), up to 80% of some Asian and African populations utilize traditional medicine for primary health care. In fact, even in developed countries, the use of alternative or complementary medicine is very prevalent. Herbal and other natural products were reportedly taken by 1 in every 5 American adults [72]. A follow-up study in the United States that evaluated the use of herbal and natural products revealed that the use was lowest among African Americans, compared to Hispanics and non-Hispanic whites, with Hispanics using the largest number of products [73].

In Asia, the use of traditional medicine systems dates back to the twelfth century B.C. Almost every nation in this region has its own use of traditional medicine, and the practice of some of these systems has spread worldwide. The traditional systems use many remedial methods, and for this chapter, only remedies using herbs and natural products will be mentioned. Some of the systems used in Asia include [74]:

- Traditional Chinese medicine (TCM)
- Ayurveda, from India
- Siddha, from south Tamil India
- Unani medicine, from Persia/Middle East, popular in India
- Kampo, Japanese herbal medicine

The Southeast Asian region shares many common herbs for medicinal uses, possibly because of the natural distribution of the available plants. This regions' use of herbal medicine is also greatly influenced by Ayurveda, Unani, Siddha, and TCM, and thus certain similarities in the use of herbs are often seen. However, it is also important to note that use of herbal medicine is usually specific to ethnicity [75]; for example, the Chinese ethnic groups from various countries tend to use herbs from TCM, rather than Ayurvedic herbs, and Indians tend to use herbs from Ayurvedic/Siddha and Unani medicines, while the Malays tend to use more herbs from the Unani system, apart from the use of folk medicine herbs unique to the ethnic Malays, such as the Tongkat Ali (*Eurycoma longifolia*) [76].

Many of the herbal medicines commonly used have been shown to have significant drug interactions. St. John's wort has been traditionally used in the treatment of depression, concomitantly with other antidepressants such as the selective serotonin reuptake inhibitors (SSRIs) and monoamine oxidase inhibitors (MAOIs), and has been implicated in the incidence of serotonin syndrome by additive effect and also CYP3A4 induction [77,78]. *In vivo* human studies have also reported a reduction in plasma concentration of drugs such as amitriptyline, cyclosporine, digoxin and fexofenadine, indinavir, methadone, midazolam, nevirapine, phenprocoumon, simvastatin, tacrolimus, theophylline, and warfarin, possibly due to CYP3A4 and P-gp induction [79]. Gingko biloba has been used to improve cognitive functions in Alzheimer's patients. A report by Galluzzi [80] highlighted a case of an Alzheimer's patient who became comatose after she was started on trazodone, an antidepressant that enhances release of GABA, and gingko biloba. In view of reversal of the patient's clinical condition by flumazenil, a specific benzodiazepine (BDZ) antagonist, it was postulated that her condition was possibly due to a drug interaction as a result of an increase in GABAergic activity mediated directly by the BDZ receptor. The usually subclinical increase in GABAergic activity of ginkgo biloba became clinically enhanced through an interaction with trazodone. A list of reported herb–drug interactions is given in Table 10.6.

It is important to keep in mind that a significant number of people regardless of ethnicity and geographical locations use herbal and natural products as remedies or daily supplements and that ethnic groups and subpopulations tend to use different types of herbal and natural products—based medicines. These medicines have the potential to interact with prescribed drugs and thus play a significant factor in determining outcome to treatment and should be given due consideration.

Differences in Medical Practice

An important extrinsic factor in determining outcome to treatment, as well as whether foreign clinical data can be extrapolated to a new region, is the evaluation of whether there are significant differences in medical practices between two regions. Common examples in this respect would be the difference between medical practices in Japan and those in the United States and Europe. Studies have shown that there are differences in drug dosing between the United States, Europe, and Japan [81,82,83]. For 32% of drugs approved between 2001 and 2007, the maximum recommended dose in the United States was at least twice as high as in Japan [83]. However, it is not certain whether the differences were due to differences in intrinsic factors or due to differences in interpretation

TABLE 10.6 Examples of herb–drug interactions and their mechanisms

Herb type	Drug interaction	Mechanism of interaction	Note on use and primary users	References
Hypericum perforatum (St. John's wort; Seiyo-otogiri-so)	Interacts with selective serotonin reuptake inhibitors and duloxetine by additive effect	Induces $CYP3A4$, P-glycoprotein membrane transporters	Used as antidepressants; Western traditional medicine; also in Japan	[171–173]
Allium sativum (Garlic)	Saquinavir, Ritonavir, Warfarin, Chlorpropamide	Induction of $CYP3A4$ and P-gp Additive effect Platelet dysfunction	Used as antidepressants; Western traditional medicine	[174–178]
Glycyrrhiza glabra (Licorice)	Prednisolone, Hydrocortisone	Potentiation of oral and topical corticosteroids by inhibition of 11β hydrogenase of its metabolite (decreasing clearance)	Widely used in Western traditional medicine	[179]
Ginkgo biloba (Ginkgo)	Thiazide diuretic, Trazodone, Warfarin, Aspirin, Digoxin	Induces $CYP2C19$ Metabolic inhibition Increase of GABAergic activity Inhibition of $CYP3A4$	Widely used in TCM, U.S., Europe	[180–187]
Panax spp (Panax ginseng)	Alcohol (ethanol), Phenelzine, Warfarin	Delayed gastric emptying and enzyme induction Additive effect	Widely used in TCM, East Asia, U.S.	[184–187]
Silybum Adans (Milk Thistle)	Indinavir	Modulation of $CYP3A$ and P-gp	Widely used in the Mediterranean, Northern Africa as liver tonic	[188]
Angelica sinensis (Dong Quai)	Warfarin	Contains coumarin	Widely used in TCM	[179]
Ephedra (Ma huang)	Monoamine oxidase inhibitors, Caffeine, Decongestants, Stimulants	Enhanced sympathomimetic effects when used with other similar drugs	Used in TCM for respiratory ailments	[189]
Danshen (Salvia miltiorrhiza)	Warfarin	Increases bleeding tendencies by decreasing clearance and increasing bioavailability	Used in TCM for chronic renal failure, coronary heart disease	[190]

(Continued)

TABLE 10.6   Examples of herb—drug interactions and their mechanisms—cont'd

Herb type	Drug interaction	Mechanism of interaction	Note on use and primary users	References
Eurycoma longifolia (Tongkat ali, Asian Viagra)	Unknown; speculated to adversely interact with immunosuppressive drugs	No available data	Aphrodisiac, anti-malarial, anti-diabetic; used mainly by Malays in Malaysia	[191]
Labisia pumila (Kacip fatimah)	No data available	Inhibits CYP2C8	Postpartum medication, treats, menstrual irregularities; used mainly by Malays in Malaysia	[192,193]
Andrographis paniculata (Hempedu bumi)	No data available	Inhibits CYP2C8, weak inhibitor of CYP2C19	Treatment of infections, diabetes mellitus; widely used in Asia-South India, Sri Lanka, Malaysia, Indonesia & South China, mainly by the Indians and Malays	[192,194]
Orthosiphon stamineus (Misai kucing)	Possible interactions with CYP2C19 substrates: Omeprazole, Citalopram, Proguanil, Diazepam	Strong inhibitor of CYP2C19	Kidney and urinary disorders; used widely in Thailand, Malaysia & Indonesia	[194]
Asparagus racemosus (Satavari)	Potential interaction with drugs interacting with cholinesterase and monoamine oxidase enzymes	Nonselective competitive inhibitor for cholinesterase and monoamine oxidase enzymes	Widely used in Ayurvedic medicine as galactagogue, aphrodisiac, diuretic, antispasmodic, nerve tonic	[195,196]
Commiphora mukul (Guggul)	Enhanced the efficacy of erlotinib, Cetuximab, and Cisplatin (*in vivo* & *in vitro*)	Induces decreased expression of both phosphotyrosine and total signal transducer and activator of transcription (STAT)-3	Used in Ayurvedic medicine to promote heart and vascular health and treat obesity & rheumatism among others	[197,198]
Agricus brazei Murill (Ji song rong; Kawariharatake)	Diltiazem and other CYP 3A4 substrates	Inhibits CYP 3A4	Used in TCM and Japan	[173]

between risk-benefit balances in the regions. Differences in patient–physician relationships between Western and Asian culture have also been highlighted, in which the Japanese/Asian culture has been viewed as being more hierarchical and paternalistic [84]. The doctor is held in great respect to the point that patients sometimes do not report adverse effects so as to not be offensive to the doctor [85]. However, it should also be noted that recent findings in Japan show that the Japanese patients now prefer the mutual "Western"-style relationship, which may also be considered ideal in other Asian cultures [84].

Conclusion

Genetic variations causing differences in drug response are well established, accounting for some of the ethnic differences in drug response. Moreover, genomics on its own is as yet unable to account for all population differences in drug response, and in many cases, despite its ambiguous definition, ethnicity—with related differences in extrinsic factors—is an important consideration for many of the differences.

ACCEPTABILITY OF FOREIGN CLINICAL DATA

According to the ICH E5 [1], it is possible to use foreign clinical data in drug registration, subject to the completeness of the data package, which should include:

- Adequate characterization of pharmacokinetics, pharmacodynamics, dose response, efficacy, and safety in the population of the foreign region(s).
- Clinical data of efficacy, dosing and safety from trials conducted according to regulatory standards (Good Clinical Practice (GCP) standards), well controlled with appropriate endpoints as well as appropriate medical and diagnostic definitions acceptable to the new region.
- The foreign population in whom the clinical trials were conducted should be representative of the populations in the new region.

Once the clinical data package fulfills the local regulatory requirements, extrapolation of foreign clinical data to the local population should be considered. If a drug is deemed ethnically sensitive, some amount of pharmacokinetic data from local subjects would be required in order to "bridge" the two sets of data from different regions or ethnic populations.

Figure 10.2, taken from the ICH E5 Appendix B, demonstrates an overview of the assessment of the clinical data package (CDP).

As can be seen from the figure, fulfillment of local regulatory requirements is mandatory; failing this fulfillment, additional clinical trials may be requested for this purpose, as well as a bridging study, should the drug be deemed ethnically sensitive. The outcome of the CDP assessment could then be:

- No bridging studies required
  - If drug is not ethnically sensitive
  - If drug is ethnically sensitive but the population is ethnically similar to ensure that the drug will behave similarly in the two populations

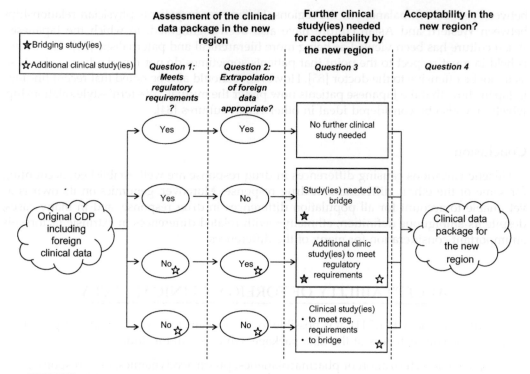

FIGURE 10.2   Assessment of the clinical data package for acceptability of foreign clinical data [1].

- Bridging studies required
  - If drug is ethnically different but with similar extrinsic factors
  - Usually requires a pharmacodynamic study (e.g., dose response) using acceptable endpoints to ensure safety, efficacy, dose, and dosage regimen are applicable to new region
  - Pharmacokinetic measurements for data support
- Controlled clinical trials (CCT) required
  - Dosage is uncertain
  - Limited CCT in the new region
  - Difference in medical practice
  - Drug is not familiar in the new region
- Safety bridging studies
  - If there are safety concerns despite adequate foreign data addressing safety and efficacy issues, for example concerns regarding possibility of higher occurrence rate of adverse events in the new region; can be done together with efficacy studies, with adequate power
  - Separate safety study may be needed if there are no efficacy bridging studies or if efficacy study is not powered for this purpose

If pharmacodynamic data from bridging studies indicate that there is a difference in drug response between the two regions, a CCT will be required. If there is a difference in pharmacokinetics, a dose adjustment may be all that is needed, without the need for a CCT.

Ethnically Sensitive or Insensitive?

The ICH E5 guideline indicates that a bridging study is necessary for drugs that may be ethnically sensitive. Table 10.7 lists some of the factors that may be used to evaluate whether a drug would have a high likelihood of being ethnically sensitive.

Due to the complex interaction among the drugs' pharmacological class, indication, and demography of the population [86], the ICH E5 guideline does not provide definitive criteria for evaluation of a drug's ethnic sensitivity in terms of the evaluating the complete clinical data package or assessing the similarity of clinical results between regions [87]. Various statistical models and strategies have been proposed to assess sensitivities and similarities in ethnicities [87,88,89]; however, no gold standard has yet been established. Specific methodology for extrapolating foreign clinical data is not provided, either. This issue has resulted in marked heterogeneity in the conduct of bridging studies in many regions, notably

TABLE 10.7 Factors affecting a drug's sensitivity to ethnic factors [1]

Factors	Ethnically sensitive	Ethnically insensitive
Pharmacokinetics	Nonlinear	Linear
Pharmacodynamics	Steep curve for efficacy and safety in the range of recommended dosage	Flat effect—concentration curve for both efficacy and safety in the range of recommended dosage
Therapeutic dose range	Narrow	Wide
Metabolism	High, especially through a single pathway Enzymes known to show genetic polymorphism Administered as a prodrug; possible ethnically variable enzymatic conversion	Minimal or distributed along multiple pathways
Bioavailability	Low; susceptible to dietary absorption effects High intersubject variation in bioavailability	High; less susceptible to dietary absorption effects
Potential for protein binding	High	Low
Potential for interactions	High; use in multiple co-medications	Low for drug–drug, drug–diet and drug–disease
Potential for inapproproiate use	Low	High (e.g., analgesics and tranquilizers)
Mode of action	Systemic	Nonsystemic

heterogeneity in the criteria for bridging evaluation, trial procedure, and statistical methods adopted.

Acceptability of Foreign Clinical Data and Drug Regulatory Procedures in East and Southeast Asia

Concerns about foreign clinical data from the drug regulatory agencies' perspective vary by region. The ICH guidelines were initially intended to harmonize regulations governing drug registration in the ICH regions, but not all regions have adopted the guidelines. Some non-ICH countries such as Taiwan and Korea have fully adopted and integrated the bridging concept; others have not. Southeast Asian countries have not adopted the ICH E5 guidelines, although some technical aspects of evaluating the ability of foreign data to be extrapolated to their regions as outlined in the guidelines have been incorporated. The East Asian and Southeast Asian perspectives on and experiences with acceptance of foreign clinical data is reviewed in the following section.

East Asia

JAPAN

In Japan, the Pharmaceuticals and Medical Devices Agency (PMDA), under the Ministry of Health, Labour, and Welfare (MHLW), is the regulatory authority responsible for the scientific review of marketing authorization application of pharmaceutical and medical devices [90]. Japan has a huge pharmaceutical market, second only to the United States, and third if the European Union (EU) is included. Various reports have been put forth to highlight the significant drug lag, which refers to the delay in the time it takes for a drug to be approved in Japan as compared to the United States and the European Union. A survey of the 100 top-selling drugs in 2004 reported a drug lag difference of 2.5 years between the United States and Japan. Another report stated that Japan took an average of 3.9 years, compared to 1.1 years in the United States and the European Union, for drugs approved between 1999 and 2005 [91].

Japan has developed many strategies to improve the drug lag; one is based on the use of bridging studies. Figure 10.3 demonstrates Japan's adoption of bridging strategies in the hope to expedite the drug approval process.

In Japan, foreign Phase I results may be used to estimate the Japanese Phase I studies to enable an abbreviated study beginning with a dose lower than the maximum tolerated

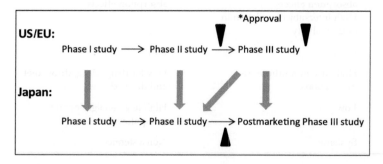

FIGURE 10.3 Overview of extrapolation of foreign clinical data to Japan. Gray arrows indicate possible routes of data extrapolation. Line arrows indicate developmental effort flows. Solid triangles indicate points of possible approval [92].

dose (MTD) in the foreign region but higher than the starting dose. Foreign Phase II data using the new drug as a single agent may replace the requirement for one of the two late Phase II studies. At least one of the Phase II studies must be conducted in Japan, in addition to Phase I studies, although studies conducted elsewhere may be considered [92]. In this case, the foreign Phase II study must be of adequate size, with dose, route, and schedule used in the study similar to those used in the Japanese studies. Otherwise, it will be necessary to prove that the difference will not give rise to a different clinical effect based on pharmacokinetic/pharmacodynamic studies conducted in Japanese subjects (local or abroad). Phase III studies conducted abroad may also be submitted to support a reexamination application, with the provision that one of the studies must be conducted in Japan [92].

Hirai et al. [91] performed a detailed study to analyze factors contributing to the drug lag and found that one of the major contributors to this was the significant delay in initiating clinical trials in Japan; that is, drugs developed in the United States and the European Union had longer lags in Japan. In about 60% of approved drugs in the United States and the European Union, a clinical development phase had not even been developed in Japan [91,93]. Apart from this, Japan's review of their bridging experience highlighted several facts that support their meticulous procedure of acceptability of foreign data, including their bridging approach. This process includes several examples of final drug dose approved in Japan that is different than doses approved in the United States, occurrence of higher adverse events in the Japanese population, as exemplified by induced interstitial lung disease with use of certain chemotherapeutic agents, as well as differences in pharmacokinetic profile for tolterodine between Japanese and Koreans. Thus, the PMDA's issuance of the Notification of Basic Principles on global clinical trials (GCTs) [94], which strongly recommends that clinical studies be done prior to or in parallel with global studies [96], was an effort to abolish the drug lag without compromising Japanese data. Japanese initiatives also resulted in a revision of the ICH E5 guideline at the sixth ICH Conference, with a set of 10 questions and answers to facilitate the implementation of the E5 guideline. The set of questions and answers outlines concepts for planning and implementing GCTs or multiregional clinical trials (MRCTs).

Subsequent to the publication of the guideline, there has been a marked increase in the number of GCTs that include Japan, with total numbers of GCTs conducted in 2007 (17) more than doubling as compared to those conducted in 2008 and 2009 (both 48) [96]. Data from GCTs conducted did highlight the fact that although there were differences in pharmacokinetics, efficacy, and safety, there were also undoubtedly similarities in the data obtained across several populations. Losartan Phase III trials showed superior effect when compared to placebo in overall population (Europe, Latin America, New Zealand, and North America—by region) including Japan, while almost no effect was seen in the U.S. population [96,97]. Global PK studies of tolterodine tartrate also showed some interesting values in the average ratio of the AUC between Japanese and Koreans, as well as the ratio between Japanese and Caucasians, which were 0.72 (95% CI: 0.62, 0.83) and 0.90 (95% CI: 0.78, 1.03), respectively. These results showed that the Japanese pharmacokinetic values were similar to those of the Caucasians and different from those of the Koreans [96]. This finding exemplifies the complexity involved in understanding the interethnic issues and in attempting to equate drug responses based on superficial ideas of ethnic differences or similarities.

KOREA

The Korea Food and Drug Administration (KFDA) is responsible for monitoring safety, assessing clinical data, and granting approvals for pharmaceuticals. Korea adopted the ICH bridging concept of extrapolating foreign clinical data in 2001. The KFDA defines bridging studies as "a trial conducted in Koreans in Korea, for the purpose of obtaining bridging data, in case it is difficult to directly apply the foreign clinical data due to differences in ethnic factors related to safety and efficacy of a drug." Bridging data, on the other hand, refers to "data of trials conducted on Koreans living in Korea or abroad, which are excerpted or selected from the clinical data package or obtained from the bridging study." Bridging data may have already been included as part of the original drug approval application package of the drug and can be used to extrapolate the foreign data. Otherwise, a bridging study must be carried out unless the drug falls into one of the seven waiver categories [98]:

1. Orphan drugs (or used to be orphan drugs)
2. Drugs for life-threatening diseases or AIDS
3. Anticancer therapy for the following
   - No standard therapy
   - Therapy after failure of a standard therapy
4. New drugs for which clinical trials were conducted on Koreans
5. Diagnostic or radioactive drugs
6. Topical drugs with no systemic effect
7. Drugs that have no ethnic differences

Basically, bridging studies must be carried out when there is absence of or inadequate bridging data or if bridging data shows ethnic differences between Koreans and non-Koreans. Figure 10.4 summarizes Korea's bridging concept.

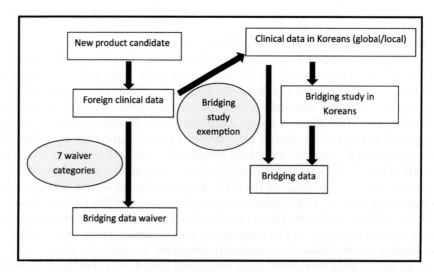

FIGURE 10.4   Overview of bridging concept in Korea [99].

TAIWAN

In Taiwan, the Center for Drug Evaluation (CDE) under the commission of the Department of Health (DOH) evaluates and reviews all new drug applications (NDAs). It was one of the first non-ICH regions to embrace the ICH E5 guideline. The bridging strategy was implemented in stages, beginning with inclusion of local (Taiwanese) clinical trials in 1993 in the 7th of July "double-seven announcement" [98]. Subsequently, the "double-twelve announcement" on December 12, 2000, recommended that sponsors first apply for a Bridging Study Evaluation (BSE) to assess the necessity of carrying out a bridging study in Taiwan, which was fully implemented in 2004. The nine waiver categories requiring no verification of ethnic sensitivity are [98]:

1. Drugs for the treatment of AIDS
2. Drugs for organ transplantation
3. Topical agents
4. Nutritional supplements
5. Cathartics used prior to surgery
6. Radio-labeled diagnostic pharmaceuticals
7. The only available treatment for a serious disease
8. Drugs with demonstrated breakthrough efficacy for a life-threatening disease
9. Drugs for the treatment of rare diseases for which it is difficult to enroll enough subjects for a trial

Should the application not fall into the waiver category, in principle, Taiwan will accept all Asian data for consideration of NDA approval, including PK/PD study data that enable a reasonable estimation of efficacy and safety of the drug. Unless the data indicated that there are ethnic sensitivities that make extrapolation of data impossible, a bridging study would then be requested.

CHINA

It is well known that China's current and future pharmaceutical market is substantial. China's 1.3 million people make up about 20% of the global population. It has grown to be the third largest pharmaceutical market and is still growing rapidly. The State Food and Drug Administration (SFDA) is in charge of drug registration and evaluation, and the Center for Drug Evaluation (CDE) is responsible for the evaluation of chemical drugs, traditional Chinese medicines, and biologic products [100]. Article 11 in the Drug Registration Regulation [101] defines five types of drug registration application: (1) new drug application, (2) generic drug application, (3) imported drug application, (4) supplemental application, and (5) renewal application. A foreign applicant shall make application according to the imported rule. For import drugs with a Certificate of Pharmaceutical Product (CPP) issued by the exporting country with a patent certificate and established Good Manufacturing Practice (GMP) status [102], foreign clinical data would then be assessed for completeness, in line with GMP requirements as well as the Chinese regulatory requirements and ethnic sensitivity assessment. Subsequently, a pharmacokinetic study and a clinical trial of 100 Chinese subjects (per arm) will be required [85,102]. Alternatively, the foreign applicant could submit a clinical trials application, in accordance with Article 44 of the Drug Registration Regulation [101], which states that the drug should already be registered in a foreign country or in Phase

II development (however, this does not apply to applications for new vaccines, which are not registered in any country). Furthermore, the SFDA may also request that the applicant first conduct a local Phase I trial. There is no explicit definition of "Chinese" given in biological or geographical terms, other than it must be local.

CHINA-KOREA-JAPAN TRIPARTITE

The China-Korea-Japan Tripartite cooperation was formed in 2007, following the ICH E5 revision as well as Japan's Notification of Basic Principles on Global Clinical Trials. The tripartite's cooperation involves research into ethnic differences in pharmacokinetics/pharmacodynamics and genetic polymorphisms affecting them, information sharing to promote regulatory framework understanding, and creating a regional clinical guidelines protocol [103]. The pioneer activity of this group was a comparative pharmacokinetic study between China, Korea, Japan, and Caucasians for three drugs—moxifloxacin, simvastatin, and meloxicam—under the Kawai Project. The study demonstrated similar pharmacokinetics between all comparator populations (moxifloxacin), similar pharmacokinetics between some comparator populations (simvastatin between Japanese and Caucasians; meloxicam between Japanese and Chinese), and also differences among comparator populations [104]. This cooperation is at an early stage of development, which may be seen as an initial move towards genomics-based bridging as opposed to ethnic-based bridging.

Southeast Asia

SINGAPORE

Singapore is regarded as the country with the most well-developed drug regulations in the Association of Southeast Asian Nations (ASEAN) region [105], as well as the country with the best capacity for new drug development [106]. Nevertheless, it accepts foreign clinical data with no ICH E5 requirements, just like all the other countries in Southeast Asia.

In Singapore, the Health Sciences Authority (HSA) administers the provisions under the Medicines Act, which requires all western medicinal products to be registered by HSA before they can be marketed in Singapore. HSA has adopted an evidence- and risk-based approach in the evaluation processes for registration of western medicinal products. Although HSA performs independent review of all applications, HSA also leverages on the assessment by other regulatory agencies.

ASEAN COUNTRIES

The ASEAN has 10 member countries: Indonesia, Malaysia, Philippines, Singapore, Thailand, Brunei Darussalam, Vietnam, Laos, Myanmar, and Cambodia. The population is about 598.5 million, with a combined gross domestic product of US$1,850,855 and a total trade of US$ 2,042,788 [107]. The pharmaceutical market in Southeast Asia is relatively small, but the region remains attractive to the pharmaceutical industry due to its growth potential. Among the ten ASEAN members, the five founding member countries (Singapore, Malaysia, Thailand, Philippines, and Indonesia) are more progressive with drug registration and drug development clinical trial activities. All ASEAN member countries are net pharmaceutical importers except for Singapore.

The ASEAN's Pharmaceutical Product Working Group (PPWG) was set up in 1999 to "harmonize pharmaceutical regulations of the ASEAN member countries to complement and facilitate the objective of ASEAN Free Trade Area (AFTA), particularly the elimination of technical barriers to trade posed by these regulations, without compromising on drug quality, safety and efficacy" [108]. The topics selected for harmonization by the PPWG were safety, quality, efficacy, and administration data, which reflects the basis for drug registration approval. PPWG was instrumental in preparing key drug regulatory harmonization documents [106]:

- The ASEAN Common Technical Requirements (ACTR) for pharmaceutical product registration
- The ASEAN Common Technical Documents (ACTD) for pharmaceutical drug registration
- ASEAN guidelines on analytical validation, bioavailability and bioequivalence studies, process validation, and stability study

Each set of guidelines gives cross-references to relevant ICH guidelines or pharmacopeia.

It should be noted that although many of the ICH guidelines were adopted by PPWG, it was decided that the ICH E5 was not going to be adopted due to lack of experience and resources and to first make a scientific justification for the need for local clinical trials and to subsequently verify the actual efficacy of drugs in local situations [109]. Instead, the ASEAN countries were strongly encouraged to participate in the Global Drug Development Programs [105]. In general, ASEAN capacity for evaluating and assessing drug quality, safety, and efficacy are limited. For this reason, they require CPPs issued by the reference country as a surrogate assurance of the product reliability [105]. Furthermore, most of the drug applications reviewed by ASEAN drug regulatory authorities are generic drugs, thus much more emphasis is given to evaluations relating to quality issues such as bioavailability/bioequivalence and stability studies. It should also be noted that Brunei Darussalam is the only ASEAN country that does not require regulatory approval for drugs to be marketed in the country [106].

The recent trend in the shift of clinical trials to Asian emerging regions, especially in Korea, Taiwan, China, Thailand, Singapore, the Philippines, and Malaysia [110], is an opportunity to provide a platform in addressing the issues of ethnic differences more objectively.

Global Drug Development and Pharmacogenomics

Currently, efforts are concentrated on developing biomarkers and pharmacogenomics information from clinical trial inception. Japan has proposed MRCT model, which incorporates special consultation on pharmacogenomics/biomarker qualification to facilitate utilization of this information for regulatory decisions [96]. This development was pioneered by the efforts from the FDA and the European Medicines Agency (EMEA) by first encouraging voluntary submission of genetic data (VGDS) (the scope was then expanded to include nongenomic biomarkers; hence, VGDS was renamed "voluntary exploratory data submission," or VXDS, for inclusion of more diverse biomarkers) by pharmaceutical industries to allow for nonthreatening discussion between industry and regulatory authorities. For this purpose, the FDA and the EMEA issued the Guidance for Industry: Pharmacogenomic Data Submissions in 2005 [111]. This platform has encouraged novel pharmacogenomics and biomarker integration in drug development. Subsequently, a harmonized submissions guideline was drafted by the ICH E16 working group for genomic biomarker qualifications in 2010 [112].

A pharmacogenomics-based (biomarker) success through the conduct of MRCTs is perhaps best exemplified by trastuzumab. Trastuzumab is a monoclonal antibody targeting the extracellular domain of the HER2 protein, an epidermal growth factor receptor gene [113], which was found to be amplified 25 to 30 times in patients with an aggressive form of breast cancer, along with an increase in the expression of its protein in the malignant cells [114]. This biomarker identification, coupled with reliable laboratory testing using fluorescence in situ hybridization (FISH) assay, has enabled a more successful treatment of this subgroup of women with breast cancer, with improved disease-free survival as well as overall survival [115,116]. The design of the trial, which preselected women who were HER2-positive, has saved much time and patient numbers to provide the statistically significant benefit of trastuzumab [117,118]. Table 10.8 lists examples of drugs that have been approved in Japan based on the use of biomarkers in MRCTs.

All of these developments have led to exciting prospects for better risk-benefit judgment for regulatory authorities, as well as making drug therapeutic effects more predictable, effective, and safe for patients.

Pharmacogenomics and Ethnicity in Global Drug Development

The use of pharmacogenomics in global drug development may very well be the turning point for the realization of actual translational medicine. The adoption of the use of pharmacogenomics as a tool to evaluate differences in population groups has added tremendously to its initial value of merely looking at interindividual differences. Invariably, characterization of each and every causative factor and quantitative relationship of their combinations of variability in pharmacologic treatment outcome [120] would be needed to truly personalize medical treatment. Pharmacogenomics and biomarkers are undoubtedly significant parts of this understanding. Nevertheless, it cannot be ignored that there are other factors that constitute an integral part of drug response at a population level that cannot be defined merely by looking at genetics and biomarkers.

In the context of using MRCTs as bridging studies, one of the key points to address is a sample size calculation to enable extrapolation of the overall trial results to the particular

TABLE 10.8 Drugs approved in Japan based on the use of pharmacogenomics (biomarkers) in multiregional clinical trials [119]

Drug name	Indication	Biomarker
Tolterodine	Overactive bladder with symptoms of urinary incontinence, urgency, and frequency	CYP2D6
Trastuzumab	Adjuvant therapy for HER2-positive breast cancer	HER2/neu
Panatimumab	Metastatic colorectal carcinoma with wild-type KRAS tumors	KRAS
Nilotinib	Newly diagnosed chronic myeloid leukemia in chronic phase	Ph chromosome
Trastuzumab	Her2-positive metastatic gastric cancer	

region. The ICH E5 (Q&A) 11 [1] emphasizes this point, and the Japanese MHLW has also provided a guideline on how to demonstrate drug efficacy in a particular region [121]. Two methods were proposed in this guideline for determination of sample size [121]:

- Method 1: D = difference between placebo and study group; D_{all} = difference in the overall study population across regions; D_{Japan} = difference within the Japanese subpopulation. The sample size is determined so that $D_{Japan}/D_{all} > 0.5$ will achieve a probability of 80% or more.
- Method 2: D_{all} = difference between placebo and entire study group across regions, assuming inclusion of three regions; D_1, D_2, and D_3 = difference between placebo and study groups in regions 1, 2, and 3. The sample size is determined so that D for each region will show similar tendencies. In the case in which $D > 0$, the number of subjects is determined so that D1, D2, and D3 will exceed 0 with a probability of 80% or higher.

It has been argued that genetic clustering (which is being used in genetic ancestry) defines a population in a more robust manner than do ethnicity and geographical approaches [120]. However, Risch et al.[122] pointed out a very important point about how data analyses evaluating genetic clusters in isolation and ignoring race and ethnicity may lead to conclusions that are seriously confounded. As an example, they illustrated how, in a study comparing the efficacy of Angiotensin Converting Enzyme (ACE) inhibitors between African American patients and Caucasian patients, there was a significantly better outcome to treatment in white patients. Had the study used a genetic clustering method, the direct inference made from this study would be that this difference was due to the difference in the genetic clusters between the two groups. Although it has been shown in other studies that genetic clustering is highly correlated with self-identified ethnicity/race [123], a direct inference as shown in this example could lead to a grossly confounded conclusion simply because the difference in treatment may very well have been other extrinsic factors that are related more to the ethnicity rather than to the actual genetics. Thus, it should be highlighted that one should not be blinded to ethnicity information while carrying out studies with genetic or even nongenetic biomarkers in order to have a more complete understanding of the given scenario.

The approval of the isosorbide dinitrate and hydralazine combination for the treatment of heart failure in African Americans by FDA in 2005 illustrates that race and ethnicity are indeed very much relevant. The initial application for the marketing of the isosorbide dinitrate/hydralazine combination for all patients was rejected as the original trial, Vasodilator Heart Failure Trial I and II (V-HeFT), failed to demonstrate required statistical significance [124]. Subsequently, when the data was reanalyzed, it was found that the drug may be selectively effective in the black population. Subsequently, the FDA recommended a new trial, named the A-HeFT trial (African American Heart Failure Trial), which demonstrated a 43% reduction in the rate of death from any cause, 33% relative reduction in the rate of first hospitalization for heart failure, and an improvement in the quality of life [125]. This study demonstrates how the inclusion of different specific subpopulation identification and definition can result in a different outcome.

The potential for deriving less expensive, more effective, and safer drugs using pharmacogenomic stratification certainly has a special appeal for developing countries that are in desperate need of more cost-effective health care strategies. Thus, close cooperation between nations for amassing and sharing genotyping data can be significantly beneficial. The HUGO

Pan-Asian SNP consortium is an example of such a cooperative effort [126] to provide a platform for disease population studies or pharmacogenomics research for investigators that can also be leveraged on by regulators.

CONCLUSION

Although ethnicity is very challenging to define, it is of utmost importance that ethnicity be defined in a standardized manner so that accurate scientific conclusions can be derived from any analysis that uses ethnicity/race as a variable. It remains a useful tool for regulatory authorities, practitioners, and researchers for providing a certain degree of insight into the risk-benefit consideration of outcomes of pharmacological treatment. Bridging strategies have been useful in addressing some of the concerns about variability in drug response, as well as expediting drug approval in some countries. The advent of bridging strategies using MRCTs has made it possible to address the issue of variability in drug response in a more global manner. However, to be able to answer specific ethnic-related questions, population selection must be clearly defined with protocols specifically catered to the target population. The incorporation of pharmacogenomics and biomarkers in drug development may allow the stratification of patient populations in a more objective manner and further characterize some of the variability observed in different ethnic populations.

DISCUSSION QUESTIONS

1. Discuss the challenges of incorporating different ethnic groups in pharmacogenetic and pharmacogenomic research.
2. Describe how foreign clinical trial data are utilized for drug approval in other parts of the world.
3. Describe culturally related extrinsic factors that might influence the design of pharmacogenomic studies and the interpretation of study data.

References

[1] ICH E5. International Conference on Harmonization of Technical Requirements for Registration of Pharmaceuticals for Human Use. ICH Harmonized Tripartite Guideline: Ethnic Factors in the Acceptability of Foreign Clinical Data, E5. Retrieved 20 April 2012 from, http://www.ema.europa.eu/docs/en_GB/document_library/Scientific_guideline/2009/09/WC500002842.pdf; 1998.

[2] Collins FS. What we do and don't know about "race", "ethnicity", genetics and health at the dawn of the genome era. Nat Genet 2004;36:S13−5.

[3] Editorial. Census, race and science. Nat Genet 2000;24:97−8.

[4] Statistics Bureau of Japan. 2010 Japan Census; 2010.

[5] Sato T, Amano T, Ono H, Ishida H, Kodera H, Matsumura H, et al. Allele frequencies of the ABCC11 gene for earwax phenotypes among ancient populations of Hokkaido, Japan. J Hum Genet 2009;54:409−13.

[6] Omoto K, Saitou N. Genetic origins of the Japanese: A partial support for the dual structure hypothesis. American Journal of Physical Anthropology 1997;102:437−46.

[7] Sato T, Amano T, Ono H, Ishida H, Kodera H, Matsumura H, et al. Origins and genetic features of the Okhotsk people, revealed by ancient mitochondrial DNA analysis. J Hum Genet 2007;52:618−27.

[8] Ruckle J, Gad SC. Bridging studies in pharmaceutical safety assessment. In: Pharmaceutical Sciences Encyclopedia. John Wiley & Sons, Inc.; 2010. Online ISBN: 9780470571224, doi:10.1002/9780470571224.

[9] Zhang F, Su B, Zhang Y-p, Jin L. Genetic studies of human diversity in East Asia. Philosophical Transactions of the Royal Society B: Biological Sciences 2007;362:987–96.

[10] Chen J, Zheng H, Bei J-X, Sun L, Jia W-h, Li T, et al. Genetic structure of the Han Chinese population revealed by genome-wide SNP variation. Am J Hum Gen 2009;85:775–85.

[11] Yao Y-G, Kong Q-P, Bandelt H-J, Kivisild T, Zhang Y-P. Phylogeographic differentiation of mitochondrial DNA in Han Chinese. Am J Hum Gen 2002;70:635–51.

[12] Yao Y-G, Lü X-M, Luo H-R, Li W-H, Zhang Y-P. Gene admixture in the Silk Road region of China: evidence from mtDNA and melanocortin 1 receptor polymorphism. Genes Genet Syst 2000;75:173–8.

[13] Yao Y-G, Kong Q-P, Wang C-Y, Zhu C-L, Zhang Y-P. Different matrilineal contributions to genetic structure of ethnic groups in the Silk Road region in China. Mol Biol Evol 2004;21:2265–80.

[14] Department of Statistics Malaysia. Population and housing census Malaysia 2010: Corrigendum; 2010.

[15] Paul W. The Golden Khersonese: Studies in the Historical Geography of the Malay Peninsula before AD 1500. Kuala Lumpur: University of Malaya Press; 1961.

[16] Driver S, editor. World and Its Peoples: Eastern and Southern Asia: Malaysia, Singapore, Brunei and Philippines, Vol. 9. New York: Marshall Cavendish Coorperation; 2008.

[17] Hatin WI, Nur-Shafawati AR, Zahri M-K, Xu S, Jin L, Tan S-G, et al. Population genetic structure of peninsular Malaysia Malay sub-ethnic groups. PLoS One 2011;6:e18312.

[18] Laws of Malaysia. Federal Constitution (The Commisioner of Law Revision Malaysia, ed.). Percetakan Nasional Malaysia Berhad; 2006.

[19] Karafet TM, Hallmark B, Cox MP, Sudoyo H, Downey S, Lansing JS, et al. Major east–west division underlies Y chromosome stratification across Indonesia. Mol Biol Evol 2010;27:1833–44.

[20] Suryadinata L, Arifin EN, Ananta A. Indonesia's Population: Ethnicity and Religion in a Changing Political Landscape. Indonesia's population series No. 1. Singapore: Institute of Southeast Asian Studies; 2003.

[21] Johnston T. Chinese diaspora: Indonesia. In: BBC. Jakarta: BBC News; 2005.

[22] MacKerras C, editor. Ethnicity in Asia. Ethnicity and the politics of ethnic classification in Thailand. Edited by Laungaramsri, P. New York: Routledge Curzon; 2003.

[23] Statistics Office. Preliminary report: The 2000 Population & Housing Census (Statistical Data Bank & Information Dissemination, ed.). Bangkok, Thailand: National Statistics Office; 2000.

[24] United Nations. United Nations Thailand. In: Thailand Info; 2008. Vol. 2012.

[25] Premsrirat S. Ethnolinguistic maps of Thailand. J Lang Cult 2002;21:5–35.

[26] Kutanan W, Kampuansai J, Fuselli S, Nakbunlung S, Seielstad M, Bertorelle G, et al. Genetic structure of the Mon-Khmer speaking groups and their affinity to the neighbouring Tai populations in Northern Thailand. BMC Genetics 2011;12:56.

[27] U.S. Department of Health and Human Services Food and Drug Administration. Guideline for industry: collection of race and ethnicity data in clinical trials. Retrieved 20 April 2012 from, http://www.fda.gov/downloads/RegulatoryInformation/Guidances/ucm126396.pdf; 2005.

[28] Kahn J. Genes, race, and population: avoiding a collision of categories. Am J Public Health 2006;96:1965–70.

[29] Schultz J. FDA guidelines on race and ethnicity: obstacle or remedy? J Natl Cancer Inst 2003;95:425–6.

[30] Ma Q, Lu AYH. Pharmacogenetics, pharmacogenomics, and individualized medicine. Pharmacol Rev 2011;63:437–59.

[31] Wolf CR, Smith G. Cytochrome P450 CYP2D6. IARC Sci Publ 1999:209–29.

[32] Wong JY, Seah ES, Lee EJ. Pharmacogenetics: the molecular genetics of CYP2D6 dependent drug metabolism. Ann Acad Med Singapore 2000;29:401–6.

[33] Jann MW, Cohen LJ. The influence of ethnicity and antidepressant pharmacogenetics in the treatment of depression. Drug Metabol Drug Interact 2000;16:39–67.

[34] Shah RR. Pharmacogenetics in drug regulation: promise, potential and pitfalls. Philos Trans R Soc Lond B Biol Sci 2005;360:1617–38.

[35] Balant-Gorgia AE, Balant LP, Garrone G. High blood concentrations of imipramine or clomipramine and therapeutic failure: a case report study using drug monitoring data. Ther Drug Monit 1989;11:415–20.

[36] Yasuda SU, Zhang L, Huang SM. The role of ethnicity in variability in response to drugs: focus on clinical pharmacology studies. Clin Pharmacol Ther 2008;84:417–23.

[37] Gardiner SJ, Begg EJ. Pharmacogenetics, drug-metabolizing enzymes, and clinical practice. Pharmacol Rev 2006;58:521—90.

[38] Man M, Farmen M, Dumaual C, Teng CH, Moser B, Irie S, et al. Genetic variation in metabolizing enzyme and transporter genes. J Clin Pharmacol 2010;50:929—40.

[39] Myrand SP, Sekiguchi K, Man MZ, Lin X, Tzeng RY, Teng CH, et al. Pharmacokinetics/genotype associations for major cytochrome p450 enzymes in native and first- and third-generation Japanese populations: comparison with Korean, Chinese, and Caucasian populations. Clin Pharmacol Ther 2008;84:347—61.

[40] Kubota T, Yamaura Y, Ohkawa N, Hara H, Chiba K. Frequencies of CYP2D6 mutant alleles in a normal Japanese population and metabolic activity of dextromethorphan O-demethylation in different CYP2D6 genotypes. Brit J Clin Pharmacol 2000;50:31—4.

[41] Zackrisson AL, Holmgren P, Gladh AB, Ahlner J, Lindblom B. Fatal intoxication cases: cytochrome P_{450} 2D6 and 2C19 genotype distributions. Eur J Clin Pharmacol 2004;60:547—52.

[42] Zackrisson AL, Lindblom B, Ahlner J. High frequency of occurrence of CYP2D6 gene duplication/multi-duplication indicating ultrarapid metabolism among suicide cases. Clin Pharmacol Ther 2010;88:354—9.

[43] Sachse C, Brockmoller J, Bauer S, Roots I. Cytochrome P450 2D6 variants in a Caucasian population: allele frequencies and phenotypic consequences. Am J Hum Genet 1997;60:284—95.

[44] Kurose K, Sugiyama E, Saito Y. Population differences in major functional polymorphisms of pharmacokinetics/pharmacodynamics-related genes in eastern Asians and Europeans: implications in the clinical trials for novel drug development. Drug Metab Pharmacokinet 2012;27:9—54.

[45] Limdi NA, Veenstra DL. Warfarin pharmacogenetics. Pharmacotherapy 2008;28:1084—97.

[46] Mahajan P, Meyer K, Wall G, Price H. Clinical applications of pharmacogenomics guided warfarin dosing. Pharmacy World & Science 2010;1—10.

[47] Mitchell C, Gregersen N, Krause A. Novel CYP2C9 and VKORC1 gene variants associated with warfarin dosage variability in the South African black population. Pharmacogenomics 2011;12:953—63.

[48] Shrif NE, Won HH, Lee ST, Park JH, Kim KK, Kim MJ, et al. Evaluation of the effects of VKORC1 polymorphisms and haplotypes, CYP2C9 genotypes, and clinical factors on warfarin response in Sudanese patients. Eur J Clin Pharmacol 2011;67:1119—30.

[49] Suriapranata IM, Tjong WY, Wang T, Utama A, Raharjo SB, Yuniadi Y, et al. Genetic factors associated with patient-specific warfarin dose in ethnic Indonesians. BMC Med Genet 2011;12:80.

[50] Vorob'eva NM, Panchenko EP, Dobrovolskii AB, Titaeva EV, Khasanova ZB, Konovalova NV, et al. Polymorphisms of genes CYP2C9 and VKORC1 in patients with venous thromboembolic complications in Moscow population: effects on stability of anticoagulant therapy and frequency of hemorrhage. Ter Arkh 2011;83:59—65.

[51] Wu AH, Wang P, Smith A, Haller C, Drake K, Linder M, Valdes Jr R. Dosing algorithm for warfarin using CYP2C9 and VKORC1 genotyping from a multi-ethnic population: comparison with other equations. Pharmacogenomics 2008;9:169—78.

[52] Finkelman BS, Gage BF, Johnson JA, Brensinger CM, Kimmel SE. Genetic warfarin dosing: tables versus algorithms. J Am Coll Cardiol 2011;57:612—8.

[53] Gage BF, Eby C, Johnson JA, Deych E, Rieder MJ, Ridker PM, et al. Use of pharmacogenetic and clinical factors to predict the therapeutic dose of warfarin. Clin Pharmacol Ther 2008;84:326—31.

[54] Lenzini PA, Grice GR, Milligan PE, Dowd MB, Subherwal S, Deych E, et al. Laboratory and clinical outcomes of pharmacogenetic vs. clinical protocols for warfarin initiation in orthopedic patients. J Thromb Haemost 2008;6:1655—62.

[55] Wu AHB, Wang P, Smith A, Haller C, Drake K, Linder M, et al. Dosing algorithm for warfarin using CYP2C9 and VKORC1 genotyping from a multi-ethnic population: comparison with other equations. Pharmacogenomics 2008;9:169—78.

[56] Mega JL, Close SL, Wiviott SD, Shen L, Walker JR, Simon T, et al. Genetic variants in ABCB1 and CYP2C19 and cardiovascular outcomes after treatment with clopidogrel and prasugrel in the TRITON-TIMI 38 trial: a pharmacogenetic analysis. Lancet 2010;376:1312—9.

[57] Johnson JA, Roden DM, Lesko LJ, Ashley E, Klein TE, Shuldiner AR. Clopidogrel: a case for indication-specific pharmacogenetics. Clin Pharmacol Ther 2012;91:774—6.

[58] Paré G, Mehta SR, Yusuf S, Anand SS, Connolly SJ, Hirsh J, et al. Effects of CYP2C19 genotype on outcomes of clopidogrel treatment. New Engl J Med 2010;363:1704—14.

[59] Holmes MV, Perel P, Shah T, Hingorani AD, Casas JP. CYP2C19 genotype, clopidogrel metabolism, platelet function, and cardiovascular events—a systematic review and meta-analysis. J Am Med Assoc 2011;306:2704—14.

[60] Mega JL, Close SL, Wiviott SD, Shen L, Hockett RD, Brandt JT, et al. Cytochrome P-450 polymorphisms and response to clopidogrel. New Engl J Med 2009;360:354—62.

[61] Shuldiner AR, O'Connell JR, Bliden KP, Gandhi A, Ryan K, Horenstein RB, et al. Association of cytochrome P450 2C19 genotype with the antiplatelet effect and clinical efficacy of clopidogrel therapy. J Am Med Assoc 2009;302:849—57.

[62] Cavallari LH, Schumock GT. Cost is not a barrier to implementing clopidogrel pharmacogenetics. Pharmacotherapy 2012;32:299—303.

[63] Bhardwaj RK, Glaeser H, Becquemont L, Klotz U, Gupta SK, Fromm MF. Piperine, a major constituent of black pepper, inhibits human P-glycoprotein and CYP3A4. J Pharmacol Exp Ther 2002;302:645—50.

[64] Saw CL-L, Cintrón M, Wu T-Y, Guo Y, Huang Y, Jeong W-S, et al. Pharmacodynamics of dietary phytochemical indoles I3C and DIM: induction of Nrf2-mediated phase II drug metabolizing and antioxidant genes and synergism with isothiocyanates. Biopharm Drug Dispos 2011;32:289—300.

[65] Arora A, Seth K, Kalra N, Shukla Y. Modulation of P-glycoprotein-mediated multidrug resistance in K562 leukemic cells by indole-3-carbinol. Toxicol Appl Pharmacol 2005;202:237—43.

[66] USDA Foreign Agricultural Service Gain Report: Global Agricultural Network. USDA; 2010. KS1027. 5/11/2010.

[67] Boriss H, Keith M. Cabbage profile. In: Agricultural Marketing Resource Center, vol. 2012. Iowa State University; 2008.

[68] Lundahl J, Regardh CG, Edgar B, Johnsson G. Effects of grapefruit juice ingestion—pharmacokinetics and haemodynamics of intravenously and orally administered felodipine in healthy men. Eur J Clin Pharmacol 1997;52:139—45.

[69] Seden K, Dickinson L, Khoo S, Back D. Grapefruit-drug interactions. Drugs 2010;70:2373—407.

[70] Honda Y, Ushigome F, Koyabu N, Morimoto S, Shoyama Y, Uchiumi T, et al. Effects of grapefruit juice and orange juice components on P-glycoprotein- and MRP2-mediated drug efflux. Br J Pharmacol 2004;143:856—64.

[71] Egashira K, Sasaki H, Higuchi S, Ieiri I. Food-drug interaction of tacrolimus with pomelo, ginger, and turmeric juice in rats. Drug Metab Pharmacokinet 2012;27:242—7.

[72] Kelly JP, Kaufman DW, Kelley K, Rosenberg L, Anderson TE, Mitchell AA. Recent trends in use of herbal and other natural products. Arch Intern Med 2005;165:281—6.

[73] Kelly JP, Kaufman DW, Kelley K, Rosenberg L, Mitchell AA. Use of herbal/natural supplements according to racial/ethnic group. J Altern Complement Med 2006;12:555—61.

[74] Chaudury RR, Rafei UM, editors. Traditional Medicine in Asia. New Delhi: World Health Organization; 2001.

[75] Dole EJ, Rhyne RL, Zeilmann CA, Skipper BJ, McCabe ML, Low Dog T. The influence of ethnicity on use of herbal remedies in elderly Hispanics and non-Hispanic whites. J Am Pharm Assoc (Wash) 2000;40:359—65.

[76] Saw JT, Bahari MB, Ang HH, Lim YH. Herbal use amongst multiethnic medical patients in Penang Hospital: patterns and perceptions. Med J Malaysia 2006;61:422—32.

[77] Parker V, Wong AH, Boon HS, Seeman MV. Adverse reactions to St John's wort. Can J Psychiatry 2001;46:77—9.

[78] Dannawi M. Possible serotonin syndrome after combination of buspirone and St John's wort. J Psychopharmacol 2002;16:401.

[79] Zhou S, Chan E, Pan SQ, Huang M, Lee EJ. Pharmacokinetic interactions of drugs with St. John's wort. J Psychopharmacol 2004;18:262—76.

[80] Galluzzi S, Zanetti O, Binetti G, Trabucchi M, Frisoni GB. Coma in a patient with Alzheimer's disease taking low dose trazodone and gingko biloba. J Neurol Neurosurg Psychiatry 2000;68:679—80.

[81] Malinowski HJ, Westelinck A, Sato J, Ong T. Same drug, different dosing: differences in dosing for drugs approved in the United States, Europe, and Japan. J Clin Pharmacol 2008;48:900—8.

[82] Uyama Y, Shibata T, Nagai N, Hanaoka H, Toyoshima S, Mori K. Successful bridging strategy based on ICH E5 guideline for drugs approved in Japan[ast]. Clin Pharmacol Ther 2005;78:102—13.

[83] Arnold FL, Kusama M, Ono S. Exploring differences in drug doses between Japan and western countries. Clin Pharmacol Ther 2010;87:714—20.

[84] Ishikawa H, Yamazaki Y. How applicable are western models of patient-physician relationship in Asia?: changing patient-physician relationship in contemporary Japan. Int J Jpn Sociol 2005;14:84−93.

[85] Kudrin A. Challenges in the clinical development requirements for the marketing authorization of new medicines in Southeast Asia. J Clin Pharmacol 2009;49:268−80.

[86] Liu JP, Chow SC. Bridging studies in clinical development. J Biopharm Stat 2002;12:359−67.

[87] Chow S-C, Shao J, Hu OY-P. Assessing sensitivity and similarity in bridging studies. J Biopharm Stat 2002;12:385−400.

[88] Shih WJ. Clinical trials for drug registrations in Asian-Pacific countries: proposal for a new paradigm from a statistical perspective. Control Clin Trials 2001;22:357−66.

[89] Glasbrenner M, Rosenkranz G. A note on ethnic sensitivity studies. J Biopharm Stat 2006;16:15−23.

[90] Pharmaceuticals and Medical Devices Agency. Pharmaceuticals and Medical Devices Agency Japan, vol. 2012; 2012. Tokyo.

[91] Hirai Y, Kinoshita H, Kusama M, Yasuda K, Sugiyama Y, Ono S. Delays in new drug applications in Japan and industrial R&D strategies. Clin Pharmacol Ther 2009;87:212−8.

[92] Fujiwara Y, Kobayashi K. Oncology drug clinical development and approval in Japan: the role of the Pharmaceuticals and Medical Devices Evaluation Center (PMDEC). Cr Rev Oncol-Hem 2002;42:145−55.

[93] Tsuji K, Tsutani K. Drug lag of clinical importance. Jpn J Clin Pharmacol Ther 2008;39(suppl):S266.

[94] Ministry of Health and Welfare, Japan. Notification on basic principles on global clinical trials (Pharmaceutical and Food Safety Bureau Notification No 0928010); 2007.

[95] Fukunaga S, Kusama M, Arnold FL, Ono S. Ethnic differences in pharmacokinetics in new drug applications and approved doses in Japan. J Clin Pharmacol 2011;51:1237−40.

[96] Ichimaru K, Toyoshima S, Uyama Y. Effective global drug development strategy for obtaining regulatory approval in Japan in the context of ethnicity-related drug response factors. Clin Pharmacol Ther 2010;87:362−6.

[97] Brenner BM, Cooper ME, de Zeeuw D, Grunfeld J-P, Keane WF, Kurokawa K, et al, RENAAL Study Investigators. The losartan renal protection study—rationale, study design and baseline characteristics of RENAAL (Reduction of Endpoints in NIDDM with the Angiotensin II Antagonist Losartan). J Renin-Angiotensin-Aldosterone Sys 2000;1:328−35.

[98] Chow S-C, Hsiao C-F. Bridging diversity: extrapolating foreign data to a new region. Pharmaceut Med 2010;24:349−62.

[99] Kim I-B. Review policies for global drug development: Korea's perspective. In: East Asian Pharmaceutical Regulatory Symposium 2008; 2008.

[100] Lu D, Huang W. Overview of drug evaluation system in China. Sci Res Essays 2010;5:514−8.

[101] The State Food and Drug Administration (SFDA). Regulations for Implementation of the Drug Administration Law of the People's Republic of China ((SFDA), ed.); 2007.

[102] Zhao N, Yao C, Chen J. On the amendments of China's provisions for drug registration. Drug Info J 2008;42:467−75.

[103] Ando Y. Overview of China-Korea-Japan Tripartite cooperation & research on ethnic factors, http://www.pmda.go.jp/english/presentations/pdf/presentations_20120327-28-4.pdf; 2012.

[104] Uzu S. Update on Korea China Japan Tripartite clinical trial collaboration in Japan, http://www.apec-ahc.org/files/tp201105/Shinobu_Uzu_Apr26Plenary_1pm.pdf; 2011.

[105] Latzel R. Development of the ASEAN Pharmaceutical Harmonization Scheme—an example of regional integration. Der Rheinischen Friedrich-Wilhelms-Universitat Bonn; 2007.

[106] S Ratanawijitrasin. Drug Regulation and Incentives for Innovation: The Case of ASEAN, WHO, 2005. http://www.who.int/intellectualproperty/studies/Sauwakon%20Ratanawijitrasin.pdf (accessed April 2012).

[107] ASEAN. ASEAN statistics leaflet: selected key indicators 2011; 2011.

[108] ASEAN. ACCSQ Pharmaceutical Product Working Group; 2011.

[109] The impact of implementation of ICH Guidelines in Non-ICH Countries. (2001). WHO.

[110] U.S. National Institutes of Health. ClinicalTrials.gov 2012;vol. 2012.

[111] Goodsaid F, Papaluca M. Evolution of biomarker qualification at the health authorities. Nat Biotech 2010;28:441−3.

[112] ICH. Biomarkers related to drug or biotechnology product development: context, structure and format of qualification submissions. E16. International Conference on Harmonization of Technical Requirements for Registration of Pharmaceuticals for Human Use. In: ICH Harmonized Tripartite Guideline; 2010.

[113] King CR, Kraus MH, Aaronson SA. Amplification of a novel V-ERBB-related gene in a human mammary-carcinoma. Science 1985;229:974—6.

[114] Slamon DJ, Godolphin W, Jones LA, Holt JA, Wong SG, Keith DE, et al. Studies of the HER-2/NEU proto-oncogene in human-breast and ovarian-cancer. Science 1989;244:707—12.

[115] Slamon DJ, Leyland-Jones B, Shak S, Fuchs H, Paton V, Bajamonde A, et al. Use of chemotherapy plus a monoclonal antibody against HER2 for metastatic breast cancer that overexpresses HER2. New Engl J Med 2001;344:783—92.

[116] Romond EH, Perez EA, Bryant J, Suman VJ, Geyer CE, Davidson NE, et al. Trastuzumab plus adjuvant chemotherapy for operable HER2-positive breast cancer. New Engl J Med 2005;353:1673—84.

[117] Peták I, Schwab R, őrfi L, Kopper L, Kéri G. Integrating molecular diagnostics into anticancer drug discovery. Nat Rev Drug Discov 2010;9:523—35.

[118] Eichler H-G, Abadie E, Breckenridge A, Flamion B, Gustafsson LL, Leufkens H, et al. Bridging the efficacy—effectiveness gap: a regulator's perspective on addressing variability of drug response. Nat Rev Drug Discov 2011;10:495—506.

[119] Ando Y, Hamasaki T. Practical issues and lessons learned from multi-regional clinical trials via case examples: a Japanese perspective. Pharm Stat 2010;9:190—200.

[120] Wilson JF, Weale ME, Smith AC, Gratrix F, Fletcher B, Thomas MG, et al. Population genetic structure of variable drug response. Nat Genet 2001;29:265—9.

[121] Ministry of Health Labour and Welfare Japan. Basic Principles on Global Clinical Trials, http://www.pmda.go.jp/english/service/pdf/notifications/0928010-e.pdf; 2007. p. 7.

[122] Risch N, Burchard E, Ziv E, Tang H. Categorization of humans in biomedical research: genes, race and disease. Genome Biol 2002;3. comment 2007.1—2007.12.

[123] Tang H, Quertermous T, Rodriguez B, Kardia SL, Zhu X, Brown A, et al. Genetic structure, self-identified race/ethnicity, and confounding in case-control association studies. Am J Hum Genet 2005;76:268—75.

[124] Kahn J. From disparity to difference: how race-specific medicines may undermine policies to address inequalities in health care. South Calif Interdiscip Law J 2005;15:105—30.

[125] Taylor AL, Ziesche S, Yancy C, Carson P, D'Agostino R, Ferdinand K, et al. Combination of isosorbide dinitrate and hydralazine in blacks with heart failure. New Engl J Med 2004;351:2049—57.

[126] Daar AS, Singer PA. Pharmacogenetics and geographical ancestry: implications for drug development and global health. Nat Rev Genet 2005;6:241—6.

[127] Han Y, Tan TMC, Lim L-Y. Effects of capsaicin on P-gp function and expression in Caco-2 cells. Biochem Pharmacol 2006;71:1727—34.

[128] Choi BH, Kim CG, Lim Y, Shin SY, Lee YH. Curcumin down-regulates the multidrug-resistance mdr1b gene by inhibiting the PI3K/Akt/NFκB pathway. Cancer Lett 2008;259:111—8.

[129] Zhang W, Lim L-Y. Effects of spice constituents on P-glycoprotein-mediated transport and CYP3A4-mediated metabolism in vitro. Drug Metab Dispos 2008;36:1283—90.

[130] Engdal S, Nilsen OG. Inhibition of P-glycoprotein in Caco-2 cells: effects of herbal remedies frequently used by cancer patients. Xenobiotica 2008;38:559—73.

[131] Jodoin J, Demeule M, Béliveau R. Inhibition of the multidrug resistance P-glycoprotein activity by green tea polyphenols. Biochimica et Biophysica Acta (BBA)—Molecular Cell Research 2002;1542:149—59.

[132] Mei Y, Qian F, Wei D, Liu J. Reversal of cancer multidrug resistance by green tea polyphenols. J Pharm Pharmacol 2004;56:1307—14.

[133] Sadzuka Y, Sugiyama T, Sonobe T. Efficacies of tea components on doxorubicin induced antitumor activity and reversal of multidrug resistance. Toxicol Lett 2000;114:155—62.

[134] Takanaga H, Ohnishi A, Matsuo H, Sawada Y. Inhibition of vinblastine efflux mediated by P-glycoprotein by grapefruit juice components in caco-2 cells. Biol Pharm Bull 1998;21:1062—6.

[135] Ikegawa T, Ushigome F, Koyabu N, Morimoto S, Shoyama Y, Naito M, et al. Inhibition of P-glycoprotein by orange juice components, polymethoxyflavones in adriamycin-resistant human myelogenous leukemia (K562/ADM) cells. Cancer Lett 2000;160:21—8.

[136] Egashira K, Fukuda E, Onga T, Yogi Y, Matsuya F, Koyabu N, et al. Pomelo-induced increase in the blood level of tacrolimus in a renal transplant patient. Transplantation 2003;75:1057.

[137] Egashira K, Sasaki H, Higuchi S, Ieiri I. Food-drug interaction of tacrolimus with pomelo, ginger, and turmeric juice in rats. Drug Metab Pharmacokinet 2012;27:242—7.

[138] Motohashi N, Kurihara T, Wakabayashi H, Yaji M, Mucsi I, Molnar J, et al. Biological activity of a fruit vegetable, "Anastasia green, " a species of sweet pepper. In Vivo 2001;15:437—42.

[139] Takanaga H, Ohnishi A, Yamada S, Matsuo H, Morimoto S, Shoyama Y, et al. Polymethoxylated flavones in orange juice are inhibitors of P-glycoprotein but not cytochrome P450 3A4. J Pharm Exp Ther 2000;293:230—6.

[140] Dresser GK, Kim RB, Bailey DG. Effect of grapefruit juice volume on the reduction of fexofenadine bioavailability: possible role of organic anion transporting polypeptides[ast]. Clin Pharmacol Ther 2005;77:170—7.

[141] Satoh H, Yamashita F, Tsujimoto M, Murakami H, Koyabu N, Ohtani H, et al. Citrus juices inhibit the function of human organic anion-transporting polypeptide OATP-B. Drug Metab Dispos 2005;33:518—23.

[142] Lilja J, Raaska K, Neuvonen P. Effects of orange juice on the pharmacokinetics of atenolol. Eur J Clin Pharmacol 2005;61:337—40.

[143] Greenblatt DJ. Analysis of drug interactions involving fruit beverages and organic anion-transporting polypeptides. J Clin Pharmacol 2009;49:1403—7.

[144] Tan H-L, Thomas-Ahner J, Grainger E, Wan L, Francis D, Schwartz S, et al. Tomato-based food products for prostate cancer prevention: what have we learned? Cancer Metast Rev 2010;29:553—68.

[145] Malhotra S, Bailey DG, Paine MF, Watkins PB. Seville orange juice-felodipine interaction: comparison with dilute grapefruit juice and involvement of furocoumarins[ast]. Clin Pharmacol Ther 2001;69:14—23.

[146] Harris KE, Jeffery EH. Sulforaphane and erucin increase MRP1 and MRP2 in human carcinoma cell lines. J Nutr Biochem 2008;19:246—54.

[147] Kamath AV, Yao M, Zhang Y, Chong S. Effect of fruit juices on the oral bioavailability of fexofenadine in rats. J Pharm Sci 2005;94:233—9.

[148] Backman JT, Maenpaa J, Belle DJ, Wrighton SA, Kivisto KT, Neuvonen PJ. Lack of correlation between in vitro and in vivo studies on the effects of tangeretin and tangerine juice on midazolam hydroxylation[ast]. Clin Pharmacol Ther 2000;67:382—90.

[149] Yoo HH, Lee M, Chung HJ, Lee SK, Kim D-H. Effects of diosmin, a flavonoid glycoside in citrus fruits, on P-glycoprotein-mediated drug efflux in human intestinal Caco-2 cells. J Agric Food Chem 2007;55:7620—5.

[150] Nowack R. Review article: cytochrome P450 enzyme, and transport protein mediated herb—drug interactions in renal transplant patients: grapefruit juice, St. John's wort—and beyond! (Review article). Nephrology 2008;13:337—47.

[151] Piver B, Berthou F, Dreano Y, Lucas D. Inhibition of CYP3A, CYP1A and CYP2E1 activities by resveratrol and other nonvolatile red wine components. Toxicol Lett 2001;125:83—91.

[152] Chan WK, Delucchi AB. Resveratrol, a red wine constituent, is a mechanism-based inactivator of cytochrome P450 3A4. Life Sci 2000;67:3103—12.

[153] Izzo AA. Herb—drug interactions: an overview of the clinical evidence. Fundam Clin Pharmacol 2005;19:1—16.

[154] Pham DQ, Pham AQ. Interaction potential between cranberry juice and warfarin. Am J Health-Syst Ph 2007;64:490—4.

[155] Ushijima K, Tsuruoka S-i, Tsuda H, Hasegawa G, Obi Y, Kaneda T, et al. Cranberry juice suppressed the diclofenac metabolism by human liver microsomes, but not in healthy human subjects. Brit J Clin Pharmacol 2009;68:194—200.

[156] Zhou S, Chan E, Pan S-Q, Huang M, Lee EJD. Pharmacokinetic interactions of drugs with St John's wort. J Psychopharmacol 2004;18:262—76.

[157] Hidaka M, Okumura M, Fujita K-i, Ogikubo T, Yamasaki K, Iwakiri T, et al. Effects of pomegranate juice on human cytochrome P450 3A (CYP3A) and carbamazepine pharmacokinetics in rats. Drug Metab Dispos 2005;33:644—8.

[158] Saruwatari A, Okamura S, Nakajima Y, Narukawa Y, Takeda T, Tamura H. Pomegranate juice inhibits sulfoconjugation in Caco-2 human colon carcinoma cells. J Med Food 2008;11:623—8.

[159] Rodeiro I, Donato MT, Lahoz A, Garrido G, Delgado R, Gomez-Lechon MJ. Interactions of polyphenols with the P450 system: possible implications on human therapeutics. Mini Rev Med Chem 2008;8:97—106.

[160] Rodeiro I, Donato MT, Jimenez N, Garrido G, Molina-Torres J, Menendez R, et al. Inhibition of human P450 enzymes by natural extracts used in traditional medicine. Phytother Res 2009;23:279—82.

[161] Chieli E, Romiti N, Rodeiro I, Garrido G. In vitro effects of Mangifera indica and polyphenols derived on ABCB1/P-glycoprotein activity. Food Chem Toxicol 2009;47:2703—10.

[162] Junyaprasert VB, Soonthornchareonnon N, Thongpraditchote S, Murakami T, Takano M. Inhibitory effect of Thai plant extracts on P-glycoprotein mediated efflux. Phytother Res 2006;20:79—81.

[163] Kim H, Yoon Y-J, Shon J-H, Cha I-J, Shin J-G, Liu K-H. Inhibitory effects of fruit juices on CYP3A activity. Drug Metab Dispos 2006;34:521—3.

[164] Pohl C, Will F, Dietrich H, Schrenk D. Cytochrome P450 1A1 expression and activity in Caco-2 cells: modulation by apple juice extract and certain apple polyphenols. J Agric Food Chem 2006;54:10262—8.

[165] Fimognari C, Lenzi M, Hrelia P. Interaction of the isothiocyanate sulforaphane with drug disposition and metabolism: pharmacological and toxicological implications. Curr Drug Metab 2008;9:668—78.

[166] Anwar-Mohamed A, El-Kadi AOS. Sulforaphane induces CYP1A1 mRNA, protein, and catalytic activity levels via an AhR-dependent pathway in murine hepatoma Hepa 1c1c7 and human HepG2 cells. Cancer Lett 2009;275:93—101.

[167] Telang U, Ji Y, Morris ME. ABC transporters and isothiocyanates: potential for pharmacokinetic diet—drug interactions. Biopharm Drug Dispos 2009;30:335—44.

[168] Leclercq I, Desager J-P, Horsmans Y. Inhibition of chlorzoxazone metabolism, a clinical probe for CYP2E1, by a single ingestion of watercress[ast]. Clin Pharmacol Ther 1998;64:144—9.

[169] Platt KL, Edenharder R, Aderhold S, Muckel E, Glatt H. Fruits and vegetables protect against the genotoxicity of heterocyclic aromatic amines activated by human xenobiotic-metabolizing enzymes expressed in immortal mammalian cells. Mutat Res-Gen Tox En 2010;703:90—8.

[170] Veeramachaneni S, Ausman LM, Choi SW, Russell RM, Wang X-D. High dose lycopene supplementation increases hepatic cytochrome P4502E1 protein and inflammation in alcohol-fed rats. J Nutr 2008;138:1329—35.

[171] British Medical Association and Royal Pharmaceutical Society of Great Britain. British National Formulary No. 59, London; 2010.

[172] Marchetti S, Mazzanti R, Beijnen JH, Schellens JHM. Concise review: clinical relevance of drug—drug and herb—drug interactions mediated by the ABC transporter ABCB1 (MDR1, P-glycoprotein). Oncologist 2007;12:927—41.

[173] Ohnishi N, Yokoyama T. Interactions between medicines and functional foods or dietary supplements. Keio J Med 2004;53:137—50.

[174] Yin OQ, Tomlinson B, Waye MM, Chow AH, Chow MS. Pharmacogenetics and herb-drug interactions: experience with ginkgo biloba and omeprazole. Pharmacogenet Genomics 2004;14:841—50.

[175] Piscitelli SC, Burstein AH, Welden N, Gallicano KD, Falloon J. The effect of garlic supplements on the pharmacokinetics of saquinavir. Clin Infect Dis 2002;34:234—8.

[176] Gallicano K, Foster B, Choudhri S. Effect of short-term administration of garlic supplements on single-dose ritonavir pharmacokinetics in healthy volunteers. Brit J Clin Pharmacol 2003;55:199—202.

[177] Chen XW, Sneed KB, Pan SY, Cao C, Kanwar JR, Chew H, Zhou SF. Herb-drug interactions and mechanistic and clinical considerations. Curr Drug Metab 2012;13:640—51.

[178] Aslam M, Stockley IH. Interaction between curry ingredient (karela) and drug (chlorpropamide). Lancet 1979;313:607.

[179] Fugh-Berman A. Herb-drug interactions. Lancet 2000;355:134—8.

[180] Shaw D, Leon C, Kolev S, Murray V. Traditional remedies and food supplements. A 5-year toxicological study (1991-1995). Drug Saf 1997;17:342—56.

[181] Matthews Jr MK. Association of ginkgo biloba with intracerebral hemorrhage. Neurology 1998;50:1933—4.

[182] Rosenblatt M, Mindel J. Spontaneous hyphema associated with ingestion of ginkgo biloba extract. New Engl J Med 1997;336. 1108—1108.

[183] Mauro VF, Mauro LS, Kleshinski JF, Khuder SA, Wang Y, Erhardt PW. Impact of ginkgo biloba on the pharmacokinetics of digoxin. Am J Ther 2003;10:247—51.

[184] Lee FC, Ko JH, Park JK, Lee JS. Effects of panax ginseng on blood alcohol clearance in man. Clin Exp Pharmacol Physiol 1987;14:543—6.

[185] Shader RI, Greenblatt DJ. Phenelzine and the dream machine—ramblings and reflections. J Clin Psychopharmacol 1985;5:65.

[186] Jones BD, Runikis AM. Interaction of ginseng with phenelzine. J Clin Psychopharmacol 1987;7:201—2.

[187] Janetzky K, Morreale A. Probable interaction between warfarin and ginseng. Am J Health-Syst Pharm 1997;54:692—3.

[188] Piscitelli SC, Formentini E, Burstein AH, Alfaro R, Jagannatha S, Falloon J. Effect of milk thistle on the pharmacokinetics of indinavir in healthy volunteers. Pharmacotherapy 2002;22:551—6.

[189] Chen J, Chen T. Chinese Medical Herbology and Pharmacology. Art of Medicine Press CA, USA; 2004.

[190] Chiu PY, Wong SM, Leung HY, Leong PK, Chen N, Zhou L, et al. Long-term treatment with danshen-gegen decoction protects the myocardium against ischemia/reperfusion injury via the redox-sensitive protein kinase C-epsilon/mK(ATP) pathway in rats. Rejuvenation Res 2011;14:173—84.

[191] Purwantiningsih, Hussin AH, Chan KL. Phase I drug metabolism study of the standardised extract of eurycoma longifolia (TAF 273) in rat hepatocytes. Int J Pharm Pharm Sci 2010;2(3S):147—52.

[192] Muthiah YD. Cytochrome P450 2C8: an investigation of types and frequencies of CYP2C8 polymorphism in Malaysia and in vitro analysis of catalytic activity. Malays J Med Sci 2010;17:116—7.

[193] Fazliana M, Ramos NL, Lüthje P, Sekikubo M, Holm Å, Wan Nazaimoon WM, et al. Labisia pumila var. alata reduces bacterial load by inducing uroepithelial cell apoptosis. J Ethnopharmacol 2011;136:111—6.

[194] Pan Y, Abd-Rashid BA, Ismail Z, Ismail R, Mak JW, Pook PCK, et al. In vitro modulatory effects of Andrographis paniculata, Centella asiatica and Orthosiphon stamineus on cytochrome P450 2C19 (CYP2C19). J Ethnopharmacol 2011;133:881—7.

[195] Meena J, Ojha R, Muruganandam AV, Krishnamurthy S. Asparagus racemosus competitively inhibits in vitro the acetylcholine and monoamine metabolizing enzymes. Neurosci Lett 2011;503:6—9.

[196] Bopana N, Saxena S. Asparagus racemosus—ethnopharmacological evaluation and conservation needs. J Ethnopharmacol 2007;110:1—15.

[197] Ojha S, Bhatia J, Arora S, Golechha M, Kumari S, Arya DS. Cardioprotective effects of Commiphora mukul against isoprenaline-induced cardiotoxicity: a biochemical and histopathological evaluation. J Environ Biol 2011;32:731—8.

[198] Leeman-Neill RJ, Wheeler SE, Singh SV, Thomas SM, Seethala RR, Neill DB, et al. Guggulsterone enhances head and neck cancer therapies via inhibition of signal transducer and activator of transcription-3. Carcinogenesis 2009;30:1848—56.

Beyond ELSIs: Where to from Here? From "Regulating" to Anticipating and Shaping the Innovation Trajectory in Personalized Medicine

Vural Özdemir[*,†,‡], *Yann Joly*[*], *Emily Kirby*[*], *Denise Avard*[*], *Bartha M. Knoppers*[*]

[*]Centre of Genomics and Policy, Department of Human Genetics, Faculty of Medicine, McGill University, Montreal, Quebec, Canada
[†]Group on Complex Collaboration, Faculty of Management, McGill University, Montreal, Quebec, Canada
[‡]Data-Enabled Life Sciences Alliance International (DELSA Global), Seattle, Washington, USA

OBJECTIVES

1. Describe the conceptual and applied analytical tools to independently understand and evaluate the societal, ethical, and policy aspects of genomics and personalized medicine in a global world.

2. Explain the need for the concept of "reflexivity" in pharmacogenomics: that is, how our own personal and professional values and unchecked assumptions—whether as scientists or bioethicists—might affect the conclusions we draw from scientific inquiry.

3. Describe how the purview of bioethics has broadened markedly in the postgenomics era and why scientists should also collaboratively take part in societal analysis of novel technologies and knowledge-based innovations with humanists and social scientists.

Pharmacogenomics
http://dx.doi.org/10.1016/B978-0-12-391918-2.00011-1

If bioethics is to address the broad range of ethical issues raised by public health policy, including targeted interventions, effectively it must place discussions of choice, consent and autonomy in the context of a wider range of ethical issues. **Onora O'Neill, 2011 [1]; Nuffield Council on Bioethics, Twentieth Anniversary Lecture, UK**

Today most discoveries are made by scientist-clinicians who are funded to generate data, build a model or hypothesis, provide a validation of their idea, and then share the results as a paper in a peer-reviewed journal. But as the scale of the data needed to make insights grows . . . the power of coordinated team approaches will grow. By analogy to physics, astronomy, and the writing of the software, the benefits of dynamic teams sharing data and ideas in real time will multiply. The logical extension is to start considering a "commons" where omics data, projects, and models can be evolved in a shared manner. **Stephen Friend, 2011 [2]**

Pharmacogenetics is now an "old" science that has evolved over the past five decades from early studies of monogenic variations in drug metabolism. Yet progress in pharmacogenetics science is not driven, nor influenced, by technology alone. In the course of the past few years, there have been seismic shifts in the way scientific knowledge is produced, not only in the traditional laboratory bench space but in hitherto unprecedented "locales" and by new stakeholders such as citizen scientists and patient advocacy groups contributing to pharmacogenetics study design and data collection. These changes have also brought to the fore the social, ethical, and policy dimensions of the very process of such knowledge coproduction by a multitude of stakeholders in- and outside academia. This chapter addresses the latter theme under the rubric of "collective innovation" for postgenomics personalized medicine.

PHARMACOGENETICS: AN OLD SCIENCE THAT STOOD THE TEST OF TIME

Pharmacogenetics is the study of the role of human genetic variation in person-to-person and population differences in drug pharmacokinetics and pharmacodynamics. Pharmacogenetics is typically considered under the rubric of personalized medicine and rational choice of medical interventions [3]. The origin of twenty-first-century personalized medicine initiatives can be traced back to the first formal comparative trial in the eighteenth century by James Lind. A naval surgeon, Lind demonstrated in 1747 that citrus juice, not the other leading remedies recommended by physicians of the day, cured scurvy [4]. At the turn of the twentieth century, British physician Archibald Garrod wrote presciently on the topic of chemical individuality [5].

As with personalized medicine, pharmacogenetics is not a new science and predates to early studies of monogenic variations in drug metabolism in the 1950s. In October 1957, Arno G. Motulsky proposed in a seminal article the idea of genetic contribution to adverse drug reactions (ADRs) [6]. Two years later in Heidelberg, Germany, Friedrich Vogel proposed the term "pharmacogenetics" [7]. Werner Kalow published the first book on pharmacogenetics in 1962 in Toronto, Ontario [8]. The *New York Times* ran both an article and an editorial on the subject that same year [9]. A salient question germane to pharmacogenetics—the *size* of genetic components of pharmacological traits—was answered thereafter by Elliot S. Vesell in Hershey, Pennsylvania, in a programmatic series of studies in monozygotic and dizygotic twins [10]. These twin studies demonstrated a strong genetic basis for individual differences in drug metabolism and pharmacokinetics. Subsequently, the sparteine/debrisoquine

(*CYP2D6*) monogenic polymorphism was identified by Eichelbaum in Germany [11,12] and Smith in England [13].

In parallel to these formative events, another impetus for personalized medicine was the recognition of ADRs for population health. The early reports of aplastic anemia in patients exposed to chloramphenicol, followed by the thalidomide disaster in 1961, led to searches for at-risk subpopulations or conditions (e.g., hepatic failure) and drug–drug interactions that predispose to drug toxicity. These developments and the accompanying scientific scrutiny culminated in the publication, for the first time, of formal principles for individualization of drug treatment based on disease, genetic, and environmental chemical influences [14].

PHARMACOGENOMICS: AN OLD SCIENCE BECOMES LARGE-SCALE AND GLOBAL

Pharmacogenomics is a newer term introduced in late 1990s and has a broader scope than pharmacogenetics. Pharmacogenomics is the study of variability in drug safety and efficacy using information from the *entire* genome of a patient without *a priori* hypotheses specific to a candidate gene. Variations in both gene sequence and expression are of interest to pharmacogenomics inquiries. Despite differences in their scope, there is also interdependency between the two fields: when a genetic biomarker relevant to drug pharmacokinetics or pharmacodynamic is identified through genome-wide pharmacogenomics research and development (R&D), each biomarker requires further validation by hypothesis-driven pharmacogenetics candidate gene approaches before it can be introduced at point of care in the clinic and public health practice. In this chapter, we hereafter use the term "pharmacogenomics," but many of the presented concepts and examples are also applicable to pharmacogenetics.

Scientific folklore has changed since the pioneers Lind, Garrod, Motulsky, Vogel, Kalow, Eichelbaum, Smith, and others first introduced the important, seminal ideas of personalized medicine over the past three centuries. Chief among these changes are:

1. Globalization of R&D and greater interdependency among nation-states for human development and sustainability, not to mention infectious and noncommunicable diseases that do not recognize the traditional Westphalian model of independent sovereign nation states
2. Enormous increase in the scale of science and data production not only in pharmacogenomics and personalized medicine but literally every aspect of daily life [2,15]
3. Emergence of infrastructure science such as population biobanks, databases, and cloud computing in parallel to traditional discovery science [16]. As we describe later in the section "Beyond Classic ELSIs," these transformative changes in postgenomics science and technology are not without ethical significance.

Insofar as global science is concerned, the annual investment in global R&D has doubled to $1.1 trillion since 1996 [17]. A recent analysis of the scientific publication data from the Web of Science over the past 30 years (1980–2009) found that Asia's share of the world scientific output grew stupendously by 155%, while the Middle East, as a region, displayed a growth four times faster than the world level [18]. Many countries large and small are embracing the concept of the knowledge society, whether because of hopeful expectations for a prosperous "knowledge economy," tighter economic interdependence, or decreasing cost of genomics analysis and the next generation (e.g., whole exome) sequencing technologies [19,20,21]. China, India, Korea, Singapore, and Qatar are among the notable nations substantially investing in science and engineering research, education, and workforce [17]. Consistent with this finding, a recent study on the globalization of science found, using quantitative indicators, that the current growth rate for state-level research funding in China greatly exceeds that in the United States and the European Union (EU) [22]. Moreover, Africa has become the emerging epicenter of genomics research investments: the Human Heredity and Health in Africa (H3Africa) Initiative (http://h3africa.org) is building momentum to accelerate the study of genomics and environmental determinants of common diseases with the goal of improving the health of African populations [23]. H3Africa is a consortium of African scientists, enabled by international partners such as the Wellcome Trust in the UK and the National Institutes of Health (NIH) in the US, that aims to bridge the research, expertise, and infrastructural genomics gap that Africa currently faces. As of early 2012, peer-reviewed and selected projects are being funded under the H3Africa theme.

As referenced in the quote in the introduction by Stephen Friend [2], postgenomics personalized medicine science is experiencing a "data deluge" from emerging high-throughput technologies and a myriad of sensors and exponential growth in electronic records. Indeed, data-intensive science was named the "fourth paradigm of science," preceded by the third (last few decades: *computational branch*, modeling, and simulating complex phenomena), the second (last few hundred years: *theoretical branch*, using models leading to generalizations), and the first paradigm (a thousand years ago: *empirical description of natural phenomena*) [24].

Together with large volumes of data, crosscutting collaborations are now essential to enable discoveries in recent data-intensive "big science" projects in the postgenomics era. Commencement in September 2010 of the Human Proteome Project (HPP) in the health sector, and the Sloan Digital Sky Survey completed in 2008 in astronomy ("Cosmic Genome Project") are two well-known recent examples of the Fourth Paradigm Science [24,25]. In the case of the HPP, efforts for the development of proteomics tests of drug safety and efficacy bring about a much-needed dynamic vision to the biomarker field because proteome-based biomarkers are more sensitive to, and thus able to capture, host−environment interactions germane to pharmacogenomics and personalized medicine [26].

With data-intensive fourth paradigm of science, collaborative governance of novel technologies among an "extended community of peers"—networks of networks—that span governments, academia, industry, and various end-users of knowledge in society has become a *sine qua non* for personalized medicine R&D [2,15,20,27]. This development signals both structural and functional transformations in the architecture of twenty-first century scientific inquiry: the hypothesis-driven tradition of science ("first hypothesize then experiment") is shifting to one that is typified by a "first experiment then hypothesize" mode of highly data-driven discovery [25]. Open science and access to large volumes of data, infrastructure science, and global governance of data-intensive "omics" technologies are now key factors that shape pharmacogenomics scientific practice in the beginning of the twenty-first century. Surprisingly, such important drivers of postgenomics personalized medicine have been thus far inadequately examined within the pharmacogenomics bioethics discourse—a gap this chapter aims to address. Moreover, the shifts in the architecture of twenty-first-century pharmacogenomics and personalized medicine science call for rethinking twenty-first-century bioethics so as to proactively steer the science and technology innovation trajectory, instead of the more narrowly framed "enabler," "protector," or "regulator" roles hitherto assigned to bioethics in the twentieth century.

The aim of this chapter is to introduce the scientific readership to emerging innovative approaches to twenty-first-century bioethics for data-intensive fourth-paradigm science that forms the central pillar of postgenomics personalized medicine R&D. Although there has been much written on the bioethics aspects of pharmacogenomics and personalized medicine that is already available elsewhere [28,29,30], herein we focus on the rapidly emerging ways in which bioethics discourse is reshaping by the data-intensive fourth-paradigm science. Chief among these new approaches is a *broadening* in bioethics frameworks in the twenty-first century to take on new roles well beyond the protectionist paradigm that has prevailed post-Nuremberg. Through awareness of these novel conceptual frameworks that are currently reshaping twenty-first-century bioethics, we trust the reader will be well poised to embark on independent personal reflection and critical analyses on issues of bioethics significance for global pharmacogenomics and personalized medicine.

RETHINKING BIOETHICS IN THE TWENTY-FIRST CENTURY

Bioethics Post-Nuremberg: The Protectionist Framework and Its Corollaries

It might perhaps come across as a surprise to some readers that biomedical ethics—as field of scholarly inquiry—has a relatively recent history. It rose to prominence in the second half of the twentieth century in response to crimes against humanity and breaches in research ethics in postwar United States [31,32]. Over the past seven decades post-Nuremberg, decision making in bioethics has primarily relied on the theme of "protection" and normative analyses (e.g., ethical/unethical technology) using moral principles such as autonomy, beneficence, nonmaleficence, and justice. These principles have been collectively termed the "Four Principles Approach," first presented in 1979 in the works of Beauchamp and Childress in their textbook *Principles of Medical Bioethics* [33]. This approach and its embodying four principles have often been utilized as guidance for setting the bioethics norms in research and clinical practice post-Nuremberg:

- *Respect for autonomy:* respecting the decision-making capacities of autonomous persons
- *Beneficence:* balancing benefits against the risks and costs
- *Nonmaleficence:* avoiding the causation of harm
- *Justice:* distributing benefits, risks, and costs fairly

Among these principles, autonomy and individual choice have historically prevailed in twentieth-century bioethics frames, particularly in the Western developed countries. Although protection and autonomy of research subjects are no doubt crucial, the idea that medical interventions that target a given person occur in a vacuum is inherently false. Consider a population in which an infectious disease is highly prevalent: treatment and preventive chemoprophylaxis of a person will have impacts, no doubt, beyond that individual. It decreases the reserve of the infectious pathogens that can be transmitted to other persons in the same population or across national borders between different populations. Similarly, vaccines benefit not only the persons who are vaccinated but also others by achieving herd immunity for the population. Both developed and developing countries are facing an alarming increase in the prevalence of noncommunicable diseases [34]. Hence, even for medical interventions ostensibly targeted on individuals, the autonomy-based narrow ethics frameworks grossly neglect such broader "connectivity" and interdependence among members of the global society in late modernity in 2012, and the population-level impact of a health intervention that "reaches out" well beyond any given person.

The narrow framing of biomedical ethics around individual choice throughout the twentieth century has bracketed out bioethics in developing or less affluent countries—a concern to be born in mind, especially in light of the current globalization of pharmacogenomics and personalized medicine R&D discussed previously. In a context of global bioethics, a much broader set of moral principles is seriously in need of consideration in order to respond to uncertainties raised by new genomics and personalized medicine applications in developed and developing countries.

In addition, ethics analyses and scholarship in twenty-first century are seriously in need of greater "symmetry": both *protection* from risks and *benefits* of global science have to be considered in tandem so as to develop a nuanced and in-depth understanding of the ethical issues in pharmacogenomics and personalized medicine. Moreover, bioethics analyses have hitherto concentrated on norms and standards concerning how things (e.g., a certain scientific practice, technology, persons, or institutions) "ought to be" drawing from moral theories and philosophical principles but have not always adequately appreciated how "things actually are" in practice or the social, political, economic, scientific, and technical contexts in which the bioethics dilemmas emerge [35]. The greater recognition of the idealized norms over their practice, and the attendant social context in a real-life setting, however, isolates bioethics from the actual practice of science and medicine. The intent of this critique is not to say that the principlism approach described above does not have instrumental utility—it does—but neglecting the real-life context of science and medicine results in a prescriptive and top-down hegemony in bioethical reasoning and eliminates the possibility of designing effective policy interventions to bring together the lived practice of science and medicine with the bioethics norms when the practice and theory are not aligned (and often they are not) [35,36].

Others have further argued that the protection pendulum has swung so far in the other direction, resulting in bioethics "mission creep" compromising the salient research ethics board mandates. For example, it has been suggested that the current ethics review system is rewarding the wrong behaviors such as mechanical, bureaucratic, and stylized documentation instead of substantive review, misdirecting precaution to research that poses little risk to subjects, which altogether can discourage innovative research that is already ethical [37,38], not to mention diverting the limited resources away from situations where genuine research oversight and protections are needed.

Taken together, a great deal of the ethical, social, legal, and public policy issues emerging from, and having an impact on, global pharmacogenomics and personalized medicine R&D cannot be adequately identified, nor resolved effectively, with the individual choice and autonomy-based protectionist ethics frames that we have inherited from the twentieth-century bioethics discourse [1,36,39]. There is a serious need to balance autonomy and individual choice-based bioethics frameworks with collective action-oriented principles such as solidarity and citizenship. The principle of proportionality is one such safeguard in domestic administrative law and international humanitarian law to balance the powers of the state or administrative bodies against the rights and autonomy of individuals. In its European formulation, the proportionality principle embraces "the express articulation and explicit weighing of the specific aims of a measure in relation to its impact on a right or interest invoked by the applicant" [40].

Bioethics in the Postgenomics Era: Responding to the Fourth Paradigm of Science and Public Health Ethics

In 1990, the U.S. Department of Energy and the National Institutes of Health allocated 3 to 5% of their annual Human Genome Project budgets toward studying the ethical, legal, and social implications (ELSIs) surrounding the availability of genetic information. With these origins more than two decades ago, the U.S. ELSI program had a number of priority areas, such as fairness in the use of genetic information (e.g., by insurers and employers), privacy and confidentiality of genetic information, societal impacts and stigmatization due to human genetic differences, adequate informed consent, clinical genetic testing, and commercialization of products. In Europe, the European Commission established a similar transdisciplinary program on ethical, legal, and social aspects (ELSA) in 1994 [19]. Unlike the U.S. ELSI framework, the European ELSA program aimed to address not only genetics/genomics but the broader field of life sciences and technologies [41].

Over the decade since the completion of the first draft of the Human Genome in 2001, the professional scope of bioethics, whether in the form of ELSI or ELSA activities, began to increasingly overlap with and extend into public health ethics, in part owing to a largely global science that demands a population focus and the necessity of collective innovation for personalized medicine. Distinct from the individual autonomy-driven bioethics discourse in the second half of the twentieth century, public health ethics issues are guided by a broader focus on the population rather than a narrow focus in a context of a given person as noted in the quote in the introduction [1]. Additionally, public health ethics aims to address issues emergent from components of health care that do not necessarily target individuals.

A case in point are the population health *infrastructures* (e.g., longitudinal biobanks and databases) that carry substantive ethical significance and where participants contribute as "citizens" with no personal or immediate benefit [16]. Indeed, the ethical concern over individual choice, autonomy, and informed consent should be understood as being embedded within such broader health infrastructures that together constitute public health and public health ethics. Such public health infrastructures are not distributable goods, nor do they represent health interventions targeted for a given person that can be subject to individual choice and consent. Public health infrastructures sustain, and are sustained by, the global or regional populations and thus often raise different sets of ethical issues that relate to "collective action" [20,42]. These ethical issues that have been traditionally omitted and marginalized in biomedical ethics are instead shaped by principles such as solidarity and citizenship [43].

Insofar as collective action related to public health ethics is concerned, we know very little about how individuals and stakeholder groups are organizing themselves for collective action in the practice of data-intensive sciences; ways in which collaborations (e.g., cooperation, competition, precompetitive collaboration, postcompetitive collaboration) are developed, sustained, or devolve; what human values guide and inform the knowledge ecologies emergent from collective innovation; and ultimately, how knowledge is created, transformed, and transferred between innovators and users of personalized medicine. As we move from a science where data collection was an essential locus of the scientific endeavor to the fourth paradigm of science where data collection is automated or at least made easier and is available in digital form, there is a need to understand the ethical principles and values that underpin collective action. In this regard, we suggest that traditional modes of knowledge sharing, such as international meetings, preplanned teams, and the traditional peer review, are possibly becoming constraints on the collaboration needed to bring the fourth paradigm of scientific inquiry to fruition. Seen in this light, the bioethics dimension of the postgenomics personalized medicine R&D is increasingly gaining a "knowledge lens" whereby *scientific knowledge itself*, not to mention its production, sharing, access, analysis, and utilization, are gaining *bona fide* bioethics and political science significance, in much the same way as the classic ELSIs such as informed consent.

BEYOND CLASSIC ELSIs

Technology Governance and Knowledge Production

The newer roles and responsibilities assigned to, and acquired by, twenty-first century bioethics in a context of the scientific knowledge trajectory and its governance—in addition to the previous limited focus on protection of research subjects and rules governing the doctor and patient relationships—make it clearer that both bioethics and personalized medicine science are in need of serious reexamination so as to respond to the actual nuances of current twenty-first-century data-intensive fourth paradigm science [24]. The solutions to these challenges in part rest on new ways of governance of science and technology as well as knowledge production [20,42,44,45,46,47,48,49,50].

Governance in a context of health can be defined as "the attempts of governments or other actors to steer communities, countries or groups of countries in the pursuit of health as integral to well-being through both a whole-of-government and a whole-of-society approach" [51]. Novel technology governance is increasingly recognized as a crucial component of responsible innovation that is closely attuned to societal values and thus socially contextualized and robust [52,53,54,55].

A crucial question in technology governance is: *precisely when should such governance efforts commence?* In this regard, recognition of the importance of early "upstream" engagement with innovations has led to anticipatory approaches to technology governance [20,56,57], including new personalized medicine-related fields, such as vaccinomics [15,58].

As an integral part of such anticipatory approaches, frequent contacts between knowledge generators (e.g., scientists, technology designers) and end users (e.g., policy makers, publics), as well as creating mechanisms between supply and demand of knowledge and technology, are necessary to move towards responsible innovation in democratic societies [59]. Asymmetry of knowledge (or divergent attitudes towards new information) among stakeholders is also an occasion to proactively foster deliberative learning processes among the constituents of an innovation ecosystem [60]. Indeed, researchers engaged in social studies of science and technology have long emphasized that the social and economic benefits of biotechnologies such as genomics are *not* automatic (i.e., "they do not flow inevitably from the marriage of biology and technology") [55,61]. Lavis et al. [62] recommend that:

> Researchers (and research funders) should create more opportunities for interactions with the potential users of their research. They should consider such activities as part of the "real" world of research, not a superfluous add-on (Page 146).

Conversely, waiting to adopt a technology until its future trajectory is "locked" into a certain path might result in greater certainty on the attendant social impacts. But attempts to modify the technology at this late stage become difficult as it is then *entrenched* in a complex nexus of sociotechnical, economic, and political dependencies [63]. That is, negotiation of technology future(s) between science and society is feasible primarily at early phases of innovations when ideas are just that—cognitive constructs open for debate and deliberation, not beliefs that are difficult to reframe in light of new evidence [63].

Anticipatory governance with participatory foresight has emerged over the past decade for policy-relevant decision making under uncertainty while the science and technology are still in the making [20,49,52,54,64]. Anticipatory governance maps the human values embedded on a volatile and nonlinear technology trajectory across an innovation ecosystem. As a contrast to narrowly framed "downstream impact assessments" for emerging technologies, anticipatory governance is a broad capacity that runs through the stakeholders of an innovation ecosystem with a view to understanding the social shaping of innovation and technology design. This approach creates a knowledge platform for collective iterative learning among the members of an innovation ecosystem. By early upstream engagement with technology designers, producers, and users, anticipatory governance builds a broad capacity for "extended peer review" and social embedding of novel technologies and innovation. Guston (54) describes anticipatory governance as:

> A broad-based capacity extended through society that can act on a variety of inputs to manage emerging knowledge-based technologies while such management is still possible (Page 4).

This upstream shift of the governance discourse from a consumer "product uptake" focus to "participatory technology design" on the innovation trajectory is an appropriately radical and necessary departure in the field of bioethics, especially given that considerable public funds are dedicated to innovations. Anticipatory governance underscores shared governance, the coproduction of knowledge by science and society, and the inseparable nature of "facts" and "values," where both of these elements need to be made explicit and deliberated to achieve innovation in governance (Table 11.1). This broader approach to "knowledge" (beyond expert opinions) allows an examination of the unchecked assumptions, values, and power systems that crucially and collectively shape visions of technical future(s).

TABLE 11.1   Options for the Governance of Data-Intensive Fourth Paradigm Science that Crucially Underpins Postgenomics Personalized Medicine R&D

By "Technological Determinism"	By "Anticipatory Governance"
Driven by technological determinism or the "science push" model of innovation and knowledge translation	Driven by the notion of "coproduction" of vaccinomics knowledge by science and society
Envisions a singular deterministic future for an innovation trajectory	Envisions possible multiplex future(s) for an innovation trajectory
Focuses on prediction of the future	Focuses on anticipation of the future(s)
Front-loaded assessment: claims to be *'future-proof'*	Adaptive and incremental responses to future(s)' uncertainties: *'future-engaged'*
Employs "regulatory" tools such as the precautionary principle or quantitative risk assessment to *control* technology and innovation	Employs "governance" instruments such as participatory foresight to *steer* technology and innovation trajectory
Based on expert knowledge	Based on "extended peer review" via expert knowledge as well as public engagement and tacit/locally situated knowledge
Scientific evidence is the only authority that can justify policy action	Recognizes the need to make policy-relevant decisions under uncertainty; public policy issues have dimensions beyond technology such as social benefits and justice as well as science; uncertainty is "sociotechnical"
Assess downstream "impacts" of technology and innovation	Influence technology upstream at "design" stage and thereby shape innovation trajectory by sociotechnical integration

Most important, anticipatory governance embraces an important theme that is not yet evident in the growth of pharmacogenomics over the past two decades: the focus on *change management* to deliver improved health. Through a process of multistakeholder engagement (for example, with professional experts, patient groups, voluntary organizations, and policy makers in various fields), this participatory knowledge is used to develop and implement strategies for improved health. The emerging field of anticipatory governance of science and technology embraces this commitment to change management on a day-to-day basis in real-life societal settings, beyond the laboratory benches and ivory towers.

Recent examples of demands by research funding agencies to put data in the public domain and to anticipate the broad impacts of proposed research at the time of research funding application suggest that anticipatory governance may be one way in which postgenomics scientific practice might transform in the future toward responsible innovation [55,65].

Taken together, technology governance and knowledge production processes are being recognized as legitimate components of the new and expanded bioethics agenda in the twenty-first century. Technology governance is also transforming its own identity in this process of inclusion in postgenomics bioethics: governance questions such as "Do we adopt/reject a technology, given that it is now well developed and mature?" are being replaced with "How can a new technology and its applications be codesigned and governed collaboratively early on, together with innovators and anticipated end users?"

Open Science and Valorization of Postgenomics Knowledge

Despite the promise and selected success stories brought about by the fourth paradigm of data-intensive scientific discovery and development, there is a need for novel mechanisms for valorization of postgenomics knowledge. Promising avenues could be based on the open source model implemented through collaborative knowledge innovation platforms (e.g., crowdsourcing, expert sourcing, cloud computing, social networking technologies). Their potential benefits, though too soon to empirically evaluate, are considerable and include the following [66]:

- Promoting the interoperability and harmonization of research platforms and tools
- Reducing financial risks
- Democratizing access to innovation
- Preventing harmful duplication of research
- Facilitating the technology transfer process

Nonetheless, shifting from a classic commercialization and individual-entrepreneur model of innovation to one that stresses collective innovation, transparency, and collaboration requires an acknowledgement of more than just the ensuing (and obvious) technical and economic changes. It also requires an exploration of the ability of different stakeholders in a global science to have access to openly shared data and actively participate in collaborations, the role of ownership and commercialization of genomic data, and the ability to uphold the duty of confidentiality and respect of participants' privacy. As postgenomic era science advances in its march towards collective innovation to create novel diagnostics, democratize "omics" research, and move from silos to systems, we need to consider that open source, along with collaborative knowledge innovation platforms, will play an essential role.

Interest in new forms of knowledge production and valorization is not limited to governments, research funding agencies, and academic investigators. In recent years, pharmaceutical companies have increasingly been willing to experiment with new business models. This apparent turn towards "open science" in the industry was largely shaped, however, by undeniable externalities such as the end of the blockbuster era, the growing innovation gap in pharmaceuticals, and public and stakeholder responses to lack of innovative products and "me-too" drugs that do not offer added clinical value. These factors have collectively invited the industry to reconsider its traditional proprietary, secretive business model centered on patenting and licensing.

The industry now considers mini-busters and the remarketing of previously withdrawn drugs to be reassigned for different conditions or for specific population subgroups who are unlikely to incur serious ADRs. Open collaborative models could offer a strategic alternative for pharmaceutical companies, health care payers, and ultimately the public. In this context, "open collaborative model" is used to refer to a broad category of potential business models such as patent pools, patent clearinghouses, open innovation, open source, open access, and patent covenants, among others [66].

The most obvious and least controversial use of open collaborative models in pharmacogenomics R&D is at the *upstream* stage. At this precompetitive exploratory stage of the research, which requires significant collaborations with and dependence on academic research, it is easy to make an argument for openness and collaboration. The best example of one such project was the Single Nucleotide Polymorphism Consortium, a not-for-profit foundation organized with the objective of producing and sharing genomic data on single-nucleotide polymorphisms (SNPs) via a publicly accessible Internet database. This new type of open collaboration between pharmaceutical companies in the context of pharmacogenomics could be a sign that new business models might be emerging in the near future, though their long-term trajectory requires empirical examination as they evolve.

On the other hand, *downstream* open collaborative models still occupy only a marginal space in the business strategy of the big pharmaceutical companies. Indeed, their adoption could have more to do with public relation strategies than with a belief that openness can constitute a viable business model. This concept is in part justified by the traditional reasoning that clinical trials are extremely expensive and that their cost must be recouped via patenting and licensing activities. It is certainly true in part: clinical trials are expensive and R&D expenditures need to be recouped somehow under the current pharmaceutical development business models. However, it is not at all clear that pharmacogenomics patents are the best way to achieve this. Pharmaceutical patents are expensive to obtain and to maintain and defend, making them unaffordable for most small and medium entities. Moreover, genetic patents are controversial and strongly contested on the legal front. Gene patents and patents on diagnostic tests based on gene mutations are often contested at the U.S. Supreme Court and thus can be invalidated in the near future. Given this prospect, the popular new strategy for pharmaceutical companies involved in pharmacogenomics research to patent drug—diagnostic combination could also eventually face legal challenges.

In all, pharmaceutical companies have much to gain in using business models that are less dependent on patenting and licensing. Brand fidelity, strategic public—private alliances, and open collaborative models will all likely play increasingly important roles in new innovation frames brought about by pharmacogenomics. It would certainly be beneficial for sustainability of pharmacogenomics to create a public forum in which all stakeholders could meet to discuss issues of commercialization of and access to drugs and diagnostics as well as open business models.

Infrastructure Science

Building the Resource Commons for the Fourth Paradigm of Twenty-First-Century Science

A hallmark of the fourth paradigm of data-intensive science is that it amasses data that are public goods (i.e., creates a "commons") that can further be creatively mined for various applications in different sectors [15,24]. In the postgenomics era, infrastructure science sets the building blocks for discovery science, in particular by providing high-quality biospecimens of different tissue types and statistically robust sample sizes required for research in personalized medicine and in rare diseases [67,68]. Numerous population biobanks are presently being constructed in countries small and large to answer this growing need for large amounts of data and high-quality biospecimens driving personalized medicine research. Many view the creation of these emerging clinical and population health infrastructures and new methods of knowledge generation as crucial enablers of postgenomics R&D.

The population biobanks provide both samples as well as related environmental, social, economic metadata and phenotypic variation data in the population, which collectively help reveal gene—environment interactions, while also providing, in some cases, controls for replication, comparison, and validation of personalized genomic discoveries [67]. It is interesting to note that population biobanks are currently being developed in many parts of the world. Although these efforts need to be further translated into applications in discovery research and the clinic to illuminate gene—environment contributions to disease risk and health intervention outcomes such as drug therapy, the ELSIs pertaining to, for example, consent, confidentiality, and oversight, as well as incorporation of genetic/genomic data into electronic records, cannot be examined using a one-size-fits-all approach; the particularities of each biobank must be taken into account. Additionally, "novel ELSIs" are also emerging, such as incidental findings from upstream genomics research such as genome-wide association studies (GWASes). In many instances, it is noteworthy that such findings do not neatly fall within the classic definition of "results" but rather fit within the new notion of "data deluge," often with unknown clinical significance. Seen in this light, it is essential to reconsider the term "return of research results" as distinct from "return of data." Moreover, such novel ELSIs emergent from discovery genomics research as well as classic ELSIs (e.g., informed consent) in the novel context of population biobanks are best addressed through empirical research to map and effectively respond to them. Future social science and policy research will help inform whether and to what extent the current governance approaches are adequate to effectively respond to these ELSIs.

Given that one of the ultimate goals of pharmacogenomics is to provide individually targeted therapies, knowledge of the underlying biological pathways is a prerequisite to clinical translation. To this end, GWASes and other sources of personalized genomic information may one day enable clinical decisions based on individual genomic makeup. Population-based screening can provide raw data and tools for understanding the biology behind genomic variants and identifying potential drug targets and biomarkers [68].

The growing reliance of postgenomics science on large infrastructures has led to both a shift in the locus of interest, from individual to population health, and has brought about the necessity of preserving these infrastructures for future use and research.

Biobanks: From Individuals to Populations, and Back Again

The ever-increasing role of infrastructure in genomic research, particularly in personalized medicine, is giving rise to a broadening of the current ethical framework of medical research, a change that may yet shape the trajectory of the long-term sustenance of these resources. On the one hand, the field of personalized medicine and pharmacogenomics has typically been interested in achieving patient-centered and individualized benefits, with the attendant ethical issues focused on individual-centered perspectives, such as questions of autonomy, consent, and privacy as discussed above. On the other hand, ethical issues related to large population biobanking initiatives (increasingly international and collaborative), though initially focused on these "individual" ethics, are now increasingly addressing broader issues relating to population health, community, solidarity, and citizenry [43]. The ethical impetus behind large research infrastructures has been broadening into considering collectivist approaches to health, in view of the realization that a narrow focus on the health of individuals may be insufficient to promote the health of populations [1,16,39]. While not setting aside or abandoning discussions about individual choice, consent, and autonomy, the shift towards population sampling and public health has set traditional ethical questions in the context of a wider range of issues [1]. Ultimately, achieving the promise of individualized and personalized medicine will depend on the willingness and trust of the public, and scientific community, to participate in the larger biobanking efforts and to provide biospecimens for translational research [69]. In addition, there is growing support in the literature and international normative documents in viewing genomic resources, and data contained in biorepositories, as having a "communal value" [70], and constituting the "common heritage of humanity" or "global public goods" [71,72]. Still, given the concern and debate over ELSIs surrounding individual autonomy, consent, and privacy, one concrete way forward for public health ethics is to be cognizant of the historical background and analyses provided in this chapter so as to recognize that these concerns in fact build on past emphasis on protectionist bioethics, rather than the *actual* needs of public health applications of genomics and personalized medicine. Following on this expansion of research ethics from individuals' autonomy to population-level considerations, it has been argued that the bioresource commons are evolving from a structure in which origin, ownership, and materials are in the hands of the same community, to one where scientific progress relies on efficient sharing of bioresources [16]. Moreover, a recent report commissioned by the Nuffield Council [73] has renewed interest in this debate. While acknowledging that "research biobanking is embedded into a rhetoric of personalized medicine, suggesting individual benefit [which] renders research biobanks an embodiment of a promise that they might struggle to fulfill" [73], the report emphasizes that given that research biobanks are typically long-term research projects that do not necessarily produce results for a certain time, only altruistic values and those prioritizing collective benefits can be used to justify individual investments [73,74]. While biobanks and attendant ELSIs continue to be mapped, a consensus within the broader bioethics community awaits the resolution of the tensions between the emerging field of public health ethics and the individual autonomy-based protectionist ethics.

Sustaining the Bioresource Commons

Complementary to the emergence of solidarity and community as new building blocks of research ethics is the idea that collective resources and benefits must also be passed on to, and shared with, future generations. Indeed, as donors have given the initial "push" to the creation of these large repositories [75], we may very soon witness the emergence of an ethical duty to "use the resource in a peaceful and responsible way, keeping it accessible to all and considering the interests of future generations" [76], and to promote the long-term sustenance of these infrastructures. Although repositories of high-quality specimens are currently being built for future research in medicine (including personalized medicine) [74], next steps will require maintaining these biobanks in order to follow through on the promises and public expectations fueled by their creation.

Sustainability of infrastructures requires both viable quality parameters (e.g., high-quality biosamples, harmonized governance structures, transparent oversight, access structure, etc.)

and long-term durability (e.g., sustainable funding source, biobank networking and growth). The biobanking community has already invested great efforts and resources in collaborating on issues related to data standardization, sharing, and acknowledgement of the data commons. In particular, at the international level, because of the complexity of managing large and longitudinal projects, much of the emphasis relating to population and other large-scale biobanks has focused on attempting to harmonize practices (e.g., consent forms, information questionnaires, data access procedures, measurements, sample treatment, etc.) in order to cultivate collaboration at an international scale. In addition, as biobanks can be costly to build and may take many years to mature to their full value, there is a call for funders to recognize these infrastructures as core platforms to support the science community, as is already the case in some regions [77].

Indeed, one of the current challenges in the development of biorepositories has been the "lack of long-term secure funding for developing and sustaining biobanks and biobank research" [69]. In global regions or countries where consistent public funding has been more difficult to secure, interesting new issues related to biobanking sustainability are emerging. For example, the study of what some have advocated as "biobankonomics" [69], which can be more broadly described as an attempt to estimate the economic, social, and health-care benefits of biobanking infrastructures, has shed new light on the necessity to broaden the "time-delimited," hypothesis-driven medical research [16].

Although the promise of translational research is nearing, current lack of data on the economic and scientific value of biobanking and biospecimen resources makes it difficult to assess the current state of return on (public or private) investments in such population-level scales [69]. This economic assessment is particularly pressing, as funding approaches and requirements can directly impact biospecimen quality (e.g., sample collection, storage conditions, data analysis, etc.) and may also shape other issues such as how access to samples and data is granted and controlled [75]. Questions of sustainability and viability may eventually provide the impetus for the creation of an independent field of "biospecimen science" to answer the important questions of how to "fund, organize and produce a more robust knowledge base" [75], and efficiently use the increasing amount of data on human populations to eventually be able to translate this knowledge for individualized health care and drug development.

AN INVITATION TO REFLEXIVE AND GLOBAL BIOETHICS

The changes in pharmacogenetics and personalized medicine R&D over the past five decades are not merely limited to technology breakthroughs such as next-generation sequencing and an expansion in the scope of scientific inquiry from candidate genes to a broad genome-wide scale. The very processes of pharmacogenomics knowledge production, synthesis, dissemination, and utilization have also transformed, with greater emphasis now placed on technology governance, large-scale collective innovation, infrastructure science, population biobanks, and data/biocommons [2].

To be sure, twenty-first-century bioethics will be (and is) much different than the narrow (bioethics) frames of the twentieth century [15,35,41,43]. As should be clear to the reader at this point, a much broader set of issues—well beyond consent and individual autonomy—are affecting the day-to-day practice of postgenomics personalized medicine R&D as discussed in this chapter. Chief among these novel ELSIs is the postgenomics scientific knowledge itself [2,78]. This development signals a knowledge turn for the new bioethics discourse in the twenty-first century whereby all segments of the scientific trajectory—from data production to data sharing, storage, analysis, and knowledge translation—are recognized to have bioethics significance. In contrast to the "normal science" [79], some have referred to the current mode of knowledge production as postnormal science, in which knowledge is produced in highly diverse and globally distributed locales far exceeding the laboratory benches and the ivory towers in developed countries and in the face of uncertainties in which the "facts are uncertain, values in dispute, stakes high and decisions urgent and where no single one of these dimensions can be managed in isolation from the rest" [44,80]. Others have named the similar forms of complex and socially embedded knowledge production processes "Mode 2" [45] or P5 Medicine [81]. Mode 2 refers to "distributed" forms of knowledge production in a manner that is inclusive from a geographical and disciplinary standpoint and one that recognizes the role of stakeholders including but beyond academia, such as citizens, nongovernmental organizations, industry, patients, and advocacy groups. Mode 2 knowledge contrasts with Mode 1 knowledge that is typically attributed to technology elites or experts in ivory towers, academia, governments, or elsewhere. Rise of Mode 2 knowledge calls for the recognition of P5 medicine, noted above, to merge predictive, preventive, personalized, and participatory (i.e., P4) medicine with an integrated study of the political science aspects of knowledge societies and innovations.

Although ELSIs have often been understood as impacts of scientific progress, bioethics is not simply a consequence of, but is rather embedded in, science and technology practice and is thus "context-emergent." Given the highly porous nature of the boundaries between technology and the social systems (e.g., human values that underpin the data-to-knowledge trajectory, not to mention the choice of scientific hypotheses by scientists), the new global bioethics is best understood as being "coproduced" with science rather than subscribing to the banal false idea that science and technology inevitably outpace the corresponding bioethics, legal, and regulatory response. Such lags in bioethics, legal, and policy responses to science and technology innovations are by no means inevitable in an era increasingly marked by collaborative governance [20,27,82]. Indeed, insofar as the much-needed coevolution of science and law is concerned, Dove (83) has recently observed that:

> This would produce regulations or laws in the science and technology domain that are not only appropriately drafted and sufficiently broad to anticipate possible future events, but that are also an accurate reflection of consensus among all community members (page 259).

This chapter asserts that both twenty-first-century bioethics and postgenomics personalized medicine science are in need of self-examination. Such self-examination can be usefully cultivated through a sound understanding of the concept of reflexivity: how our own values—as individuals or communities of scientists, social scientists, humanists, and publics—affect *how* we know *what* we know.

As a concept, reflexivity is important for the twenty-first-century bioethics discourse because social systems such as human values and ways of knowing—*what we choose to know and how we know it*—expressly influence what gets to be produced as scientific knowledge [84,85]. The choice and framing of scientific hypotheses, experimental methodology, and interpretation of data can all be influenced by experts and their institutions' value systems that often remain *implicit* in scientific decision making. Reflexivity can play a key role to explicate these value systems and, by extension, offers a new promise for modern bioethics to render postgenomics knowledge production more accountable, globally distributed, equitable, and thus sustainable.

We suggest that reflexivity ought to be cultivated within twenty-first-century bioethics so as to recognize and map the practice context of postgenomics science from which many of the bioethics issues emerge (Figures 11.1 and 11.2). Without a keen empirical knowledge of the scientific practice in the postgenomics era, bioethics analyses are at risk of being remote

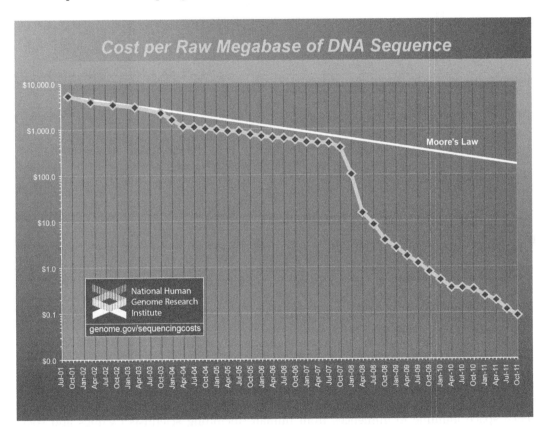

FIGURE 11.1 **Declining cost of sequencing a human genome.** *Moore's law* is the concept that the number of transistors (and by extension, computation and data *storage and analysis* capacity) on a computer chip doubles every 24 months, allowing 'chip scale' to be reduced proportionately. However, in the years from 2007 to 2011, the rate of reduction in human genome sequencing cost has far exceeded the data analysis and storage capacity defined by Moore's law. Consequently, this shift in 'bioinformatics bottleneck' also shifts the current 'bioethics bottleneck' to that of addressing the ethical, legal, social, and policy issues pertaining to *data analysis and storage* (e.g., consider the rise of cloud computing and data storage) instead of the past long-standing narrow bioethics emphasis on sequencing alone.

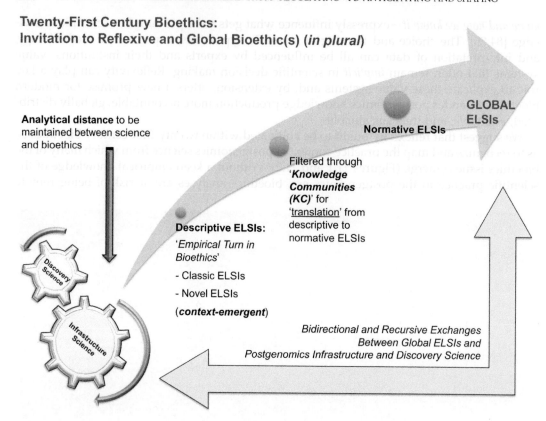

FIGURE 11.2   An invitation to reflexive bioethics. Global bioethics aims to capture both classical ELSIs and novel ELSIs. In the face of a highly globalized and dynamic postgenomics science, novel ELSIs will be 'context-emergent' and demand real-time monitoring. These factors require 'context-engagement' and empirical description of the scientific 'practice milieu' through bioethics research. The descriptive ELSI data to be generated will then filter through different 'knowledge (epistemic) communities' for *translation* into normative frames and global ELSIs in a highly nuanced and context-aware manner so as to steer the innovation and technology trajectory towards responsible innovation.

(and removed) from the key mandates of bioethics—protection of research subjects and steering scientific practice towards responsible innovation.

For too long, we approached scientific innovation in a piecemeal and linear fashion in which innovation was conceived to occur in a vacuum within the confines of the laboratory bench space immune to influences from human (including scientists') values and social systems and invariably moving down a linear pipeline of product development. There are alternatives, however, to such narrow views on scientific knowledge and practice, such as the innovation ecosystems approach.

An innovation ecosystem is comprised of two qualitatively different sets of constituents: *actors* who carry out the research and development and *narrators* (or observatories) who examine the evolution of data from knowledge to action (Figure 11.3). An innovation ecosystem recognizes the processes of interactive collective learning and knowledge

Innovations as Knowledge Ecosystems

(of *Actors* and *Narrators*)

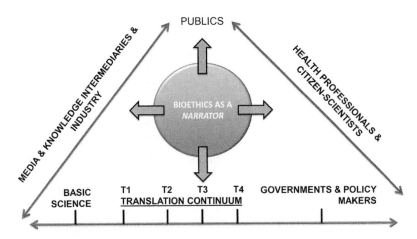

FIGURE 11.3 **Building a Global Positioning System (GPS) for personalized medicine.** In a knowledge and innovation ecosystem (e.g., global personalized medicine R&D) comprised of *actors* of scientific practice, bioethics could serve the role of an *independent narrator* to steer the bidirectional exchanges and contacts between technology designers, knowledge generators, and users for 21st century collective innovation. *Phase 1* translation (T1) aims to advance a basic genome-based discovery into a candidate health application (e.g., metagenomics test). *Phase 2* translation (T2) concerns the development of evidence-based guidelines for an "omics" application. *Phase 3* translation (T3) aims to connect evidence-based guidelines with health practice through delivery, dissemination, and diffusion research. *Phase 4* translation (T4) evaluates the "real world" health outcomes of a genomic application. Inclusion of T1-T4 translation research stakeholders allows a "high resolution" innovation ecosystem, instead of lumping all research under one category of undifferentiated research.

translation between actors such as basic, translational (see 86 for T1-T4 translation research), and clinical scientists, policy makers, governments, publics, health professionals, industry, and media. The proposed reflexive and global bioethics for personalized medicine above (Figure 11.2) is ideally positioned in an independent observatory structure within the knowledge ecosystem wherein the personalized medicine innovations currently materialize (Figure 11.3).

CONCLUSION

Twenty-first-century bioethics will be served well by closer proximity to the practice of postgenomics science as well as the materiality of "omics" technologies (genomics versus proteomics) while also maintaining a crucial analytical distance so as to secure independence from science and, thus, assure credibility in bioethics advice (Figure 11.2). By mapping the actual practice of postgenomics science, twenty-first-century bioethics will be poised to anticipate and develop the diverse set of globally distributed criteria in order to make normative decisions (e.g., ethical/unethical technology) in a highly contextualized and nuanced manner (Figures 11.2 and 11.3).

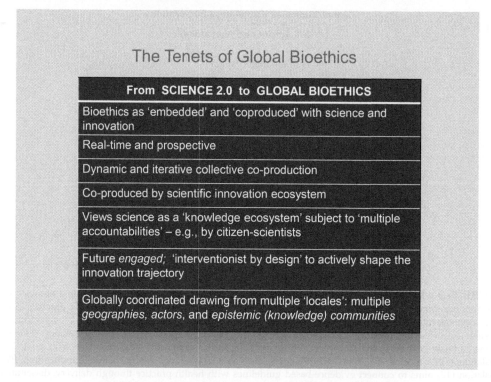

FIGURE 11.4   **The tenets of global and reflexive bioethics.**

Indeed, the ELSI research throughout the past two decades since the inception of the Human Genome Project (HGP) has created a large body of useful but static ELSI data (in much the same way as the genome sequence data hitherto created by the HGP itself). Now is the time to further contextualize and examine its functional impact and relevance (in much the same way functional genomics aims to achieve this for genomes and proteins in the postgenomics era). The tenets of such dynamic global bioethics are further summarized in Figure 11.4.

In the face of a global science that undeniably harbors highly dynamic, diverse, and versatile knowledge communities in developed and developing countries, the new global bioethics will be in a position to best serve its long-touted mandates, responding to both classical and novel ELSIs [87,88].

DISCUSSION QUESTIONS

1. Explain the limitations of the individual autonomy-based bioethics frameworks in regards to genomics sciences and their applications for population health.
2. Provide an overview of the emerging subspecialty of public health ethics that is cognizant of the novel ethical issues attendant to new public health infrastructures for personalized medicine, such as biobanks and large-scale collective innovation consortia.

3. Discuss the nuanced difference between classic ELSIs versus novel ELSIs.
4. What conceptual approaches and frameworks should be considered in addressing postgenomics ELSIs, including population biobanks?

Acknowledgments

The analysis, concepts, and work reported herein were supported by the following grants to the authors: FRQ-S research scholar salary award for science-in-society research in personalized medicine and -omics data-intensive health technologies (VÖ), and open science and innovation (YJ); a research grant from the Social Sciences and Humanities Research Council (231644) on foresight research (VÖ) and; Canada Research Chair in Law and Medicine (BMK). The views expressed in this article are the personal opinions of the authors and do not necessarily represent the positions of their affiliated institutions or the funding agencies. We thank Edward S. Dove, Samer A. Faraj, and Erik Fisher for helpful and spirited discussions on reflexivity, sociomateriality of scientific knowledge, and social studies of science and technology.

References

[1] O'Neill O. Broadening bioethics: clinical ethics, public health and global health. Twentieth Anniversary Lecture. Nuffield Council on Bioethics 2011. May 19.
[2] Friend S. Thinking outside the genome. The Scientist 2011 (October). Available from: http://the-scientist.com/2011/10/01/opinion-thinking-outside-the-genome/ [Accessed February 29, 2012].
[3] Nuffield Council on Bioethics. Medical Profiling and Online Medicine: The Ethics of Personalized Healthcare in a Consumer Age. Oxford: Nuffield Council on Bioethics 2.
[4] James Lind Library. Available from: http://www.jameslindlibrary.org/about_us.html 2003 [Accessed February 29, 2012].
[5] Garrod AE. The incidence of alcaptonuria: a study in chemical individuality. Lancet 1902;2:1616−20.
[6] Motulsky AG. Drug reactions, enzymes and biochemical genetics. J Am Med Assoc 1957;165:835−7.
[7] Vogel F. Moderne problem der humangenetik. Ergeb Inn Med U Kinderheilk 1959;12:52−125.
[8] Kalow W. Pharmacogenetics. Heredity and the Response to Drugs. Philadelphia: W.B. Saunders Co.; 1962.
[9] Schmeck HM. Heredity linked to drug effects. New York Times October 10, 1962.
[10] Vesell ES, Page JG. Genetic control of drug levels in man: antipyrine. Science 1968;161:72−3.
[11] Eichelbaum M, Spannbrucker N, Dengler HJ. Noxidation of sparteine in man and its interindividual differences. Arch. Pharmacol 1975;287. R94.
[12] Eichelbaum M, Spannbrucker N, Steincke B, et al. Defective N-oxidation of sparteine in man: a new pharmacogenetic defect. Eur J Clin Pharmacol 1979;16:183−7.
[13] Mahgoub A, Idle JR, Dring LG, et al. Polymorphic hydroxylation of debrisoquine in man. Lancet 1977;2:584−6.
[14] Reidenberg MM. Individualization of drug therapy. Med Clin North Am 1974;58:905−1162.
[15] Ozdemir V, Pang T, Knoppers BM, et al. Vaccines of the twenty-first century and vaccinomics: data-enabled science meets global health to spark collective action for vaccine innovation. OMICS 2011a;15(9):523−7.
[16] Schofield PN, Eppig J, Huala E, et al. Research funding. Sustaining the data and bioresource commons. Science 2010;330(6004):592−3.
[17] Suresh S. Moving toward global science. Science 2011;333(6044). 802.
[18] Science Metrix. 30 years in science. Secular movements in knowledge creation. Science Metrix. Available from: http://www.science-metrix.com/30years-Paper.pdf 2010 [Accessed March 7, 2012].
[19] European Commission. Ethical, Legal and Social Aspects of the Life Science and the Technologies. Available from: http://cordis.europa.eu/elsa-fp4/home.html 1994 [Accessed February 29, 2012].

[20] Ozdemir V, Knoppers BM. From government to anticipatory governance. Responding to challenges set by emerging technologies and innovation. In: Kickbusch I, editor. Governance for Health in the Twenty-first Century. New York: Springer; 2012. in press.

[21] World Economic Forum. Global Risks. seventh ed, http://reports.weforum.org/global-risks-2012/; 2012 [Accessed February 29, 2012].

[22] Hather GJ, Haynes W, Higdon R, et al. The United States of America and scientific research. PLoS One 2010;5(8):e12203.

[23] Rotimi CN. Conceptual Framework of the Human Heredity and Health in Africa. Ethics and Genomics Research in Africa (EAGER-AFRICA) Conference. Abuja: Nigeria. Available from: http://eager-africa.com/ea/slide.php; Nov 28-29, 2011 [Accessed February 29, 2012].

[24] Hey T, Tansley S, Tolle K. The Fourth Paradigm: Data-Intensive Scientific Discovery. Redmond, WA: Microsoft Research; 2009.

[25] Ozdemir V, Smith C, Bongiovanni K, et al. Policy and data intensive scientific discovery in the beginning of the 21st century. OMICS 2011b;15(4):221−5.

[26] Ozdemir V, Armengaud J, Dubé L, et al. Nutriproteomics and proteogenomics: cultivating two novel hybrid fields of personalized medicine with added societal value. Curr Pharmacogenomics Person Med 2010;8(4):240−4.

[27] Kaye J. From single biobanks to international networks: developing e-governance. Hum Genet 2011;130(3): 377−82.

[28] Knoppers BM. Consent to "personal" genomics and privacy. Direct-to-consumer genetic tests and population genome research challenge traditional notions of privacy and consent. EMBO Rep 2010;11(6):416−9.

[29] Howard HC, Knoppers BM, Borry P. Blurring lines. The research activities of direct-to-consumer genetic testing companies raise questions about consumers as research subjects. EMBO Rep 2010;11(8):579−82.

[30] Knoppers BM, Avard D. "Principled" personalized medicine? Personalized Medicine 2009;6:663−7.

[31] Nuremberg Code. Trials of war criminals before the Nuremberg Military Tribunals under Control Council Law. Washington: US Government Printing Office; 1949. 181−2.

[32] Belmont Report. Ethical principles and guidelines for the protection of human subjects of research. National Commission for the Protection of Human Subjects in Biomedical and Behavioral Research. Department of Health, Education, and Welfare (DHEW) Publication No. (OS); 1978. 78−0012.

[33] Beauchamp TL, Childress JF. Principles of Biomedical Ethics. sixth ed. Oxford: Oxford University Press; 2008.

[34] Probst-Hensch N, Tanner M, Kessler C, et al. Prevention—a cost-effective way to fight the non-communicable disease epidemic: an academic perspective of the United Nations High-level NCD Meeting. Swiss Med Wkly 2011:141. w13266.

[35] Ozdemir V. What to do when the risk environment is rapidly shifting and heterogeneous? Anticipatory governance and real-time assessment of social risks in multiply marginalized populations can prevent IRB mission creep, ethical inflation or underestimation of risks. Am J Bioeth 2009;9(11):65−8.

[36] Schicktanz S, Schweda M, Wynne B. The ethics of "public understanding of ethics"—why and how bioethics expertise should include public and patients' voices. Med Health Care Philos 2011 Mar 30 [Epub ahead of print].

[37] Illinois White Paper. Improving the System for Protecting Human Subjects: Counteracting Institutional Review Board (IRB). "Mission Creep". Urbana−Champagne, IL: Center for Advanced Study, University of Illinois; 2005.

[38] Gunsalus CK, Bruner EM, Burbules NC, et al. Mission creep in the institutional review board (IRB) world. Science 2006;312(5779). 1441.

[39] Knoppers BM. Genomics and policymaking: from static models to complex systems? Hum Genet 2009;125(4): 375−9.

[40] Burca G. Proportionality and wednesbury unreasonableness: the influence of european legal concepts on UK law. European Public Law 1997;561.

[41] Lunshof JE. From genetic privacy to open consent. Person Med 2006;3:187−94.

[42] Ostrom E. Collective action and the evolution of social norms. Journal Econ Perspectives 2000;14(3):137−58.

[43] Knoppers BM, Chadwick R. Human genetic research: emerging trends in ethics. Nat Rev Genet 2005;6(1):75−9.

[44] Funtowicz SO, Ravetz JR. A new scientific methodology for global environmental issues. In: Costanza R, editor. Ecological Economics: The Science and Management of Sustainability. New York: Columbia University Press; 1991. p. 137–52.

[45] Gibbons M, Limoges C, Nowotny H, et al. The new production of knowledge. London: Sage; 1994.

[46] Nowotny H, Scott P, Gibbons M. "Mode 2" revisited: the new production of knowledge. Minerva 2003;41: 179–94.

[47] Faraj S, Kwon D, Watts S. Contested artifact: technology sensemaking, actor networks, and the shaping of the Web browser. Information Technology & People 2004;17(2):186–209.

[48] Fisher E, Mahajan R, Mitcham C. Midstream modulation of technology: governance from within. Bulletin of Science. Technology & Society 2006;26(6):485–96.

[49] Douma KF, Karsenberg K, Hummel MJ, et al. Methodology of constructive technology assessment in health care. International Journal of Technology Assessment in Health Care 2007;23(2):162–8.

[50] Denis JL, Lamothe L, Langley A, et al. Governance and health: the rise of the managerialism in public sector reform. Rev Salud Publica (Bogota) 2010;12(Suppl 1):105–22.

[51] Kickbusch I. Governance for health in the 21st century. A study conducted for the WHO Regional Office for Europe. Presented at the First meeting of the European Health Policy Forum for High-Level Government Officials. Andorra la Vella: Andorra; March 2011. 9–11.

[52] Guston DH, Sarewitz D. Real-time technology assessment. Technol Soc 2002;24:93–109.

[53] Fisher E. Editorial overview: public science and technology scholars: engaging whom? Sci Eng Ethics 2011;17(4):607–20.

[54] Guston DH. Anticipatory Governance: A Strategic Vision for Building Reflexivity into Emerging Technologies. Presentation. Resilience 2011. Tempe, AZ: Arizona State University; March 14, 2011.

[55] Ommer R, Wynne B, Downey R, et al. Pathways to Integration. Vancouver: Genome British Columbia GSEAC Subcommittee on Pathways to Integration. Available from: http://www.genomebc.ca/index.php/download_file/view/611/910/; 2011 [Accessed on February 29, 2012].

[56] Barben D, Fisher E, Selin C, et al. Anticipatory governance of nanotechnology: foresight, engagement and integration. In: Hackett EJ, Amsterdamska O, Lynch M, Wajcman J, editors. The Handbook of Science and Technology Studies. Cambridge, MA: MIT Press; 2008. p. 979–1000.

[57] Owen R, Goldberg N. Responsible innovation: a pilot study with the U.K. Engineering and Physical Sciences Research Council. Risk Anal 2010;30(11):1699–707.

[58] Joly Y, McClellan K, Knoppers BM. Personalized vaccines and public health genomics: anticipating and monitoring the ELSIs. Curr Pharmacogenomics Person Med 2010;8:5–8.

[59] Landry R, Amara N, Pablos-Mendes A, et al. The knowledge-value chain: a conceptual framework for knowledge translation in health. Bull World Health Organ 2006;84(8):597–602.

[60] Daudelin G, Lehoux P, Abelson J, et al. The integration of citizens into a science/policy network in genetics: governance arrangements and asymmetry in expertise. Health Expect 2011;14:261–71.

[61] Lehoux P. Moving beyond our mutual ignorance. Or, how would engaging the public benefit the personalized medicine community? Curr Pharmacogenomics Person Med 2011;9(2):76–9.

[62] Lavis JN, Ross SE, Hurley JE, et al. Examining the role of health services research in public policymaking. Milbank Q 2002;80:125–54.

[63] Collinridge D. The Social Control of Technology. New York: St. Martin's Press; 1980.

[64] Guston D. Toward anticipatory governance. Available from: http://nanohub.org/resources/3270 2007 [Accessed February 29, 2012].

[65] Reddy PJ, Jain R, Paik YK, et al. Personalized medicine in the age of pharmacoproteomics: a close up on India and need for social science engagement for responsible innovation in post-proteomic biology. Curr Pharmacogenomics Person Med 2011;9(1):67–75.

[66] Joly Y. Open source approaches in biotechnology: utopia revisited. Maine Law Review 2007;59:386–431.

[67] Knoppers BM, Zawati MH, Kirby ES. Sampling populations of humans across the world: ELSI issues. Annual Review of Genomics and Human Genetics 2012 Mar 8 [Epub ahead of print].

[68] Hewitt RE. Biobanking: the foundation of personalized medicine. Current Opinion in Oncology 2011;23(1): 112–9.

[69] Vaught J, Rogers J, Carolin T, et al. Biobankonomics: developing a sustainable business model approach for the formation of a human tissue biobank. JNCI Monographs 2011;42:24–31.

[70] World Health Organization (WHO). Genetic Databases Assessing the Benefits and the Impact on Human and Patient Rights. Geneva: Switzerland; 2003.
[71] Knoppers BM. Biobanking: international norms. J Law Med Ethics 2005;33(1):7—14.
[72] Knoppers BM. Of genomics and public health: building public "goods"? Can Med Assoc J 2005;173(10):1185—6.
[73] Prainsack B, Buyx A. Solidarity: reflection on an emerging concept in bioethics. Swindon, UK: Nuffield Council on Bioethics; 2011.
[74] Vegvari A, Welinder C, Lindberg H, et al. Biobank resources for future patient care: developments, principles and concepts. J Clin Res Bioeth 2011;1(1):24.
[75] Betsou F, Rimm DL, Watson PH, et al. What are the biggest challenges and opportunities for biorepositories in the next three to five years? Biopreserv Biobanking 2010;8(2):81—8.
[76] Meslin E, Garba I. Biobanking and public health: is a human rights approach the tie that binds? Hum Genet 2011;130(3):451—63.
[77] Murtagh M, Demir I, Harris J, et al. Realizing the promise of population biobanks: a new model for translation. Hum Genet 2011;130(3):333—45.
[78] Carlile PR. Transferring, translating, and transforming: an integrative framework for managing knowledge across boundaries. Organ Sci 2004;15(5):555—68.
[79] Kuhn T. The Structure of Scientific Revolutions. second ed. Chicago: University of Chicago Press; 1962.
[80] Ravetz JR. Usable knowledge, usable ignorance: incomplete science with policy implications. Sci Commun 1987;9:87—116.
[81] Ozdemir V, Fisher E, Dove ES, et al. End of the beginning and public health pharmacogenomics: knowledge in "mode 2" and p5 medicine. Curr Pharmacogenomics Person Med 2012;10(1):1—6.
[82] Dandara C, Adebamowo C, de Vries J, et al. An idea whose time has come? An African foresight observatory on genomics medicine and data-intensive global science. Curr Pharmacogenomics Person Med 2012;10(1):1—6.
[83] Dove ES. The genetic privacy carousel: a discourse on proposed genetic privacy bills and the co-evolution of law and science. Curr Pharmacogenomics Person Med 2011;9(4):252—63.
[84] Bourdieu P. Outline of a Theory of Practice. Cambridge: Cambridge University Press; 1977.
[85] Bourdieu P, Wacquant L. An invitation to reflexive sociology. Chicago: University of Chicago Press; 1992.
[86] Khoury MJ, Gwinn M, Yoon PW, Dowling N, Moore CA, Bradley L. The continuum of translation research in genomic medicine: how can we accelerate the appropriate integration of human genome discoveries into health care and disease prevention? Genet Med 2007;9(1):665—74.
[87] Dove ES, Faraj SA, Kolker E, Ozdemir V. Designing a post-genomics knowledge ecosystem to translate pharmacogenomics into public health action. Genome Med 2012;4(11):91.
[88] Ozdemir V. OMICS 2.0: A Practice Turn for 21(st) Century Science and Society. OMICS 2013;17(1):1—4.

12

Pharmacoeconomics of Pharmacogenetics within the Context of General Health Technology Assessments

Maarten J. Postma*, Cornelis Boersma*, S. Vegter*,
Dominique Vandijck†, L. Annemans**

*Department of Pharmacy, University of Groningen, Netherlands,
†University of Ghent & University of Hasselt, Belgium,
**University of Ghent & University of Brussels, Belgium

OBJECTIVES

1. Describe the position of health economics within the broader context of health technology assessment (HTA) processes.

2. Define the position of genetic/genomic strategies within the broader context of HTAs.

3. Identify the issues concerning reimbursement of genetic/genomic strategies, with the focus on health economics.

INTRODUCTION

Health care markets are becoming more and more complex, and the number of stakeholders is generally increasing. The latter change can be explained by either more regional decision making or increased privatization leading to more health care insurers. For example, in the Netherlands, decentralization and ongoing privatization of health care has increased the independent role in financing and reimbursement of health care insurance companies. In Germany, further decentralization is ongoing by limiting the influence of central bodies such as the Institut für Qualität und Wirtschaftlichkeit im Gesundheitswesen (IQWiG, or Institute for Quality and Efficiency in Health Care), whereas decentralized procedures have been in place for many decades with major decisive authorities for the individual regions. Purchasers of health care (technologies) in general (also including sickness funds, managed care organizations, patient bodies, and private clinics) are becoming concerned about getting value for money, demanding increased evidence of proven added value before granting even market access, let alone reimbursement of technologies. Within the context of a gross budget for health care—for example, around 10% of gross domestic product (GDP)—a diabetes patient body, an individual health care insurer, hospital formulary committees, and cooperative organizations of general practitioners (GPs) and pharmacists may all serve as examples here. Next to central levels, this is also the case for the decentralized levels, where considerations on evidence, costs, and cost-effectiveness are becoming more eminent. Decisions on reimbursement for drugs more often require economic evidence in terms of cost-effectiveness in many European countries. An alternative methodology has been applied from a German perspective (IQWiG) with valuation of net costs and additional benefits of an intervention in terms of monetary values. Often these evidence, costs, and cost-effectiveness considerations start with drugs, which are relatively easy to control, and then rapidly extend to other areas in health care.

Health technology assessment (HTA) is essentially a comparative analysis of two or more interventions with respect to benefits and costs. Demanding structured and generic HTA is considered an option to achieve the combined goals of increasing quality of care and access to care while controlling expenses. HTA uses evidence-based medicine techniques and is based on sets of guidelines for methodologies to be used (www.ispor.org). As a result, the reimbursement environment is often stringent, both at the central as well as local health care insurance and hospital levels. For instance, in the Netherlands, the implementation of technology assessment, including pharmacoeconomic analysis, has resulted in the denial of various potentially innovative pharmacotherapeutic technologies for reimbursement (taken from the website www.cvz.nl, HPV vaccine, rotavirus vaccine, ivabradine, oseltamivir, entecavir, and rasagiline may serve as some examples). Further examples of recent innovative therapies that have gone through HTA in western countries include various new antithrombotic therapies, antimicrobial agents, psychotropic drugs, and orphan drugs [1–3].

Such stringent rules/procedures for technology approval have various consequences. On the one hand, stringent rules/procedures hamper the dissemination of new technologies, particularly if this leads to denial of reimbursement (Netherlands, Belgium, Sweden, etc.) or negative guidance on use (National Institute of Clinical Excellence [NICE] and Scottish Medicines Consortium [SMC] in the UK). On the other hand, despite being stringent, procedures and the set of requirements they encompass provide suppliers of technologies with rather clear-cut pathways to follow for market access and reimbursement. Indeed, in many western economies the pathways for manufacturers to get new drugs on the market are straightforward, generally including clear steps and decision criteria. Also, and most clearly at the central level, timelines generally exist in the technology assessment, reassuring that the evaluation will be finished within an adequate amount of time both regarding market access and reimbursement, although clock stops (extending this legislation timeline) are obviously more the rule than the exception. In many western economies, such procedures exist for drugs with processes and criteria and specific requirements on how to report evidence with regard to some of the decision criteria (for example, guidelines for pharmacoeconomic analysis). In some areas, however, lags in this respect still exist and ample potentials for further development of rational procedures, criteria, and requirements also exist. This is the case for vaccines and donated blood and its derived products (www.gr.nl), for example.

Sheer absence of implemented procedures and criteria is identified in genetic screening and testing. Obviously, ideas have been proposed; however, applications in the real world beyond scarce pilot projects have yet to be started. Yet we can drawn on the experiences gathered in assessing drug technologies within the context of new drugs having to prove their cost-effectiveness in many countries nowadays. The trend towards personalizing medicine therefore provides a new challenge for HTAs [4].

As mentioned earlier, this chapter specifically focuses on health investment decisions for genetic tests for diagnosis and screening, whereby it is acknowledged that health care decisions on genetic tests can be expected on several decision levels (e.g., country, regional, hospital) [5]. For uniformity reasons, it is highly relevant to identify a set of procedures and criteria to develop a tool for decision-making procedures on health investments for genetic tests. For this purpose, genetic tests are classified into the following four categories:

1. Genetic testing in *population-based screening* programs, which are defined as tests aimed at individuals (children or adults) who belong to a stratified population (sub)group (e.g., age, race/ethnicity, gender) without clinical signs of disease. Prenatal screening for Down syndrome is an example of this category.
2. Genetic testing for *diagnostic/predictive testing* on an individual basis, which is defined as diagnostic testing (for confirmation of clinical symptoms) or predictive testing (presymptomatic but existing risk of developing a condition because of family history) for predisposition to either a common disease or a rare diseases (rare disease being defined as occurring <1/2,500 population). Screening for hereditary breast cancer is an example of this category.
3. Genetic testing for *carrier* testing, which is defined as testing for the "carrier" of an inherited (recessive or X-linked) disorder, which does not affect the person but could eventually affect his or her relatives. An example is testing for cystic fibrosis.
4. Genetic testing to assess the *response* to specific therapies (*pharmacogenetics* per se), which is defined as the study of interindividual variations in DNA sequence related to drug response (efficacy/effectiveness and/or safety (e.g., adverse drug reactions)); the variation in drug response could be due to inherited changes in drug absorption, distribution, metabolism, or excretion or at the drug target site; a distinction between common and rare disease can be made again. Examples may be testing for potential response to trastuzumab in breast cancer and for adverse effects in azathioprine therapy.

In the following sections, we discuss and propose criteria for decision making related to investments in genetic screening and diagnosis (as defined by the four categories listed above) as well as processes and requirements. Additionally, one illustrative case study is described in an appendix to the chapter.

CRITERIA FOR DECISION MAKING

Introduction

In our current approach, we define criteria as various items to be considered within the whole tool kit to assess a new genetic screening test. In particular, the tool kit comprises the process of deciding which steps to take in an assessment, who actually performs these (national, local, or meso level decision makers/advisors), specific procedures to perform the steps, and the consideration of methodological standards or even guidelines for conducting specific parts within the tool kit.

Notably, many HTAs have been developed for drugs and vaccines in various EU countries, and HTA criteria and tools have been designed and used for their assessment. In our building of such criteria and tools for assessing genetic screening tests, we have therefore drawn on the experience with developments in the areas of drugs and vaccines. Obviously, similarities exist between drugs and vaccines and the area of screening and diagnostics. For example, genetic tests may be directly linked to drug treatment (fluorescence *in situ* hybridization, or FISH, test and trastuzumab; or imatinib, MammaPrint® in breast cancer, and chemotherapy). Another similarity exists in that screening (and some diagnostics) are used in general populations rather than in specific patient groups. Therefore, criteria and tools developed for drugs and vaccines, if combined and selected, should enable the construction of an approach and set of methods that are useful for genetic screening and diagnostics. On the other hand, unique characteristics are associated with the field of genetic screening and diagnosis: the difficulties in obtaining robust evidence on clinical utility of the tests, the ethical issues especially related to false-positive and false-negative tests, the fact that tests for which reimbursement is requested are often already used in the market, and often clinical utility (and cost-effectiveness) of tests is directly linked to utility (and cost-effectiveness) of specific pharmacotherapy.

Previously, the EUnetHTA Core Model for HTA was developed within an EU project (www.EUnetHTA.eu). The model employs 10 domains of criteria within a procedure that can be applied for assessing a new screening test or diagnostic (e.g., a genetic test). These domains can be summarized as:

- Current use of the technology (dissemination so far)
- Epidemiology of the relevant disease(s)
- The exact technology and its characteristics
- Safety/toxicity of the technology
- Diagnostic accuracy
- Clinical effectiveness/efficacy
- Costs and economic evaluation
- Ethical aspects
- Organizational aspects
- Psychosocial and legal aspects

These ten criteria are discussed in detail in the following subsections.

Current Use

WHAT KNOWLEDGE IS REQUIRED?

It is obviously important to assess the current use of the new technology and the environment in which it is/might be adopted. The environment relates to alternatives, such as other tests (e.g., older or less advanced technologies) or drug treatments irrespective of screening/test outcomes. Current use may be limited if reimbursement is not (yet) adequately provided or market access is still in the process of being acquired.

Hence, an important distinction must be made between tests that are already on the market and tests that have not yet reached the market. This distinction has consequences for what information regarding the other criteria can be requested and how it can be obtained, as will be discussed later.

WHICH SOURCES ARE AVAILABLE?

For the previous information, various sources are available: published literature, market authorization authorities' reports, registries, routine statistical data by census bureaus (if relevant, such as EUROSTAT), experts in the field, and the manufacturers. Finally, data may be gathered by the decision maker, such as the Ministry of Health; however, this approach will be time consuming.

Epidemiology and Management of the Health Condition

WHAT KNOWLEDGE IS REQUIRED?

The epidemiology and burden of disease for which the technology is intended should be analyzed. Aspects to be considered involve prevalence of disease, symptoms, natural progression, influence of risk factors, mortality and life-years lost, expected potential impact of screening or diagnosis, current treatment practice, clinical guidelines in the disease area, alternative treatment(s) (comparator[s]), positioning and decisions on market access, and reimbursement in other countries and with competitor drugs. Additional factors are any competitor technologies expected to be off patent soon (with potential subsequent price reductions) and regulatory status of both the technology and competitor technologies in the upcoming years.

WHICH SOURCES ARE AVAILABLE?

Again, various data sources are available. In particular, these include reports and published literature, but databases in health care can also be helpful. Examples of databases include hospital admissions, general practitioner visits, and pharmacy prescriptions. Generally, these databases are standalone; however, within targeted projects databases can be linked anonymously—for example, using trusted third parties for linkage.

The Exact Technology and Its Characteristics

WHAT KNOWLEDGE IS REQUIRED?

In this stage, various types of questions have to be asked and answered. These questions relate not only to the technology itself but also to material and immaterial requirements for its use. Examples of such questions are: Are investments in specific equipment needed? Are highly skilled staff required to operate the technology?

WHICH SOURCES ARE AVAILABLE?

The basis for answering these types of questions would generally be literature research using review articles from MEDLINE, EMBASE, Cochrane, and potential HTA reports already available in other countries. Notably, the United Kingdom and the Scottish Medicines Consortium may sometimes be early in their HTA assessments, providing very useful information (for some technologies, "early" reports exist for other countries as well; however, they have limited application due to the use of local language).

Besides public sources, it might be advisable to directly contact the manufacturer for additional (potentially classified) information. Involvement of the manufacturer in the HTA process may have disadvantages because of potential conflicts of interest. However, advantages may vastly outweigh disadvantages in that manufacturers (1) have the most up-to-date and detailed information and (2) realize the importance of dissemination of objective assessment. One technique used could be to invite the industry for targeted lectures on the technology along the lines of predefined formats (the Dutch Health Council uses this technique). Also, interviewing some potential early adopters of the technology and frontline specialists in the area should be undertaken. At this stage, it could already be considered whether the design of a registry on its use would be recommended.

Safety

WHAT KNOWLEDGE IS REQUIRED?

Two types of harm or safety may be considered for diagnostics and screening tests: (1) morbidity or even mortality due to radiation, toxic contrast media, or invasiveness; and (2) harm due to wrong or misleading diagnosis (or even rightful diagnosis). For most diagnostics and screening tests, the latter aspect might be the most relevant. Wrong or misleading diagnosis based on genetic testing technologies might lead to suboptimal treatment or even mistreatment in addition to emotional distress for the patient and family members. Rightful diagnosis will facilitate optimal treatment; however, genetic knowledge may affect areas of the individual's and his or her family's life.

WHICH SOURCES ARE AVAILABLE?

National and international data banks on toxicity and safety information can be used to investigate this aspect of the new technology. The HTA can therefore draw on web-based information from the U.S. Food and Drug Administration (FDA; www.fda.gov) and the section on medical devices of the UK Medicines and Healthcare Products Regulatory Agency (devices.mhra.gov.uk).

Accuracy

WHAT KNOWLEDGE IS REQUIRED?

The basic question regarding diagnostic accuracy regards whether the diagnostic or screening technology correctly distinguishes diseased/at-risk populations from nondiseased/low-risk populations. Often, for this purpose sensitivity and specificity are used or, alternatively, predictive values can be estimated and reported. The following definitions may be used:

- Accuracy = proportion of subjects that the test correctly identifies as positive/negative
- Sensitivity = probability of a positive test result in persons with disease/risk
- Specificity = probability of a negative test result in persons without disease/risk
- False positivity = probability of a positive test result in persons without disease/risk
- False negative = probability of a negative test result in persons with disease/risk
- Predictive value = proportion of persons with risk present/absent in those with a positive/negative result

WHICH SOURCES ARE AVAILABLE?

Although studies in the literature evaluating such technology usually report measurements related to accuracy, it is important to understand the caveats regarding the interpretation for use in daily clinical practice. For example, one should consider whether those persons receiving the diagnostic/test are similar to those studied; that is, is the latter group representative for populations of relevance? Is the reference test used to assess the performance of the index technology the gold standard and highly likely to correctly identify disease/risk? Were adequate sample sizes used to study the index technology, and were variations reported across observers, instruments, and patient subgroups? The Receiver Operating Characteristics (ROC) curve, in which pairs of sensitivity are plotted versus the estimated probability of false positivity ($1 - $ specificity) is primarily used for analyzing these type of studies [6].

With respect to potential inaccuracy of the technology and the consequences of inaccuracies, particularly false-positive and false-negative outcomes, it is also important to evaluate characteristics of available alternative technologies, such as how any potential harm of the index technology compares to any evidence on harm of alternative technologies. Finally, even if sensitivity and specificity of the test is relatively high, rendering a high analytical validity, in the end what actually counts is clinical validity, that is, how useful the test is in daily clinical practice.

Clinical Effectiveness

WHAT KNOWLEDGE IS REQUIRED?

Rather than evidence on accuracy only, one might argue that evidence on actually changing medical practice and improving patient outcomes is required in diagnose-and-treat or screen-and-treat settings, potentially showing statistically significant improvements in serious morbidity and mortality, or at least surrogate markers. However, such trials combining the diagnostic/test technology with subsequent health care interventions to demonstrate clinical utility are scarce. Also, specific outcomes of the test may differ between sites due to different quality assurance between labs and tests that are used, which further complicates issues.

Ideally, direct evidence is gathered on the new diagnostic/screening technology by randomizing patients in clinical trials to receive either the new technology or standard care and assess their respective therapeutic outcomes. Notably, the old technology or comparator in this case might be absent any screening or testing or it might be just standard care treatment if the outcome of the test would involve an actual treatment. If not available, some indirect evidence on effectiveness may be derived from safety studies (although generally underpowered for effectiveness, but possibly showing trends), observational cohorts, and time series analyses. (Please refer to Chapter 9a for a more in-depth review and discussion of the various study designs used in evaluating different aspects of pharmacogenomics diagnostic tests.)

WHICH SOURCES ARE AVAILABLE?

Notably, in the area of economic evaluation within HTAs for diagnostics/tests (see the next criterion) the issue of (cost-)efficacy versus (cost-)effectiveness plays a crucial role. Given the type of evidence, morbidity and mortality outcomes are likely lacking for these technologies and only intermediate performances are available (e.g., sensitivity, specificity, predictive values). Decision analytic models can be used to estimate the trade-offs between safety, harms, accuracy, and effectiveness in a potentially complex model integrating all these aspects [7]. If cost considerations are important, then economic evaluation is needed.

Costs and Economic Evaluation

Economic evaluation has become paramount in the last decades to help prioritize health care spending, that is, to spend health care budgets optimally to ensure the highest health gains for limited resources. It is undertaken to inform health care decision makers with the explicit goal of enhancing rational decision making. Guidelines are used as explicit tools for designing, executing, and judging economic evaluations. Health economic/pharmacoeconomic guidelines exist for various countries all over the world (www.ispor.org). Setting the methodological standards, most guidelines require the transparent presentation of cost-effectiveness planes, cost-effectiveness acceptability curves, value-of-information analysis, uncertainty analysis using both sensitivity and scenario analysis, and explicit probabilistic sensitivity analysis reporting averages, credibility intervals surrounding the cost-effectiveness estimates, and sometimes explicit thresholds for willingness to pay [8,9]. These guidelines and quality requirements will be addressed in detail in the next section of this chapter.

Ethical Aspects

Ethical analysis should be considered a separate domain in HTA, and a structured analysis of both situations with and without the new diagnostic/testing technology should be embarked upon. It is obvious that genetic testing is especially prone to misuse and abuse, and it is generally considered a major ethical concern associated with implementation. For example, knowledge on genetic information may on one hand enhance optimal treatment possibilities; however, on the other hand, it may provide undesirable information that might have an impact beyond health care—for example, life insurance.

Notably, health economic methodologies also involve ethical issues, which could be addressed in this paragraph of the HTA. In particular, quality-adjusted life year (QALY) measurement, weighting of life years, inclusion of indirect costs in the societal perspective, and discounting of future costs, savings, and health gains all involve ethical issues that have been listed in textbooks (chapters 4 and 6 in [10]) but have not yet been awarded the attention that they deserve. Notably, health economists merely do the calculations, including indicating potential ethical issues on equity in health care between generations, social classes, and gender. Yet the ethical issues of time preference and QALY measurements, including production losses in economic evaluations, should always be kept in mind.

Organizational Aspects

With the assessment focus on the technology itself and not considering the environment for implementation, organizational issues are often neglected in HTA. Neglecting organizational aspects of HTA may seriously endanger the achievement of optimal health gains while controlling costs. Some of the issues include whether the screening should be implemented in an opportunistic or systematic approach and whether the diagnostic is most suitable to be applied in an outpatient or inpatient setting. Specifically, who do we target and how do we target them? For a screening program, a specific subgroup registered with a general practitioner could be targeted either opportunistically when they come for a visit or systematically by sending all within a specific age range an invitation for testing. Obviously, both approaches will also have differing economic impacts, with systematic approaches potentially having higher up-front costs but opportunistic approaches potentially resulting in less widespread savings [11,12].

Drugs are generally developed, registered, and used in clear-cut patient populations, and the environments for their use are often inherent to the technology considered, such as intravenous chemotherapy in hospitals and general analgesics in the outpatient setting. However, this may not be the case for diagnosis/screening. Populations considered may even comprise (major parts of) the general population at large or groups with elevated risk *a priori*. So, the question "whom do we target" is often less straightforward to answer, let alone how to target them.

Finally, an organizational aspect may be listed as to whether the HTA should involve the local use of a diagnostic/testing technology or an application and assessment at the national level. Different aspects may be relevant in either of both cases, including size and location of trials, specific patient groups and geographic areas, and national cost structures to be considered.

Psychosocial and Legal Aspects

The psychosocial impact of the use of HTA needs to be assessed not only for patients but for family, friends, employers, and employees as well. Also, impacts on leisure time should be considered here if not taken into account explicitly in the economic evaluation, which likely would not have been done if the third-payer party perspective was emphasized in the economic analysis (as in NICE in the UK; www.ispor.org). Notably, in health economic evaluations, the QALY impacts of technologies should be evaluated beyond the index case, for example, including only partners of those diseased, parents of sick children, and sexual contacts in the area of sexually transmitted diseases.

Broader areas to be considered comprise again ethical issues, distrust, dilemmas, stigmatization and even humiliation, taboos, and legal aspects. Legal issues should be sorted out *a priori* before recommendations on use can be considered. Legislation is both at the national and supranational level (for example, the EU's legislation on medical devices available at ec.europa.eu/enterprise/medical_devices).

Notably, assessment of the criteria listed above could and should be done at several levels of decision. Obviously, some criteria are to be assessed at central levels, whereas others would primarily be considered at decentralized levels. In particular, aspects regarding the exact technology, its general safety, and clinical effectiveness profile would represent considerations ideally performed at the national level, but potentially partly at the EU level. Where specific regional/national aspects come in, for example, epidemiology of the underlying disease, resistance patterns if microbes are targeted, and accuracy of testing in specific circumstances, national assessments should be done. In particular, as local circumstances play a bigger role for screening/testing than for drug treatments, we might expect a relatively strong focus on the national or even regional process for assessing new pharmacotherapies.

Obviously, health economic and cost considerations as well as organizational issues are generally targeted at the national level. This approach would be fully in line with how new drugs are economically assessed currently but is still fully opposed to the current practices for screening and testing, with a strong focus generally on the local or even individual health care center level. Ethical issues would generally be rather similar over various localities and therefore the majority of considerations could be listed at a supranational level. Differences in cultures in the different member states might lead to specific adaptations of the general assessment in individual countries.

Generally, we would advocate a process primarily at the national level, building on an overall assessment at the EU level on common aspects such as technology description, safety, and clinical effectiveness. At the national level, a clear process should be installed on which steps to be taken, who takes the initiative, who assesses, and what timelines apply.

PHARMACOECONOMICS/HEALTH ECONOMICS WITHIN THE HTA CONTEXT

Introduction

We will now discuss in more detail which tools, methodological standards, and guidelines exist for assessing the above recommended criteria and the quality of the available data. In this respect, our focus will be on health and economics, that is, in particular the first ten domains discussed above. Notably, this will, for example, concern cost of illness and assessment of adequate comparator (section 1; *current use*), budget impact analysis (section 2; *epidemiology*), cost consequences of the technology under consideration (section 3; *technology and characteristics*), cost and QALY impacts of toxicity/safety and efficacy/effectiveness characteristics of the technology (sections 4 & 6; *safety/toxicity* and *efficacy/effectiveness*), health economic consequences of (in)accuracy of the technology (section 5; *accuracy*), and actual economic evaluation (section 7). Again, in specifying criteria into quality assessment, technologies regarding screening tests and diagnostics involve various specificities that were analyzed before [13] and are revisited next. This approach will clarify in which aspects HTAs of screening tests and diagnostics might differ from HTAs in other areas.

Assessment Tools

Regarding tools to judge and structure health economic parts of HTAs, guidelines for health economic or pharmacoeconomic evaluations can be extremely helpful. As an illustration, we list the Dutch guidelines for economic evaluation (Table 12.1) as presenting such a set of tools. This specific set of guidelines reflects one of many possible sets of guidelines that are now available all over the world and can be accessed at www.ispor.org. Generally, these sets of guidelines are comparable between countries and institutions (for example, the Dutch guidelines resemble the Belgian ones, and those of NICE resemble those of the Scottish Medicines Consortium) [14]. Yet, notable differences exist—for example, where the first Dutch guideline prescribes the societal perspective, NICE generally recommends the third-party payer (National Health Service [NHS]) perspective be adopted. In particular, the Dutch guidelines consist of a set of 11 guidelines with several subheadings specified [15]. Table 12.1 lists this set of items that can be applied to evaluating any HTA, including those concerning screening test and diagnostics.

Notably, these guidelines are used for assessing/evaluating files on drugs that are prepared by the *manufacturers*, including pharmacoeconomic analysis (generally outsourced to a consultancy specialized in this area but under strict supervision of the company/producer). Ideally, the process of HTA should be a fruitful cooperation between authorities and the manufacturer; however, most analyses and assessments are performed primarily by independent bodies (such as NICE).

Some general aspects are reflected in this set of tools. In particular, transparency in reporting is an important issue, as the economic evaluations are often performed by (or under the strict supervision of) the manufacturer. Also, effectiveness versus efficacy (safety vs. toxicity) is important, with the goal of analyzing population/real world level rather than at clinical trial level; that is, we prefer cost-effectiveness analysis versus cost-efficacy analysis. For genetic tests, this means that we prefer to see clinical utility rather than analytical validity or clinical validity. A related issue is the desire to measure effects in "hard end points" such as morbidity and mortality, rather than intermediate endpoints (such as blood pressure, albuminuria, HB1Ac, and cholesterol levels) that may have been used in Phase III registration trials. This approach often requires the use of models, which is allowed according to the guidelines if adequately designed. In practice, most economic evaluation HTAs involve some level of modeling to infer from efficacy (toxicity) to effectiveness (safety) and from intermediate to "hard" end points. As such, this set of tools seems adequate to apply to screening tests and diagnostics, although some specific aspects should be considered and slight revisions may be appropriate. This topic will be discussed in detail in the next section.

Specific Issues in Assessment of Genetic Testing

We recently published a literature review on economic evaluations of pharmacogenetic and pharmacogenomic technologies [13,16] using the same set of tools and pharmacoeconomic guidelines available at www.ispor.org. For example, all guidelines refer to discounting issues, choice of the comparator, and measurement of QALYs. In this respect, discounting refers to how to deal with costs, savings, and health gains that all may have different timings and are thus subject to time preference. Guidelines are generally in agreement but differ on

TABLE 12.1 Dutch guidelines for pharmacoeconomics as set of tools and subcriteria (items), adapted from Hoomans et al. [15]

1. Perspective	The analysis should be done and reported from the societal perspective, including costs, savings, and health effects irrespective of who benefits and who loses.
2.1 Comparator	The new technology should be compared with the standard technology recommended by health care professionals or the technology usually provided to patients (as derived from registries).
2.2 Indications	The indication should be clearly specified and the population considered in the economic analysis should be same as the population for which the new technology is considered.
2.3 Subgroups	All relevant subgroups should be specified and explicitly considered in the analysis; all subgroups should be clearly defined.
3.1 Cost-utility	If quality of life is important, a cost-utility analysis should be performed.
3.2 Cost-effectiveness	If mortality dominates the health effects of the disease considered, a cost-effectiveness study can be performed, using life years gained as the preferred clinical outcome.
4 Analytical period	The analytical period should be such to allow valid inferences to be drawn on consequences of the old comparator and the new technology regarding costs, savings, and health.
5.1 Identification of costs	All relevant costs should be identified: direct medical, direct nonmedical, indirect nonmedical, and indirect medical and individually included/excluded in sensitivity analysis.
5.2 Measurement of costs	Resource use must be measured validly and adequately.
5.3 Valuing resources	Adequate unit cost prices should be used to value resource use.
6.1 Efficacy vs. effectiveness	Relevant measures of effect (morbidity and mortality) should all be identified validly and with highest achievable certainty.
6.2 Quality of life	Quality of life should be measured validly and reliably.
6.3 Utilities	Appropriate methods should be used to measure utilities.
6.4 QALYs	Calculations of QALYs should be done correctly and reported transparently.
7.1 Modeling	Use of a model is allowed for analyzing costs and effects of treatments if the necessity of modeling is sufficiently justified.
7.2 Reporting the model	Transparent reporting of the model is required, including structure, cycles, transition probabilities, estimates, and assumptions.
7.3 Model validity	The validity of the model should be adequately investigated, tested, and reported.
8.1 Reporting totals	Totals for costs and effects of both the index drug and the comparator should be provided.
8.2 Reporting incrementals	Incremental costs and effects between index and comparator should be reported in detail.

TABLE 12.1   Dutch guidelines for pharmacoeconomics as set of tools and subcriteria (items), adapted from Hoomans et al. [15]—cont'd

9 Discounting	All relevant future effects (at 1.5%) and costs (at 4%) should be adequately discounted to current values.
10.1 Deterministic SA*	To analyze uncertainty, all relevant deterministic uni- and multivariate SAs should be conducted adequately and transparently.
10.2 Probabilistic SA*	To analyze uncertainty, all relevant probabilistic SAs should be conducted adequately and transparently.
11.1 Expert panel	If adequately motivated, expert panel opinions can be used.
11.2 Consensus	It should be transparent how consensus was arrived at by the expert panel.

* SA = sensitivity analysis

details—for example, what exact number to use for discounting (generally, 4%, 3%, or 1.5%). Studies also varied regarding disease, involving HIV, breast cancer, thromboembolism, nephropathy, psychiatric disorders, rheumatoid arthritis, duodenal ulcer, atrial fibrillation, hypertension, cystic fibrosis, lung cancer, inflammatory bowel disorder, hepatitis C, and leukemia. The review focused on genetic technologies linked to treatment to enhance effectiveness or safety profiles, with treatments comprising selective serotonin reuptake inhibitors, angiotensin-converting enzyme (ACE) inhibitors, proton pump inhibitors, antithrombotics, trastuzumab, and antiviral therapies. Obviously, the pharmacogenetic or pharmacogenomic biomarker is a specific group of screening tests; however, issues identified may be typical for all four groups of tests: (1) population-based screening programs, (2) diagnostic/predictive testing on an individual basis, (3) carrier testing, and (4) assessing the response to specific therapies (pharmacogenetics per se), as mentioned in the introduction to this chapter.

Pertinent factors for pharmacoeconomic evaluations of pharmacogenetic or pharmacogenomics diagnostic tests are discussed further in the following sections.

Disease under Study

In general, diagnostic tests and diagnostics related to high prevalence and/or severe complications are much more likely to result in favorable cost-effectiveness. Also, severe side effects of medications that can be dosed on genetic information enhance the cost-effectiveness profile. Therefore, the epidemiological aspects of the disease and associated complications should be considered much more than what has been currently performed in a pharmacoeconomic assessment. Also, side effects are often lacking in economic evaluations of drugs but may be crucial here. So specific attention is required for disease epidemiology and medication side effects if screening tests and diagnostics are involved, including the characteristics of the study population. In the end one could then validly estimate the number needed to test to identify one person (NNT), a crucial measure in epidemiology of screening. The NNT reflects the number that one needs to screen to actually identify one person eligible for intervention, for example, with pharmacotherapy.

Association Genotype–Phenotype

Furthermore, inconsistent results were identified for the association between genotype and phenotype in most studies. In general, for screening tests and diagnostics, the study design may not be optimal, and randomized controlled trials (RCTs) and meta-analyses may be lacking. It is therefore crucial that all evidence is gathered, including conflicting results, case control studies, and large observational cohort studies. Special care should be taken with respect to allele frequencies. Percentages may differ among various ethnic populations. In the area of screening tests and diagnostics, inferences from one set of populations to another may be much less straightforward than for drug treatments; that is, a careful evidence synthesis is required, without deleting any (potentially conflicting) information!

Predictive Values

The major challenge in the area of screening tests and diagnostics—which differs from the area of drug treatments—is how to integrate sensitivity, specificity, and predictive values adequately in the health economic model. For specific studies measuring the pertinent factors, various caveats regarding their interpretation for use in daily clinical practice exist. This issue again relates to the difference between efficacy and effectiveness or analytical validity versus clinical validity in this case. High *analytical validity* (ability to detect a clinical marker) reflects good performance in laboratory/research circumstances with high ability to detect, whereas *clinical validity* (ability to identify a disease/condition/response) reflects high performance in daily use and would correspond with high *clinical utility* (improvement in clinical management and outcomes). For example, one should consider whether those persons receiving the diagnostic/test are indeed similar to those studied in the clinical studies; that is, is the clinical trial group representative for populations of relevance, and what is the prevalence in the target population? Is the reference test used to assess the performance of the index technology indeed the gold standard and highly likely to correctly identify disease/risk?

Decision Model

Another difference for tests and diagnostics compared to drug treatments involves the requirement of an explicit analysis of the advantages of knowing the test/diagnostic outcomes. This analysis should ideally lead to a valid design of a decision-analytic model involving the options "screening/testing" versus "no screening/no testing." One specific example can be found in the appendix. Notably, a decision-analytic model could be developed next to an RCT to support the findings in the RCT, which could be limited in size due to cost issues and to enable extrapolation of estimates beyond the time horizon of the RCT.

Type of Analysis

The type of economic analysis should be chosen based on the aim of the study and available data. A cost-utility analysis is preferred but is sometimes not practical. Formal cost-effectiveness analyses present a good second option, especially when using life-years gained as the outcome measurement. Still, if using QALYs, the analysts should always report life years gained (LYGs)—in absolute terms and weighted for utilties—separately in the analysis, in case the concerned condition is life threatening.

Study Perspective

Costs of screening tests and diagnostics, medications, and all relevant events, such as adverse reactions to tests, diagnostics, and related drug treatments, should be carefully assessed. A third-party payer perspective is sufficient, as adopting a societal perspective may be impossible or impractical in the field of screening and diagnosis. However, if possible, a societal perspective should also be adopted. Notably, clinical or even patient perspectives are not taken. The time horizon of an analysis should be sufficient to capture all the differential costs and effects between the testing and nontesting strategies (ideally even lifelong), including adequate discounting for time preference using country-specific numbers.

Sensitivity Analysis

Additional parameters with inherent uncertainties need to be investigated in sensitivity analysis (SA). In particular, uncertainty around test performances should be included, as well as the epidemiology of the underlying disease and characteristics of the tested in the population. Preferably, both equal (for example, 3%, 3% and 4%, 4%) and differential discounting (notably lower discount rate for health—for example, 1.5% [17]) is investigated in *deterministic scenario analysis*, as several EU countries have now adopted the latter approach (www.ispor.org/peguidelines/comp1.asp). Notably, lower discount rates for health improve the economic profiles of preventive strategies, such as vaccination, but also screening. Most of the discussions on discounting for time preference have actually come from the vaccination strategies [18].

Vegter et al. [16] presented additional findings with a larger set of published literature on the topic. Over time, increasingly cost-utility analysis (CUA) was used, longer time windows were employed, and more extensive SA had been conducted. So (also) in the area of screening tests and diagnostics, studies seem to improve. Yet considerable and difficult-to-explain differences were seen in costs of screening tests (for example, tests for polymorphisms such as in *CYP*, *VKORC1*, and *TPMT* genes). In particular, cost prices varied between and within countries. As an example, costs for a HER-2 test varied between €283 for France and €548 in Sweden, and costs for ACE I/D polymorphism were estimated at €7 and €50 in two different Dutch studies. Notably, most studies were conducted in academia or the hospital setting without direct association with the pharmaceutical or diagnostic manufacturers. Transparency in reporting is certainly an issue where further improvement is needed; however, in general the quality of economic analyses in genetic screening seems to have improved over time. Finally, it is noted that the economic analyses seem to be conducted to increase awareness of cost-effective possibilities and perspectives of (genetic) testing rather than to influence policy decisions on reimbursement. With budget impact analyses and full HTAs being added, the economic analyses may serve the specific purpose of underpinning reimbursement decisions.

The latter is not strange because often the process for policy decisions on screening tests and diagnostics is not as clear cut as it is for new pharmacotherapies in many countries. In the following section, we suggest a process that could be followed to standardize such decisions and enhance a clear positioning of such health economic studies outlined above.

In general, we conclude that it would be a good approach to follow the health economic/ pharmacoeconomic guidelines as a set of standards, although slight changes should possibly be made in focus regarding epidemiology, evidence synthesis of associations between disease and characteristic tested for, analysis of the related pharmacotherapy, accuracy and predictive values (NNT), clinical utility, type of analysis, study perspective, and scope of SA—all of which have been described earlier.

PROCESS

The procedures taken to arrive at a decision on health care investments consist of a sequence of steps. Figure 12.1 reflects a graphical representation of an algorithm for the decision-making process for different genetic tests at different levels. Also, this general approach involves steps that apply to all types of genetic tests. The procedure can be initiated by the manufacturer (e.g., for commercial tests), clinician (e.g., laboratory tests for hospital

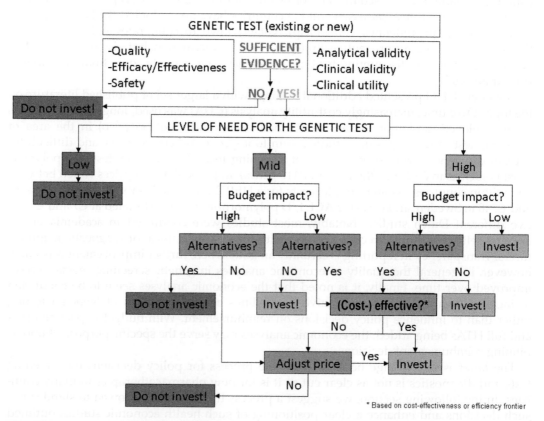

FIGURE 12.1   Potential algorithm to assess the impact of health investments for genetic tests (Jaime Caro, presented at ISPOR 2011, Madrid). *Source: Adapted from Care et al (presented during ISPOR 2010).*

use and coverage from the hospital budget), or HTA entities. During the process, initiators (e.g., manufacturers) should be involved in all stages (hearings, presentations, etc.), but the HTA should primarily be conducted by a single party, either a manufacturer-independent body (e.g., NICE) or health care payer.

First, the genetic test will be evaluated based on evidence related to the quality, efficacy/effectiveness, and safety/toxicity of the test or related (pharmaceutical) intervention (e.g., registration studies or other literature on clinical studies). The next step involves the assessment of the analytical validity or accuracy of the test (e.g., acceptable test characteristics and ability to detect the marker of interest). After proven analytical validity, the clinical validity—ability to identify a disease/condition/response—should be evaluated.

If all these requirements for the genetic test under consideration are found to be positive, the next step is to evaluate the clinical utility of the test. Relevant questions here are:

1. Does the genetic test make a difference in clinical management and outcomes?
2. Does the genetic test add to optimal health care (e.g., dependent on epidemiology of the relevant disease and standard care)?

If both these questions are answered in the affirmative, it is then important to identify the exact need for the test. Here, the following options "reflecting" the need for access to the specific test are possible:

1. High need; extremely high risk of high disease burden or death.
2. In between low/high need; lower risk of death, risk of major permanent malfunction or major symptomatic condition.
3. Low need; minor symptomatic condition and factors indicating risk of future problems.

If the need for the genetic test is low, investing in the test is not recommended from a payer's perspective (e.g., public money). If the need for the genetic test is high, and the budget impact on national health care spending is low, investing in the test is recommended. For in-between options with or without alternatives and different budget impacts, recommendations are based on efficiency, cost-effectiveness, and price criteria. Budget impact implicitly also includes number of patients and cost-effectiveness-based decision making and requires guidance in terms of the cost-effectiveness ratios of testing strategies. In general, a decision to invest (e.g., low budget impact and high medical need) may be linked to a conditional reimbursement trajectory requiring reassessment and including cost-effectiveness analysis if alternative tests become available.

The obvious example in the high-need category would be trastuzumab treatment and testing for overexpression of HER2 proteins. For this purpose, the FISH test is often used. In the absence of overexpressed protein, treatment would have no effect and only the side effects are potentially experienced. Trastuzumab treatment relevantly prolongs survival and thus reflects crucial life-saving treatment. The high-need category certainly has cost-effectiveness analysis as one of the potential outcomes of the decision tree in Figure 12.1. For example, in the Netherlands, trastuzumab was one of the first drugs with cost-effectiveness formally analyzed within the framework of reimbursement (www.cvz.nl). An example from the "in-between low/high" category—or even in the high-need category—is the 70-gene signature in the treatment of breast cancer [19]. It is extensively addressed as an example in the appendix. Another example in this category might be CYP2C19 genotyping for clopidogrel treatment. An example of in-between to even low need might be nortriptyline for depression and testing for CYP2D6, where adverse effects in slow metabolizers for CYP2D6 can be minimized. However, in practice, dose adjustments with empirical and genotype-guided therapy may be similar, although potentially with some time delays with empiric treatment. So some gain can be achieved with CYP2D6 testing but this will be limited and will potentially not affect mortality. Yet given the size of the patient population, budget impact can be extensive. Therefore, formal cost-effectiveness analysis is potentially not required to underpin investment decisions; however, negotiations on the price are recommended.

CONCLUSION

We note that guidelines for pharmacoeconomic evaluations address issues that were common between clinical pharmacology and personalized medicine. For example, all guidelines refer to discounting issues, choice of comparator, and measurement of QALYs [13,17]. However, in contrast to population-based innovative drug assessments, various aspects of the guidelines require specific attention for application to personalized medicine—for example, potential negative effects resulting from false-positive outcomes of testing warrant consideration, including QALY impacts. Moreover, ethical concerns may not be in line with economic arguments seeking optimal sensitivity, specificity, and cost combinations. Notably, the level of economic (but also technical) evidence may differ from what is generally experienced in clinical pharmacology, thus stressing the need to include all evidence in the economic analysis, including potentially conflicting results from case control and observational settings. Given genetic variability, the question of whether the patients in the studies on testing and diagnostic technologies are representative of the target groups for personalized medicine is one of utmost importance. In addition to clinical validity, the test should show a high clinical utility in practice, translating into an acceptable cost-effectiveness that is robust in extensive sensitivity analysis regarding uncertainty in test characteristics such as accuracy and predictive value. Finally, it is noted that the economic analyses often appear to be conducted to increase awareness of cost-effective possibilities and perspectives of (genetic) testing rather than to influence policy decisions on reimbursement.

In general, we conclude that an adequate approach for evaluating personalized medicine would be to follow the economic guidelines developed for clinical pharmacology as a set of standards, although some modifications and specific foci should be made to optimize applicability in testing strategies. These specific foci should comprise optimal evidence synthesis of associations between disease and characteristic tested for, analysis of the related pharmacotherapy, accuracy and predictive values, clinical utility, representativeness of the available studies on the technology, study perspective, and scope of the SA. Also, recent use of HTAs in personalized medicine are increasing—for example, regarding genetic profiling and self-testing [19,20]. If specific steps listed are adopted in the coming years with further applications, HTAs in personalized medicine may greatly benefit from the abundance of experience that has been gathered with clinical pharmacological HTAs [21].

QUESTIONS FOR DISCUSSION

1. Describe some of the health technology assessments (HTAs) that can be applied to pharmacogenomics.
2. Apply the concept of HTA with specific drug examples.
3. Are there limitations and/or challenges when applying HTAs to pharmacogenomics?

References

[1] Postma MJ. Public health aspects of vaccines in The Netherlands: methodological issues and applications. J Public Health 2008;16:267−73.

[2] Vegter S, Rozenbaum MH, Postema R, Tolley K, Postma MJ. Review of regulatory recommendations for orphan drug submissions in the Netherlands and Scotland: focus on the underlying pharmacoeconomic evaluations. Clinical Therapeutics 2010;9:1651−61.

[3] Blankart CR, Stargardt T, Schreyögg J. Availability of and access to orphan drugs. Pharmacoeconomics 2011;29:63−82.

[4] Trusheim MR, Berndt ER, Douglas FL. Stratified medicine and economic implications of combining drugs and clinical biomarkers. Nat Rev Drug Discov 2007;6:287−93.

[5] Flowers CR, Veenstra D. The role of cost-effectiveness analysis in the era of pharmacogenomics. Pharmacoeconomics 2004;22:481−93.

[6] Haddix AC, Teutsch SM, Corso PS. Prevention effectiveness: a guide to decision analysis and economic evaluation. 2nd ed. Oxford: Oxford University Press; 2003.

[7] Reese ES, Daniel Mullins C, Beitelshees AL, Onukwugha E. Cost-effectiveness of cytochrome P450 2C19 genotype screening for selection of antiplatelet therapy with clopidogrel or prasugrel. Pharmacotherapy 2012;32:323−32.

[8] Barton GR, Briggs AH, Fenwick EAL. Optimal cost-effectiveness decisions: the role of the cost-effectiveness acceptability curve (CEAC), the cost-effectiveness frontier (CEAF), and the expected value of perfect information (EVPI). Value in Health 2008;11:886−97.

[9] Boersma C, Broere A, Postma MJ. Quantification of the potential impact of cost-effectiveness thresholds on Dutch drug expenditures using retrospective analysis. Value in Health 2010:853−6.

[10] Drummond MF, Sculpher MJ, Torrance GW, O'Brien BJ, Stoddart GL. Methods for the economic evaluation of health care programmes. 3rd ed. Oxford: Oxford University Press; 2005.

[11] Knies S, Ament AJHA, Evers SMAA, Severens JL. The transferability of economic evaluations. Value in Health 2009;12:730−8.

[12] Welte R, Feenstra F, Jager H, Leidl R. A decision chart for assessing and improving the transferability of economic evaluation results between countries. Pharmacoeconomics 2004;22:857−76.

[13] Vegter S, Boersma C, Rozenbaum M, Wilffert B, Navis G, Postma MJ. Pharmacoeconomic evaluation of pharmacogenetic and genomic screening programmes: a systematic review on content and adherence to guidelines. Pharmacoeconomics 2008;26:569−87.

[14] Tolley K, Postma MJ. Pharmacoeconomics and market access in Europe: case studies in Scotland and The Netherlands. ISPOR Connections 2006;6(10):3−6.

[15] Hoomans T, van der Roer N, Severens JL, Delwel GO. Kosteneffectiviteit van Nieuwe Geneesmiddelen; van belang bij geneesmiddelenvergoeding, maar voor verbetering vatbaa (in Dutch). Nederlands Tijdschrift voor Geneeskunde 2010;154. A958.

[16] Vegter S, Jansen E, Postma MJ, Boersma C. Pharmacoeconomic evaluation of pharmacogenetic and genomic screening programs: an update of the literature. Drug Develop Res 2010;71:492−501.

[17] Gravelle H, Brouwer W, Niessen L, Postma M, Rutten F. Discounting in economic evaluations: stepping forward towards optimal decision rules. Health Econ 2007:307−17.

[18] Westra TA, Parouty M, Brouwer WB, Beutels PH, Rogoza RM, Rozenbaum MH, et al. On discounting of health gains from human papillomavirus vaccination: effects of different approaches. Value in Health 2012;15:562−7.

[19] Retèl VP, Hummel MJM, van de Vijver MJ, Douma KFL, Karsenberg K, van Dam FSAM, et al. Constructive Technology Assessment (CTA) as a tool in coverage with evidence development: the case of the 70-gene prognosis signature for breast cancer diagnostics. Int J Technol Assess in Health Care 2009;25:73—83.

[20] Polonsky WH, Fisher L, Schikman CH, Hinnen DA, Parkin CG, Jelsovsky Z, et al. Structured self-monitoring of blood glucose significantly reduces A1C levels in poorly controlled, noninsulin-treated type 2 diabetes. Diabetes Care 2011;34:262—7.

[21] Postma MJ, Boersma C, Vandijck D, Vegter S, Le HH, Annemans L. Health technology assessments in personalized medicine: illustrations for cost-effectiveness analysis. Expert Rev Pharmacoeconomics Outcomes Res 2011;11:367—9.

Appendix: Case Study in the Netherlands

In the Netherlands, clear procedures exist for the HTA for inpatient and outpatient drugs, yet for diagnostics, screening tests, and devices, the analysis is not straightforward, is still in development, and currently lacks health economic assessment. However, two examples of structured HTAs on diagnostic tests—inclusive health economic assessment—exist related to Thyrotropin Alpha (Thyrogen®) and the 70-gene signature (MammaPrint®) [1,2]. The first was evaluated in 2005 within the (then newly developed) system of HTA for new innovative drugs in the Netherlands, whereas it actually was a test rather than a therapeutic drug. An extensive report on the use of technology, a pharmacotherapeutic and a pharmacoeconomic report, and a cost-consequence (budget impact) evaluation became available. On this basis of this, the technology was reimbursed for scintigraphics with iodine in combination with measurement of thyreoglobulin levels for the detection of thyroid cancer. Since then, no other test-like technology has been evaluated in this way within the Dutch drug reimbursement system.

The case of the technology assessment of the 70-gene prognosis signature for breast cancer in the Netherlands may serve as an example of the procedures and tools suggested. This technology was evaluated both clinically and economically within a pilot project financed by the Dutch Foundation for Health Care Insurance [1-5]. Although the exact position of the assessment and potential impact of the signature regarding reimbursement were not immediately clear cut, some aspects of a generic assessment as we suggest can be identified. Also, the assessment of the 70-gene signature is known as the only example of a consistent central assessment of a genetic diagnostic technology to support pharmacotherapeutic decisions in the Netherlands.

The 70-gene prognosis signature for breast cancer was developed by the Netherlands Cancer Institute. This new genomic technology (MammaPrint®) was cleared by the FDA as a diagnostic medical device and ISO-17025 was accredited. The technology was claimed to outperform the use of clinical factors in predicting disease progression and overall survival. Better prediction of these aspects enhances targeted and adequately personalized chemotherapy and averts over- or undertreatment. In particular, using the 70-gene signature, the selection of those patients benefiting most from adjuvant systemic treatment could be done most accurately.

To support the evidence on the noneconomic but clinical aspects, the RASTER (MicroarRAy PrognoSTics in Breast CancER) study was developed as a multicenter, prospective observational study. The main aim was to evaluate the differences between adjuvant systemic treatment advice based on the Dutch clinical (CBO) guideline and the 70-gene signature. The clinical analysis of the 70-gene signature within the RASTER design has also been labeled Constructive TA rather than HTA to emphasize the relatively early assessment of the technology: in some other cases, more evidence on the performance of the technology is already available. The main outcome of the study referred to the fact that the prognosis signature was discordant with CBO risk assessment in 30% of cases, often resulting in treatment changes. Discordant cases were defined as patients clinically at low risk but genetically at high risk or the other way around [6,7].

The cost-effectiveness analysis built on the clinical findings by economically comparing the 70-gene signature with St. Gallen guidelines (SG) and Adjuvant Online software (AO) on pathological test results. Both SG and AO may be considered rather comparable to CBO guidelines. Previous cost-effectiveness analyses on the 70-gene signature concluded that additional data had to be gathered on clinical utility and efficiency [8], which was done in the meantime [9], or were directed towards the competitor Oncotype DX, a 21-gene RP_PCR assay [10,11].

In particular, a Markov model for progression in breast cancer was developed and combined with test outcomes and treatment modalities. In the model, all patients received endocrine treatment, tamoxifen, and aromatase inhibitors. Subsequent treatment depended on HER2neu status (trastuzumab therapy), estrogen receptor positivity, European guidelines, and the risk assessment with either the 70-gene signature or SG/AO. Sensitivity and specificity for the three latter diagnostics were calculated in a pooled analysis. Quality of life was based on EQ-5D, and costs were estimated from various sources. Cost savings could occur if the 70-gene signature would indicate mild treatment where SG/AO would indicate aggressive treatment. Obviously, risks for false-positive (negative) results were explicitly taken into account in the model. Extensive sensitivity analysis was performed, although discount rates were not explicitly mentioned in this respect. Reporting was done using novel health economic tools such as CE planes, efficiency frontiers (incremental analysis), and CE acceptability curves. Value of Information (VoI) analysis, which would have been helpful to target further research, was not undertaken. Per QALY and for willingness to pay above approximately €10,000, the 70-gene signature had the highest probability of being cost-effective. The 70-gene dominated St. Gallen and had an incremental CE over AO of €5650 per life year gained and €4700 per QALY.

Appendix 1 References

[1] Retèl VP, Hummel MJM, van de Vijver MJ, et al. Constructive technology assessment (CTA) as a tool in coverage with evidence development: the case of the 70-gene prognosis signature for breast cancer diagnostics. Int J Technol Assessment in Health Care 2009;25:73—83.

[2] www.cvz.nl

[3] Van de Vijver MJ, He YD, van 't Veer LJ, et al. A gene-expression signature as a predictor of survival in breast cancer. New Engl J Med 2002;347:1999—2009.

[4] Retèl VP, Hummel MJ, van Harten WH. Early phase technology assessment of nanotechnology in oncology. Tumori 2008;94:284—90.

[5] Retèl VP, Joore M, Knauer M, et al. Cost-effectiveness of the 70-gene signature versus St. Gallen guidelines and Adjuvant Online for early breast cancer. Eur J Cancer 2010;46:1382—91.

[6] Bueno de Mesquita JM, Linn SC, Keijzer R, et al. Validation of 70-gene prognosis signature in node-negative breast cancer. Breast Cancer Res Treat 2009 Oct;117(3):483−95. doi: 10.1007/s10549-008-0191-2. Epub 2008 Sep 26.

[7] Bueno de Mesquita JM, van Harten W, Retèl V, et al. Use of 70-gene signature to predict prognosis of patients with node-negative breast cancer: a prospective community-based feasibility study (RASTER). Lancet Oncol 2007;8:1078−87.

[8] Oestreicher N, Ramsey SD, Linden HM, et al. Gene expression profiling and breast cancer care: what are the potential benefits and policy implications? Genet Med 2005;7:380−9.

[9] Buyse M, Loi S, van 't Veer LJ, et al. Validation and clinical utility of a 70-gene prognostic signature for women with node-negative breast cancer. J Natl Cancer Inst 2006;98:1183−92.

[10] Hornberger J, Cosler LE, Lyman GH, et al. Economic analysis of targeting chemotherapy using a 21-gene RT-PCR assay in lymph-node-negative, estrogen-receptor-positive, early-stage breast cancer. Am J Manag Care 2005;11:313−24.

[11] Lyman GH, Cosler LE, Kuderer NM, et al. Impact of a 21-gene RT-PCR assay on treatment decisions in early-stage breast cancer: an economic analysis based on prognostic and predictive validation studies. Cancer 2007;109:1011−8.

13

Integrating Genomics into Pharmacy Education and Practice: Pharmacogenomics Is Not Enough

Daniel A. Brazeau*, Gayle A. Brazeau†

*Genomics, Analytics and Proteomics Core,
University of New England, Biddeford, Maine, USA

†College of Pharmacy, University of New England, Biddeford, Maine, USA

OBJECTIVES

1. Identify the case for genomics in the education of current and future pharmacists and other health care professionals.

2. Describe generally the technologies that have and will continue to drive genomic sciences and their importance in future patient care.

3. Discuss the current status of genomics in pharmacy education and the gap between what is available and what will be needed for contemporary pharmacy practice in which genomics will play a key role in patient care.

4. Identify and describe four key elements for a contemporary genomics course in pharmacy education.

INTRODUCTION: THE CASE FOR GENOMICS

The nature of health care has always been one of constant change and advancement; however, with the completion of the human genome and, more important, the ongoing advances in genomic technologies, the delivery of appropriate health care is witnessing an extraordinary revolution. The application of genetic or genomic principles to our understanding of human health has and will continue to dramatically transform the nature of and expectations for the standards of practice for health care providers at all levels and

Pharmacogenomics
http://dx.doi.org/10.1016/B978-0-12-391918-2.00013-5

451

professional settings. This is particularly relevant given the complementary emphasis on optimizing patient care and minimizing health care disparities while at the same time reducing health care costs.

Genomics has become integral to all fields of research in the life sciences. The genomic tools now available to scientists allow for a more precise understanding of disease processes and therapeutic agents. It is becoming possible to assess the cellular response to perturbations to homeostasis in living systems as well as to determine the cellular mechanisms that underlie the success of therapeutic interventions. Advances in the genomic sciences are bringing about the ultimate realization of "personalized medicine." In a broader scope, personalized medicine is not simply the right therapy or drug for the right patient, but the "individualization of disease," encompassing the characterization of human disease based upon individual-specific biological mechanisms rather than a phenotypic description.

Determining cellular—and sometimes genetic—markers is now becoming routine in guiding clinical decisions. For example, the parsing of breast cancers into subtypes is becoming customary in clinical diagnosis. One such subtype is defined in part by the overexpression of the oncogene *ERBB2*. *ERBB2* codes for the human epidermal growth factor receptor 2 protein (HER2). Overexpression of this gene has been observed in approximately 30% of breast cancers and results in notably aggressive tumors and generally poor prognosis [1]. *ERBB2* (for a complete description, see the online database Mendelian Inheritance in Man, OMIM #164870) is a member of a family of human epidermal growth factor receptors that when activated results in anti-apoptosis and increased cell proliferation. With an understanding of the underlying molecular mechanisms of this specific subclass of tumors, new drugs have been developed to target and block the HER2 receptor. The monoclonal antibody trastuzumab is one of the first drugs approved by the Food and Drug Administration (FDA) based upon genomic individualization of disease.

Similarly, recent work has identified a subgroup of cystic fibrosis patients whose disease is caused by a specific missense mutation in the Cystic Fibrosis Transmembrane conductance Regulator (CFTR) protein. Cystic fibrosis (CF) is caused by mutations in the *CFTR* gene (OMIM #602421) that codes for a transmembrane chloride ion channel. Cystic fibrosis in most patients (~70%) is caused by a three base pair deletion in *CFTR* for which there is no cure. However, in those 4 to 5% of patients whose particular subtype of cystic fibrosis is caused by a different mutation, the missense mutation, G551D (a substitution of a glycine in normal individuals with an aspartic acid in CF patients at the 551st amino acid), a new drug is in development that augments the activity of p. G551D-impaired CFTR [2]. This event represents the first possible cure for selected cystic fibrosis patients. Interestingly, CF can be caused independently by any number of mutations in the *CFTR* gene, a common genetic phenomenon known as allelic heterogeneity.

Allelic heterogeneity is an important concept in understanding the pharmacogenetics of drug response. There are many cases in which different mutations (alleles) in the same gene (a drug metabolizing enzyme for example) result in the same phenotype (poor metabolizer). For example, within the cytochrome P450 gene, 2D6 (*CYP2D6*), there are a number of mutations, all of which result in the poor metabolizer phenotype (Table 13.1). Often, these genetic variants are unique to specific human populations. This often unknown allelic heterogeneity is one of the main complicating factors to the development and interpretation of genetic tests.

TABLE 13.1 Allelic heterogeneity in human cytochrome P450 2D6 (CYP2D6)

Allele	Phenotype	Comments
CYP2D6*1	Extensive metabolizer	Reference allele
CYP2D6*2	Extensive metabolizer	Nonsynonymous SNP
CYP2D6*3	Poor metabolizer	Frame shift mutation
CYP2D6*5	Poor metabolizer	Gene deletion
CYP2D6*6	Poor metabolizer	Frame shift mutation

For more examples within *CYP2D6* see www.pharmGKB.org. To date there are at least 12 important alleles identified in CYP2D6.

TECHNOLOGY AS THE DRIVER OF GENOMIC SCIENCE

The foundations of genomic medicine are based upon more than 50 years of research in physiology, molecular, and cellular biology. Recently, the development of technologies that led to the completion of the human genome has allowed an unprecedented access to the host of cellular players that underlie human health and disease. Three technologies warrant special attention in pharmacogenomics, as they will further advance our potential to enhance future patient care.

Quantitative Polymerase Chain Reaction (Q-PCR)

Quantitative Polymerase Chain Reaction (Q-PCR) — an advancement in standard polymerase chain reaction (PCR) that allows for both the detection and quantification of very rare DNA targets by assessing the amplification process with each PCR cycle (in real time), this quantification is target specific from samples derived from the mRNA pool of a few hundred cells (assessing the transcriptome) or from genomic DNA from a buccal swab (genetic testing). Q-PCR is the method of choice for assessing the expression levels of any gene or genes in the human genome. Because the technology is based upon the polymerase chain reaction quantification of gene expression, it can be accomplished from minute quantities of tissue, thus allowing for the localization of gene expression. Q-PCR is routinely used to validate the findings of larger microarray or sequencer-based genome expression studies.

Silencing RNA (siRNA)

Silencing RNA (siRNA) - synthetically produced small RNAs (~21 base pairs) that inhibit the translation of the complementary mRNA for virtually any human protein—coding gene. Depending upon the vector being used and the cell type, silencing of any human gene can be accomplished in a matter of hours to weeks, thus providing researchers with a very precise molecular switch to assess the function of a gene. siRNA technologies are now standard in research laboratories and have great therapeutic potential [3,4].

Next-Generation Sequencing-The Newest Iteration

The newest iteration of DNA sequencing technology is based upon a number of different "chemistries," but all use massively parallel sequencing reactions, producing hundreds of thousands to millions of sequences in a single run. These technologies have increased the capacity to sequence an entire human genome from years to weeks and will ultimately cost no more than many standard medical tests. Readily available, and inexpensive, sequence data will allow health scientists and clinicians to assess and evaluate the role of underlying human genetic variation in disease [5,6] or to explicitly define tumors in order to guide both therapy and predict outcomes [7,8]. In addition, massively parallel sequencing of cDNAs reverse-transcribed from mRNA pools are replacing "older" microarrays as a means to assess gene expression patterns [9] and accurately characterize the transcriptome to identify causative agents or useful biomarkers. With the advent of digital medical records that include patient genome data, decisions concerning the use of "pharmacogenetically" appropriate drugs can be made immediately at the point of care. Finally, genome data, including an enumeration of patient-specific disease risk alleles, will greatly facilitate preventative care.

THE CRITICAL GAP IN PHARMACY EDUCATION

These genomic technologies provide an unprecedented ability to explore the nature of human disease at the cellular and subcellular level. Their importance to health care professionals lies in the fact that this information will change the very nature of how we define human disease. Increasingly, disease will be characterized by the language of genomics or cellular/molecular biology rather than "symptoms." To date, although the need for coverage of genomics in the education of health care professionals is widely acknowledged, appropriate content has been slow to enter the training of health care professionals [10–21]. It is interesting to note that in the health education literature outside of pharmacy, the discussion is of the need to integrate "genomics" and "genomics medicine" into curricula. Within the field of pharmacy, in contrast, the talk is of "pharmacogenomics" or "pharmacogenetics." Although these terms are often defined broadly to include some aspects of the field of genomics, all too often the discussion is more narrowly defined as the contributions of human genetic variation to drug response involving single genes (pharmacogenetics) or many genes—perhaps the entire genome (pharmacogenomics).

Pharmacy education, like other contemporary health profession educational programs, needs to move forward to integrate the entirety of genomic sciences into the pharmacy curricula early on in the professional program. Enhancing and optimizing patient care is going "genomic," as the above examples in personalized medicine demonstrated in this chapter (breast cancer and cystic fibrosis) and throughout this book. In order for society to realize the benefits of our increased understanding of the nature of human disease, all health care practitioners in interprofessional teams must become conversant in genomics [21].

Status of Genomics in Pharmacy Education

Of course, in many respects pharmacy has been leading the way in incorporating many aspects of pharmacogenetics into the curricula of colleges and schools of pharmacy. This development is due in large part to the immediate application of genetic data in some aspects of drug therapy. The fact that some drugs are metabolized largely by a product of a single gene makes for a simple monogenic trait whose principles of inheritance have been known since Mendel. For example, codeine is metabolized in the liver into the active agent morphine. A single enzyme, *CYP2D6*, is largely responsible for this conversion. Thus, the metabolism of codeine to morphine in humans can be treated as a monogenic trait and thus polymorphisms in this gene (Table 13.1) may result in clinically relevant outcomes. In fact the term "pharmacogenetics" was first coined by Vogel [22] in 1959 following an earlier summary by Motulsky [23] concerning interindividual drug response and its relationship to genetics (as understood to be the study of patterns of inheritance).

Interestingly, a paper prompted by the American Association of Colleges of Pharmacy early in the genomics era presented the case for the need for pharmacogenomics education in schools and colleges of pharmacy [11]. This paper includes suggested core competencies in pharmacogenetics and pharmacogenomics. These competencies were quite broad and included basic genetic concepts; the nature of genetic variation and disease (genotype to phenotype); the role of behavioral, social, and environmental factors in disease and drug response (multifactorial or quantitative genetics); and ethical and economic implications of genetic information. Yet, despite this early call for a more broad-based presentation of pharmacogenomics in schools and colleges, in practice what is being offered is more narrowly focused, though there are attempts to take a broader approach [18].

The need for inclusion of pharmacogenetics and pharmacogenomics in pharmacy school curricula is essential and has been included in the recent Accreditation Council for Pharmacy Education (ACPE) Standards [24]. In a 2003 survey of 377 community pharmacists in the United States, less than 50% were satisfied with their knowledge of the implications of the Human Genome Project, genetic testing, and pharmacogenetics [25]. Thus, among practicing pharmacists, there is the realization of the need for a better background (foundation) in genetics and its application to pharmaceutical principles. In a 2004 survey of 85 schools of pharmacy in the United States, of the 41 respondents only 16 schools (39%) provided any content on pharmacogenetics or pharmacogenomics to their professional students [26]. At that time, only five schools had a standalone pharmacogenetics or pharmacogenomics course. A 2010 assessment of the coverage of pharmacogenomics in the curricula of colleges or schools of pharmacy in the United States showed an increase to 89% (69 schools among 75 respondents) of schools that now included some aspect of pharmacogenomics in their curricula [27]. Approximately 22% of the schools offered a standalone course in pharmacogenomics. A greater percentage (35%) offered pharmacogenomics as an elective. At the time of this survey, fewer than half of the schools had plans to increase coverage in the following three years. For a discipline that is becoming foundational to modern health care, this coverage seems inadequate and could limit pharmacists' ability to successfully engage in interprofessional patient care.

Ultimate integration of genomics into clinical practice will require of health care professionals a much more sophisticated understanding of the underlying genomics principles and technologies. Here we present a brief outline of a standalone pharmacogenomics course designed to be offered early in the curriculum, as it is the foundation for future courses such as pharmacokinetics/pharmacodynamics, pharmacology and pharmacotherapeutics, and clinical pharmacokinetics. In that the principles presented in this course are largely basic sciences in molecular biology and genetics, it is believed that introducing the students early in their education to these basic principles will provide them with a foundation in genomics that they can then apply as they encounter specific topics in which genomics and therapeutics intersect. The outcomes of this course are adapted from competencies first suggested by the American Society of Human Genetics (Table 13.2).

A two-credit-hour, semester-long course is suited for the first professional year of the doctor of pharmacy program. The course goal is to provide students with an understanding of the principles and applications of human genetics and genomics in drug discovery, drug therapy optimization, and patient care and counseling. The course is divided into four sections (Table 13.3), including the basic science that underlies genomics, a section on the nature of genomes and the genomic technologies including bioinformatics that are driving the science, a presentation of pharmacogenomics and therapeutics, and the ethical issues novel to the genomics revolution in health care.

TABLE 13.2   Educational outcomes for a standalone genomics course

The educational outcomes that a student should be able to achieve at the end of the course are:

1) Discuss and apply the basis science of genomics and how this relates to the role that pharmacists play in optimizing patient care.
2) Identify and discuss how genomic technologies and bioinformatics will affect the role that pharmacists can play in patient care.
3) Identify and demonstrate how pharmacogenetics will have an impact on drug pharmacokinetics and pharmacogenomics on pharmacotherapeutics.
4) Identify and discuss the ethical issues associated in the genome revolution and how pharmacy practice and patient care is affected in contemporary practice.

The course objectives include:

- Describe the structure and function of DNA and RNA.
- The role of mutations in biological variation.
- What genes are, how they are organized and regulated.
- The role of genome evolution in biological systems.
- How alleles segregate in and among populations.
- Environmental and genetic factors that affect development of the phenotype, including drug response.
- The multifactorial nature of most human traits, including drug response and the principles of multifactorial inheritance.
- The role of gene–gene interaction in phenotypic variation.
- How polymorphisms arise and are maintained in human populations, and how gene linkage and human gene mapping are used to identify candidate genes.
- How human genetic variation affects drug metabolism, activation, and disposition.
- The advantages, limitations, and dangers of predictive testing for genetic disease and drug response.
- How to navigate among the many comprehensive genomic databases and resources on the Internet.
- The genomic technologies employed in drug discovery and development.
- Legal and ethical issues in genetic testing and patient stratification in clinical trials.

The specific learning objectives of the course were based upon the American Society of Human Genetics (ASHG) guidelines for a medical school core curriculum in genetics.

TABLE 13.3 Basic outline for a pharmacogenomics course

1) Basic Science of Genomics
The Case for Pharmacogenomics
The History of Genetics and Pharmacogenetics

1.1) Information Flow in Biological Systems
    Review of Cell and Molecular Biology
    Gene Expression — Regulation of Transcription and Translation

1.2) Information Transmission/Inheritance
    Mendelian Transmission Patterns
    Dominance/Recessive Expression Patterns
    Sex-Linkage

1.3) Population Genetics and Evolution
    Gene and Allele Frequencies
    Hardy-Weinberg Equilbria
    Factors Affecting Gene Frequencies
    Selection, Genetic drift
    Population Structure/Admixture
    Race/Ethnicity and Ancestral/Geographic Origin Of Alleles

1.4) Multifactorial Inheritance - Quantitative Genetics
    Complex Traits
    Locus and Allelic Heterogeneity
    Phenocopies
    Polygenic Traits and Environmental Factors
    Genetic Background
    Genetic Mapping
    Genetic Markers and Linkage Mapping
    Pedigree Analysis
    Haplotypes
    Genome Wide Association Studies (GWAS)

2) The Human Genome and Genomic Technologies

2.1) Genomes — Diversity, Size and Structure

2.2) Genome Evolution
    Mechanisms of Gene Duplication and Development of Gene Families
    Paralogs versus Orthlogs

2.3) Genomic Technologies
    Microarrays
    Quantitative PCR
    Next Generation SequencingSilencing RNA

2.4) Bioinformatics
    Genetic Data and The Internet
    The National Center for Biotechnology Information
    The Pharmacogenetics and Pharmacogenomics Knowledge Base
    Electronic Medical Records

(Continued)

TABLE 13.3   Basic outline for a pharmacogenomics course—cont'd

3) Pharmacogenetics

3.1) Application of basic principles
  Drug Transporters
  Drug Metabolizing Enzymes
  Cellular Signaling Pathways
  Drug Target Pharmacogenetics

3.2) Pharmacogenomics and Pharmacotherapeutics
  Cancer Genomics and Oncology
  Hematology
  Cardiovascular Diseases
  Transplantation
  Central Nervous System and Psychiatry

4) Ethics and the Genome Revolution

4.1) Ethics and Privacy in Research
  The Belmont Report

4.2) Ethical, Legal and Social Implications of Pharmacogenomics

4.3) Economics of Pharmacogenomics
  Regulatory Issues

Basic Science of Genomics

Fully one half of the course is devoted to coverage of the science that underlies genomics. This coverage is crucial in that most undergraduate programs provide only rudimentary exposure to the advanced genetic principles. It is in some sense unfortunate that pharmacogenomics encompasses three distinct disciplines of genetics (Figure 13.1), two of which (Population Genetics and Quantitative Genetics) are generally not part of most undergraduate programs—even for biology majors. Following an introduction to the history of pharmacogenetics and pharmacogenomics and the fields' importance in modern therapeutics [15,21], two to three lectures are spent reviewing DNA and RNA structure and synthesis, ultimately leading to the mechanisms of gene expression and regulation. Throughout these discussions, the cellular machinery involved in all these essential cellular functions is described in the context of its function and potential as a drug target [27,28].

Basic Mendelian principles are reviewed, including discussions of the standard monogenic transmission patterns (autosomal dominance/recessive, sex linkage). Slight variations on these basic patterns (penetrance, codominance) are then introduced as simple modifications from Mendelian traits, though still encompassed in single gene systems. In all cases, examples are drawn from the pharmacogenetics literature. For example, the role of the phase two enzyme thiopurine-S-methyltransferase (TPMT) in the metabolism of thiopurine drugs provides an excellent case study in variable enzyme activity among patients displaying classic monogenic patterns of phenotypic expression [29]. That assessment of TPMT activity is now routine prior to treatment helps make this topic relevant to students.

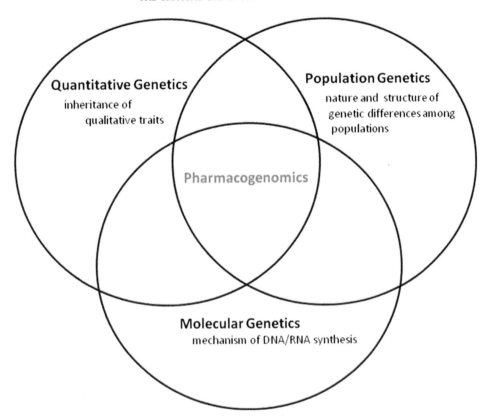

FIGURE 13.1 Pharmacogenomics represents the intersection of three distinct fields of modern genetics. Quantitative genetics is the study of qualitative genetic traits whose expression is due to the interaction of many genes and the environment. Population genetics is the study of the genetic differences among populations. Molecular genetics encompasses the cellular mechanisms of DNA and RNA synthesis, transcription, and translation. Molecular processes underlie the revolution in genomic technologies.

The calculation of gene and allele frequencies is demonstrated in a class exercise in which the students are genetically "tested" for two traits whose expression follows simple Mendelian rules. Students are polled to assess the number of students who can or cannot roll their tongues (Table 13.4) and can or cannot taste the compound phenylthiocarbamide (PTC). Based upon the frequencies of the homozygous recessive alleles, genotype frequencies are calculated. Finally, as a proof of concept, the number of students who can neither roll their neither tongues nor taste PTC is predicted and compared to the observed. Chi-square analysis can then be introduced as a statistical tool for evaluating deviations from Hardy–Weinberg equilibrium. This approach also provides a useful introduction of genetic linkage because the calculations assume the two traits are not linked (they are not). This exercise leads to a discussion of the nature of linkage in genetic systems, preparing the groundwork for coverage of genetic mapping. Finally, with an understanding of allele frequencies, a discussion of the nature of the genetic structure of populations and its role in pharmacogenetics [30,31] is presented. Factors that result in changes in allele frequencies within and among populations (genetic drift, selection) can now be introduced.

Critical to the student's ultimate understanding of pharmacogenomics is the realization that nearly all human traits, including most human disease and drug response, do not behave in a simple Mendelian fashion. A significant number of lectures are devoted to a discussion of the science of multifactorial inheritance or quantitative genetics. As noted earlier, the field of quantitative genetics will be new material for most students. Of all the concepts that underlie an understanding of pharmacogenomics, this issue is perhaps the most critical and warrants considerable effort in order for students to acquire an understanding of how human traits (disease risk, drug response, intelligence, height) that are determined by multiple, interacting genes (polygenic) can be further subject to considerable modification by environmental factors. Here genetic mapping techniques, including genome-wide association studies (GWASes) are used as a means to introduce the concepts of genetic background, locus and allelic heterogeneity, phenocopies, and gene—environment interactions [32]. An awareness

TABLE 13.4 In-class exercise to demonstrate the calculation of genotype and allele frequencies

Consider the trait: "tongue rolling"
  1. Two-allele system
      a. "R" allele → the ability to roll tongue, **dominant**
      b. "r" allele → cannot roll tongue, **recessive**
  2. Autosomal

Example for a class with 100 students			
Phenotype	**Genotype**	**Number in class**	**Phenotypic frequencies**
Roller	RR, Rr, rR	80	80%
Non-roller	rr	20	20%
		Total 100	100%

To calculate allele frequencies:
Look at recessive cases. Individuals who cannot roll their tongue genotype must be → "rr"
  If frequency of "rr" individuals $= r^2 = 20/100 = 0.20$
  Then, frequency of $r = \sqrt{0.20} = $ **0.447**
  Because the frequencies of R and r must $= 1$, then frequency of $R = 1 - 0.447 = $ **0.553**

To calculate genotypic frequencies:

Genotype	**Frequency**	**Number in class**
rr (r^2)	$(0.447)(0.447) = 0.20$	0.20 * 100 students $= 20.0$
RR (R^2)	$(0.553)(0.553) = 0.306$	0.306 * 100 students $= 30.6$
Rr and rR (2Rr)	$2 * (0.553)(0.447) = 0.494$	0.494 * 100 students $= 49.4$
Total	1.0	100.0

TABLE 13.4 In-class exercise to demonstrate the calculation of genotype and allele frequencies—cont'd

Note: Given Hardy—Weinberg equilibrium, allele frequencies will follow standard probability theory (beanbag genetics!).

Remember: Binomial expansion

$$R + r = 1$$
$$(R + r)^2 = 1^2 = 1$$
$$(R + r)^2 = R^2 + 2Rr + r^2 = 1$$

Now, consider a second gene:

Trait: Phenylthiocarbamide (PTC) tasters

1. Two-allele system

    a. "T" allele → can taste PTC, **dominant**
    b. "t" allele → cannot taste PTC, **recessive**

2. Autosomal

Phenotype	Genotype	Number in class	Phenotypic frequencies
Taster	TT, Tt, tT	60	60%
Non-taster	tt	40	40%
		Total 100	100%

Therefore:

	Allelic frequencies		Genotypic frequencies		
	T allele	**t allele**	**TT**	**Tt**	**tt**
Frequency	0.37	0.63	0.14	0.46	0.40
Number	–	–	14	46	40

Finally,

We can now calculate the number of students who can *neither* roll their tongues *nor* taste PTC

Non-roller, non-taster genotype: **rr,tt**

Frequency: 0.20 * 0.40 = 0.08

Number of students: 0.08 * 100 = 8 students

Important: This assumes that there is no LINKAGE—the genes (and their alleles) assort independently.

If the calculated number is significantly different (observed ≠ expected)

Then → **1)** Sampling errors (usually small sample sizes)
        2) Phenocopies
        3) Linkage

of the nature of gene—gene and gene—environment interactions is essential in understanding why most of the therapeutically relevant human traits (disease risk, drug response) do not behave in a straightforward Mendelian fashion. A more sophisticated knowledge of the factors that underlie the complexity of most human traits is essential if health professionals are to bring the benefits of genomics to patient care. It will be essential for pharmacists and other health care professionals to have and employ a sophisticated knowledge that underlies the complexity of most human traits in order to optimize the benefits of genomics to the care of their patients. The educational components proposed in this type of course provide the foundation for pharmacists and other health care providers to achieve this goal for their professional practice.

The Human Genome and Genomic Technologies

The human genome comprises approximately 3 billion base pairs arranged on 23 chromosomes, of which we have two sets. Within this DNA sequence are scattered 23,000 to 26,000 genes representing slightly more than 10,000 gene families. The complexity of the human genome not only complicates the discovery of genes but also confounds the understanding of gene function including the role specific genes play in drug response. For example, the cytochrome P450 (CYP) gene family includes genes that code for important drug metabolizing enzymes. Humans have 57 functioning CYP genes and 58 members of the CYP family that are nonfunctional in all humans. Many of these "pseudogenes" in the human genome are still functional in other organisms. For example, mice have 102 functioning members of this CYP family, many of which are the homologues of pseudogenes in humans. The relevance to pharmacy is that many of the members of this family have very similar DNA sequences and therefore have significant overlap in the compounds metabolized and tissues in which they are expressed. With so many very similar genes performing similar, sometimes overlapping functions, it is difficult to identify the specific genes responsible for a given drug's action. Importantly, pharmacogenetically important genes are often (if not entirely) those genes that act alone at a crucial metabolic step. Hence, the activation or inactivation of the drug is monogenic—determined by a single gene. Thus polymorphisms in this crucial gene may have clinical importance. This fundamental property of the genome—many genes with overlapping substrates—underlies one of the major limitations to the widespread utility and adoption of pharmacogenetics in health care practice. Finally, an understanding of the evolution of genomes among species is crucial in choosing the correct gene in the correct model organism to study drug action.

Also included in the section is a brief discussion of the most important genomic technologies involved in drug discovery, drug development, and genotyping, including next-generation DNA sequencing technologies, microarrays, silencing RNA, and the quantitative polymerase chain reaction (QPCR). These technologies are covered in the specific context of drug discovery and development, and in human genotyping of genes involved in drug metabolism and transport. Finally, the database resources (Table 13.5) now available to health professionals are introduced. Most important, students are shown how to navigate among the information to be found at the National Center for Biotechnology Information (NCBI) and the Pharmacogenomics Knowledge Base (PharmGKB, described in detail in chapter 3). PharmGKB in particular represents the ultimate repository and the most current source of information on drugs and human genetics. In many respects, the entire course is designed so that students are capable of competently querying this database for information on the pharmacogenomics of any drug or gene.

TABLE 13.5 Freely available internet resources in genomics

1	ClinVar	Human sequence variation and its relationship to health.	www.ncbi.nlm.nih.gov/clinvar/
2	DNA Learning Center Cold Spring Harbor Laboratory	Laboratory exercises and explorations for students and educators.	www.dnalc.org
3	Evaluation of Genomic Applications in Practice and Prevention	Evidence-based evaluation of genetic tests and applications of genomic technologies.	www.egappreviews.org
4	Gene Tests NCBI	Current information on genetic testing and its use in diagnosis, management, and genetic counseling.	www.ncbi.nlm.nih.gov/sites/GeneTests/?db=GeneTests
5	GeneCards: The Human Gene Compendium	Searchable, integrated database of human genes.	www.genecards.org
6	Genetics Home Reference NIH	Consumer-friendly information about the effects of genetic variations on human health.	ghr.nlm.nih.gov
7	Global Genetics and Genomics Community	Interactive case studies simulating patient encounters.	g-3-c.org/resources.php
8	National Cancer Institute	Part of the National Institutes of Health (NIH) dealing with cancer research and training.	www.cancer.gov/cancertopics/understandingcancer/targetedtherapies
9	National Center for Biotechnology Information	Ultimate repository for biomedical and genetic information.	www.ncbi.nlm.nih.gov
10	National Coalition for Health Professional Education in Genetics	Promotes health professional education and access to information about advances in human genetics.	www.nchpeg.org
11	National Human Genome Research Institute	Educational materials in genetics and genomics.	www.genome.gov/education/
12	Online Mendelian Inheritance in Man	Comprehensive, authoritative, and timely compendium of human genes and genetic phenotypes.	www.ncbi.nlm.nih.gov/omim
13	Pharmacogenomics Education Program, University of California San Diego	Evidence-based pharmacogenomics education program designed for pharmacists and physicians, pharmacy and medical students, and other health care professionals.	pharmacogenomics.ucsd.edu/home.aspx

(Continued)

TABLE 13.5 Freely available internet resources in genomics—cont'd

14	Pharmacogenomics Knowledge Base	Comprehensive resource that curates knowledge about the impact of genetic variation on drug response for clinicians and researchers.	www.pharmgkb.org
15	Your Genome: Wellcome Trust Sanger Institute	Explanatory animations, downloadable classroom activities, and accurate information about genetics and genomics.	www.yourgenome.org

Pharmacogenetics and Pharmacotherapeutics

With a foundation established in the basics of genomics, genomes, genomic technologies, and genomic databases, students are now introduced to case studies in pharmacogenetics and drug response. Included are discussions of the pharmacogenetics of drug metabolizing enzymes, and drug transporters, drug target pharmacogenetics, cancer therapeutics, organ transplantations, gene therapies, and multiple sclerosis. Ideally, guest faculty members are invited with specific areas of specialization, including cardiovascular disease, respiratory disease, oncology, psychiatry, and infectious diseases. The emphasis here is on in-depth discussions incorporating broad genomic principles and technologies in characterizing and treating disease.

Ethics and the Genome Revolution

Genomic medicine presents novel ethical challenges that pharmacists and other health care professionals will face. Pharmacists in particular face many ethical issues and dilemmas because of their interactions with patients through individualized counseling, medication therapy management programs, medication reconciliation programs, transitional care programs, and consultation activities in the community and institutional setting. Genomic patient data present unique privacy issues, risk of discrimination (social risk), and implications for other family members not incurred by standard medical tests. In addition, pharmacogenomics introduces a new type of stratification among treatment populations and perhaps a new class of therapeutic orphans based upon allelic status at crucial genes. Also covered here are ethical and privacy issues arising in pharmaceutical research and the role of stratification in drug trials.

EDUCATIONAL COMPETENCIES

The key to any genomics educational program is the competencies demonstrated by the student, whether it is a pharmacy student, other health care professional student, pharmacist, or other health care professional. In the current two-hour required course, competencies are evaluated using written essay examinations, which enables assessment at the higher levels of thinking such as application, analysis, synthesis, and evaluation. Student assessments incorporate a relative grading approach as these concepts and their applications can be challenging even to students who have had a previous genetics course.

Students are also asked to integrate their learning through a two- to three-page written summary and critique of a contemporary peer-reviewed publication in the area of pharmacogenomics. Students must choose a paper from the primary literature that has been published in the past year (this prevents students from using others' work from past years). The incorporation of this written assessment, with feedback provided to those students who request assistance in looking at draft papers, is an effective means of identifying areas where key concepts and applications of these concepts may still require time in the lecture setting.

The most challenging aspect in offering this course is to demonstrate the relevancy of the many complex genetic concepts to the practice of pharmacy. Many of our students work or have had experiences in pharmacy settings, and they know that genomic concepts are seldom encountered. That this is a new, emerging field that will come to dominate the practice of medicine is often not convincing. Thus, it is essential to incorporate clinically relevant examples throughout the course. It is also helpful that articles describing recent genomic discoveries in health care appear nearly weekly in the general press.

CONCLUSION

In conclusion, the practice of medicine in the immediate future will require of health professionals the ability to accurately communicate and interpret pharmacogenomic information to patients. Pharmacists, given their education and expertise in aspects related to interpatient variability in drug metabolism, disposition, and pharmacodynamics, are well placed to assume the leading role as the "learned intermediary" to consumers and patients about pharmacogenomic information and its relationship to optimizing and enhancing their health care. Furthermore, the easy accessibility of pharmacists in a variety of professional settings will most likely result in pharmacists being the primary point of contact for patients and customers seeking help to understand these concepts and issues. This development will be extremely important given the increasing genetic information and services, some of questionable validity, available from many sources both within and outside of the health care system. Genomics will contribute greatly to improved therapeutics, but realization of its benefits will require a more realistic understanding of the roles that environmental factors, multiple genes with multiple variants, and human population genetic structure play in predicting individual therapies and therapeutic outcomes. The appropriate competency in the genomic sciences is essential to ensure that our students have the basic understandings to be lifelong learners in this field as it progresses.

QUESTIONS FOR DISCUSSION

1. What are examples of drug therapies that were developed based on an understanding of individual-specific mechanisms of a disease?

2. What are examples of technologies that are driving genetic and pharmacogenetic discoveries?
3. How does a thorough understanding of human genetics—including an understanding of population genetics, quantitative genetics, genetic technologies, and ethics—complement typical pharmacogenetic content included in the pharmacy curricula?

References

[1] Tan M, Yu D. Molecular mechanisms of erbB2-mediated breast cancer chemoresistance. Adv Exp Med Biol 2007;608:119–29.

[2] Ramsey, B. W., Davies, J., McElvaney, N. G., Tullis, E., Bell, S. C., Ďevínek, P., Griese, M., McKone, E. F., Wainwright, C. E., Konstan, M. W., Moss, R., Ratjen, F., Sermet-Gaudelus, I., Rowe, S. M., Dong, Q., et-al, and. for the VX08-770-102 Study Group. (2011). A CFTR potentiator in patients with cystic fibrosis and the G551D mutation. New Engl J Med 365, 1663–72.

[3] Takeshita F, Ochiya T. Therapeutic potential of RNA interference against cancer. Cancer Sci 2006;97:689–96.

[4] Singh SK, Hajeri PB. siRNAs: their potential as therapeutic agents, II. Methods of delivery. Drug Discov Today 2009;14:859–65.

[5] Tucker T, Marra M, Friedman JM. Massively parallel sequencing: the next big thing in genetic medicine. Am J Hum Genet 2009;85:142–54.

[6] Voelkerding KV, Dames SA, Durtschi JD. Next-generation sequencing: from basic research to diagnostics. Clin Chem 2009;55:641–58.

[7] Ley TJ, Mardis ER, Ding L, Fulton B, McLellan MD, Chen K, et al. DNA sequencing of a cytogenetically normal acute myeloid leukaemia genome. Nature 2008;456:66–72.

[8] Aparicio SA, Huntsman DG. Does massively parallel DNA resequencing signify the end of histopathology as we know it? J Pathol 2010;220:307–15.

[9] Wang Z, Gerstein M, Snyder M. RNA-Seq: a revolutionary tool for transcriptomics. Nat Rev Genet 2009;10:57–63.

[10] Emery J, Hayflick S. The challenge of integrating genetic medicine into primary care. Brit Med J 2001;322:1027–30.

[11] Johnson JA, Bootman JL, Evans WE, Hudson RA, Knoell D, Simmons L, et al. Pharmacogenomics: a scientific revolution in pharmaceutical sciences and pharmacy practice. Report of the 2001–2002 Academic Affairs Committee. A J Pharm Ed 2002;66:12s–5s.

[12] Gurwitz D, Weizman A, Rehavi M. Education: teaching pharmacogenomics to prepare future physicians and researchers for personalized medicine. Trends Pharmacol Sci 2003;24:122–5.

[13] Gurwitz D, Lunshof JE, Dedoussis G, Flordellis CS, Fuhr U, Kirchheiner J, et al. Pharmacogenomics education: International Society of Pharmacogenomics recommendations for medical, pharmaceutical, and health schools deans of education. Pharmacogenomics J 2005;5:221–5.

[14] Guttmacher AE, Porteous ME, McInerney JD. Educating health-care professionals about genetics and genomics. Nat Rev Genet 2007;8:151–7.

[15] El-Ibiary SY, Cheng C, Alldredge B. Potential roles for pharmacists in pharmacogenetics. J Am Pharm Assoc 2008;48:e21–9.

[16] Gaff CL, Williams JK, McInerney JD. Genetics in health practice and education special issue. J Genet Couns 2008;17:143–4.

[17] McInerney JD. Genetics education for health professionals: a context. J Genet Couns 2008;17:145–51.

[18] O'Brien TJ, Goodsaid F, Plack M, Harralson A, Harrouk W, Hales TG, et al. Development of an undergraduate pharmacogenomics curriculum. Pharmacogenomics 2009;10:1979–86.

[19] Daack-Hirsch S, Dieter C, Quinn Griffin MT. Integrating genomics into undergraduate nursing education. J Nurs Scholarsh 2011;43:223–30.

[20] Lea DH, Skirton H, Read CY, Williams JK. Implications for educating the next generation of nurses on genetics and genomics in the 21st century. J Nurs Scholarsh 2011;43:3–12.

[21] Frueh FW, Gurwitz D. From pharmacogenetics to personalized medicine: a vital need for educating health professionals and the community. Pharmacogenomics 2004;5:571—9.

[22] Vogel F. Probleme der Humangenetik. Ergebnisse der inneren Medizin und Kinderheilkunde 1959;12:65—126.

[23] Motulsky AG. Drug reactions, enzymes, and biochemical genetics. J Am Med Assoc 1957;165:835—7.

[24] American Council on Pharmaceutical Education (ACPE). Accreditation standards and guidelines for the professional program in pharmacy leading to the doctor of pharmacy degree. Chicago, Ill. 2011. http://www.acpe-accredit.org/standards/default.asp.

[25] Sansgiry SS, Kulkarni AS. The Human Genome Project: assessing confidence in knowledge and training requirements for community pharmacists. Am J Pharm Educ 2003;67(2). article 39.

[26] Latif DA, McKay AB. Pharmacogenetics and pharmacogenomics instruction in colleges and schools of pharmacy in the United States. Am J Pharm Educ 2005;69(2). article 23.

[27] Papavassiliou AG. Transcription-factor-modulating agents: precision and selectivity in drug design. Mol Med Today 1998;4:358—66.

[28] Ghosh D, Papavassiliou AG. Transcription factor therapeutics: long-shot or lodestone. Curr Med Chem 2005;12:691—701.

[29] Weinshilboum R. Pharmacogenetics of methyl conjugation and thiopurine drug toxicity. Bioessays 1987;7:78—82.

[30] Wilson JF, Weale ME, Smith AC, Gratrix F, Fletcher B, Thomas MG, et al. Population genetic structure of variable drug response. Nat Gen 2001;29:265—9.

[31] O'Donnell PH, Dolan ME. Cancer pharmacoethnicity: ethnic differences in susceptibility to the effects of chemotherapy. Clin Cancer Res 2009;15:4806—14.

[32] Schork NJ. Genetics of complex disease: approaches, problems, and solutions. Am J Respir Crit Care Med 1997;156:S103—9.

Index

Color Plates

PharmGKB
The Pharmacogenomics
Knowledgebase

Pharmacogenomics. Knowledge. Implementation.

PharmGKB is a comprehensive resource that curates knowledge about the impact of genetic variation on drug response for clinicians and researchers.

About Us ▼   News & Events   Projects   Search ▼   Download   Help

Search PharmGKB: [   ]   Submit

What is the PharmGKB?

Find out how we go from extraction of gene-drug relationships in the literature to implementation of pharmacogenomics in the clinic...

Find out more →

Pharmacogenomics Knowledge Implementation
- Clinical Implementation
- Clinical Interpretation
- Knowledge Annotation, Aggregation & Integration
- Knowledge Extraction
- Primary Pharmacogenomic Literature

CPIC Simvastatin/SLCO1B1 Guideline

New VDR VIP Publication

New CYP2A6 VIP Publication

PharmGKB Knowledge Pyramid

CPIC: PGx Drug Dosing Guidelines

Clinically-Relevant PGx

- Well-known PGx associations
- Clinically relevant PGx summaries
- PGx drug dosing guidelines
- Drug labels with PGx info
- Genetic tests for PGx
- Star (*) allele translations

PGx-Based Drug Dosing Guidelines

- SLCO1B1/simvastatin: article and supplement
- HLA-B/abacavir: article and supplement
- more guidelines...

CPIC Gene-Drug Pairs

CPIC: Implementing PGx
a *PharmGKB* & PGRN collaboration

PGx Research

- **VIP:** Very Important PGx gene summaries
- PharmGKB pathways
- Annotated SNPs by gene
- Drugs with genetic information

find interpretations [   ]
hint: enter a gene, drug, rsid, disease

find PGx Research [   ]
hint: enter a gene, rsid, drug, disease

Follow us on:

Get your PGx fix:

 Curators' Favorite Papers

 Download PharmGKB Data

PharmGKB is a partner of the

 Pharmacogenomics Research Network
PGRN

COLOR PLATE 1   The Pharmacogenomics Knowledge Base home page. *Copyright to PharmGKB with permission given by PharmGKB and Stanford University for reproduction.*

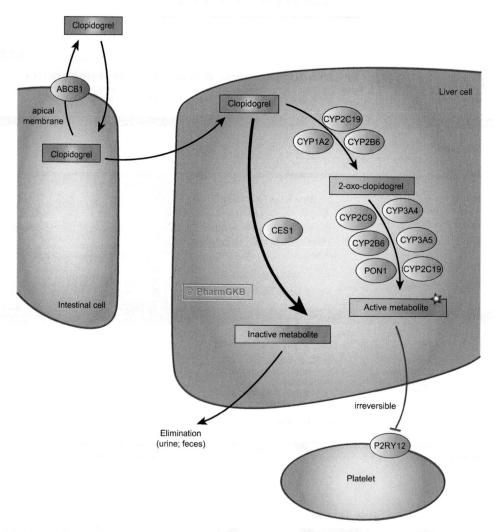

COLOR PLATE 2   Clopidogrel pathway on the Pharmacogenomics Knowledge Base. *Copyright to PharmGKB with permission given by PharmGKB and Stanford University for reproduction.*

COLOR PLATE 3 Estrogen Receptor Staining of Breast Tumor Tissue Section. Image of ER-positive breast cancer tissue section with immunohistochemical staining for ER (brown) and hematoxylin counterstaining (blue).

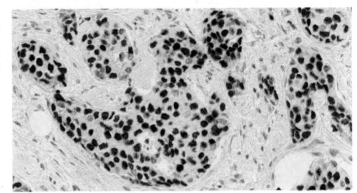

COLOR PLATE 4 Progesterone Receptor Staining of Breast Tumor Tissue Section. Image of PR-positive breast cancer tissue section with immunohistochemical staining for PR (brown) and hematoxylin counterstaining (blue).

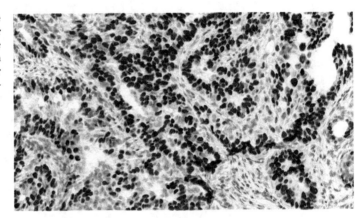

COLOR PLATE 5 HER2 Staining of Breast Tumor Tissue Section. Image of HER2-positive breast cancer tissue section with immunohistochemical staining for HER2 (brown) and hematoxylin counterstaining (blue).

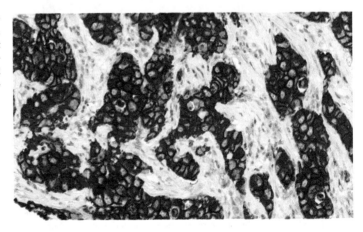

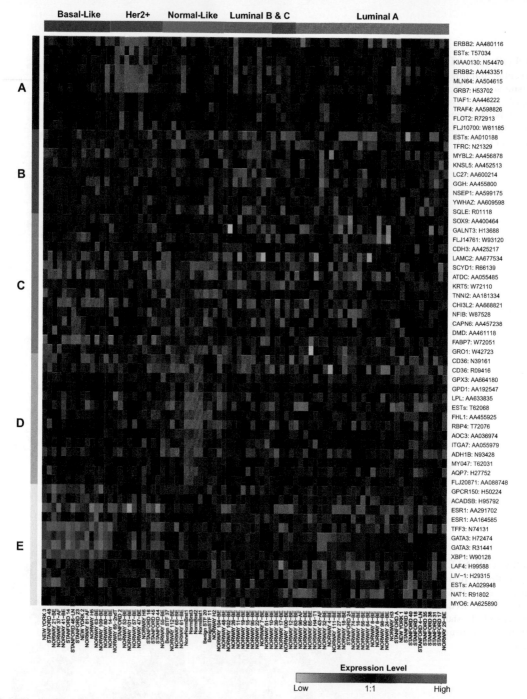

COLOR PLATE 6 Heat Map of Breast Cancer Gene Expression Data. Heat map illustrating intrinsic breast cancer gene signature identified by Sørlie et al. by profiling 78 breast tumor samples, 3 benign tumor samples, and 4 normal breast tissue samples. Each row in the heatmap represents a gene and is annotated with the Entrez Gene gene symbol, when available, and Genbank accession number. Each sample depicted as a column and annotated with a unique sample identifier. Gene expression levels are depicted as a color gradient relative to the median expression ratio per row. Based on distinct expression signatures, breast tumors can be grouped into five broad classes: Basal-like, Her2, Normal-like, or one of the Luminal subtypes (typically now reported as either "Luminal A" or "Luminal B"). These expression patterns have been reproduced across several independent studies and reflect unique drivers of tumor growth. The genes shown are representative of a large set of 456 intrinsic genes previously reported and are arranged by distinct functional groups noted by labels A-E. Note two considerable division in the data derive from expression of ER-related genes (gene group E) and Her2-related genes (gene group A). *This figure was created in the software package from original data deposited in the Stanford Microarray Database and gene lists adapted from Sørlie et al.*

Printed and bound by CPI Group (UK) Ltd, Croydon, CR0 4YY

03/10/2024

01040312-0001